W9-AQB-172

INSTRUCTOR'S SOLUTIONS MANUAL

to accompany

MULTIVARIABLE CALCULUS

William G. McCallum
University of Arizona

Deborah Hughes-Hallett
Harvard University
et al.

Andrew M. Gleason
Harvard University

John Wiley & Sons, Inc.

New York   Chichester   Weinheim   Brisbane   Singapore   Toronto

This project was supported, in part,
by the
National Science Foundation
Opinions expressed are those of the authors
and not necessarily those of the Foundation

Grant No. DUE-9352905

ISBN 0-471-17357-6

Printed in the United States of America

10 9 8 7 6 5 4 3 2 1

Printed and bound by Bradford & Bigelow, Inc.

CONTENTS

CHAPTER ELEVEN

Solutions for Section 11.1

1. (a) 80-90°F (b) 60-72°F (c) 60-100°F

2.

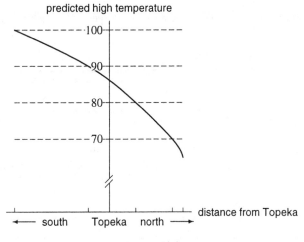

predicted high temperature

Figure 11.1

3.

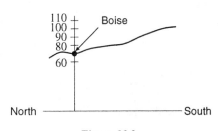

Figure 11.2

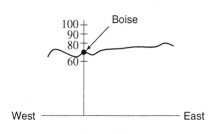

Figure 11.3

4. The amount of money spent on beef equals the product of the unit price p and the quantity of beef consumed:

$$M = pC = pf(I,p).$$

Thus, we multiply each entry in Table 11.1 on page 3 by the price at the top of the column. This yields Table 11.1.

TABLE 11.1   *Amount of money spent on beef ($/household/week)*

income\price	3.00	3.50	4.00	4.50
20	7.95	9.07	10.04	10.94
40	12.42	14.18	15.76	17.46
60	15.33	17.50	19.88	21.78
80	16.05	18.52	20.76	22.82
100	17.37	20.20	22.40	24.89

5.   (a)   Beef consumption by households making $20,000/year is given by Row 1 of Table 11.1

TABLE 11.2

p	3.00	3.50	4.00	4.50
$f(20, p)$	2.65	2.59	2.51	2.43

For households making $20,000/year, beef consumption decreases as price goes up.

(b)   Beef consumption by households making $100,000/year is given by Row 5 of Table 11.1

TABLE 11.3

p	3.00	3.50	4.00	4.50
$f(100, p)$	5.79	5.77	5.60	5.53

For households making $100,000/year, beef consumption also decreases as price goes up.

(c)   Beef consumption by households when the price of beef is $3.00/lb is given by Column 1 of Table 11.1

TABLE 11.4

I	20	40	60	80	100
$f(I, 3.00)$	2.65	4.14	5.11	5.35	5.79

When the price of beef is $3.00/lb, beef consumption increases as income increases.

(d)   Beef consumption by households when the price of beef is $4.00/lb is given by Column 3 of Table 11.1

TABLE 11.5

I	20	40	60	80	100
$f(I, 4.00)$	2.51	3.94	4.97	5.19	5.60

When the price of beef is $4.00/lb, beef consumption increases as income increases.

6. If the price of beef is held constant, beef consumption for households with various incomes can be read from a fixed column in Table 11.1. For example, the column corresponding to $p = 3.00$ gives the function $h(I) = f(I, 3.00)$; it tells you how much beef a household with income I will buy at \$3.00/lb. Looking at the column from the top down, you can see that it is an increasing function of I. This is true in every column. This says that at any fixed price for beef, consumption goes up as household income goes up—which makes sense. Thus, f is an increasing function of I for each value of p.

7. Table 11.7 gives the amount M spent on beef per household per week. Thus, the amount the household spent on beef in a year is $52M$. Since the household's annual income is I thousand dollars, the proportion of income spent on beef is

$$P = \frac{52M}{1000I} = 0.052\frac{M}{I}.$$

Thus, we need to take each entry in Table 11.7, divide it by the income at the left, and multiply by 0.052. Table 11.6 shows the results.

TABLE 11.6 *Proportion of annual income spent on beef.*

Income	Price of Beef ($)			
($1,000)	3.00	3.50	4.00	4.50
20	0.021	0.024	0.026	0.028
40	0.016	0.018	0.020	0.023
60	0.013	0.015	0.017	0.019
80	0.010	0.012	0.013	0.015
100	0.009	0.011	0.012	0.013

TABLE 11.7 *Amount of money spent on beef ($/household/week)*

income\price	3.00	3.50	4.00	4.50
20	7.95	9.07	10.04	10.94
40	12.42	14.18	15.76	17.46
60	15.33	17.50	19.88	21.78
80	16.05	18.52	20.76	22.82
100	17.37	20.20	22.40	24.89

8. In the answer to Problem 7 we saw that

$$P = 0.052\frac{M}{I},$$

and in the answer to Problem 4 we saw that

$$M = pf(I, p).$$

Putting the expression for M into the expression for P, gives:

$$P = 0.052\frac{pf(I, p)}{I}.$$

9. We have $M = f(B,t) = B(1.05)^t$.

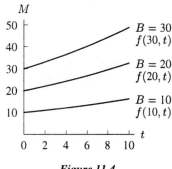

Figure 11.4

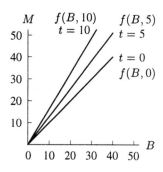

Figure 11.5

Figure 11.4 gives the graphs of f as a function of t for B fixed at 10, 20, and 30. For each fixed B, the function $f(B,t)$ is an increasing function of t. The larger the fixed value of B, the larger $f(B,t)$ is.

Figure 11.5 gives the graphs of f as a function of B for t fixed at 0, 5, and 10. For each fixed t, $f(B,t)$ is an increasing (and in fact linear) function of B. The larger t is, the larger the slope of the line.

10. (a) The daily fuel cost is calculated:

$$\text{Cost} = \text{Price per gallon} \times \text{Number of gallons}$$

TABLE 11.8 *Cost vs. Price & Gallons*

Number of Gallons	Price per gallon (dollars)						
	1.00	1.05	1.10	1.15	1.20	1.25	1.30
5	5.00	5.25	5.50	5.75	6.00	6.25	6.50
6	6.00	6.30	6.60	6.90	7.20	7.50	7.80
7	7.00	7.35	7.70	8.05	8.40	8.75	9.10
8	8.00	8.40	8.80	9.20	9.60	10.00	10.40
9	9.00	9.45	9.90	10.35	10.80	11.25	11.70
10	10.00	10.50	11.00	11.50	12.00	12.50	13.00
11	11.00	11.55	12.10	12.65	13.20	13.75	14.30
12	12.00	12.60	13.20	13.80	14.40	15.00	15.60

Note: Table entries may vary depending on the price increase interval chosen.

(b)

$$\text{Number of gallons} = (\text{Distance in miles}) / (30 \text{ miles per gallon})$$

$$\text{Cost} = \text{Price per gallon} \times \text{Number of gallons}$$

TABLE 11.9 *Cost vs. Price & Distance*

Distance (miles)	Price per gallon (dollars)						
	1.00	1.05	1.10	1.15	1.20	1.25	1.30
100	3.33	3.50	3.67	3.83	4.00	4.17	4.33
150	5.00	5.25	5.50	5.75	6.00	6.25	6.50
200	6.67	7.00	7.33	7.67	8.00	8.33	8.67
250	8.33	8.75	9.17	9.58	10.00	10.42	10.83
300	10.00	10.50	11.00	11.50	12.00	12.50	13.00
350	11.67	12.25	12.83	13.42	14.00	14.58	15.17
400	13.33	14.00	14.67	15.33	16.00	16.67	17.33
450	15.00	15.75	16.50	17.25	18.00	18.75	19.50

Note: Table entries may vary depending on the price increase interval chosen.

11. (a) The acceleration due to gravity decreases as h increases, because the gravitational force gets weaker the farther away you are from the planet. (In fact, g is inversely proportional to the square of the distance from the center of the planet.)

    (b) The acceleration due to gravity increases as m increases. The more massive the planet, the larger the gravitational force. (In fact, g is proportional to m.)

12. (a) According to Table 11.3, it feels like $-31°$F.

    (b) A wind of 10 mph, according to Table 11.3.

    (c) About 5.5 mph. Since at a temperature of 25°F, when the wind increases from 5 mph to 10 mph, the temperature adjusted for wind-chill decreases from 21°F to 10°F, we can say that a 5 mph increase in wind speed causes an 11°F decrease in the temperature adjusted for wind-chill. Thus, each 0.5 mph increase in wind speed brings *about* a 1°F drop in the temperature adjusted for wind-chill.

    (d) With a wind of 15 mph, approximately 23.5°F would feel like 0°F. With a 15 mph wind speed, when air temperature drops five degrees from 25°F to 20°F, the temperature adjusted for wind-chill drops 7 degrees from 2°F to $-5°$F. We can say that for every 1°F decrease in temperature there is *about* a 1.4°F ($= 7/5$) drop in the temperature you feel.

13.

TABLE 11.10 *Temperature adjusted for wind-chill at 20°F*

°F\mph	5	10	15	20	25
20 °F	16	3	−5	−10	−15

TABLE 11.11 *Temperature adjusted for wind-chill at 0°F*

°F\mph	5	10	15	20	25
0 °F	−5	−22	−31	−39	−44

14.

TABLE 11.12   *Temperature adjusted for wind-chill at 5 mph*

(mph)\°F	35	30	25	20	15	10	5	0
5 mph	33	27	21	16	12	7	0	−5

TABLE 11.13   *Temperature adjusted for wind-chill at 20 mph*

(mph)\°F	35	30	25	20	15	10	5	0
20 mph	12	4	−3	−10	−17	−24	−31	−39

15. Distance-wise, this wave has half the period of the original wave, that is, the distance from crest to trough at any one moment is halved. Time-wise it also has half the period of the original wave, that is, the time it takes for any one person to complete one cycle is halved.

One way to find the speed of the wave is to compare the position of the crest at time t and at time $t + 1$. At any time, t, the crest of the wave is at an x that makes the quantity

$$\cos(x - 2t) = 1$$

So

$$x - 2t = 2k\pi,$$

where k is an integer such that $2t + 2k\pi$ is a non negative integer. Thus at any time t the position of the crest is at

$$x = 2t + 2k\pi$$

where k is an integer such that $2t + 2k\pi$ is a non negative integer. For $t = 0$, the crest is at $x = 0$; for $t = 1$ the crest is at $x = 2$. Thus the wave is moving at 2 seats/second.

16. Since the wave in the text has formula $h(x, t) = 5 + \cos(0.5x - t)$, a wave moving in the opposite direction has formula:

$$h(x, t) = 5 + \cos(0.5x + t).$$

17. (a)   For $t = 0$, we have $y = f(x, t) = \sin x, \ 0 \leq x \leq \pi$

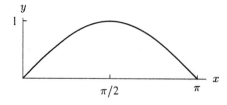

Figure 11.6

For $t = \pi/4$, we have $y = f(x,t) = \frac{\sqrt{2}}{2} \sin x, 0 \le x \le \pi$

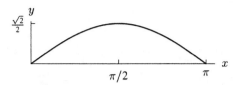

Figure 11.7

For $t = \pi/2$, we have $y = f(x,t) = 0$

Figure 11.8

For $t = 3\pi/4$, we have $y = f(x,t) = \frac{-\sqrt{2}}{2} \sin x, 0 \le x \le \pi$

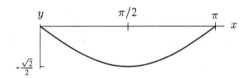

Figure 11.9

For $t = \pi$, we have $y = f(x,t) = -\sin x, 0 \le x \le \pi$

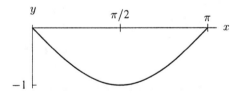

Figure 11.10

(b) The graphs show an arch of a sine wave which is above the x-axis, concave down at $t = 0$, is straight along the x-axis at $t = \pi/2$, and below the x-axis, concave up at $t = \pi$, like a guitar string vibrating up and down.

18. The function $y = f(x,0) = \cos 0 \sin x = \sin x$ gives the displacement of each point of the string when time is held fixed at $t = 0$. The function $f(x,1) = \cos 1 \sin x = 0.54 \sin x$ gives the displacement of each point of the string at time $t = 1$. Graphing $f(x,0)$ and $f(x,1)$ gives in each case an arch of the sine curve, the first with amplitude 1 and the second with amplitude 0.54. For each different fixed value of t, we get a different snapshot of the string, each one a

sine curve with amplitude given by the value of $\cos t$. The result looks like the sequence of snapshots shown in Figure 11.11.

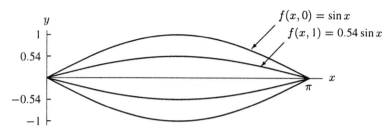

Figure 11.11

19. The function $f(0, t) = \cos t \sin 0 = 0$ gives the displacement of the left end of the string as time varies. Since that point remains stationary, the displacement is zero. The function $f(1, t) = \cos t \sin 1 = 0.84 \cos t$ gives the displacement of the point at $x = 1$ as time varies. Since $\cos t$ oscillates back and forth between 1 and -1, this point moves back and forth with maximum displacement of 0.84 in either direction. Notice the maximum displacements are greatest at $x = \pi/2$ where $\sin x = 1$.

20. (a) For $g(x, t) = \cos 2t \sin x$, our snapshots for fixed values of t are still one arch of the sine curve. The amplitudes, which are governed by the $\cos 2t$ factor, now change twice as fast as before. That is, the string is vibrating twice as fast.

   (b) For $y = h(x, t) = \cos t \sin 2x$, the vibration of the string is more complicated. If we hold t fixed at any value, the snapshot now shows one full period, i.e. one crest and one trough, of the sine curve. The magnitude of the sine curve is time dependent, given by $\cos t$. Now the center of the string, $x = \pi/2$, remains stationary just like the end points. This is a vibrating string with the center held fixed, as shown in Figure 11.12.

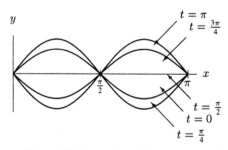

Figure 11.12: Another vibrating string:
$y = h(x, t) = \cos t \sin 2x$

Solutions for Section 11.2

1. The distance of a point $P = (x, y, z)$ from the yz-plane is $|x|$, from the xz-plane is $|y|$, and from the xy-plane is $|z|$. So, B is closest to the yz-plane, since it has the smallest x-coordinate in absolute value. B lies on the xz-plane, since its y-coordinate is 0. B is farthest from the xy-plane, since it has the largest z-coordinate in absolute value.

2. The distance of a point $P = (x, y, z)$ from the yz-plane is $|x|$, from the xz-plane is $|y|$, and from the xy-plane is $|z|$. So A is closest to the yz-plane, since it has the smallest x-coordinate in absolute value. B lies on the xz-plane, since its y-coordinate is 0. C is farthest from the xy-plane, since it has the largest z-coordinate in absolute value.

3. Your final position is $(1, -1, -3)$. Therefore, you are in front of the yz-plane, to the left of the xz-plane, and below the xy-plane.

4. Your final position is $(1, -1, 1)$. This places you in front of the yz-plane, to the left of the xz-plane, and above the xy-plane.

5. The graph is a plane parallel to the yz-plane, and passing through the point $(-3, 0, 0)$. See Figure 11.13.

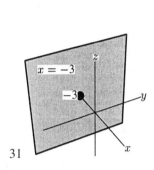

Figure 11.13

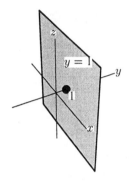

Figure 11.14

6. The graph is a plane parallel to the xz-plane, and passing through the point $(0, 1, 0)$. See Figure 11.14.

7. The graph is all points with $y = 4$ and $z = 2$, i.e., a line parallel to the x-axis and passing through the points $(0, 4, 2); (2, 4, 2); (4, 4, 2)$ etc. See Figure 11.15.

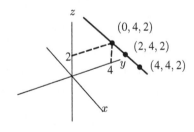

Figure 11.15

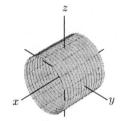

Figure 11.16

8. The coordinates of points on the y-axis are $(0, y, 0)$. The distance from any such point $(0, y, 0)$ to the point (a, b, c) is $d = \sqrt{a^2 + (b - y)^2 + c^2}$. Therefore, the closest point will have $y = b$ in order to minimize d. The resulting distance is then: $d = \sqrt{a^2 + c^2}$.

9. The equation for the points whose distance from the x-axis is 2 is given by $\sqrt{y^2 + z^2} = 2$, i.e. $y^2 + z^2 = 4$. It specifies a cylinder of radius 2 along the x-axis. See Figure 11.16.

10. The distance of any point with coordinates (x, y, z) from the x-axis is $\sqrt{y^2 + z^2}$. The distance of the point from the yz-plane is $|x|$. Since the condition states that these distances are equal, the equation for the condition is

$$\sqrt{y^2 + z^2} = |x| \quad \text{i.e.} \quad y^2 + z^2 = x^2.$$

This is the equation of a cone whose tip is at the origin and which opens along the x-axis with a slope of 1 as shown in Figure 11.17.

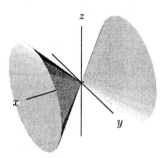

Figure 11.17

11. The point P is $\sqrt{1^2 + 2^2 + 1^2} = \sqrt{6} = 2.45$ units from the origin, and Q is $\sqrt{2^2 + 0^2 + 0^2} = 2$ units from the origin. Since $2 < \sqrt{6}$, the point Q is closer.

12. The distance formula: $d = \sqrt{(x_2 - x_1)^2 + (y_2 - y_1)^2 + (z_2 - z_1)^2}$ gives us the distance between any pair of points (x_1, y_1, z_1) and (x_2, y_2, z_2). Thus, we find

$$\text{Distance from } P_1 \text{ to } P_2 = 2\sqrt{2}$$
$$\text{Distance from } P_2 \text{ to } P_3 = \sqrt{6}$$
$$\text{Distance from } P_1 \text{ to } P_3 = \sqrt{10}$$

So P_2 and P_3 are closest to each other.

13. By drawing the top four corners, we find that the length of the edge of the cube is 5. See Figure 11.18. We also notice that the edges of the cube are parallel to the coordinate axis. So the x-coordinate of the the center equals

$$-1 + \frac{5}{2} = 1.5.$$

The y-coordinate of the center equals

$$-2 + \frac{5}{2} = 0.5.$$

The z-coordinate of the center equals

$$2 - \frac{5}{2} = -0.5.$$

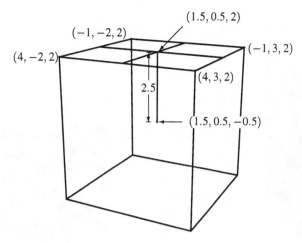

Figure 11.18

14. The length corresponds to the y-axis, therefore the y-coordinates of the corners must be $1 \pm \dfrac{13}{2} = -5.5, 7.5$. See Figure 11.19. The height corresponds to the z-axis, therefore the z-coordinates of the corners must be $-2 \pm \dfrac{5}{2} = 0.5, -4.5$. The width corresponds to the x-axis, therefore the x-coordinates of the corners must be $1 \pm 3 = 4, -2$. The coordinates of those eight corners are therefore

$$(4, 7.5, 0.5), \ (-2, 7.5, 0.5), (-2, -5.5, 0.5), (4, -5.5, 0.5),$$
$$(4, 7.5, -4.5), \ (-2, 7.5, -4.5), (-2, -5.5, -4.5), (4, -5.5, -4.5).$$

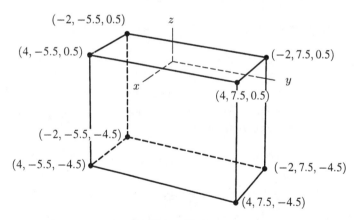

Figure 11.19

15. Using the distance formula, we find that

$$\text{Distance from } P_1 \text{ to } P = \sqrt{206}$$
$$\text{Distance from } P_2 \text{ to } P = \sqrt{152}$$
$$\text{Distance from } P_3 \text{ to } P = \sqrt{170}$$
$$\text{Distance from } P_4 \text{ to } P = \sqrt{113}$$

So $P_4 = (-4, 2, 7)$ is closest to $P = (6, 0, 4)$.

16.   An example is the line $z = -x$ in the xz-plane. See Figure 11.20.

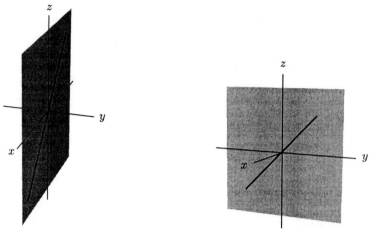

Figure 11.20   *Figure 11.21*

17.   An example is the line $y = z$ in the yz-plane. See Figure 11.21.

18.   The equation is $x^2 + y^2 + z^2 = 25$

19.   The equation is $(x - 1)^2 + (y - 2)^2 + (z - 3)^2 = 25$

20.   (a)   To find the intersection of the sphere with the yz-plane, substitute $x = 0$ into the equation of the sphere:

$$(-1)^2 + (y + 3)^2 + (z - 2)^2 = 4,$$

therefore

$$(y + 3)^2 + (z - 2)^2 = 3$$

This equation represents a circle of radius $\sqrt{3}$.
On the xz-plane $y = 0$:

$$(x - 1)^2 + 3^2 + (z - 2)^2 = 4,$$

therefore

$$(x - 1)^2 + (z - 2)^2 = -5$$

The negative sign on the right side of this equation shows that the sphere does not intersect the xz-plane, since the left side of the equation is always non-negative.
On the xy-plane, $z = 0$:

$$(x - 1)^2 + (y + 3)^2 + (-2)^2 = 4,$$

therefore

$$(x - 1)^2 + (y + 3)^2 = 0.$$

This equation has the unique solution $x = 1, y = -3$, so the xy-plane intersects the sphere in the single point $(1, -3, 0)$.

(b)   Since the sphere does not intersect the xz-plane, it cannot intersect the x or z axes. On the y-axis, we have $x = z = 0$. Substituting this into the equation for the sphere we get

$$(-1)^2 + (y + 3)^2 + (-2)^2 = 4,$$

therefore

$$(y + 3)^2 = -1.$$

This equation has no solutions because the right hand side is negative, and the left-hand side is always non-negative. Thus the sphere does not intersect any of the coordinate axes.

Solutions for Section 11.3

1. (a) The value of z decreases as x increases. See Figure 11.22.
    (b) The value of z increases as y increases. See Figure 11.23.

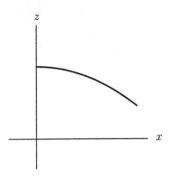

Figure 11.22

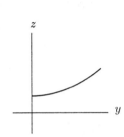

Figure 11.23

2. (a) In this company success only increases when money increases, so success will remain constant along the work axis no matter how much work is put in. However, as money increases so does success, which is shown in Graph (III).
    (b) As both work and money increase, success never increases, so we have a flat plane with no success, which corresponds to Graph (II).
    (c) If the money doesn't matter, then regardless of how much the money increases success will be constant along the money axis. However, success increases as work increases. This is best represented in Graph (I).
    (d) This company's success increases as both money and work increase, which is demonstrated in Graph (IV).

3. (a) This is a bowl; z increases as the distance from the origin increases, from a minimum of 0 at $x = y = 0$.
    (b) Neither. This is an upside-down bowl. This function will decrease from 1, at $x = y = 0$, to arbitrarily large negative values as x and y increase due to the negative squared terms of x and y. It will look like the bowl in part (a) except flipped over and raised up slightly.
    (c) This is a plate. Solving the equation for z gives $z = 1 - x - y$ which describes a plane whose x and y slopes are -1. It is perfectly flat, but not horizontal.
    (d) Within its domain, this function is a bowl. It is undefined at points at which $x^2 + y^2 > 5$, but within those limits it describes the bottom half of a sphere of radius $\sqrt{5}$ centered at the origin.
    (e) This function is a plate. It is perfectly flat and horizontal.

4. (a)

(i)

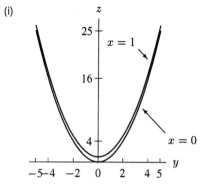

Figure 11.26: Cross-sections of $z = x^2 + y^2$

(ii)

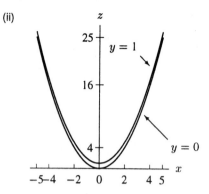

Figure 11.27: Cross-sections of $z = x^2 + y^2$

(b)

(i)

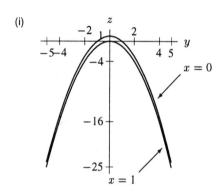

Figure 11.30: Cross-sections of $z = 1 - x^2 - y^2$

(ii)

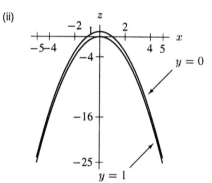

Figure 11.31: Cross-sections of $z = 1 - x^2 - y^2$

(c)

(i)

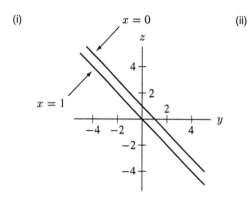

(ii)

Figure 11.34: Cross-sections of $x + y + z = 1$ *Figure 11.35:* Cross-sections of $x + y + z = 1$

(d)

(i)

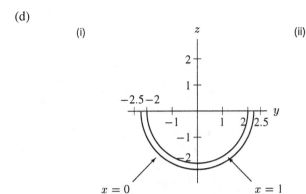

Figure 11.38: Cross-sections of
$z = -\sqrt{5 - x^2 - y^2}$

(ii)

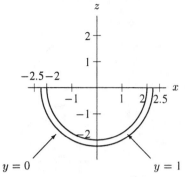

Figure 11.39: Cross-sections of
$z = -\sqrt{5 - x^2 - y^2}$

(e)

(i)

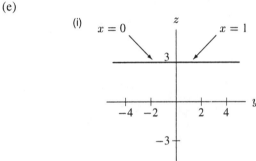

(ii)

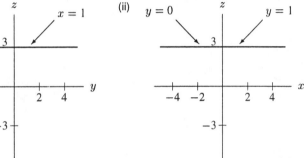

Figure 11.42: Cross-section of $z = 3$ **Figure 11.43:** Cross-section of $z = 3$

5. (a) The value of z only depends on the distance from the point (x, y) to the origin. Therefore the graph has a circular symmetry around the z-axis. There are two such graphs among those depicted in Figure 5: I and V. The one corresponding to $z = \frac{1}{x^2+y^2}$ is I since the function blows up as (x, y) gets close to $(0, 0)$.

   (b) For similar reasons as in part (a), the graph is circularly symmetric about the z-axis, hence the corresponding one must be V.

   (c) The graph has to be a plane, hence IV.

   (d) The function is independent of x, hence the corresponding graph can only be II. Notice that the cross-sections of this graph parallel to the yz-plane are parabolas, which is a confirmation of the result.

   (e) The graph of this function is depicted in III. The picture shows the cross-sections parallel to the zx-plane, which have the shape of the cubic curves $z = x^3 -$ constant.

6. (a) If we have iron stomachs and can consume cola and pizza endlessly without ill effects, then we expect our happiness to increase without bound as we get more cola and pizza. Graph (IV) shows this since it increases along both the pizza and cola axes throughout.

   (b) If we get sick upon eating too many pizzas or drinking too much cola, then we expect our happiness to decrease once either or both of those quantities grows past some optimum value. This is depicted in graph (I) which increases along both axes until a peak is reached, and then decreases along both axes.

   (c) If we do get sick after too much cola, but are always able to eat more pizza, then we expect our happiness to decrease after we drink some optimum amount of cola, but continue to increase as we get more pizza. This is shown by graph (III) which increases continuously along the pizza axis but, after reaching a maximum, begins to decrease along the cola axis.

7.

(a)

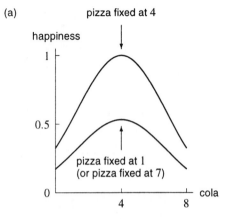

Figure 11.44: Cross-sections of graph I

(b)

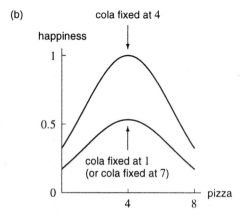

Figure 11.45: Cross-sections of graph I

(a)

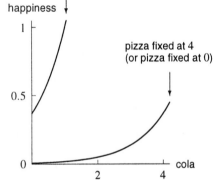

Figure 11.46: Cross-sections of graph II

(b)

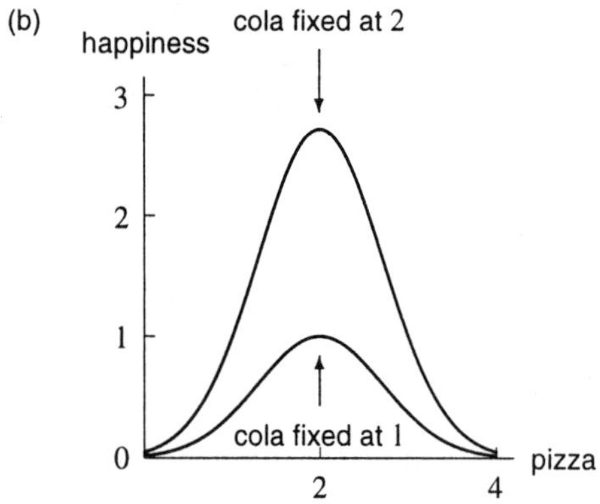

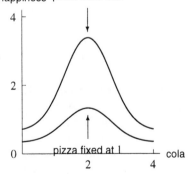

Figure 11.47: Cross-sections of graph II

(a)

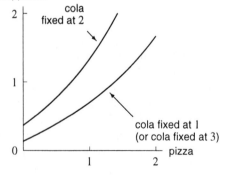

Figure 11.48: Cross-sections of graph III

(b)

Figure 11.49: Cross-sections of graph III

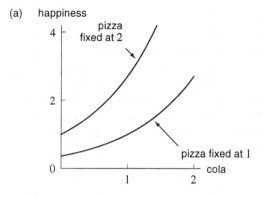

(a)

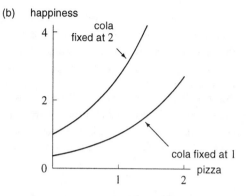

(b)

Figure 11.50: Cross-sections of graph IV

Figure 11.51: Cross-sections of graph IV

8. Planes perpendicular to the positive y-axis should yield the graphs of upright parabolas $f(x, y)$, which widen as y decreases (giving $f(x, 2)$ and $f(x, 1)$). When $y = 0$, the parabola flattens out, creating a horizontal line for $f(x, 0)$. The graphs then turn downward, creating the parabolas $f(x, -1)$ and $f(x, -2)$ which become narrower as y decreases. So the graph (IV) bests fits this information.

9. (a)

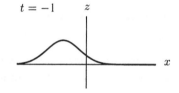

Figure 11.52

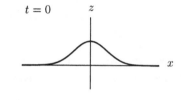

Figure 11.53

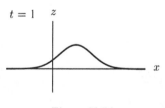

Figure 11.54

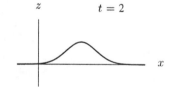

Figure 11.55

(b) Increasing x

(c) The graph in Figure 11.56 represents a wave traveling in the opposite direction.

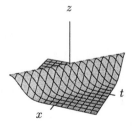

Figure 11.56

10. The cross-sections with t fixed describe snapshots of the string at different instants. Graphs of these cross-sections can be seen in the graph of f shown in Figure 11.57. Every plane perpendicular to the t-axis intersects the surface in one arch of a sine curve. The magnitude of the arch changes with t as a cosine curve.

The cross-sections with x fixed show how a single point on the string moves as time goes by. Notice in Figure 11.57 that the cross-sections with $x = 0$ and $x = \pi$ are flat lines since the endpoints of the string don't move. The cross-section with $x = \pi/2$ is a cosine curve with amplitude 1, because the midpoint of the string oscillates back and forth. Cross-sections with x fixed between 0 and $\frac{\pi}{2}$ and between $\frac{\pi}{2}$ and π are cosine curves with amplitude between 0 and 1, representing the fact that these points on the string oscillate back and forth with the same period as $x = \frac{\pi}{2}$, but a smaller amplitude.

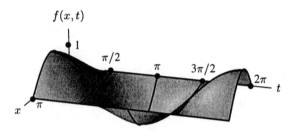

Figure 11.57: The graph of the vibrating string function

11. The cross-sections perpendicular to the t-axis are sine curves of the form $g(x, b) = (\cos b) \sin 2x$; these have period π. The cross-sections perpendicular to the x-axis are cosine curves of the form $g(a, t) = (\sin 2a) \cos t$; these have period 2π.

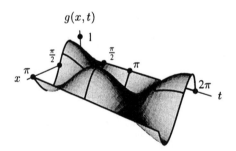

Figure 11.58: Graph $g(x, t) = \cos t \sin 2x$

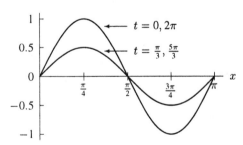

Figure 11.59: Cross-section
$g(x, b) = (\cos b) \sin 2x$,
with $b = 0, \pi/3, 5\pi/3, 2\pi$

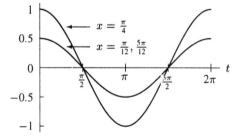

Figure 11.60: Cross-section
$g(a, t) = (\sin 2a) \cos t$
with $a = \pi/12, \pi/4, 5\pi/12$

12.

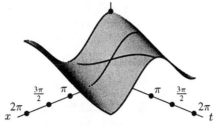

Figure 11.61: Graph of
$h(x,t) = 3 + \cos(x - 0.5t)$

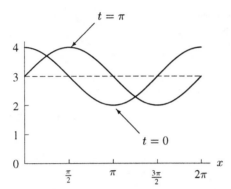

Figure 11.62: Cross-section of
$h(x, b) = 3 + \cos(x - 0.5b)$
with $b = 0, \pi$

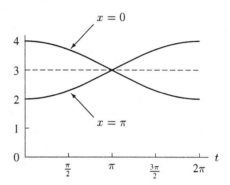

Figure 11.63: Cross-section of
$h(a, t) = 3 + \cos(a - 0.5t)$,
with $a = 0, \pi$

The cross-sections perpendicular to the t-axis are cosine curves with period 2π. The cross-sections perpendicular to the x-axis are also cosine curves with period 4π.

13. (a) Cross-sections with x fixed at $x = b$ are in Figure 11.64.

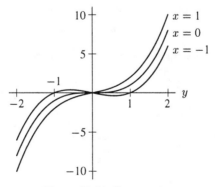

Figure 11.64: Cross-section
$f(a, y) = y^3 + ay$, with $a = -1, 0, 1$

(b)   Cross-section with y fixed at $y = 6$ are in Figure 11.65.

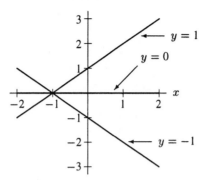

Figure 11.65: Cross-section
$f(x, b) = b^3 + bx$, with $b = -1,\ 0,\ 1$

14.   (a)   (i)   If $x = c$, then

$$E = 1 - \cos c + \frac{y^2}{2}.$$

This is a parabola opening upwards, symmetric about the E-axis with a nonnegative E-intercept since $1 - \cos c \geq 0$.

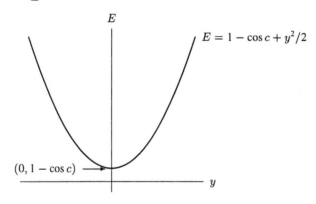

Figure 11.66

(ii)   If $y = c$, then

$$E = 1 + \frac{c^2}{2} - \cos x.$$

This is a the cosine curve flipped over and moved up by $1 + c^2/2$.

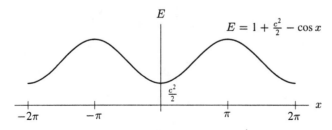

Figure 11.67

(b)

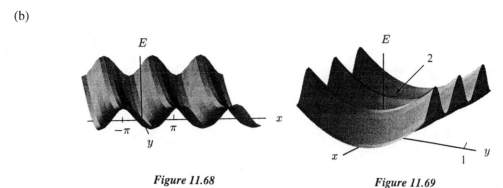

Figure 11.68   *Figure 11.69*

Solutions for Section 11.4

1. The contour where $f(x, y) = x + y = c$, or $y = -x + c$, is the graph of the straight line with slope -1 as shown in Figure 11.70. Note that we have plotted the contours for $c = -3, -2, -1, 0, 1, 2, 3$. The contours are evenly spaced.

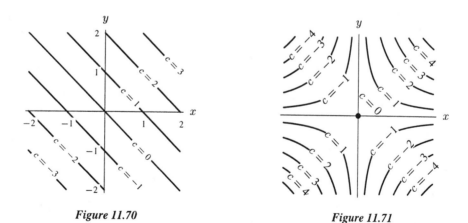

Figure 11.70   *Figure 11.71*

2. The contour where $f(x, y) = xy = c$, is the graph of the hyperbola $y = c/x$ if $c \neq 0$ and the coordinate axes if $c = 0$, as shown in Figure 11.71. Note that we have plotted contours for $c = -5, -4, -3, -2, -1, 0, 1, 2, 3, 4, 5$. The contours become more closely packed as we move further from the origin.

3. The contour where $f(x, y) = x^2 + y^2 = c$, where $c \geq 0$, is the graph of the circle centered at $(0, 0)$, with radius $\sqrt{c}$ as shown in Figure 11.72. Note that we have plotted the contours for $c = 0, 1, 2, 3, 4$. The contours become more closely packed as we move further from the origin.

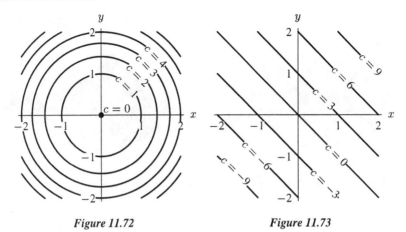

Figure 11.72   Figure 11.73

4. The contour where $f(x, y) = 3x + 3y = c$ or $y = -x + c/3$ is the graph of the straight line of slope -1 as shown in Figure 11.73. Note that we have plotted the contours for $c = -9, -6, -3, 0, 3, 6, 9$. The contours are evenly spaced.

5. The contour where $f(x, y) = -x^2 - y^2 + 1 = c$, where $c \leq 1$, is the graph of the circle centered at $(0, 0)$, with radius $\sqrt{1 - c}$ as shown in Figure 11.74. Note that we have plotted the contours for $c = -3, -2, -1, 0, 1$. The contours become more closely packed as we move further from the origin.

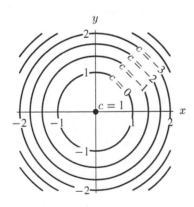

Figure 11.74

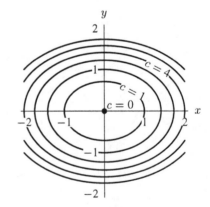

Figure 11.75

6. The contour where $f(x, y) = x^2 + 2y^2 = c$, where $c \geq 0$, is the graph of the ellipse with focuses $(-\sqrt{\frac{c}{2}}, 0)$, $(\sqrt{\frac{c}{2}}, 0)$ and axes lying on x- and y-axes as shown in Figure 11.75. Note that we have plotted the contours for $c = 0, 1, 2, 3, 4$. The contours become more closely packed as we move further from the origin.

7. The contour where $f(x,y) = \sqrt{x^2 + 2y^2} = c$, where $c \geq 0$, is the graph of the ellipse with focuses $(-\frac{c\sqrt{2}}{2}, 0)$, $(\frac{c\sqrt{2}}{2}, 0)$ and axes lying on x- and y-axes as shown in Figure 11.76. Note that we have plotted the contours for $c = 0, 1, 2, 3, 4$.

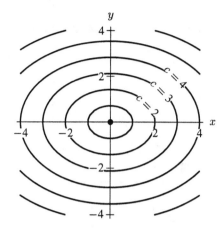

Figure 11.76

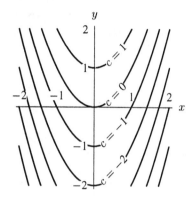

Figure 11.77

8. The contour where $f(x,y) = y - x^2 = c$ is the graph of the parabola with vertex $(0, c)$ and symmetric about the y-axis, shown in Figure 11.77. Note that we have plotted the contours for $c = -2, -1, 0, 1$. The contours become more closely packed as we move further from the y-axis.

9. The contour where $f(x,y) = \cos(\sqrt{x^2 + y^2}) = c$, where $-1 \leq c \leq 1$, is a set of circles centered at $(0,0)$, with radius $\cos^{-1} c + 2k\pi$ with $k = 0, 1, 2, ..$ and $-\cos^{-1} c + 2k\pi$, with $k = 1, 2, 3, ...$ as shown in Figure 11.78. Note that we have plotted contours for $c = 0, 0.2, 0.4, 0.6, 0.8, 1$.

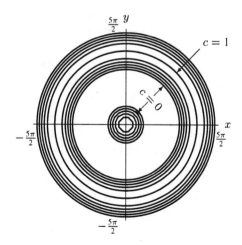

Figure 11.78

10. (a) The point representing 13% and \$6000 on the graph lies between the 120 and 140 contours. We estimate the monthly payment to be about \$137.

(b)   Since the interest rate has dropped, we will be able to borrow more money and still make a monthly payment of $137. To find out how much we can afford to borrow, we find where the interest rate of 11% intersects the $137 contour and read off the loan amount to which these values correspond. Since the $137 contour is not shown, we estimate its position from the $120 and $140 contours. We find that we can borrow an amount of money that is more than $6000 but less than $6500. So we can borrow about $250 more without increasing the monthly payment.

(c)   The entries in the table will be the amount of loan at which each interest rate intersects the 137 contour. Using the $137 contour from (b) we make table 11.14.

TABLE 11.14   *Amount borrowed at a monthly payment of $137.*

Interest Rate (%)	0	1	2	3	4	5	6	7
Loan Amount ($)	8200	8000	7800	7600	7400	7200	7200	6800
Interest rate (%)	8	9	10	11	12	13	14	15
Loan Amount ($)	6650	6500	6350	6250	6100	6000	5900	5800

11.   (a)   The contour lines are much closer together on path A, so path A is steeper.

(b)   If you are on path A and turn around to look at the countryside, you find hills to your left and right, obscuring the view. But the ground falls away on either side of path B, so you are likely to get a much better view of the countryside from path B.

(c)   There is more likely to be a stream alongside path A, because water follows the direction of steepest descent.

12.   One possible answer follows.

(a)   If there is a city at the center of the diagram, then the population is very dense at the center, but progressively less dense as you move into the suburbs, further from the city center. This scenario corresponds to diagram (I) or (II). We pick (I) because it has the highest density, as we would expect in a city.

(b)   If the center of the diagram is a lake, and is a very busy and thriving center, where lake front property is considered the most desirable, then the most dense area will be at lakeside, and decrease as you move further from the lake in the center, as in diagram (I) or (II). We pick (II) because we expect the population density at lake front to be less than that in the center of a city.

(c)   If the center of the diagram is a power plant and if the plant is not in a densely populated area, where people can and will choose not to live anywhere near it, the population density will then be very low nearby, increasing slightly further from the plant, as in diagram (III).

In an alternative solution, if the lake were in the middle of nowhere, the entire area would be very sparsely populated, and there would be slightly fewer people living on the actual lake shore, as in diagram (III).

13. We'll set $z = 4$ at the peak.

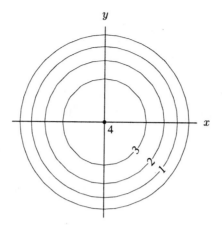

Figure 11.79

14. We will take $z = 4$ to be the flat area.

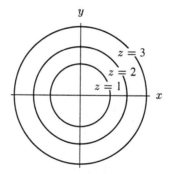

Figure 11.80

15.

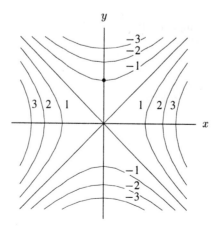

Figure 11.81

16. (a) East-west cross-section along the line $N = 50$ kilometers:

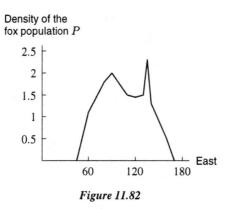

Figure 11.82

(b) East-west cross-section along the line $N = 100$ kilometers:

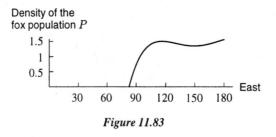

Figure 11.83

(c) North-south cross-section along the line $E = 60$ kilometers:

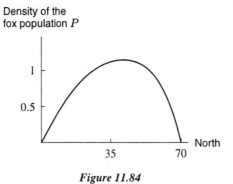

Figure 11.84

(d)   North-south cross-section along the line $E = 120$ kilometers:

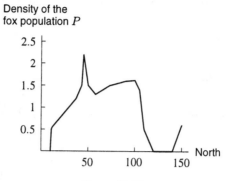

Figure 11.85

17.

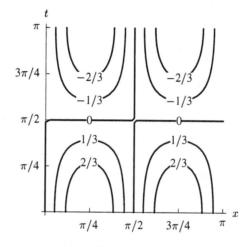

Figure 11.86

18.   The contour diagram of $h(x, t) = 5 + \cos(0.5x - t)$ is in Figure 11.87.

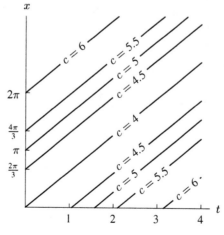

Figure 11.87

By moving vertically along a line t = constant, we find the height of the wave along a row at any given time. Alternatively, by moving horizontally, we get the height at a particular position at any time. Cross-sections of the form $h(2,t)$ correspond to horizontal lines across this graph; cross-sections of the form $h(x,5)$ correspond to vertical lines. The contours are most closely spaced near where $c = 5$ and most widely spaced near where $c = 4$ and $c = 6$.

19. (I)

cola
.1
.3
.5
.7
.9
.8 .6 .4 .2
pizzas

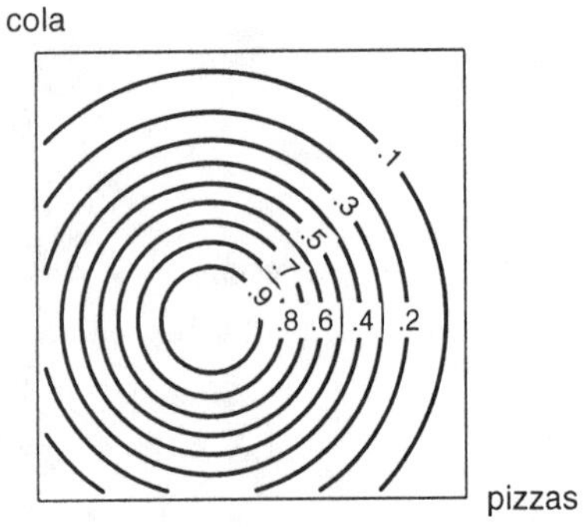

Figure 11.88

(II)

cola
1.5
1.2
1
0.9
0.8
0.7
0.6
0.5
0.4
0.3
0.2
0.1
0.05
0.01
pizzas

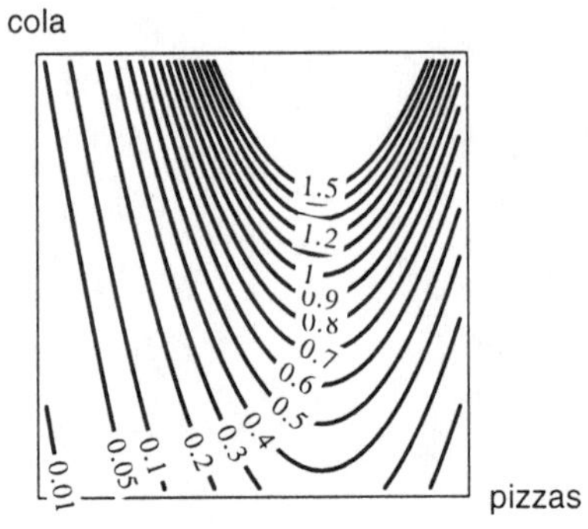

Figure 11.89

(III)

cola
2
1.5
1.2
0.9
0.7
0.5
0.4
0.3
0.2
0.1
0.05
pizzas

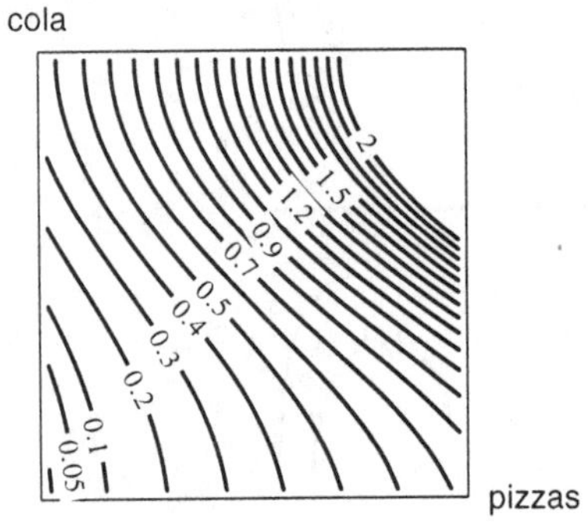

Figure 11.90

(IV)

cola
2
1.5
1.2
0.9
0.7
0.5
0.4
0.3
0.2
0.1
pizzas

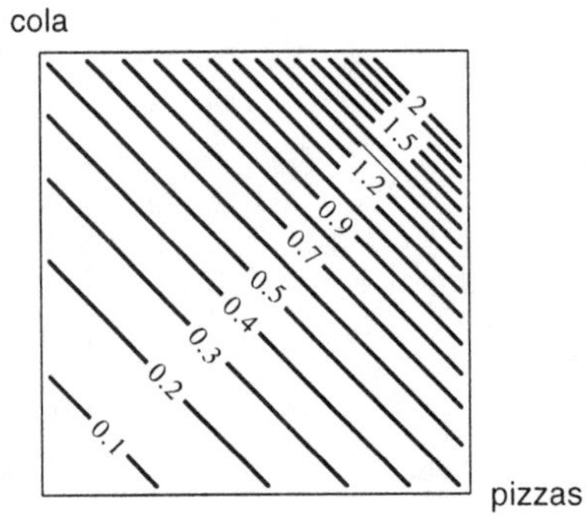

Figure 11.91

20. (a) The profit is given by the following:

$$\pi = \text{Revenue from } q_1 + \text{Revenue from } q_2 - \text{Cost}.$$

Measuring π in thousands, we obtain:

$$\pi = 3q_1 + 12q_2 - 4.$$

(b) A contour diagram of π follows. Note that the units of π are in thousands.

q_2
3
2
1
$\pi = 30$
$\pi = 20$
$\pi = 10$
$\pi = 0$
2
4
6
8
q_1

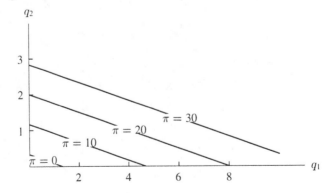

21. (a) The TMS map of an eye of constant curvature will have only one color, with no contour lines dividing the map.
     (b) The contour lines are circles, because the cross-section is the same in every direction. The largest curvature is in the center.

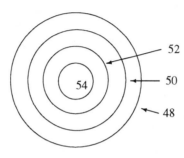

22. The values in Table 11.6 are not constant along rows or columns and therefore cannot be the lines shown in (I) or (IV). Also observe that as you move away from the origin, whose contour value is 0, the z-values on the contours increase. Thus, this table corresponds to diagram (II).

   The values in Table 11.7 are also not constant along rows or columns. Since the contour values are decreasing as you move away from the origin, this table corresponds to diagram (III).

   Table 11.8 shows that for each fixed value of x, we have constant contour value, suggesting a straight vertical line at each x-value, as in diagram (IV).

   Table 11.9 also shows lines, however these are horizontal since for each fixed value of y we have constant contour values. Thus, this table represents diagram (I).

23. (a) (III)
     (b) (I)
     (c) (V)
     (d) (II)
     (e) (IV)

24. (a) *False*. The values on the level curves are decreasing as you go northward in Canada..
     (b) *False*. It is below 100, since the values on the level curves are decreasing as you go southward in Florida and Miami is south of the 100 level curve.
     (c) *True*. In general, the contour levels are increasing from peninsulas to mainland. This is true for all three examples. For instance, the density on the Baja peninsula is mostly below 180 whereas on the mainland nearby it is 200 and up to 280.
     (d) *True*. Pick the point P where the level curves are the closest together, and pick the direction in which the values on the level curves are increasing fastest. See the arrow marked in Figure 11.92.

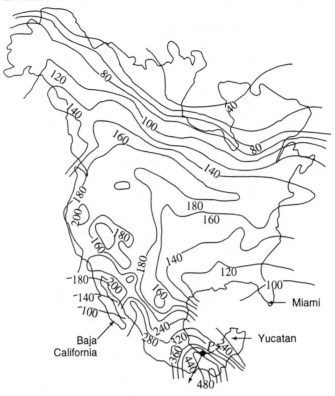

Figure 11.92

25.   (a)   To find the level curves, we let T be a constant.

$$T = 100 - x^2 - y^2$$
$$x^2 + y^2 = 100 - T,$$

which is an equation for a circle of radius $\sqrt{100 - T}$ centered at the origin. At $T = 100°$, we have a circle of radius 0 (a point). At $T = 75°$, we have a circle of radius 5. At $T = 50°$, we have a circle of radius $5\sqrt{2}$. At $T = 25°$, we have a circle of radius $5\sqrt{3}$. At $T = 0°$, we have a circle of radius 10.

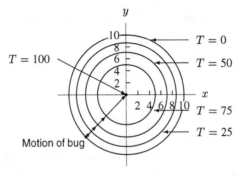

Figure 11.93

(b)   No matter where we put the bug, it should go straight toward the origin—the hottest point on the xy-plane. Its direction of motion is perpendicular to the tangent lines of the level curves, as can be seen in Figure 11.93.

26. Many different answers are possible. Answers are in degrees Celsius.
    (a) Minnesota in winter.

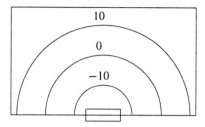

    (b) San Francisco in winter.

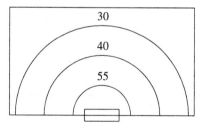

    (c) Houston in summer.

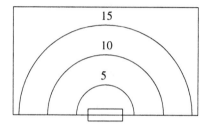

    (d) Oregon in summer.

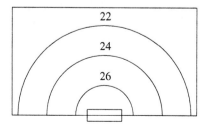

27.   For any Cobb-Douglass function $F(K, L) = bL^\alpha K^\beta$, if we increase the inputs by a factor of m, from (K, L) to (mK, mL) we get:

$$F(mK, mL) = b(mL)^\alpha (mK)^\beta$$
$$= m^{\alpha+\beta} b L^\alpha K^\beta$$
$$= m^{\alpha+\beta} F(K, L)$$

Thus we see that increasing inputs by a factor of m increases outputs by a factor of $m^{\alpha+\beta}$.

If $\alpha + \beta < 1$, then increasing each input by a factor of m will result in an increase in output of less than a factor of m. This applies to statements (a) and (E). In statement (a), $\alpha + \beta = 0.25 + 0.25 = 0.5$, so increasing inputs by a factor of $m = 4$, as in statement (E), increases output by a factor of $4^{0.5} = 2$. We can match statements (a) and (E) to graph (II) by noting that when $(K, L) = (1, 1)$, we have $F = 1$ and when we double the inputs $(m = 2)$ to $(K, L) = (2, 2)$, F increases by *less than* a factor of 2. This is called decreasing returns to scale.

If $\alpha + \beta = 1$, then increasing K and L by a factor of m will result in an increase in F by the same factor m. This applies to statements (b) and (D). In statement (b), $\alpha + \beta = 0.5 + 0.5 = 1$, and in statement (D), an increase in inputs by a factor of 3 results in an increase in F by the same factor. We match these statements to graph (I) where we see that increasing (K, L) from $(1, 1)$ to $(3, 3)$ results in an increase in F from $F = 1$ to $F = 3$. This is called constant returns to scale.

If $\alpha + \beta > 1$, then we have increasing returns to scale, i.e. an increase in K and L by a factor of m results in an increase in F by more than a factor of m. This is the case for equation (c), where $\alpha + \beta = 0.75 + 0.75 = 1.5$. Statement (G) also applies an increase in inputs by a factor of $m = 2$ results in an increase in output by *more than* 2, in this case by a factor of almost 3. We can match statements (c) and (G) to graph (III), where we see that increasing (K, L) from $(1, 1)$ to $(2, 2)$ results in a change in F by more than a factor of 2 (but less than a factor of 3). This is called increasing returns to scale.

This information is summarized in Table 11.15.

TABLE 11.15

Function	Graph	Statement
$F(L, K) = L^{0.25} K^{0.25}$	(II)	(E)
$F(L, K) = L^{0.5} K^{0.5}$	(I)	(D)
$F(L, K) = L^{0.75} K^{0.75}$	(III)	(G)

28.   If

$$P_0 = f(L_0, K_0) = 1.01 L_0^{0.75} K_0^{0.25}$$

then replacing L_0 and K_0 by $2L_0$ and $2K_0$ gives

$$f(2L_0, 2K_0) = 1.01(2L_0)^{0.75}(2K_0)^{0.25}$$
$$= 2^{0.75} 2^{0.25} \cdot 1.01 L_0^{0.75} K_0^{0.25}$$
$$= 2f(L_0, K_0)$$
$$= 2P_0.$$

So, doubling labor and capital doubles production.

29. Suppose P_0 is the production given by L_0 and K_0, so that

$$P_0 = f(L_0, K_0) = cL_0^\alpha K_0^\beta.$$

We want to know what happens to production if L_0 is increased to $2L_0$ and K_0 is increased to $2K_0$:

$$\begin{aligned} P &= f(2L_0, 2K_0) \\ &= c(2L_0)^\alpha (2K_0)^\beta \\ &= c2^\alpha L_0^\alpha 2^\beta K_0^\beta \\ &= 2^{\alpha+\beta} cL_0^\alpha K_0^\beta \\ &= 2^{\alpha+\beta} P_0. \end{aligned}$$

Thus, doubling L and K has the effect of multiplying P by $2^{\alpha+\beta}$. Notice that if $\alpha + \beta > 1$, then $2^{\alpha+\beta} > 2$, if $\alpha + \beta = 1$, then $2^{\alpha+\beta} = 2$, and if $\alpha + \beta < 1$, then $2^{\alpha+\beta} < 2$. Thus, $\alpha + \beta > 1$ gives increasing returns to scale, $\alpha + \beta = 1$ gives constant returns to scale, and $\alpha + \beta < 1$ gives decreasing returns to scale.

30. (a) About 15 feet along the wall, because that's where there are regions of cold air (55° and 65°).
     (b) Roughly between 10 am and 12 noon, and between 4 pm and 6 pm.
     (c) Roughly between midnight and 2 am, between 10 am and 1 pm, and between 4 pm and 9 pm, since that is when the temperature near the heater is greater than 80°.
     (d)

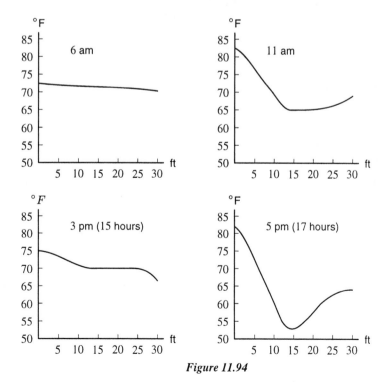

Figure 11.94

(e)

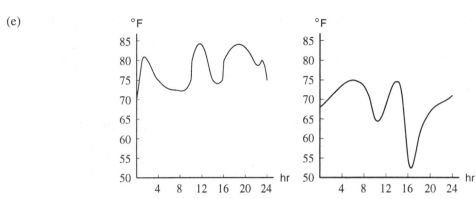

Figure 11.95: Temp vs. Time at heater

Figure 11.96: Temp vs. Time at window

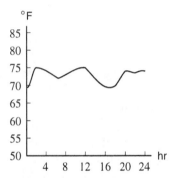

Figure 11.97: Temp vs. Time
midway between heater and window

(f)   The temperature at the window is colder at 5 pm than at 11 am because the outside temperature is colder at 5 pm than at 11 am.

(g)   The thermostat is set to roughly 70°F because the temperature stays close to 70°F a couple feet from the window.

(h)   We are told that the thermostat is about 2 feet from the window. Thus, the thermostat is either about 13 feet or about 17 feet from the wall. If the thermostat is set to 70°F, every time the temperature at the thermostat goes over or under 70°F, the heater turns off or on. Look at the point at which the vertical lines at 13 feet and 17 feet cross the 70°F contours. We need to decide which of these crossings correspond best with the times that the heater turns on and off. (These times can be seen along the wall.) Notice that the 17 foot line does not cross the 70°F contour after 16 hours (4 pm). Thus, if the thermostat were 17 feet from the wall, the heater would not turn off after 4 pm. However, the heater does turn off at about 21 hours (9 pm). Since this is the time that the 13 foot line crosses the 70°F contour, we estimate that the thermostat is about 13 feet away from the wall.

Solutions for Section 11.5

1.   (a)   Since z is a linear function of x and y with slope 2 in the x-direction, and slope 3 in the y-direction, we have:

$$z = 2x + 3y + c$$

We can write an equation for changes in z in terms of changes in x and y:

$$\Delta z = (2(x + \Delta x) + 3(y + \Delta y) + c) - (2x + 3y + c)$$
$$= 2\Delta x + 3\Delta y$$

Since $\Delta x = 0.5$ and $\Delta y = -0.2$, we have

$$\Delta z = 2(0.5) + 3(-0.2) = 0.4$$

So a 0.5 change in x and a -0.2 change in y produces a 0.4 change in z.

(b) As we know that $z = 2$ when $x = 5$ and $y = 7$, the value of z when $x = 4.9$ and $y = 7.2$ will be

$$z = 2 + \Delta z = 2 + 2\Delta x + 3\Delta y$$

where Δz is the change in z when x changes from 4.9 to 5 and y changes from 7.2 to 7. We have $\Delta x = 4.9 - 5 = -0.1$ and $\Delta y = 7.2 - 7 = 0.2$. Therefore, when $x = 4.9$ and $y = 7.2$, we have

$$z = 2 + 2 \cdot (-0.1) + 5 \cdot 0.2 = 2.4$$

2. Since

$$\begin{aligned}
0 &= c + m \cdot 0 + n \cdot 0 & c &= 0 \\
-1 &= c + m \cdot 0 + n \cdot 2 & c + 2n &= -1 \\
-4 &= c + m \cdot (-3) + n \cdot 0 & c - 3m &= -4
\end{aligned}$$

we get:

$$c = 0, m = \frac{4}{3}, n = -\frac{1}{2}.$$

Thus, $z = \frac{4}{3}x - \frac{1}{2}y$.

3. Let the equation of the plane be

$$z = c + mx + ny$$

Since we know the points: $(4, 0, 0)$, $(0, 3, 0)$, and $(0, 0, 2)$ are all on the plane, we know that they satisfy the same equation. We can use these values of (x, y, z) to find c, m, and n. Putting these points into the equation we get:

$$0 = c + m \cdot 4 + n \cdot 0 \quad \text{so } c = -4m$$

$$0 = c + m \cdot 0 + n \cdot 3 \quad \text{so } c = -3n$$

$$2 = c + m \cdot 0 + n \cdot 0 \quad \text{so } c = 2$$

Because we have a value for c, we can solve for m and n to get

$$c = 2, m = -\frac{1}{2}, n = -\frac{2}{3}.$$

So the linear function is

$$f(x, y) = 2 - \frac{1}{2}x - \frac{2}{3}y.$$

4. Figure 11.98 shows the two lines the plane must contain.

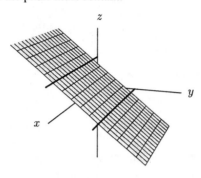

Figure 11.98

Both lines are parallel to the x-axis; thus our plane must have x-slope zero. On the other hand, the line in the xy-plane is 2 units down and one unit to the right of the line in the xz-plane; hence the y-slope of our plane must be -2. Thus the equation is

$$z = 0x - 2y + c = -2y + c,$$

for some constant c. Since the plane contains the point $(0, 0, 2)$, the value of c must be 2. So the equation is

$$z = -2y + 2.$$

5. When $y = 0$, $c + mx = 3x + 4$, so $c = 4$, $m = 3$. Thus, when $x = 0$, we have $4 + ny = y + 4$, so $n = 1$. Thus, $z = 4 + 3x + y$.

6. When $z = 0$, $mx + ny + c = 0$, $y = -\dfrac{m}{n}x - \dfrac{c}{n} = 3x + 4$.

    So, $-\dfrac{m}{n} = 3$, $-\dfrac{c}{n} = 4$.

    Since $m \cdot 0 + n \cdot 0 + c = 5$, $c = 5$.

    So $n = -\dfrac{c}{4} = -\dfrac{5}{4}$, $m = -3n = \dfrac{15}{4}$.

    Thus, $z = \dfrac{15}{4}x - \dfrac{5}{4}y + 5$.

7. The first row is linear with slope $1/0.1 = 10$. The second row is linear with slope $1.07/0.1 = 10.7$. Since the slope of the first row is not the same as the slope of the second row, the function is not linear.

8. The revenue function, R, is linear and so we may write it as:

    $$R = (p_1)t + (p_2)c$$

    where p_1 is the price of tapes and p_2 is the price of compact discs, in dollars. From the diagram, we can pick two points, such as $t = 100, c = 100, R = 2000$ and $t = 50, c = 300, R = 4000$. Using these we can then solve the following system of linear equations:

    $$2000 = 100p_1 + 100p_2$$
    $$4000 = 50p_1 + 300p_2.$$

    Thus, $p_1 = 8$ dollars and $p_2 = 12$ dollars.

9. For each column in the table, we find that as x increases by 1, $f(x, y)$ increases by 2, so the x slope is 2. For each row in the table, we find that as y increases by 1, $f(x, y)$ decreases by 0.5, so the y slope is -0.5. So the function has the form $f(x, y) = 2x - 0.5y + c$. Also note that $f(0, 0) = 1$, so $c = 1$. Therefore, the function is $f(x, y) = 2x - 0.5y + 1$.

10. For each column in the table, we find that as x increases by 100, $f(x, y)$ decreases by 1, so the x slope is -0.01. For each row in the table, we find that as y increases by 10, $f(x, y)$ increases by 3, so the y slope is 0.3. So the function has the form $f(x, y) = -0.01x + 0.3y + c$. Also note that $f(100, 10) = 3$, so $c = 1$. Therefore, the function is $f(x, y) = -0.01x + 0.3y + 1$

11.

TABLE 11.16

$x\backslash y$	0.0	1.0
0.0	-1.0	1.0
2.0	3.0	5.0

12.

TABLE 11.17

$x\backslash y$	-1.0	0.0	1.0
2.0	4.0	6.0	8.0
3.0	1.0	3.0	5.0

13. In the diagram the contours correspond to values of the function that are 2 units apart, i.e., there are contours for $-2, 0, 2$, etc. Note that moving two units in the y direction we cross three contours; i.e., a change of 2 in y changes the function by 6, so the y slope is 3. Similarly, a move of 1 in the positive x direction crosses one contour line and changes the function by -2; so the x slope is -2. Hence $f(x, y) = c - 2x + 3y$. We see from the diagram that $f(0, 1) = 6$. Solving for c gives $c = 3$. Therefore the function is $f(x, y) = 3 - 2x + 3y$.

14. In the diagram the contours correspond to values of the function that are 15 units apart, i.e., there are contours for $-90, -75, -60$, etc. An increase of 3 units in the y direction moves you from one contour to the next and changes the function by -15, so the y slope is $-15/3 = -5$. Similarly, an increase of 6 in the x direction crosses two contour lines and changes the function by 30; so the x slope is $30/6 = 5$. Hence $f(x, y) = c + 5x - 5y$. We see from the diagram that $f(8, 4) = -75$. Solving for c gives $c = -95$. Therefore the function is $f(x, y) = -95 + 5x - 5y$

15.

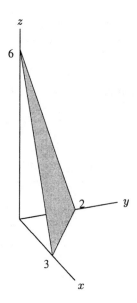

Figure 11.99

16.

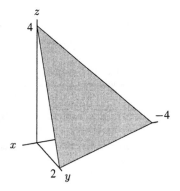

Figure 11.100

17.

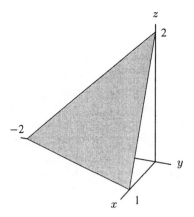

Figure 11.101

18. (a) Expenditure, E, is given by the equation:

$$E = (\text{price of raw material 1})m_1 + (\text{price of raw material 2})m_2 + C$$

where C denotes all the other expenses (assumed to be constant). Since the prices of the raw materials are constant, but m_1 and m_2 are variables, we have a linear function.

(b) Revenue, R, is given by the equation:

$$R = (p_1)q_1 + (p_2)q_2.$$

Since p_1 and p_2 are constant, while q_1 and q_2 are variables, we again have a linear function.

(c) Revenue is again given by the equation,

$$R = (p_1)q_1 + (p_2)q_2.$$

Since p_2 and q_2 are now constant, the term $(p_2)q_2$ is also constant. However, since p_1 and q_1 are variables, the $(p_1)q_1$ term means that the function is not linear.

19. (a) A student with SATs of 1050 and a GPA of 3.0 has a z-value given by

$$z = 0.003 \cdot 1050 + 0.8 \cdot 3.0 - 4 = 1.55$$

Since $1.55 < 2.3$, this student will not be admitted.

(b) A student with SATs of 1600 and GPA of y has a z-value given by

$$z = 0.003 \cdot 1600 + 0.8y - 4 = 0.8 + 0.8y = 0.8(y + 1)$$

Since $0.8(y + 1)$ may be greater than or less than 2.3, not all of the students with SAT scores of 1600 will be admitted.

(c) A student with GPA of 4.3 and SATs of x has a z-value given by

$$z = 0.003x + 0.8 \cdot 4.3 - 4 = 0.003x - 0.56$$

Since $0.003x - 0.56$ may be greater than or less than 2.3, not all of the students with a high school GPA of 4.3 will be admitted.

(d)

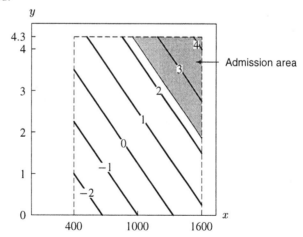

Figure 11.102

(e) If $\Delta x = 100$, then $\Delta z = 0.003\Delta x = 0.003 \cdot 100 = 0.3$.
If $\Delta y = 0.5$, then $\Delta z = 0.8 \cdot 0.5 = 0.4$.
An extra 0.5 of high school GPA increases a student's z-value by more than an extra 100 points on the SAT. Thus, the increase in GPA is more important.

20. (a) The contours of f have equation

$$k = c + mx + ny, \quad \text{where } k \text{ is a constant.}$$

Solving for y gives:

$$y = -\frac{m}{n}x + \frac{k-c}{n}$$

Since c, m, n and k are constants, this is the equation of a line. The coefficient of x is the slope and is equal to $-m/n$.

(b) Substituting $x + n$ for x and $y - m$ for y into $f(x,y)$ gives

$$f(x+n, y-m) = c + m(x+n) + n(y-m)$$

Multiplying out and simplifying gives

$$f(x+n, y-m) = c + mx + mn + ny - nm$$

$$f(x+n, y-m) = c + mx + ny = f(x,y)$$

(c) Part (b) tells us that if we move n units in the x direction and $-m$ units in the y direction, the value of the function $f(x,y)$ remains constant. Since contours are lines where the function has a constant value, this implies that we remain on the same contour. This agrees with part (a) which tells us that the slope of any contour line will be $-m/n$. Since the slope is $\Delta y/\Delta x$, it follows that changing y by $-m$ and x by n will keep us on the same contour.

Solutions for Section 11.6

1. (a) On the surface, in the corner where the hot water enters, the temperature is highest. As we move further away from this corner, the temperature of the water decreases. In the corner furthest from the hot water's entry point, the water is the coolest. A possible contour diagram is shown in Figure 11.103. Temperatures are in C°.

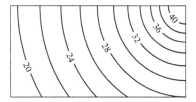

Figure 11.103: Surface temperatures

(b) One meter below the surface, the water is cooler than that at the surface, but still it is warmest in the corner where the hot water enters and gets cooler further away. The contour diagram is similar, but the temperature is lower because of the water's depth. (See Figure 11.104.)

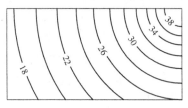

Figure 11.104: Temperatures at depth of one meter

2. The temperature is decreasing away from the window, suggesting that heat is flowing in from the window. As time goes by the temperature at at each point in the room increases. This could be caused by opening the window of an air conditioned room at $t = 0$ thus letting heat from the hot summer day outside raise the temperature inside.

3. (a) Observe that setting $f(x, y, z) = c$ gives a cylinder about the x-axis, with radius $\sqrt{c}$. These surfaces are in graph (I).
    (b) By the same reasoning the level curves for $h(x, y, z)$ are cylinders about the y-axis, so they are represented in graph (II).

4. The level surfaces are the graphs of $\sin(x + y + z) = k$ for constant k (with $-1 \le k \le 1$). This means $x + y + z = \sin^{-1}(k) + 2\pi n$, or $\pi - \sin^{-1}(k) + 2n\pi$ for all integers n. Therefore for each value of k, with $-1 \le k \le 1$, we get an infinite family of parallel planes. So the level surfaces are families of parallel planes.

5.

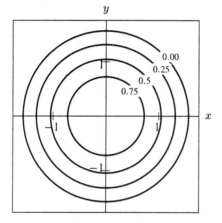

Figure 11.105: Contour diagram when $t = 0$

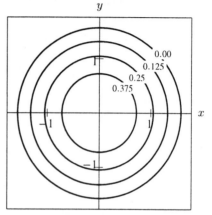

Figure 11.106: Contour diagram when $t = \pi/3$

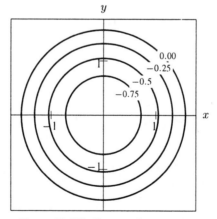

Figure 11.107: Contour diagram when $t = \pi$

6. The contour diagram consists of a family of surfaces, one for each value of t. The surfaces for $t = 0, 2\pi/3, 4\pi/3, 2\pi$ in Figure 11.108 show that the water bobs up and down, going through one cycle in 2π seconds.

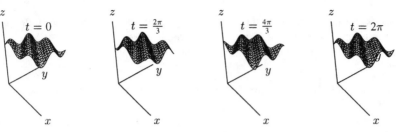

Figure 11.108

7. The level surfaces are the graphs of $g(x, y, z) = e^{-(x^2+y^2+z^2)} = k$ for constant values of k such that $0 < k \leq 1$. So $x^2 + y^2 + z^2 = -\ln k$, which is the graph of a sphere since $-\ln k \geq 0$.

8. Suppose $w = F(x, y, z) = ax + by + cz + d$.

   We get a, the x slope, by looking at the cross-section with $z = 4$. We have $a = \Delta w/\Delta x = (9 - 14)/(1 - 0) = -5$.

   We get b, the y slope, from the cross-section with $z = 1$. We have $b = \Delta w/\Delta y = (8 - 4)/(5 - 3) = 2$.

   The z slope, c, is obtained by comparing values between tables. Since when $x = 0$, $y = 5$, the value of the function changes from 8 to 14, we have $c = \Delta w/\Delta z = (14 - 8)/(4 - 1) = 2$.

   Now use the fact that $a = -5, b = 2, c = 2$ and $f(0, 3, 1) = -5(0) + 2(3) + 2(1) + d = 4$, so $d = -4$. Thus the linear function is

$$f(x, y, z) = -5x + 2y + 2z - 4$$

9. Suppose $w = f(x, y, z) = ax + by + cz + d$.

   We get b, the y slope, from the table with $z = 4$. We have $b = \Delta w/\Delta y = (9 - 14)/(5 - 3) = -5/2$.

   Since the y slope is $-5/2$, the bottom left entry in the $z = 1$ table is obtained from the 8 in the same table by adding 5 to give 13. Thus $f(1, 3, 1) = 13$. Thus the x slope is given by $a = \Delta w/\Delta x = (13-4)/(1-0) = 9$.

   We get c, the z slope, by comparing the values of w when $x = 1$, $y = 5$. Then $c = \Delta w/\Delta z = (9 - 8)/(4 - 1) = 1/3$.

   Now use the fact that $a = 9, b = -5/2, c = 1/3$ and

$$f(0, 3, 1) = 9(0) - \frac{5}{2}(3) + \frac{1}{3}(1) + d = 4$$

so $d = 67/6$. Thus

$$f(x, y, z) = 9x - \frac{5}{2}y + \frac{1}{3}z + \frac{67}{6}$$

10. If there were only three values filled in the two tables, we could not determine f because a system of three linear equations with four unknowns has more than one solution. If there were five values, we could not always determine f since a system of five linear equations with four unknowns may or may not have solutions.

11. From the graph we can read that

$$f(2, 0, 3) = 2$$

$$f(0, -1, 3) = 8$$

$$f(4, 0, 4) = 2$$

$$f(0, 0, 4) = 8$$

So we get

$$2a + 3c + d = 2$$
$$-b + 3c + d = 8$$
$$4a + 4c + d = 2$$
$$4c + d = 8$$

Solving the equations, we get $a = -\dfrac{3}{2}, b = -3, c = 3, d = -4$. Therefore, the function is

$$f(x, y, z) = -\frac{3}{2}x - 3y + 3z - 4.$$

12. $f(x, y, z) = x^2 - y^2 + z^2$ has 3 types of level surfaces depending on the values of c in the equation $x^2 - y^2 + z^2 = c$. We write this as $x^2 + z^2 = y^2 + c$ and think of what happens as we take a cross-section of the surface, perpendicular to the y-axis by holding y fixed.
    (i) For $c > 0$, the level surface is a hyperboloid of 1 sheet.
    (ii) For $c < 0$, the level surface is a hyperboloid of 2 sheets.
    (iii) For $c = 0$, the level surface is a cone.

13. A hyperboloid of two sheets.

14. A cone.

15. An elliptic paraboloid.

16. A cylindrical surface.

17. An ellipsoid.

18. A plane.

19. A sphere.

20. Let's consider the function $y = 2 + \sin z$ drawn in the yz-plane in Figure 11.109.

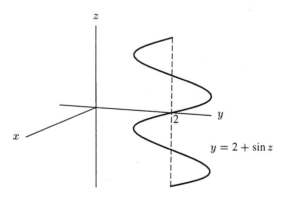

Figure 11.109

Now rotate this graph around the z-axis. Then, a point (x, y, z) is on the surface if and only if $x^2 + y^2 = (2 + \sin z)^2$. Thus, the surface generated is a surface of rotation with the profile shown in Figure 11.109.

Similarly, the surface with equation $x^2 + y^2 = (f(z))^2$ is the surface obtained rotating the graph of $y = f(z)$ around the z-axis.

Solutions for Section 11.7

1. Let us suppose that (x, y) tends to $(0,0)$ along the curve $y = kx^2$, where $k \neq -1$. We get

$$f(x, y) = f(x, kx^2) = \frac{x^2}{x^2 + kx^2} = \frac{1}{1 + k}.$$

Therefore:

$$\lim_{x \to 0} f(x, kx^2) = \frac{1}{1 + k}$$

and so for $k = 0$ we get

$$\lim_{\substack{(x,y) \to (0,0) \\ y=0}} f(x, y) = 1$$

and for $k = 1$

$$\lim_{\substack{(x,y) \to (0,0) \\ y=x^2}} f(x, y) = \frac{1}{2}.$$

Thus no matter how close they are to the origin, there will be points (x, y) where the value $f(x, y)$ is close to 1 and points (x, y) where $f(x, y)$ is close to $\frac{1}{2}$. So the limit:

$$\lim_{(x,y) \to (0,0)} f(x, y)$$

doesn't exist.

2. Let us suppose that (x, y) approaches $(0,0)$ along the line $y = x$. Then

$$f(x, y) = f(x, x) = \frac{x^3}{x^4 + x^2} = \frac{x}{x^2 + 1}.$$

Therefore

$$\lim_{\substack{(x,y) \to (0,0) \\ y=x}} f(x, y) = \lim_{x \to 0} \frac{x}{x^2 + 1} = 0.$$

On the other hand, if (x, y) approaches $(0,0)$ along the parabola $y = x^2$ we have

$$f(x, y) = f(x, x^2) = \frac{x^4}{2x^4} = \frac{1}{2}$$

and

$$\lim_{\substack{(x,y) \to (0,0) \\ y=x^2}} f(x, y) = \lim_{x \to 0} f(x, x^2) = \frac{1}{2}.$$

Thus no matter how close they are to the origin, there will be points (x, y) such that $f(x, y)$ is close to 0 and points (x, y) such that $f(x, y)$ is close to $\frac{1}{2}$. So the limit

$$\lim_{(x,y) \to (0,0)} f(x, y)$$

doesn't exist.

</>

3. (a) The graphs are shown in Figure 11.110.

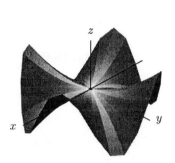
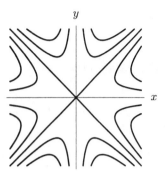

Figure 11.110

(b) Yes, it seems that if x and y are both close to 0, the values of the function are both close to $0 = f(0,0)$.

4. (a) We have $f(x,0) = 0$ for all x and $f(0,y) = 0$ for all y, so these are both continuous (constant) functions of one variable.

(b) The contour diagram suggests that the contours of f are lines through the origin. Providing it is not vertical, the equation of such a line is

$$y = mx.$$

To confirm that such lines are contours of f, we must show that f is constant along these lines. Substituting into the function, we get

$$f(x,y) = f(x,mx) = \frac{x(mx)}{x^2 + (mx)^2} = \frac{mx^2}{x^2 + m^2x^2} = \frac{m}{1 + m^2} = \text{constant.}$$

Since $f(x,y)$ is constant along the line $y = mx$, such lines are contained in contours of f.

(c) We consider the limit of $f(x,y)$ as $(x,y) \to (0,0)$ along the line $y = mx$. We can see that

$$\lim_{x \to 0} f(x,mx) = \frac{m}{1 + m^2}.$$

Therefore, if $m = 1$ we have

$$\lim_{\substack{(x,y) \to (0,0) \\ y=x}} f(x,y) = \frac{1}{2}$$

whereas if $m = 0$ we have

$$\lim_{\substack{(x,y) \to (0,0) \\ y=0}} f(x,y) = 0.$$

Thus, no matter how close we are to the origin, we can find points (x,y) where the value $f(x,y)$ is $1/2$ and points (x,y) where the value $f(x,y)$ is 0. So the limit $\lim_{(x,y)\to(0,0)} f(x,y)$ does not exist. Thus, f is not continuous at $(0,0)$, even though the one-variable functions $f(x,0)$ and $f(0,y)$ are continuous at $(0,0)$. See Figures 4 and 4

5. Since the composition of continuous functions is continuous, the function f is continuous at $(0,0)$. We have:

$$\lim_{(x,y)\to(0,0)} f(x,y) = \lim_{(x,y)\to(0,0)} (x^2 + y^2) = 0 + 0 = 0.$$

6. Since the composition of continuous functions is continuous, the function f is continuous at $(0,0)$ and we have

$$\lim_{(x,y)\to(0,0)} f(x,y) = \lim_{(x,y)\to(0,0)} e^{-x-y} = e^{-0-0} = 1$$

7. Since f doesn't depend on y we have:

$$\lim_{(x,y)\to(0,0)} f(x,y) = \lim_{x\to 0} \frac{x}{x^2+1} = \frac{0}{0+1} = 0.$$

8. Since the composition of continuous functions is continuous, the function f is continuous at $(0,0)$. We have:

$$\lim_{(x,y)\to(0,0)} f(x,y) = \lim_{(x,y)\to(0,0)} \frac{x+y}{\sin y + 2} = \frac{0+0}{0+2} = 0.$$

9. We want to compute

$$\lim_{(x,y)\to(0,0)} f(x,y) = \lim_{(x,y)\to(0,0)} \frac{\sin(x^2+y^2)}{x^2+y^2}.$$

As $r = \sqrt{x^2+y^2}$ is the distance from (x,y) to $(0,0)$ we have that $(x,y) \to (0,0)$ is equivalent to $r \to 0$. Hence the limit becomes:

$$\lim_{(x,y)\to(0,0)} f(x,y) = \lim_{r\to 0} \frac{\sin r^2}{r^2} = 1.$$

10. We want to show that f doesn't have a limit as (x,y) approaches $(0,0)$. So let us suppose that (x,y) tends to $(0,0)$ along the line $y = mx$, where the slope $m \neq 1$. Then

$$f(x,y) = f(x,mx) = \frac{x+mx}{x-mx} = \frac{(1+m)x}{(1-m)x} = \frac{1+m}{1-m}.$$

Therefore

$$\lim_{x\to 0} f(x,mx) = \frac{1+m}{1-m}$$

and so for $m=2$ we get

$$\lim_{\substack{(x,y)\to(0,0)\\y=2x}} f(x,y) = \frac{1+2}{1-2} = \frac{3}{-1} = -3$$

and for $m=3$

$$\lim_{\substack{(x,y)\to(0,0)\\y=3x}} f(x,y) = \frac{1+3}{1-3} = \frac{4}{-2} = -2.$$

Thus no matter how close they are to the origin, there will be points (x,y) where the value $f(x,y)$ is close to -3 and points (x,y) where $f(x,y)$ is close to -2. So the limit:

$$\lim_{(x,y)\to(0,0)} f(x,y) \text{ doesn't exist.}$$

11. We want to show that f doesn't have a limit as (x, y) approaches $(0, 0)$. Let us suppose that (x, y) tends to $(0, 0)$ along the line $y = mx$. Then

$$f(x, y) = f(x, mx) = \frac{x^2 - m^2 x^2}{x^2 + m^2 x^2} = \frac{1 - m^2}{1 + m^2}.$$

Therefore

$$\lim_{x \to 0} f(x, mx) = \frac{1 - m^2}{1 + m^2}$$

and so for $m = 1$ we get

$$\lim_{\substack{(x,y) \to (0,0) \\ y = x}} f(x, y) = \frac{1 - 1}{1 + 1} = \frac{0}{2} = 0$$

and for $m = 0$

$$\lim_{\substack{(x,y) \to (0,0) \\ y = 0}} f(x, y) = \frac{1 - 0}{1 + 0} = 1.$$

Thus no matter how close they are to the origin, there will be points (x, y) such that $f(x, y)$ is close to 0 and points (x, y) where $f(x, y)$ is close to 1. So the limit:

$$\lim_{(x,y) \to (0,0)} f(x, y) \text{ doesn't exist}$$

12. We want to show that f doesn't have a limit as (x, y) approaches $(0, 0)$. For this let us consider $x > 0, y > 0$, which gives

$$\lim_{\substack{(x,y) \to (0,0) \\ x > 0, y > 0}} f(x, y) = \lim_{\substack{(x,y) \to (0,0) \\ x > 0, y > 0}} \frac{xy}{|xy|} = 1.$$

On the other hand, if $x > 0, y < 0$, we get

$$\lim_{\substack{(x,y) \to (0,0) \\ x > 0, y < 0}} f(x, y) = \lim_{\substack{(x,y) \to (0,0) \\ x > 0, y < 0}} \frac{xy}{|xy|} = \lim_{\substack{(x,y) \to (0,0) \\ x > 0, y < 0}} \frac{xy}{-xy} = -1.$$

Thus no matter how close to the origin they are, there will be points (x, y) such that $f(x, y)$ is close to 1 and points (x, y) such that $f(x, y)$ is close to -1. So the limit

$$\lim_{(x,y) \to (0,0)} f(x, y) \text{ doesn't exist.}$$

13. A contour of g is the curve defined by the equation $g(x, y) = c$, where c is a constant.

$$\frac{x^2}{x^2 + y^2} = c \quad \text{or} \quad x^2 = cx^2 + cy^2 \text{ or again}$$
$$(1 - c)x^2 = cy^2.$$

Thus if $c < 0$ or $c > 1$ the only point which satisfies the above equation is $(0, 0)$ but this is not in the domain of g. For these values of c there are no contours.

   If $c = 0$ we get $x = 0$ which is the y-axis without the origin (hence two rays of undefined slope).
   If $c = 1$ we get $y = 0$ which is the x-axis without the origin (hence two rays of slope 0).
   Finally, if $0 < c < 1$, we get two lines without the origin, namely:

$$y = \pm \sqrt{\frac{1 - c}{c}} x, \quad x \neq 0.$$

Therefore we get four rays of slopes $\sqrt{\frac{1-c}{c}}$ and $-\sqrt{\frac{1-c}{c}}$ respectively.

14. We will study the continuity of f at $(a, 0)$. Now $f(a, 0) = 1 - a$. In addition:

$$\lim_{\substack{(x,y)\to(a,0)\\y>0}} f(x,y) = \lim_{x\to a}(1-x) = 1 - a$$

$$\lim_{\substack{(x,y)\to(a,0)\\y<0}} f(x,y) = \lim_{x\to a} -2 = -2.$$

If $a = 3$, then

$$\lim_{\substack{(x,y)\to(3,0)\\y>0}} f(x,y) = 1 - 3 = -2 = \lim_{\substack{(x,y)\to(3,0)\\y<0}} f(x,y)$$

and so $\lim_{(x,y)\to(3,0)} f(x,y) = -2 = f(3,0)$. Therefore f is continuous at $(3, 0)$.
On the other hand, if $a \neq 3$, then

$$\lim_{\substack{(x,y)\to(a,0)\\y>0}} f(x,y) = 1 - a \neq -2 = \lim_{\substack{(x,y)\to(a,0)\\y<0}} f(x,y)$$

so $\lim_{(x,y)\to(a,0)} f(x,y)$ doesn't exist. Thus f is not continuous at $(a, 0)$ if $a \neq 3$.
   Thus, f is not continuous along the line $y = 0$. (In fact the only point on this line where f is continuous is the point $(3, 0)$.)

15. The function f is continuous at all points (x, y) with $x \neq 3$. So let's analyze the continuity of f at the point $(3, a)$. We have

$$\lim_{\substack{(x,y)\to(3,a)\\x<3}} f(x,y) = \lim_{y\to a}(c+y) = c + a$$

$$\lim_{\substack{(x,y)\to(3,a)\\x>3}} f(x,y) = \lim_{y\to a}(5-y) = 5 - a.$$

So we need to see if we can find one value for c such that $c + a = 5 - a$ for all a. This would require that $c = 5 - 2a$, but then c would depend on a, which is exactly what we don't want. Therefore, we cannot make the function continuous everywhere.

16. The function, f is continuous at all points (x, y) with $x \neq 3$. We analyze the continuity of f at the point $(3, a)$. We have:

$$\lim_{(x,y)\to(3,a),x<3} f(x,y) = \lim_{y\to a}(c+y) = c + a$$

$$\lim_{(x,y)\to(3,a),x>3} f(x,y) = \lim_{x>3,x\to 3}(5-x) = 2.$$

We want to see if we can find one value of c such that $c + a = 2$ for all a. This would mean that $c = 2 - a$, but then c would be dependent on a. Therefore, we cannot make the function continuous everywhere.

Solutions for Chapter 11 Review

1. These conditions describe a line parallel to the z-axis which passes through the xy-plane at $(2, 1, 0)$.

2. We complete the square

$$x^2 + 4x + y^2 - 6y + z^2 + 12z = 0$$
$$x^2 + 4x + 4 + y^2 - 6y + 9 + z^2 + 12z + 36 = 4 + 9 + 36$$
$$(x + 2)^2 + (y - 3)^2 + (z + 6)^2 = 49$$

The center is $(-2, 3, -6)$ and the radius is 7.

3. The equation will be of the form $mx + ny + ez = d$, but you can divide through by d to get an equation of the form $ax + by + cz = 1$ (d can not be zero, as the origin is not in the plane). Now plug in the points: From $(0, 0, 2)$, we get $a(0) + b(0) + c(2) = 1$. From this we get $c = \frac{1}{2}$. Similarly we get $a = \frac{1}{5}$, and $b = \frac{1}{3}$. So the equation that fits these points is

$$\frac{x}{5} + \frac{y}{3} + \frac{z}{2} = 1.$$

The equation of this plane can also be obtained by calculating the normal as the cross product of two vectors lying in the plane.

4. Let the equation of the plane be $z = ax + by + c$. When $z = 0$, the line on the xy-plane is $ax + by + c = 0$. Since we know that the plane intersects the xy-plane along the line $y = 2x + 2$ we have $b \neq 0$ and

$$-\frac{a}{b} = 2 \qquad -\frac{c}{b} = 2$$

Since $(1, 2, 2)$ lies on the plane, we can use the equation $z = ax + by + c$ to get

$$2 = a + 2b + c$$

Solving the equations gives

$$a = 2,$$
$$b = -1,$$
$$c = 2.$$

Hence $z = 2x - y + 2$ and the linear function is $f(x, y) = 2x - y + 2$.

5. When h is fixed, say $h = 1$, then
$$V = f(r, 1) = \pi r^2 1 = \pi r^2$$

Similarly,
$$f(r, \frac{2}{3}) = \frac{4}{9}\pi r^2 \quad \text{and} \quad f(r, \frac{1}{3}) = \frac{\pi}{9} r^2$$

When r is fixed, say $r = 1$, then
$$f(1, h) = \pi (1)^2 h = \pi h$$

Similarly,
$$f(2, h) = 4\pi \quad \text{and} \quad f(3, h) = 9\pi h.$$

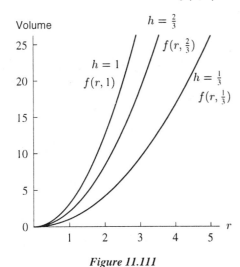

Figure 11.111

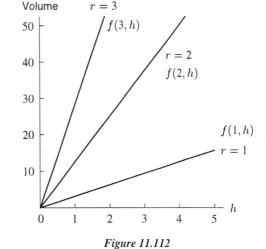

Figure 11.112

6. (a) Since $z = c$, where $-1 \le c \le 1$ is a constant, gives $\sqrt{x^2 + y^2} = \pm \cos^{-1}(c) + 2k\pi$, where k is any integer such that $\pm \cos^{-1}(c) + 2k\pi$ is non-negative, or $x^2 + y^2 = r^2$, where $r = \pm \cos^{-1}(c) + 2k\pi$, which represents a family of circles of radius r centered at $(0, 0)$, the level curves of the function are families of circles, as shown in Figure 11.113.

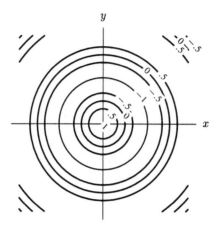

Figure 11.113

(b) The plane containing the x- and z-axes is the plane $y = 0$. Thus the cross-section is $z = \cos \sqrt{x^2 + 0^2} = \cos(|x|) = \cos x$, as shown in Figure 11.114.

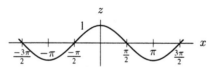

Figure 11.114

(c) Denote the line $y = x$ in the xy-plane as r-axis and put units on it such that the units on the r-axis coincide with the units on the x-axis and y-axis, namely, $r^2 = x^2 + y^2$. Thus, the cross-section is $z = \cos \sqrt{r^2} = \cos(|r|) = \cos r$, as shown in Figure 11.115.

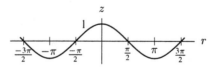

Figure 11.115

7.

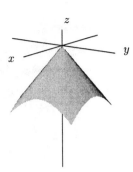

One possible equation: $z = -\sqrt{x^2 + y^2}$.

8.

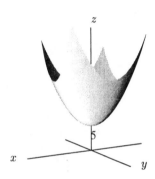

One possible equation: $z = x^2 + y^2 + 5$

9.

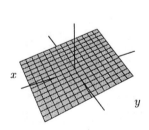

One possible equation: $x + y + z = 1$.

10.

Figure 11.116

One possible equation: $z = (x - y)^2$.

11. Could not be true. If the origin is on the level curve $z = 1$, then $z = f(0,0) = 1 \neq -1$. So $(0,0)$ cannot be on both $z = 1$ and $z = -1$.

12. Might be true. One may consider the function

$$z = f(x, y) = (x^2 + y^2 - 2)(x^2 + y^2 - 3) + 1$$

13. Might be true. The function $z = x^2 - y^2 + 1$ has this property. The level curve $z = 1$ is the lines $y = x$ and $y = -x$.

14. Not true. There are no level curves for $z > 1$ or $z \leq 0$.

15. True. For every point (x, y), compute the value $z = e^{-(x^2+y^2)}$ at that point. The level curve obtained by getting z equal to that value goes through the point (x, y).

16. Since setting $z = c$, with $-1 \leq c \leq 1$ gives $y = \sin^{-1} c + 2n\pi$ or $y = \pi - \sin^{-1} c + 2n\pi =$ constant, where n is any integer, contours are horizontal lines as shown in Figure 11.117.

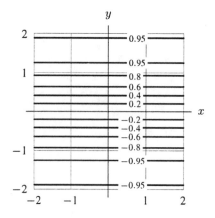

Figure 11.117

17. Contours are lines of the form $3x - 5y + 1 = c$ as shown in Figure 11.118. Note that for the regions of x and y given, the c values range from $-15 < c < 17$.

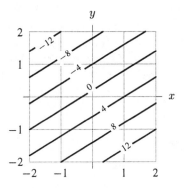

Figure 11.118

18. Contours are ellipses of the form $2x^2 + y^2 = c$ as shown in Figure 11.119. Note that for the ranges of x and y given, the range of c value is $0 \le c < 12$.

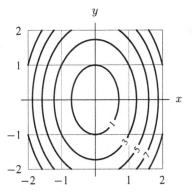

Figure 11.119

19. The contours are ellipses of the form $2x^2 + y^2 = -\ln c$ as shown in Figure 11.120. For the ranges of x and y given, the c values range from just above 0 to 1.

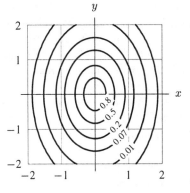

Figure 11.120

20.   The point $x = 10$, $t = 5$ is between the contours $H = 70$ and $H = 75$, a little closer to the former. So we estimate $H(10, 5) \approx 72$, i.e., it is about 72°F. Five minutes later we are at the point $x = 10$, $t = 10$, which is just above the contour $H = 75$, so we estimate that it has warmed up to 76°F by then.

21.   The line $t = 5$ crosses the contour $H = 80$ at about $x = 4$; this means that $H(4, 5) \approx 80$, and so the point $(4, 80)$ is on the graph of the one-variable function $y = H(x, 5)$. Each time the line crosses a contour, we can plot another point on the graph of $H(x, 5)$, and thus get a sketch of the graph. See Figure 11.121. Each data point obtained from the contour map has been indicated by a dot on the graph. The graph of $H(x, 20)$ was obtained in a similar way.

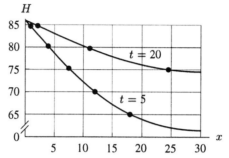

Figure 11.121: Graph of $H(x, 5)$ and $H(x, 20)$: heat as a function of distance from the heater at $t = 5$ and $t = 20$ minutes

These two graphs describe the temperature at different positions as a function of x for $t = 5$ and $t = 20$. Notice that the graph of $H(x, 5)$ descends more steeply than the graph of $H(x, 20)$; this is because the contours are quite close together along the line $t = 5$, whereas they are more spread out along the line $t = 20$. In practical terms the shape of the graph of $H(x, 5)$ tells us that the temperature drops quickly as you move away from the heater, which makes sense, since the heater was turned on just five minutes ago. On the other hand, the graph of $H(x, 20)$ descends more slowly, which makes sense, because the heater has been on for 20 minutes and the heat has had time to diffuse throughout the room.

22.   To read off the cross-sections of f with t fixed, we choose a t value and move horizontally across the diagram looking at the values on the contours. For $t = 0$, as we move from the left at $x = 0$ to the right at $x = \pi$, we cross contours of 0.25, 0.50, 0.75 and reach a maximum at $x = \pi/2$, and then decrease back to 0. That is because if time is fixed at $t = 0$, then $f(x, 0)$ is the displacement of the string at that time: no displacement at $x = 0$ and $x = \pi$ and greatest displacement at $x = \pi/2$. For cross-sections with t fixed at larger values, as we move along a horizontal line, we cross fewer contours and reach a smaller maximum value: the string is becoming less curved. At time $t = \pi/2$, the string is straight so we see a value of 0 all the way across the diagram, namely a contour with value 0. For $t = \pi$, the string has vibrated to the other side and the displacements are negative as we read across the diagram reaching a minimum at $x = \pi/2$.

The cross-sections of f with x fixed are read vertically. At $x = 0$ and $x = \pi$, we see vertical contours of value 0 because the end points of the string have 0 displacement no matter what time it is. The cross-section for $x = \pi/2$ is found by moving vertically up the diagram at $x = \pi/2$. As we expect, the contour values are largest at $t = 0$, zero at $t = \pi/2$, and a minimum at $t = \pi$.

Notice that the spacing of the contours is also important. For example, for the $t = 0$ cross-section, contours are most closely spaced at the end points at $x = 0$ and $x = \pi$ and most spread out at $x = \pi/2$. That is because the shape of the string at time $t = 0$ is a sine curve, which is steepest at the end points and relatively flat in the middle. Thus, the contour diagram shows the steepest terrain at the end points and flattest terrain in the middle.

23. The function, g, has a slope of 3 in the x direction and a slope of 1 in the y direction, so $g(x,y) = c + 3x + y$. Since $g(0,0) = 0$, the formula is $g(x,y) = 3x + y$.

24. The function h decreases as y increases: each increase of y by 2 takes you down one contour and hence changes the function by 2, so the slope in the y direction is -1. The slope in the x direction is 2, so the formula is $h(x,y) = c + 2x - y$. From the diagram we see that $h(0,0) = 4$, so $c = 4$. Therefore, the formula for this linear function is $h(x,y) = 4 + 2x - y$.

25. (a) Let $x =$ distance (microns) from center of waveguide, $t =$ time (nanoseconds) as shown in the problem, and $I =$ intensity of light as marked on the given level curves.

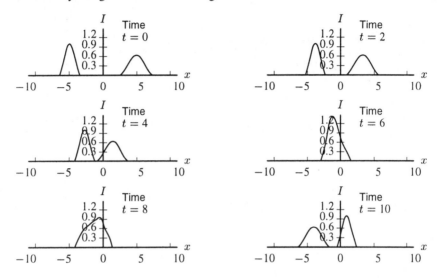

Figure 11.122

(b) Two waves would start out at opposite ends of the screen. The wave on the left would be slightly taller and narrower than the wave on the right. The waves would move toward one another, the wave on the right moving a little faster. They would meet to the left of the center and appear to merge, becoming taller. They would then proceed in the directions they were initially going, ultimately leaving the screen on the side opposite to where they began.

(c) Let $x =$ distance (microns), $t =$ time (nanoseconds), and $I =$ intensity.

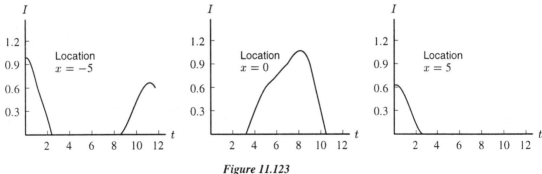

Figure 11.123

(d) Two pulses of light are traveling down a wave-guide toward one another. They meet in the center and, as they pass through one another, appear brighter. They then continue along in the wave-guide in the directions they were going.

CHAPTER TWELVE

1.

$$\vec{p} = 2\vec{w}, \quad \vec{q} = -\vec{u}, \quad \vec{r} = \vec{w} + \vec{u} = \vec{u} + \vec{w},$$
$$\vec{s} = \vec{p} + \vec{q} = 2\vec{w} - \vec{u}, \quad \vec{t} = \vec{u} - \vec{w}$$

2.

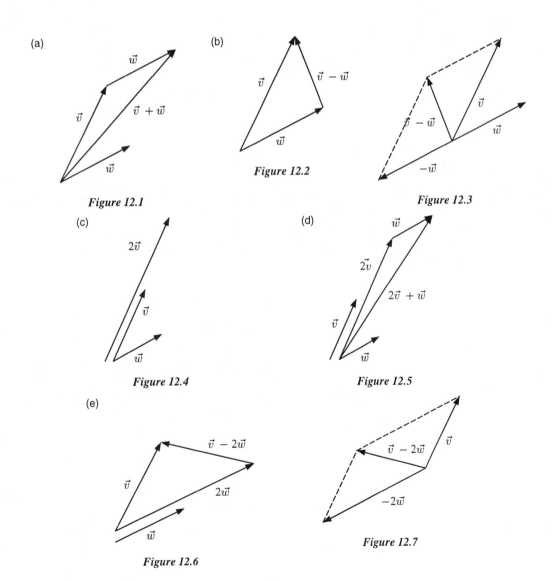

(a)

Figure 12.1

(b)

Figure 12.2

Figure 12.3

(c)

Figure 12.4

(d)

Figure 12.5

(e)

Figure 12.6

Figure 12.7

3. $\|\vec{z}\| = \sqrt{(1)^2 + (-3)^2 + (-1)^2} = \sqrt{1 + 9 + 1} = \sqrt{11}$.

4. $\vec{a} + \vec{z} = (2\vec{j} + \vec{k}) + (\vec{i} - 3\vec{j} - \vec{k}) = (0 + 1)\vec{i} + (2 - 3)\vec{j} + (1 - 1)\vec{k} = \vec{i} - \vec{j}$

5. $5\vec{b} = 5(-3\vec{i} + 5\vec{j} + 4\vec{k}) = -15\vec{i} + 25\vec{j} + 20\vec{k}$.

6. $2\vec{c} + \vec{x} = 2(\vec{i} + 6\vec{j}) + (-2\vec{i} + 9\vec{j}) = (2\vec{i} + 12\vec{j}) + (-2\vec{i} + 9\vec{j}) = (2 - 2)\vec{i} + (12 + 9)\vec{j} = 21\vec{j}$.

7. $\|\vec{y}\| = \sqrt{(4)^2 + (-7)^2} = \sqrt{16 + 49} = \sqrt{65}$.

8.

$$2\vec{a} + 7\vec{b} - 5\vec{z} = 2(2\vec{j} + \vec{k}) + 7(-3\vec{i} + 5\vec{j} + 4\vec{k}) - 5(\vec{i} - 3\vec{j} - \vec{k})$$
$$= (4\vec{j} + 2\vec{k}) + (-21\vec{i} + 35\vec{j} + 28\vec{k}) - (5\vec{i} - 15\vec{j} - 5\vec{k})$$
$$= (-21 - 5)\vec{i} + (4 + 35 + 15)\vec{j} + (2 + 28 + 5)\vec{k} = -26\vec{i} + 54\vec{j} + 35\vec{k}.$$

9. $4\vec{i} + 2\vec{j} - 3\vec{i} + \vec{j} = \vec{i} + 3\vec{j}$

10. $\vec{i} + 2\vec{j} - 6\vec{i} - 3\vec{j} = -5\vec{i} - \vec{j}$

11. $-4\vec{i} + 8\vec{j} - 0.5\vec{i} + 0.5\vec{k} = -4.5\vec{i} + 8\vec{j} + 0.5\vec{k}$

12. $(0.9\vec{i} - 1.8\vec{j} - 0.02\vec{k}) - (0.6\vec{i} - 0.05\vec{k}) = 0.3\vec{i} - 1.8\vec{j} + 0.03\vec{k}$

13. $\|\vec{v}\| = \sqrt{1^2 + (-1)^2 + 3^2} = \sqrt{11}$.

14. $\|\vec{v}\| = \sqrt{1^2 + (-1)^2 + 2^2} = \sqrt{6}$.

15. $\|\vec{v}\| = \sqrt{1.2^2 + (-3.6)^2 + 4.1^2} = \sqrt{31.21} \approx 5.6$.

16. $\|\vec{v}\| = \sqrt{7.2^2 + (-1.5)^2 + 2.1^2} = \sqrt{58.5} \approx 7.6$.

17. We get displacement by subtracting the coordinates of the origin $(0, 0, 0)$ from the coordinates of the cat $(1, 4, 0)$, giving

    Displacement $= (1 - 0)\vec{i} + (4 - 0)\vec{j} + (0 - 0)\vec{k} = \vec{i} + 4\vec{j}$.

18. We get displacement by subtracting the coordinates of the bottom of the tree, $(2, 4, 0)$, from the coordinates of the squirrel, $(2, 4, 1)$, giving:

    $$\text{Displacement} = (2 - 2)\vec{i} + (4 - 4)\vec{j} + (1 - 0)\vec{k} = \vec{k}.$$

19.

$$\text{Displacement} = \text{Cat's coordinates} - \text{Bottom of the tree's coordinates}$$
$$= (1 - 2)\vec{i} + (4 - 4)\vec{j} + (0 - 0)\vec{k} = -\vec{i}.$$

20.

$$\text{Displacement} = \text{Squirrel's coordinates} - \text{Cat's coordinates}$$
$$= (2 - 1)\vec{i} + (4 - 4)\vec{j} + (1 - 0)\vec{k} = \vec{i} + \vec{k}.$$

21.

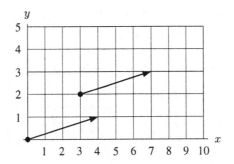

Figure 12.8: $\vec{v}$

22. Using a ruler, we measure the length of the x component of the vector to be slightly less than 0.75 inches in the $-x$ direction. Therefore the x component is $\frac{-0.725}{0.25}\vec{i} = -2.9\vec{i}$. The length of the y component is 0.95 inches in the $-y$ direction, so the y component is $\frac{-0.95}{0.25}\vec{j} = -3.8\vec{j}$. Therefore, $\vec{w} = -2.9\vec{i} - 3.8\vec{j}$.

23. The vector we want is the displacement from P to Q, which is given by

$$\vec{PQ} = (4-1)\vec{i} + (6-2)\vec{j} = 3\vec{i} + 4\vec{j}$$

24. The vector we want is the displacement from Q to P, which is given by

$$\vec{QP} = (1-4)\vec{i} + (2-6)\vec{j} = -3\vec{i} - 4\vec{j}$$

25. $\vec{a} = \vec{b} = \vec{c} = 3\vec{k}$, $\quad \vec{d} = 2\vec{i} + 3\vec{k}$, $\quad \vec{e} = \vec{j}$, $\quad \vec{f} = -2\vec{i}$

26. $\vec{u} = \vec{i} + \vec{j} + 2\vec{k}$ and $\vec{v} = -\vec{i} + 2\vec{k}$.

27. $\|\vec{u}\| = \sqrt{1^2 + 1^2 + 2^2} = \sqrt{6}, \|\vec{v}\| = \sqrt{(-1)^2 + 2^2} = \sqrt{5}$

28. The length of the vector $\vec{i} - \vec{j} + 2\vec{k}$ is $\sqrt{1^2 + (-1)^2 + 2^2} = \sqrt{6}$. We can scale the vector down to length 2 by multiplying it by $\frac{2}{\sqrt{6}}$. So the answer is $\frac{2}{\sqrt{6}}\vec{i} - \frac{2}{\sqrt{6}}\vec{j} + \frac{4}{\sqrt{6}}\vec{k}$.

29. (a) The displacement from P to Q is given by

$$\vec{PQ} = (4\vec{i} + 6\vec{j}) - (\vec{i} + 2\vec{j}) = 3\vec{i} + 4\vec{j}.$$

Since

$$\|\vec{PQ}\| = \sqrt{3^2 + 4^2} = 5,$$

a unit vector $\vec{u}$ in the direction of $\vec{PQ}$ is given by

$$\vec{u} = \frac{1}{5}\vec{PQ} = \frac{1}{5}(3\vec{i} + 4\vec{j}) = \frac{3}{5}\vec{i} + \frac{4}{5}\vec{j}.$$

(b) A vector of length 10 pointing in the same direction is given by

$$10\vec{u} = 10(\frac{3}{5}\vec{i} + \frac{4}{5}\vec{j}) = 6\vec{i} + 8\vec{j}.$$

30. To determine if two vectors are parallel, we need to see if one vector is a scalar multiple of the other one. Since $\vec{u} = -2\vec{w}$, and $\vec{v} = \frac{1}{4}\vec{q}$ and no other pairs have this property, only $\vec{u}$ and $\vec{w}$, and $\vec{v}$ and $\vec{q}$ are parallel.

31. We must check that all the points are the same distance apart, i.e., the magnitude of the displacement vectors $\overrightarrow{OA}, \overrightarrow{OB}, \overrightarrow{OC}, \overrightarrow{BA}, \overrightarrow{CB}$ and $\overrightarrow{CA}$ is the same. Here goes:

$$\|\overrightarrow{OA}\| = \|(2\vec{i} + 0\vec{j} + 0\vec{k}) - (0\vec{i} + 0\vec{j} + 0\vec{k})\| = \sqrt{2^2 + 0^2 + 0^2} = 2$$

$$\|\overrightarrow{OB}\| = \|(1\vec{i} + \sqrt{3}\vec{j} + 0\vec{k}) - (0\vec{i} + 0\vec{j} + 0\vec{k})\| = \sqrt{1^2 + (\sqrt{3})^2 + 0^2} = 2$$

$$\|\overrightarrow{OC}\| = \|(1\vec{i} + 1/\sqrt{3}\vec{j} + 2\sqrt{2/3}\vec{k}) - (0\vec{i} + 0\vec{j} + 0\vec{k})\| = \sqrt{1 + 1/3 + 4(2/3)} = 2$$

$$\|\overrightarrow{BA}\| = \|(2\vec{i} + 0\vec{j} + 0\vec{k}) - (1\vec{i} + \sqrt{3}\vec{j} + 0\vec{k})\| = \sqrt{1 + 3 + 0} = 2$$

$$\|\overrightarrow{CB}\| = \|(1\vec{i} + \sqrt{3}\vec{j} + 0\vec{k}) - (1\vec{i} + 1/\sqrt{3}\vec{j} + 2\sqrt{2/3}\vec{k})\|$$

$$= \sqrt{0^2 + (\sqrt{3} - 1/\sqrt{3})^2 + 4(2/3)} = \sqrt{3 - 2 + 1/3 + 8/3} = 2$$

$$\|\overrightarrow{CA}\| = \|(2\vec{i} + 0\vec{j} + 0\vec{k}) - (1\vec{i} + 1/\sqrt{3}\vec{j} + 2\sqrt{2/3}\vec{k})\| = \sqrt{1 + 1/3 + 4(2/3)} = 2.$$

32. The vectors $\vec{u}$, $\vec{v}$, and $\vec{w}$ are shown in Figure 12.9. Notice that $\vec{v}$ and $\vec{w}$ are in the same direction.

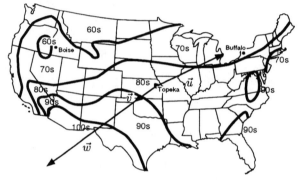

Figure 12.9

(a) Decreasing because you are moving from the 80s toward the 70s.
(b) Increasing because you are moving from the 80s toward the 90s.
(c) Increasing

33.

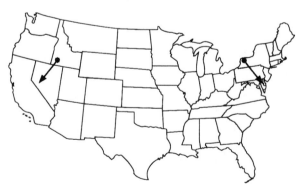

Figure 12.10

34.

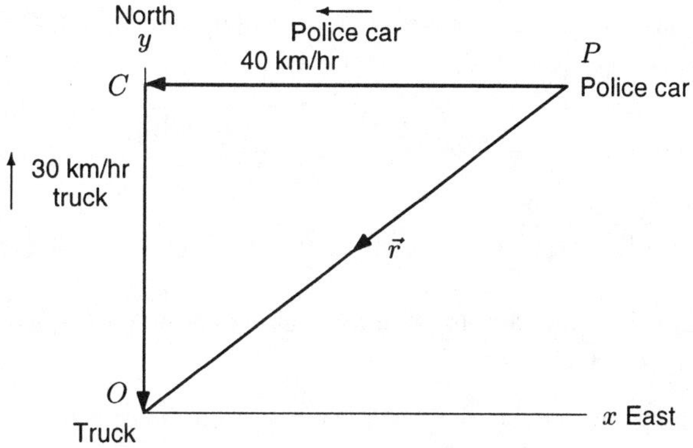

North
y
Police car
40 km/hr
P
Police car
C
30 km/hr
truck
$\vec{r}$
O
Truck
x East

Figure 12.11

Since both vehicles reach the crossroad in exactly one hour, at the present the truck is at O in Figure 12.11; the police car is at P and the crossroads is at C. If $\vec{r}$ is the vector representing the line of sight of the truck with respect to the police car.

$$\vec{r} = -40\vec{i} - 30\vec{j}$$

35. In Figure 12.12 let O be the origin, points A, B, and C be the vertices of the triangle, point D be the midpoint of $\overline{BC}$, and Q be the point in the line segment $\overline{DA}$ that is $\frac{1}{3}|DA|$ away from D.

y
C
A
Q
D
B
O
x

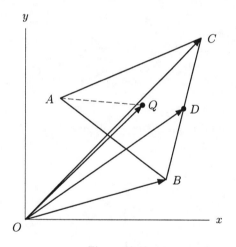

Figure 12.12

From Figure 12.12 we see that

$$\begin{aligned}
\overrightarrow{OQ} = \overrightarrow{OD} + \overrightarrow{DQ} &= \overrightarrow{OD} + \frac{1}{3}\overrightarrow{DA} \\
&= \overrightarrow{OD} + \frac{1}{3}(\overrightarrow{OA} - \overrightarrow{OD}) \\
&= \overrightarrow{OD} + \frac{1}{3}\overrightarrow{OA} - \frac{1}{3}\overrightarrow{OD} \\
&= \frac{1}{3}\overrightarrow{OA} + \frac{2}{3}\overrightarrow{OD}.
\end{aligned}$$

Because the diagonals of a parallelogram meet at their midpoint, and $2\overrightarrow{OD}$ is a diagonal of the parallelogram formed by $\overrightarrow{OB}$ and $\overrightarrow{OC}$, we have:

$$\overrightarrow{OD} = \frac{1}{2}(\overrightarrow{OB} + \overrightarrow{OC}),$$

so we can write:

$$\overrightarrow{OQ} = \frac{1}{3}\overrightarrow{OA} + \frac{2}{3}\left(\frac{1}{2}\right)(\overrightarrow{OB} + \overrightarrow{OC}) = \frac{1}{3}(\overrightarrow{OA} + \overrightarrow{OB} + \overrightarrow{OC}).$$

Thus a vector from the origin to a point $\frac{1}{3}$ of the way along median AD from D, the midpoint, is given by $\frac{1}{3}(\overrightarrow{OA} + \overrightarrow{OB} + \overrightarrow{OC})$.

In a similar manner we can show that the vector from the origin to the point $\frac{1}{3}$ of the way along any median from the midpoint of the side it bisects is also $\frac{1}{3}(\overrightarrow{OA} + \overrightarrow{OB} + \overrightarrow{OC})$. See Figure 12.13 and 12.14.

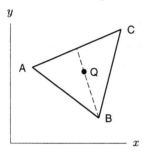

Figure 12.13

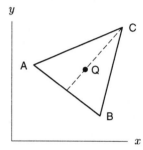

Figure 12.14

Thus the medians of a triangle intersect at a point $\frac{1}{3}$ of the way along each median from the side that each bisects.

36. We want to find an expression for a vector from the origin to a point that is $\frac{1}{4}$ of the way from a centroid to its opposite vertex.

In Figure 12.15 let O be the origin, A, B, C, and D be the vertices of a tetrahedron, P be the centroid of face BCD, and Q be the point on $\overline{PA}$ that is $\left|\frac{1}{4}\overline{PA}\right|$ away from P.

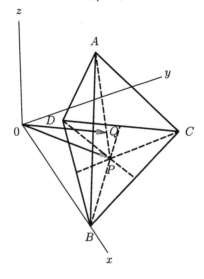

Figure 12.15

$$\overrightarrow{OQ} = \overrightarrow{OP} + \overrightarrow{PQ}$$
$$= \overrightarrow{OP} + \frac{1}{4}\overrightarrow{PA}$$
$$= \overrightarrow{OP} + \frac{1}{4}(\overrightarrow{OA} - \overrightarrow{OP})$$
$$= \overrightarrow{OP} + \frac{1}{4}\overrightarrow{OA} - \frac{1}{4}\overrightarrow{OP}$$
$$= \frac{1}{4}\overrightarrow{OA} + \frac{3}{4}\overrightarrow{OP}.$$

In Problem 35 we showed that a vector from the origin to P, the centroid of a triangle, is

$$\overrightarrow{OP} = \frac{1}{3}(\overrightarrow{OB} + \overrightarrow{OC} + \overrightarrow{OD}).$$

Substituting this into our expression for $\overrightarrow{OQ}$ gives

$$\overrightarrow{OQ} = \frac{1}{4}\overrightarrow{OA} + \frac{3}{4}\left(\frac{1}{3}\right)(\overrightarrow{OB} + \overrightarrow{OC} + \overrightarrow{OD})$$
$$= \frac{1}{4}\overrightarrow{OA} + \frac{1}{4}(\overrightarrow{OB} + \overrightarrow{OC} + \overrightarrow{OD})$$
$$= \frac{1}{4}(\overrightarrow{OA} + \overrightarrow{OB} + \overrightarrow{OC} + \overrightarrow{OD}).$$

In a similar manner we can show that a vector from the origin to a point $\frac{1}{4}$ of the way from the centroid of *any* face to its opposite vertex is $\frac{1}{4}(\overrightarrow{OA} + \overrightarrow{OB} + \overrightarrow{OC} + \overrightarrow{OD})$. Thus, lines joining the centroid of each face to its opposite vertex all meet at a single point which is $\frac{1}{4}$ of the way from any centroid to its opposite face.

Solutions for Section 12.2

1. Scalar

2. Scalar

3. The magnetic field is a vector because it has both a magnitude (the strength of the field) and a direction (the direction of the compass).

4. Temperature is measured by a single number, and so is a scalar.

5. (a) If the car is going east, it is going solely in the positive x direction, so its velocity vector is $50\vec{i}$.
   (b) If the car is going south, it is going solely in the negative y direction, so its velocity vector is $-50\vec{j}$.
   (c) If the car is going southeast, the angle between the x-axis and the velocity vector is $-45°$. Therefore

   $$\text{velocity vector} = 50\cos(-45°)\vec{i} + 50\sin(-45°)\vec{j}$$
   $$= 25\sqrt{2}\vec{i} - 25\sqrt{2}\vec{j}.$$

   (d) If the car is going northwest, the velocity vector is at a $45°$ angle to the y-axis, which is $135°$ from the x-axis. Therefore:

   $$\text{velocity vector} = 50(\cos 135°)\vec{i} + 50(\sin 135°)\vec{j} = -25\sqrt{2}\vec{i} + 25\sqrt{2}\vec{j}.$$

6. We need to calculate the length of each vector.

$$\|21\vec{i} + 35\vec{j}\| = \sqrt{21^2 + 35^2} = \sqrt{1666} \approx 40.8,$$
$$\|40\vec{i}\| = \sqrt{40^2} = 40.$$

So the first car is faster.

7. There are 1000 meters in 1 km and $60 \cdot 60 = 3600$ seconds in 1 hour. Thus,

$$\vec{v} = \frac{3600 \text{ sec}}{1 \text{ hour}} \cdot \frac{1 \text{ km}}{1000 \text{ meters}} (50\vec{i} + 20\vec{j}) \frac{\text{meters}}{\text{sec}}$$
$$= \frac{3600 \cdot 50}{1000}\vec{i} + \frac{3600 \cdot 20}{1000}\vec{j} \frac{\text{km}}{\text{hour}}$$
$$= 180\vec{i} + 72\vec{j} \quad \text{kilometers per hour.}$$

8.

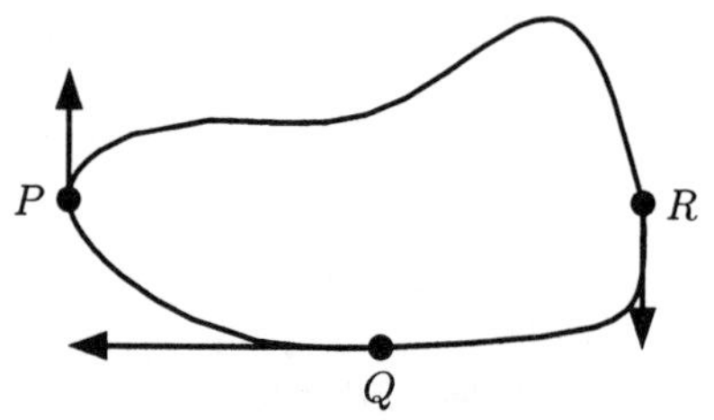

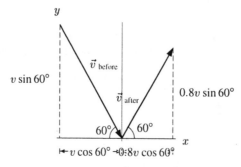

Figure 12.16

9. At the point P, the velocity of the car is changing the quickest; not in magnitude, but in direction only. The acceleration vector is therefore the longest at this point. The direction of the vector is directed in towards the center of the track because the difference in velocity vectors at nearby points is a vector pointing toward the center.

10. The speed of the particle before impact is v, so the speed after impact is $0.8v$. If we consider the barrier as being along the x-axis (see Figure 12.17), then the $\vec{i}$-component is $0.8v \cos 60° = 0.8v(0.5) = 0.4v$.

Similarly, the $\vec{j}$-component is $0.8v \sin 60° = 0.8v(0.8660) \approx 0.7v$. Thus

$$\vec{v}_{\text{after}} = 0.4v\vec{i} + 0.7v\vec{j}.$$

Figure 12.17

11. The total scores are out of 300 and are given by the total score vector $\vec{v} + 2\vec{w}$:

$$\vec{v} + 2\vec{w} = (73, 80, 91, 65, 84) + 2(82, 79, 88, 70, 92)$$
$$= (73, 80, 91, 65, 84) + (164, 158, 176, 140, 184)$$
$$= (237, 238, 267, 205, 268).$$

To get the scores as a percentage, we divide by 3, giving

$$\frac{1}{3}(237, 238, 267, 205, 268) \approx (79.00, 79.33, 89.00, 68.33, 89.33).$$

12. Let the velocity vector of the airplane be $\vec{V} = x\vec{i} + y\vec{j} + z\vec{k}$ in km/hr. We know that $x = -y$ because the plane is traveling northwest. Also, $\|\vec{V}\| = \sqrt{x^2 + y^2 + z^2} = 200$ km/hr and $z = 300$ m/min $= 18$ km/h. We have $\sqrt{x^2 + y^2 + z^2} = \sqrt{x^2 + x^2 + 18^2} = 200$, so $x = -140.8, y = 140.8, z = 18$. (The value of x is negative and y is positive because the plane is heading northwest.) Thus,

$$\vec{v} = -140.8\vec{i} + 140.8\vec{j} + 18\vec{k}.$$

13. Suppose $\vec{u}$ represents the velocity of the plane relative to the air and $\vec{w}$ represents the velocity of the wind. We can add these two vectors by adding their components. Suppose north is in the y-direction and east is the x-direction. The vector representing the airplane's velocity makes an angle of $45°$ with north; the components of $\vec{u}$ are

$$\vec{u} = 700\sin 45°\vec{i} + 700\cos 45°\vec{j} \approx 495\vec{i} + 495\vec{j}.$$

Since the wind is blowing from the west, $\vec{w} = 60\vec{i}$. By adding these we get a resultant vector $\vec{v} = 555\vec{i} + 495\vec{j}$. The direction relative to the north is the angle θ shown in Figure 12.18 given by

$$\theta = \tan^{-1}\frac{x}{y} = \tan^{-1}\frac{555}{495}$$
$$\approx 48.3°$$

The magnitude of the velocity is

$$\|\vec{v}\| = \sqrt{495^2 + 555^2} = \sqrt{553{,}050}$$
$$= 744 \text{ km/hr.}$$

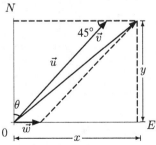

Figure 12.18: Note that θ is the angle between north and the vector $\vec{v}$

14. Let the x-axis point east and the y-axis point north. Since the wind is blowing from the northeast at a speed of 50 km/hr, the velocity of the wind is

$$\vec{w} = -50\cos 45^\circ \vec{i} - 50\sin 45^\circ \vec{j} \approx -35.4\vec{i} - 35.4\vec{j}.$$

Let $\vec{a}$ be the velocity of the airplane, relative to the air, and let ϕ be the angle from the x-axis to $\vec{a}$; since $\|\vec{a}\| = 600$ km/hr, we have $\vec{a} = 600\cos\phi\vec{i} + 600\sin\phi\vec{j}$. (See Figure 12.19.)

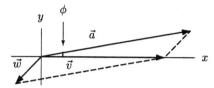

Figure 12.19

Now the resultant velocity, $\vec{v}$, is given by

$$\vec{v} = \vec{a} + \vec{w} = (600\cos\phi\vec{i} + 600\sin\phi\vec{j}) + (-35.4\vec{i} - 35.4\vec{j})$$
$$= (600\cos\phi - 35.4)\vec{i} + (600\sin\phi - 35.4)\vec{j}.$$

Since the airplane is to fly due east, i.e., in the x direction, then the y-component of the velocity must be 0, so we must have

$$600\sin\phi - 35.4 = 0$$
$$\sin\phi = \frac{35.4}{600}.$$

Thus $\phi = \arcsin(35.4/600) \approx 3.4^\circ$.

15. The velocity vector of the plane with respect to the air has the form

$$\vec{v} = a\vec{i} + 80\vec{k} \text{ where } \|\vec{v}\| = 480.$$

(See Figure 12.20.) Therefore $\sqrt{a^2 + 80^2} = 480$ so $a = \sqrt{480^2 - 80^2} \approx 473.3$ km/hr. We conclude that $\vec{v} \approx 473.3\vec{i} + 80\vec{k}$.

The wind vector is

$$\vec{w} = 100(\cos 45^\circ)\vec{i} + 100(\sin 45^\circ)\vec{j}$$
$$\approx 70.7\vec{i} + 70.7\vec{j}$$

The velocity vector of the plane with respect to the ground is then

$$\vec{v} + \vec{w} = (473.3\vec{i} + 80\vec{k}) + (70.7\vec{i} + 70.7\vec{j})$$
$$= 544\vec{i} + 70.7\vec{j} + 80\vec{k}$$

From Figure 12.21, we see that the velocity relative to the ground is

$$544\vec{i} + 70.7\vec{j}.$$

The ground speed is therefore $\sqrt{544^2 + 70.7^2} \approx 548.6$ km/hr.

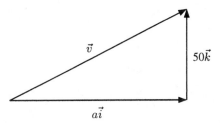

Figure 12.20: Side view

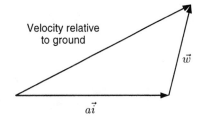

Figure 12.21: Top view

16.

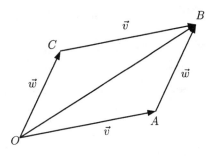

Figure 12.22

The vector $\vec{v} + \vec{w}$ is equivalent to putting the vectors $\overrightarrow{OA}$ and $\overrightarrow{AB}$ end-to-end as shown in Figure 12.22; the vector $\vec{w} + \vec{v}$ is equivalent to putting the vectors $\overrightarrow{OC}$ and $\overrightarrow{CB}$ end-to-end. Since they form a parallelogram, $\vec{v} + \vec{w}$ and $\vec{w} + \vec{v}$ are both equal to the vector $\overrightarrow{OB}$, we have $\vec{v} + \vec{w} = \vec{w} + \vec{v}$.

17. The vectors $\vec{v}$, $\alpha\vec{v}$ and $\beta\vec{v}$ are all parallel. Figure 12.23 shows them with α, $\beta > 0$, so all the vectors are in the same direction. Notice that $\alpha\vec{v}$ is a vector α times as long as $\vec{v}$ and $\beta\vec{v}$ is β times as long as $\vec{v}$. Therefore $\alpha\vec{v} + \beta\vec{v}$ is a vector $(\alpha + \beta)$ times as long as $\vec{v}$, and in the same direction. Thus,

$$\alpha\vec{v} + \beta\vec{v} = (\alpha + \beta)\vec{v}.$$

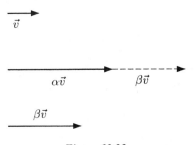

Figure 12.23

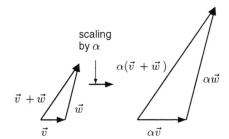

Figure 12.24

18. The effect of scaling the left-hand picture in Figure 12.24 is to stretch each vector by a factor of α (shown with $\alpha > 1$). Since, after scaling up, the three vectors $\alpha\vec{v}$, $\alpha\vec{w}$, and $\alpha(\vec{v} + \vec{w})$ form a similar triangle, we know that $\alpha(\vec{v} + \vec{w})$ is the sum of the other two: that is

$$\alpha(\vec{v} + \vec{w}) = \alpha\vec{v} + \alpha\vec{w}.$$

19.

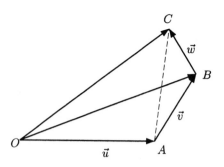

Figure 12.25

The vector $\vec{u} + \vec{v}$ is represented by $\overrightarrow{OB}$. The vector $(\vec{u} + \vec{v}) + \vec{w}$ is represented by $\overrightarrow{OB}$ followed by $\overrightarrow{BC}$, which is therefore $\overrightarrow{OC}$. Now $\vec{v} + \vec{w}$ is represented by $\overrightarrow{AC}$. So $\vec{u} + (\vec{v} + \vec{w})$ is $\overrightarrow{OA}$ followed by $\overrightarrow{AC}$, which is $\overrightarrow{OC}$. Since we get the vector $\overrightarrow{OC}$ by both methods, we know

$$(\vec{u} + \vec{v}) + \vec{w} = \vec{u} + (\vec{v} + \vec{w})$$

20. Assume α, $\beta > 0$. The vector $\beta\vec{v}$ is in the same direction and β times as long as $\vec{v}$. The vector $\alpha(\beta\vec{v})$ is in the same direction and α times as long as $\beta\vec{v}$, and so is $\alpha\beta$ times as long as $\vec{v}$ and in the same direction as $\vec{v}$. Thus,

$$\alpha(\beta\vec{v}) = (\alpha\beta)\vec{v}.$$

21. Since the zero vector has zero length, adding it to $\vec{v}$ has no effect.

22. According to the definition of scalar multiplication, $1 \cdot \vec{v}$ has the same direction and magnitude as $\vec{v}$, so it is the same as $\vec{v}$.

23. By Figure 12.26, the vectors $\vec{v} + (-1)\vec{w}$ and $\vec{v} - \vec{w}$ are equal.

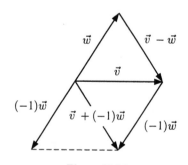

Figure 12.26

Solutions for Section 12.3

1. $\vec{c} \cdot \vec{y} = (\vec{i} + 6\vec{j}) \cdot (4\vec{i} - 7\vec{j}) = (1)(4) + (6)(-7) = 4 - 42 = -38.$
2. $\vec{a} \cdot \vec{z} = (2\vec{j} + \vec{k}) \cdot (\vec{i} - 3\vec{j} - \vec{k}) = (0)(1) + (2)(-3) + (1)(-1) = 0 - 6 - 1 = -7.$
3. $\vec{a} \cdot \vec{b} = (2\vec{j} + \vec{k}) \cdot (-3\vec{i} + 5\vec{j} + 4\vec{k}) = (0)(-3) + (2)(5) + (1)(4) = 0 + 10 + 4 = 14.$

4. Since $\vec{a} \cdot \vec{b}$ is a scalar and $\vec{a}$ is a vector, the answer to this equation is a vector parallel to $\vec{a}$. We have

$$\vec{a} \cdot \vec{b} = (2\vec{j} + \vec{k}) \cdot (-3\vec{i} + 5\vec{j} + 4\vec{k}) = 0(-3) + 2(5) + 1(4) = 14.$$

Thus,

$$(\vec{a} \cdot \vec{b}) \cdot \vec{a} = 14\vec{a} = 14(2\vec{j} + \vec{k}) = 28\vec{j} + 14\vec{k}$$

5. Since $\vec{a} \cdot \vec{y}$ and $\vec{c} \cdot \vec{z}$ are both scalars, the answer to this equation is the product of two numbers and therefore a number. We have

$$\vec{a} \cdot \vec{y} = (2\vec{j} + \vec{k}) \cdot (4\vec{i} - 7\vec{j}) = 0(4) + 2(-7) + 1(0) = -14$$

$$\vec{c} \cdot \vec{z} = (\vec{i} + 6\vec{j}) \cdot (\vec{i} - 3\vec{j} - \vec{k}) = 1(1) + 6(-3) + 0(-1) = -17$$

Thus,

$$(\vec{a} \cdot \vec{y})(\vec{c} \cdot \vec{z}) = 238$$

6. Since $\vec{c} \cdot \vec{c}$ is a scalar and $(\vec{c} \cdot \vec{c})\vec{a}$ is a vector, the answer to this equation is another scalar. We could calculate $\vec{c} \cdot \vec{c}$, then $(\vec{c} \cdot \vec{c})\vec{a}$, and then take the dot product $((\vec{c} \cdot \vec{c})\vec{a}) \cdot \vec{a}$. Alternatively, we can use the fact that

$$((\vec{c} \cdot \vec{c})\vec{a}) \cdot \vec{a} = (\vec{c} \cdot \vec{c})(\vec{a} \cdot \vec{a}).$$

Since

$$\vec{c} \cdot \vec{c} = (\vec{i} + 6\vec{j}) \cdot (\vec{i} + 6\vec{j}) = 1^2 + 6^2 = 37$$

$$\vec{a} \cdot \vec{a} = (2\vec{j} + \vec{k}) \cdot (2\vec{j} + \vec{k}) = 2^2 + 1^2 = 5,$$

we have,

$$(\vec{c} \cdot \vec{c})(\vec{a} \cdot \vec{a}) = 37(5) = 185$$

7.

$$\cos\theta = \frac{(\vec{i} + \vec{j} + \vec{k}) \cdot (\vec{i} - \vec{j} - \vec{k})}{\|\vec{i} + \vec{j} + \vec{k}\| \|\vec{i} - \vec{j} - \vec{k}\|} = \frac{(1)(1) + (1)(-1) + (1)(-1)}{\sqrt{1^1 + 1^2 + 1^2}\sqrt{1^2 + (-1)^2 + (-1)^2}}$$

$$= -\frac{1}{3}.$$

So, $\theta = \arccos(-\frac{1}{3}) \approx 1.91$ radians, or $\approx 109.5°$.

8. Since $3\vec{i} + \sqrt{3}\vec{j} = \sqrt{3}(\sqrt{3}\vec{i} + \vec{j})$, we know that $3\vec{i} + \sqrt{3}\vec{j}$ and $\sqrt{3}\vec{i} + \vec{j}$ are scalar multiples of one another, and therefore parallel.
    Since $(\sqrt{3}\vec{i} + \vec{j}) \cdot (\vec{i} - \sqrt{3}\vec{j}) = \sqrt{3} - \sqrt{3} = 0$, we know that $\sqrt{3}\vec{i} + \vec{j}$ and $\vec{i} - \sqrt{3}\vec{j}$ are perpendicular.
    Since $3\vec{i} + \sqrt{3}\vec{j}$ and $\sqrt{3}\vec{i} + \vec{j}$ are parallel, $3\vec{i} + \sqrt{3}\vec{j}$ and $\vec{i} - \sqrt{3}\vec{j}$ are perpendicular, too.

9. In general, $\vec{u}$ and $\vec{v}$ are perpendicular when $\vec{u} \cdot \vec{v} = 0$.
    In this case, $\vec{u} \cdot \vec{v} = (t\vec{i} - \vec{j} + \vec{k}) \cdot (t\vec{i} + t\vec{j} - 2\vec{k}) = t^2 - t - 2$.
    This is zero when $t^2 - t - 2 = 0$, i.e. when $(t - 2)(t + 1) = 0$, so $t = 2$ or -1.
    In general, $\vec{u}$ and $\vec{v}$ are parallel if and only if $\vec{v} = \alpha\vec{u}$ for some real number α.
    Thus we need $\alpha t\vec{i} - \alpha\vec{j} + \alpha\vec{k} = t\vec{i} + t\vec{j} - 2\vec{k}$, so we need $\alpha t = t$, and $-\alpha = t$, and $\alpha = -2$. But if $\alpha = -2$, we can't have $\alpha t = t$ unless $t = 0$, and if $t = 0$, we can't have $-\alpha = t$, so there are no values of t for which $\vec{u}$ and $\vec{v}$ are parallel.

10. A normal vector can be obtained from the coefficients: $\vec{n} = 2\vec{i} + \vec{j} - \vec{k}$.

11. Writing the equation in the form
$$3x + 4y - z = 7$$

    shows that a normal vector is
$$\vec{n} = 3\vec{i} + 4\vec{j} - \vec{k}$$

12. Rewriting the equation as
$$2x - 2z = 3x + 3y$$

    or
$$x + 3y + 2z = 0$$

    tells us that a normal vector is
$$\vec{n} = \vec{i} + 3\vec{j} + 2\vec{k}.$$

13. Since a normal vector of the plane is $\vec{n} = -\vec{i} + 2\vec{j} + \vec{k}$, an equation for the plane is
$$-x + 2y + z = -1 + 2 \cdot 0 + 2 = 1$$
$$-x + 2y + z = 1.$$

14. Since the plane is normal to the vector $5\vec{i} + \vec{j} - 2\vec{k}$ and passes through the point $(0, 1, -1)$, an equation for the plane is
$$5x + y - 2z = 5 \cdot 0 + 1 \cdot 1 + (-2) \cdot (-1) = 3$$
$$5x + y - 2z = 3.$$

15. Since the plane is normal to the vector $2\vec{i} - 3\vec{j} + 7\vec{k}$ and passes through the point $(1, -1, 2)$, an equation for the plane is
$$2x - 3y + 7z = 2 \cdot 1 - 3 \cdot (-1) + 7 \cdot 2 = 19$$
$$2x - 3y + 7z = 19.$$

16. Two planes are parallel if their normal vectors are parallel. Since the plane $2x + 4y - 3z = 1$ has normal vector $\vec{n} = 2\vec{i} + 4\vec{j} - 3\vec{k}$, the plane we are looking for has the same normal vector and passes through the point $(1, 0, -1)$. Thus the plane we want has equation:
$$2x + 4y - 3z = 2 \cdot 1 + 4 \cdot 0 + (-3) \cdot (-1) = 5$$

17. Two planes are parallel if their normal vectors are parallel. Since the plane $3x + y + z = 4$ has normal vector $\vec{n} = 3\vec{i} + \vec{j} + \vec{k}$, the plane we are looking for has the same normal vector and passes through the point $(-2, 3, 2)$. Thus, it has the equation
$$3x + y + z = 3 \cdot (-2) + 3 + 2 = -1.$$

18.   (a)   The points A, B and C are shown in Figure 12.27.

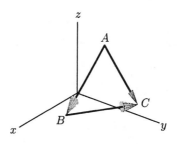

Figure 12.27

First, we calculate the vectors which form the sides of this triangle:

$$\overrightarrow{AB} = (4\vec{i} + 2\vec{j} + \vec{k}) - (2\vec{i} + 2\vec{j} + 2\vec{k}) = 2\vec{i} - \vec{k}$$
$$\overrightarrow{BC} = (2\vec{i} + 3\vec{j} + \vec{k}) - (4\vec{i} + 2\vec{j} + \vec{k}) = -2\vec{i} + \vec{j}$$
$$\overrightarrow{AC} = (2\vec{i} + 3\vec{j} + \vec{k}) - (2\vec{i} + 2\vec{j} + 2\vec{k}) = \vec{j} - \vec{k}$$

Now we calculate the lengths of each of the sides of the triangles:

$$\|\overrightarrow{AB}\| = \sqrt{2^2 + (-1)^2} = \sqrt{5}$$
$$\|\overrightarrow{BC}\| = \sqrt{(-2)^2 + 1^2} = \sqrt{5}$$
$$\|\overrightarrow{AC}\| = \sqrt{1^2 + (-1)^2} = \sqrt{2}$$

Thus the length of the shortest side of S is $\sqrt{2}$.

(b) $\cos \angle BAC = \dfrac{\overrightarrow{AB} \cdot \overrightarrow{AC}}{\|\overrightarrow{AB}\| \cdot \|\overrightarrow{AC}\|} = \dfrac{2 \cdot 0 + 0 \cdot 1 + (-1) \cdot (-1)}{\sqrt{5} \cdot \sqrt{2}} \approx 0.32$

19.   Let

$$\vec{a} = \vec{a}_{\text{parallel}} + \vec{a}_{\text{perp}}$$

where $\vec{a}_{\text{parallel}}$ is parallel to $\vec{d}$, and $\vec{a}_{\text{perp}}$ is perpendicular to $\vec{d}$. Then $\vec{a}_{\text{parallel}}$ is the projection of $\vec{a}$ in the direction of $\vec{d}$:

$$\begin{aligned}
\vec{a}_{\text{parallel}} &= \left(\vec{a} \cdot \frac{\vec{d}}{\|\vec{d}\|} \right) \frac{\vec{d}}{\|\vec{d}\|} \\
&= \left((3\vec{i} + 2\vec{j} - 6\vec{k}) \cdot \frac{(2\vec{i} - 4\vec{j} + \vec{k})}{\sqrt{2^2 + 4^2 + 1^2}} \right) \frac{(2\vec{i} - 4\vec{j} + \vec{k})}{\sqrt{2^2 + 4^2 + 1^2}} \\
&= -\frac{8}{21}(2\vec{i} - 4\vec{j} + \vec{k}) \\
&= -\frac{8}{21}\vec{d}
\end{aligned}$$

Since we now know $\vec{a}$ and $\vec{a}_{\text{parallel}}$, we can solve for $\vec{a}_{\text{perp}}$:

$$\vec{a}_{\text{perp}} = \vec{a} - \vec{a}_{\text{parallel}}$$

$$= (3\vec{i} + 2\vec{j} - 6\vec{k}) - \left(-\frac{8}{21}\right)(2\vec{i} - 4\vec{j} + \vec{k})$$

$$= \frac{79}{21}\vec{i} + \frac{10}{21}\vec{j} - \frac{118}{21}\vec{k}.$$

Thus we can now write $\vec{a}$ as the sum of two vectors, one parallel to $\vec{d}$, the other perpendicular to $\vec{d}$:

$$\vec{a} = -\frac{8}{21}\vec{d} + \left(\frac{79}{21}\vec{i} + \frac{10}{21}\vec{j} - \frac{118}{21}\vec{k}\right)$$

20. The plane cuts the x-axis where $y = z = 0$, so $x = -3/5 = -0.6$, giving the point

$$P = (-0.6, 0, 0)$$

Similarly, the plane cuts the y-axis where $x = z = 0$, so $y = 3/4 = 0.75$, so

$$Q = (0, 0.75, 0)$$

The plane cuts the z-axis at $x = y = 0$, so that $z = 3$, so

$$R = (0, 0, 3)$$

Now we have the three vertices of the triangle, P, Q, and R. The vectors along the three sides of the triangle are

$$\overrightarrow{QP} = -0.6\vec{i} - 0.75\vec{j}$$
$$\overrightarrow{RP} = -0.6\vec{i} - 3\vec{k}$$
$$\overrightarrow{QR} = -0.75\vec{j} + 3\vec{k}$$

The lengths of the sides of the triangle are

$$\|\overrightarrow{QP}\| = \sqrt{(-0.6)^2 + (-0.75)^2} \approx 0.96$$
$$\|\overrightarrow{RP}\| = \sqrt{(-0.6)^2 + (-3)^2} \approx 3.059$$
$$\|\overrightarrow{QR}\| = \sqrt{(-0.75)^2 + (3)^2} \approx 3.092$$

The angle between the vectors $\vec{v}$ and $\vec{w}$ is given by

$$\cos\theta = \frac{\vec{v} \cdot \vec{w}}{\|\vec{v}\|\|\vec{w}\|} \quad \text{so} \quad \theta = \arccos\left(\frac{\vec{v} \cdot \vec{w}}{\|\vec{v}\|\|\vec{w}\|}\right).$$

Thus,

$$\text{Angle at } P = \arccos\left(\frac{\overrightarrow{QP} \cdot \overrightarrow{RP}}{\|\overrightarrow{QP}\|\|\overrightarrow{RP}\|}\right)$$

$$\approx \arccos\left(\frac{(-0.6\vec{i} - 0.75\vec{j}) \cdot (-0.6\vec{i} - 3\vec{k})}{0.96 \cdot 3.059}\right) \approx \arccos\left(\frac{0.36}{0.96 \cdot 3.059}\right)$$

$$\approx \arccos(0.123) \approx 83.0°.$$

Angle at $Q = \arccos\left(\dfrac{\overrightarrow{QP} \cdot \overrightarrow{QR}}{\|\overrightarrow{QP}\|\|\overrightarrow{QR}\|}\right)$

$\approx \arccos\left(\dfrac{(-0.6\vec{i} - 0.75\vec{j}) \cdot (-0.75\vec{j} + 3\vec{k})}{0.96 \cdot 3.092}\right) \approx \arccos\left(\dfrac{0.5625}{0.96 \cdot 3.092}\right)$

$\approx \arccos(0.19) \approx 79.1°.$

Now we use the fact that the angles of the triangle add up to 180°. Thus

Angle at $R \approx 180° - (83.0° + 79.1°) \approx 17.9°.$

21. The angle between two planes is equal to the angle between the normal vectors of the two planes. A normal vector to the plane $5(x - 1) + 3(y + 2) + 2z = 0$ is

$$\vec{n}_1 = 5\vec{i} + 3\vec{j} + 2\vec{k},$$

and a normal vector to the plane $x + 3(y - 1) + 2(z + 4) = 0$ is

$$\vec{n}_2 = \vec{i} + 3\vec{j} + 2\vec{k}.$$

Since $\vec{n}_1 \cdot \vec{n}_2 = \|\vec{n}_1\|\|\vec{n}_2\| \cos\theta$, then

$$\cos\theta = \frac{\vec{n}_1 \cdot \vec{n}_2}{\|\vec{n}_1\|\|\vec{n}_2\|} = \frac{(5\vec{i} + 3\vec{j} + 2\vec{k}) \cdot (\vec{i} + 3\vec{j} + 2\vec{k})}{\sqrt{5^2 + 3^2 + 2^2}\sqrt{1^2 + 3^2 + 2^2}}$$

$$= \frac{18}{\sqrt{532}} = 0.78$$

Hence, $\theta \approx 38.7°.$

22. Let the room be put in the coordinate system as shown in Figure 12.28.

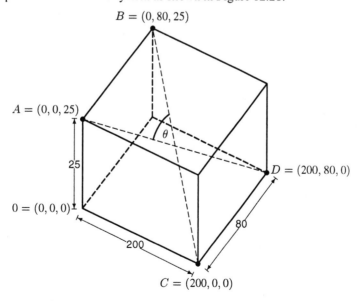

Figure 12.28

Then the vectors of the two strings are given by:

$$\overrightarrow{AD} = (200\vec{i} + 80\vec{j} + 0\vec{k}) - (0\vec{i} + 0\vec{j} + 25\vec{k}) = 200\vec{i} + 80\vec{j} - 25\vec{k}$$
$$\overrightarrow{BC} = (200\vec{i} + 0\vec{j} + 0\vec{k}) - (0\vec{i} + 80\vec{j} + 25\vec{k}) = 200\vec{i} - 80\vec{j} - 25\vec{k}.$$

Let the angle between $\overrightarrow{AD}$ and $\overrightarrow{BC}$ be θ. Then we have

$$\cos\theta = \frac{\overrightarrow{AD} \cdot \overrightarrow{BC}}{\|\overrightarrow{AD}\|\,\|\overrightarrow{BC}\|}$$

$$= = \frac{200(200) + (80)(-80) + (-25)(-25)}{\sqrt{200^2 + 80^2 + (-25)^2}\sqrt{(200)^2 + (-80)^2 + (-25)^2}}$$

$$= \frac{34225}{47025}$$

$$= 0.727804$$

23. If $\vec{x}$ and $\vec{y}$ are two consumption vectors corresponding to points satisfying the same budget constraint, then

$$\vec{p} \cdot \vec{x} = k = \vec{p} \cdot \vec{y}.$$

Therefore we have

$$\vec{p} \cdot (\vec{x} - \vec{y}) = \vec{p} \cdot \vec{x} - \vec{p} \cdot \vec{y} = 0.$$

Thus $\vec{p}$ and $\vec{x} - \vec{y}$ are perpendicular; that is, the difference between two consumption vectors on the same budget constraint is perpendicular to the price vector.

24. We need to find the speed of the wind in the direction of the track. Looking at Figure 12.29, we see that we want the component of $\vec{w}$ in the direction of $\vec{v}$. We calculate

$$\|\vec{w}_{\text{parallel}}\| = \|\vec{w}\|\cos\theta = \frac{\vec{w} \cdot \vec{v}}{\|\vec{v}\|} = \frac{(5\vec{i} + \vec{j}) \cdot (2\vec{i} + 6\vec{j})}{\|2\vec{i} + 6\vec{j}\|}$$

$$= \frac{16}{\sqrt{40}} \approx 2.53$$

$$< 5$$

Therefore, the race results will not be disqualified.

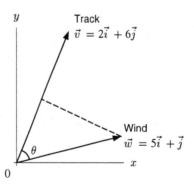

Figure 12.29

25. We have

$$\|\vec{a}_2\| = \sqrt{0.10^2 + 0.08^2 + 0.12^2 + 0.69^2} = 0.7120$$

$$\|\vec{a}_3\| = \sqrt{0.20^2 + 0.06^2 + 0.06^2 + 0.66^2} = 0.6948$$

$$\|\vec{a}_4\| = \sqrt{0.22^2 + 0.00^2 + 0.20^2 + 0.57^2} = 0.6429$$

$$\vec{a}_2 \cdot \vec{a}_3 = 0.10 \cdot 0.20 + 0.08 \cdot 0.06 + 0.12 \cdot 0.06 + 0.69 \cdot 0.66 = 0.4874$$

$$\vec{a}_3 \cdot \vec{a}_4 = 0.20 \cdot 0.22 + 0.06 \cdot 0.00 + 0.06 \cdot 0.20 + 0.66 \cdot 0.57 = 0.4322$$

The distance between the English and the Bantus is given by θ where

$$\cos\theta = \frac{\vec{a}_2 \cdot \vec{a}_3}{\|\vec{a}_2\|\|\vec{a}_3\|} = \frac{0.4874}{(0.7120)(0.6948)} \approx 0.9852$$

so $\theta \approx 9.9°$.

The distance between the English and the Koreans is given by ϕ where

$$\cos\phi = \frac{\vec{a}_3 \cdot \vec{a}_4}{\|\vec{a}_3\|\|\vec{a}_4\|} = \frac{0.4322}{(0.6948)(0.6429)} \approx 0.9676$$

so $\phi \approx 14.6°$. Hence the English are genetically closer to the Bantus than to the Koreans.

26. Suppose $\vec{v} = v_1\vec{i} + v_2\vec{j} + v_3\vec{k}$ and $\vec{w} = w_1\vec{i} + w_2\vec{j} + w_3\vec{k}$.

  • Property 1:
    We calculate both $\vec{v} \cdot \vec{w}$ and $\vec{w} \cdot \vec{v}$ using the algebraic definition of the dot product:

$$\vec{v} \cdot \vec{w} = v_1w_1 + v_2w_2 + v_3w_3$$

$$\vec{w} \cdot \vec{v} = w_1v_1 + w_2v_2 + w_3v_3$$

    But since ordinary multiplication of scalars is commutative, $v_1w_1 = w_1v_1$ and so on. Therefore

$$\vec{v} \cdot \vec{w} = \vec{w} \cdot \vec{v}.$$

  • Property 2:
    First we observe that

$$\lambda\vec{w} = \lambda(w_1\vec{i} + w_2\vec{j} + w_3\vec{k}) = (\lambda w_1)\vec{i} + (\lambda w_2)\vec{j} + (\lambda w_3)\vec{k}$$

$$\lambda\vec{v} = \lambda(v_1\vec{i} + v_2\vec{j} + v_3\vec{k}) = (\lambda v_1)\vec{i} + (\lambda v_2)\vec{j} + (\lambda v_3)\vec{k}.$$

    Now we calculate the three quantities $\vec{v} \cdot (\lambda\vec{w})$ and $\lambda(\vec{v} \cdot \vec{w})$ and $(\lambda\vec{v}) \cdot \vec{w}$

$$\vec{v} \cdot (\lambda\vec{w}) = v_1(\lambda w_1) + v_2(\lambda w_2) + v_3(\lambda w_3)$$

$$\lambda(\vec{v} \cdot \vec{w}) = \lambda(v_1w_1 + v_2w_2 + v_3w_3)$$

$$(\lambda\vec{v}) \cdot \vec{w} = (\lambda v_1)w_1 + (\lambda v_2)w_2 + (\lambda v_3)w_3$$

    Since ordinary multiplication is associative and commutative, we know that $v_1(\lambda w_1) = \lambda v_1 w_1 = (\lambda v_1)w_1$ and so on. Thus, we have $\vec{v} \cdot (\lambda\vec{w}) = (\lambda\vec{v}) \cdot \vec{w}$.
    In addition, the distributive property of ordinary multiplication tells us that

$$\lambda(v_1w_1 + v_2w_2 + v_3w_3) = \lambda v_1w_1 + \lambda v_2w_2 + \lambda v_3w_3$$

    Thus, we know that all three quantities are equal

$$\vec{v} \cdot (\lambda\vec{w}) = \lambda(\vec{v} \cdot \vec{w}) = (\lambda\vec{v}) \cdot \vec{w}$$

- Property 3:
  First we observe that

$$\vec{v} + \vec{w} = (v_1 + w_1)\vec{i} + (v_2 + w_2)\vec{j} + (v_3 + w_3)\vec{k}.$$

Next we calculate the quantities $((\vec{v} + \vec{w}) \cdot \vec{u})$ and $(\vec{v} \cdot \vec{u} + \vec{w} \cdot \vec{u})$.

$$(\vec{v} + \vec{w}) \cdot \vec{u} = (v_1 + w_1)u_1 + (v_2 + w_2)u_2 + (v_3 + w_3)u_3$$
$$\vec{v} \cdot \vec{u} + \vec{w} \cdot \vec{u} = (v_1u_1 + v_2u_2 + v_3u_3) + (w_1u_1 + w_2u_2 + w_3u_3).$$

The distributive law of ordinary multiplication shows that $(v_1 + w_1)u_1 = v_1u_1 + w_1u_1$, and so on. Thus, the dot product is distributive also:

$$(\vec{v} + \vec{w}) \cdot \vec{u} = \vec{v} \cdot \vec{u} + \vec{w} \cdot \vec{u}$$

27. Property 2 says that multiplying one of the vectors by a scalar simply multiplies the dot product by the same scalar. If $\lambda > 0$, then when one vector is multiplied by λ, the angle between the vectors does not change, but the length of one vector, and hence the dot product, is multiplied by λ. The result remains true when $\lambda < 0$. For a justification in the case when $\lambda < 0$, see Problem 31 on page 77.

28. We want to show that $(\vec{b} \cdot \vec{c})\vec{a} - (\vec{a} \cdot \vec{c})\vec{b}$ and $\vec{c}$ are perpendicular. We do this by taking their dot product:

$$((\vec{b} \cdot \vec{c})\vec{a} - (\vec{a} \cdot \vec{c})\vec{b}) \cdot \vec{c} = (\vec{b} \cdot \vec{c})(\vec{a} \cdot \vec{c}) - (\vec{a} \cdot \vec{c})(\vec{b} \cdot \vec{c}) = 0.$$

Since the dot product is 0, the vectors $(\vec{b} \cdot \vec{c})\vec{a} - (\vec{a} \cdot \vec{c})\vec{b}$ and $\vec{c}$ are perpendicular.

29. Since $\vec{u} \cdot \vec{w} = \vec{v} \cdot \vec{w}$, $(\vec{u} - \vec{v}) \cdot \vec{w} = 0$. This equality holds for any $\vec{w}$, so we can take $\vec{w} = \vec{u} - \vec{v}$. This gives

$$\|\vec{u} - \vec{v}\|^2 = (\vec{u} - \vec{v}) \cdot (\vec{u} - \vec{v}) = 0,$$

that is,

$$\|\vec{u} - \vec{v}\| = 0.$$

This implies $\vec{u} - \vec{v} = 0$, that is, $\vec{u} = \vec{v}$.

30. If $\vec{u} = \vec{0}$, then both sides of the equation are zero. If $\vec{u} \neq \vec{0}$, write $\vec{v}_{\text{parallel}}$, $\vec{w}_{\text{parallel}}$, and $(\vec{v} + \vec{w})_{\text{parallel}}$ for the components of $\vec{v}$, $\vec{w}$, and $\vec{v} + \vec{w}$ in the direction of $\vec{u}$. Then Figure 30 shows that

$$\vec{v}_{\text{parallel}} + \vec{w}_{\text{parallel}} = (\vec{v} + \vec{w})_{\text{parallel}}.$$

So

$$\left(\frac{\vec{v} \cdot \vec{u}}{\|\vec{u}\|^2}\right)\vec{u} + \left(\frac{\vec{w} \cdot \vec{u}}{\|\vec{u}\|^2}\right)\vec{u} = \left(\frac{(\vec{v} + \vec{w}) \cdot \vec{u}}{\|\vec{u}\|^2}\right)\vec{u}.$$

Thus, since $\vec{u} \neq \vec{0}$, we deduce that

$$\frac{\vec{v} \cdot \vec{u}}{\|\vec{u}\|^2} + \frac{\vec{w} \cdot \vec{u}}{\|\vec{u}\|^2} - \frac{(\vec{v} + \vec{w}) \cdot \vec{u}}{\|\vec{u}\|^2} = 0,$$

so

$$\vec{v} \cdot \vec{u} + \vec{w} \cdot \vec{u} = (\vec{v} + \vec{w}) \cdot \vec{u}.$$

31. Suppose θ is the angle between $\vec{u}$ and $\vec{v}$.

(a) By the definition of scalar multiplication, we know that $-\vec{v}$ is in the opposite direction of $\vec{v}$, so the angle between $\vec{u}$ and $-\vec{v}$ is $\pi - \theta$. (See Figure 12.30.) Hence,

$$\vec{u} \cdot (-\vec{v}) = \|\vec{u}\|\| - \vec{v}\| \cos(\pi - \theta)$$
$$= \|\vec{u}\|\|\vec{v}\|(-\cos\theta)$$
$$= -(\vec{u} \cdot \vec{v})$$

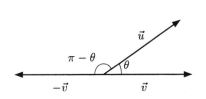

Figure 12.30

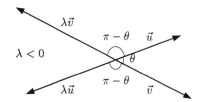

Figure 12.31

(b) If $\lambda < 0$, the angle between $\vec{u}$ and $\lambda\vec{v}$ is $\pi - \theta$, and so is the angle between $\lambda\vec{u}$ and $\vec{v}$. (See Figure 12.31.) So we have,

$$\vec{u} \cdot (\lambda\vec{v}) = \|\vec{u}\|\|\lambda\vec{v}\| \cos(\pi - \theta)$$
$$= |\lambda|\|\vec{u}\|\|\vec{v}\|(-\cos\theta)$$
$$= -\lambda\|\vec{u}\|\|\vec{v}\|(-\cos\theta) \quad \text{since } |\lambda| = -\lambda$$
$$= \lambda\|\vec{u}\|\|\vec{v}\| \cos\theta$$
$$= \lambda(\vec{u} \cdot \vec{v})$$

By a similar argument, we have

$$(\lambda\vec{u}) \cdot \vec{v} = \|\lambda\vec{u}\|\|\vec{v}\| \cos(\pi - \theta)$$
$$= -\lambda\|\vec{u}\|\|\vec{v}\|(-\cos\theta)$$
$$= \lambda(\vec{u} \cdot \vec{v})$$

32. Let $\vec{u}$ and $\vec{v}$ be the displacement vectors from C to the other two vertices. Then

$$c^2 = \|\vec{u} - \vec{v}\|^2$$
$$= (\vec{u} - \vec{v}) \cdot (\vec{u} - \vec{v})$$
$$= \vec{u} \cdot \vec{u} - \vec{v} \cdot \vec{u} - \vec{u} \cdot \vec{v} + \vec{v} \cdot \vec{v}$$
$$= \|\vec{u}\|^2 - 2\|u\|\|v\| \cos C + \|\vec{v}\|^2$$
$$= a^2 - 2ab \cos C + b^2$$

33. We substitute $\vec{u} = u_1\vec{i} + u_2\vec{j} + u_3\vec{k}$ and by the result of Problem 30, we expand as follows:

$$(\vec{u} \cdot \vec{v})_{\text{geom}} = (u_1\vec{i} + u_2\vec{j} + u_3\vec{k}) \cdot \vec{v}$$
$$= (u_1\vec{i}) \cdot \vec{v} + (u_2\vec{j}) \cdot \vec{v} + (u_3\vec{k}) \cdot \vec{v}$$

where all the dot products are defined geometrically By the result of Problem 31 we can write

$$(\vec{u} \cdot \vec{v})_{\text{geom}} = u_1(\vec{i} \cdot \vec{v})_{\text{geom}} + u_2(\vec{j} \cdot \vec{v})_{\text{geom}} + u_3(\vec{k} \cdot \vec{v})_{\text{geom}}.$$

Now substitute $\vec{v} = v_1\vec{i} + v_2\vec{j} + v_3\vec{k}$ and expand, again using Problem 30 and the geometric definition of the dot product:

$$(\vec{u} \cdot \vec{v})_{\text{geom}} = u_1 \left(\vec{i} \cdot (v_1\vec{i} + v_2\vec{j} + v_3\vec{k}) \right)_{\text{geom}}$$

$$+ u_2 \left(\vec{j} \cdot (v_1\vec{i} + v_2\vec{j} + v_3\vec{k}) \right)_{\text{geom}}$$

$$+ u_3 \left(\vec{k} \cdot (v_1\vec{i} + v_2\vec{j} + v_3\vec{k}) \right)_{\text{geom}}$$

$$= u_1v_1(\vec{i} \cdot \vec{i})_{\text{geom}} + u_1v_2(\vec{i} \cdot \vec{j})_{\text{geom}} + u_1v_3(\vec{i} \cdot \vec{k})_{\text{geom}}$$

$$+ u_2v_1(\vec{i} \cdot \vec{i})_{\text{geom}} + u_2v_2(\vec{i} \cdot \vec{j})_{\text{geom}} + u_2v_3(\vec{i} \cdot \vec{k})_{\text{geom}}$$

$$+ u_3v_1(\vec{i} \cdot \vec{i})_{\text{geom}} + u_3v_2(\vec{i} \cdot \vec{j})_{\text{geom}} + u_3v_3(\vec{i} \cdot \vec{k})_{\text{geom}}$$

The geometric definition of the dot product shows that

$$\vec{i} \cdot \vec{i} = \|\vec{i}\| \, \|\vec{i}\| \cos 0 = 1$$

$$\vec{i} \cdot \vec{j} = \|\vec{i}\| \, \|\vec{j}\| \cos \frac{\pi}{2} = 0.$$

Similarly $\vec{j} \cdot \vec{j} = \vec{k} \cdot \vec{k} = 1$ and $\vec{i} \cdot \vec{k} = \vec{j} \cdot \vec{k} = 0$. Thus, the expression for $(\vec{u} \cdot \vec{v})_{\text{geom}}$ becomes

$$(\vec{u} \cdot \vec{v})_{\text{geom}} = u_1v_1(1) + u_1v_2(0) + u_1v_3(0)$$

$$+ u_2v_1(0) + u_2v_2(1) + u_2v_3(0)$$

$$+ u_3v_1(0) + u_3v_2(0) + u_3v_3(1)$$

$$= u_1v_1 + u_2v_2 + u_3v_3.$$

34. (a) Since $q(t) = (\vec{v} + t\vec{w}) \cdot (\vec{v} + t\vec{w}) = \|\vec{v} + t\vec{w}\|^2$ and since the length of any vector is nonnegative, we must have

$$q(t) = \|\vec{v} + t\vec{w}\|^2 \geq 0$$

for all real t.

  (b) Using the distributive law

$$q(t) = (\vec{v} + t\vec{w}) \cdot (\vec{v} + t\vec{w}) = \vec{v} \cdot \vec{v} + t\vec{w} \cdot \vec{v} + \vec{v} \cdot t\vec{w} + t^2\vec{w} \cdot \vec{w}$$

$$= \|\vec{v}\|^2 + 2(\vec{v} \cdot \vec{w})t + \|\vec{w}\|^2 t^2.$$

If $\vec{w} \neq 0$, then $\|\vec{w}\| \neq 0$ and $q(t)$ is quadratic in t.

  (c) Since $q(t) \geq 0$, the quadratic has one repeated root or no roots, so the discriminant must be less than or equal to zero. Thus,

$$(2\vec{v} \cdot \vec{w})^2 - 4\|\vec{v}\|^2\|\vec{w}\|^2 \leq 0.$$

Taking square roots, we have

$$|\vec{v} \cdot \vec{w}| \leq \|\vec{v}\|\|\vec{w}\|.$$

If $\vec{w} = 0$, then $q(t)$ is no longer a quadratic. However, in that case,

$$|\vec{v} \cdot \vec{w}| = 0 = \|\vec{v}\|\|\vec{w}\|$$

so the inequality still holds.

Solutions for Section 12.4 ━━━

1. $\vec{k} \times \vec{j} = -\vec{i}$ (remember $\vec{i}, \vec{j}, \vec{k}$ are unit vectors along the axes, and you must use the right hand rule.)
2. $\vec{i} \times \vec{i}$ is a vector, $\vec{i} \cdot \vec{i}$ is a scalar. They are never equal.
3. $\vec{a} = \vec{i} + \vec{k}$, and $\vec{b} = \vec{i} + \vec{j}$

$$\vec{a} \times \vec{b} = \begin{vmatrix} \vec{i} & \vec{j} & \vec{k} \\ 1 & 0 & 1 \\ 1 & 1 & 0 \end{vmatrix} = -\vec{i} + \vec{j} + \vec{k}$$

4. $\vec{a} = -\vec{i}$, and $\vec{b} = \vec{j} + \vec{k}$

$$\vec{a} \times \vec{b} = \begin{vmatrix} \vec{i} & \vec{j} & \vec{k} \\ -1 & 0 & 0 \\ 0 & 1 & 1 \end{vmatrix} = \vec{j} - \vec{k}$$

5. $\vec{a} = \vec{i} + \vec{j} + \vec{k}$, and $\vec{b} = \vec{i} + \vec{j} - \vec{k}$

$$\vec{a} \times \vec{b} = \begin{vmatrix} \vec{i} & \vec{j} & \vec{k} \\ 1 & 1 & 1 \\ 1 & 1 & -1 \end{vmatrix} = -2\vec{i} + 2\vec{j}$$

6. $\vec{a} = 2\vec{i} - 3\vec{j} + \vec{k}$, and $\vec{b} = \vec{i} + 2\vec{j} - \vec{k}$

$$\vec{a} \times \vec{b} = \begin{vmatrix} \vec{i} & \vec{j} & \vec{k} \\ 2 & -3 & 1 \\ 1 & 2 & -1 \end{vmatrix} = \vec{i} + 3\vec{j} + 7\vec{k}$$

7.

$$\vec{a} \times \vec{b} = \begin{vmatrix} \vec{i} & \vec{j} & \vec{k} \\ 3 & 1 & -1 \\ 1 & -4 & 2 \end{vmatrix}$$

$$= \begin{vmatrix} 1 & -1 \\ -4 & 2 \end{vmatrix} \vec{i} - \begin{vmatrix} 3 & -1 \\ 1 & 2 \end{vmatrix} \vec{j} + \begin{vmatrix} 3 & 1 \\ 1 & -4 \end{vmatrix} \vec{k}$$

$$= -2\vec{i} - 7\vec{j} - 13\vec{k}.$$

Since

$$\vec{a} \cdot (\vec{a} \times \vec{b}) = 3(-2) + (-7) - (-13) = 0$$

and

$$\vec{b} \cdot (\vec{a} \times \vec{b}) = 1(-2) - 4(-7) + 2(-13) = 0,$$

$\vec{a} \times \vec{b}$ is perpendicular to both $\vec{a}$ and $\vec{b}$.

8. Since

$$\|\vec{v} \times \vec{w}\| = \|\vec{v}\| \cdot \|\vec{w}\| \sin\theta,$$

and

$$\vec{v} \cdot \vec{w} = \|\vec{v}\| \cdot \|\vec{w}\| \cos\theta,$$

so

$$\frac{\|\vec{v} \times \vec{w}\|}{\vec{v} \cdot \vec{w}} = \frac{\|\vec{v}\| \cdot \|\vec{w}\| \sin\theta}{\|\vec{v}\| \cdot \|\vec{w}\| \cos\theta} = \tan\theta,$$

so

$$\tan\theta = \frac{\|2\vec{i} - 3\vec{j} + 5\vec{k}\|}{3} = \frac{\sqrt{38}}{3} \approx 2.05.$$

9. The magnitude of $\vec{a} \times \vec{b}$ is given by

$$\|\vec{a} \times \vec{b}\| = \|\vec{a}\|\|\vec{b}\| \sin\theta = (3 \cdot 2) \sin\theta = 6\sin\theta.$$

Therefore, the maximum possible value of $\|\vec{a} \times \vec{b}\|$ occurs when $\sin\theta = 1$. This occurs when $\theta = \pi/2$; that is, when $\vec{a}$ and $\vec{b}$ are perpendicular. The maximum value of $\|\vec{a} \times \vec{b}\|$ is 6. The minimum value of $\|\vec{a} \times \vec{b}\|$ occurs when $\sin\theta = 0$ so $\theta = 0$ or π; that is, when $\vec{a}$ and $\vec{b}$ are parallel. Then $\|\vec{a} \times \vec{b}\|$ is 0.

   The direction of $\vec{a} \times \vec{b}$ will be along the positive z-axis when $\vec{b}$ is in the first or second quadrant and along the negative z-axis when $\vec{b}$ is in the third or fourth quadrant. See Figure 12.32.

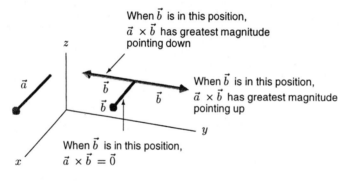

Figure 12.32: A fixed vector $\vec{a}$ and a rotating vector $\vec{b}$

10. The direction we choose should be perpendicular to both missiles, therefore parallel to

$$(3\vec{i} + 5\vec{j} + 2\vec{k}) \times (\vec{i} - 3\vec{j} - 2\vec{k}),$$

which is equal to

$$\begin{vmatrix} \vec{i} & \vec{j} & \vec{k} \\ 3 & 5 & 2 \\ 1 & -3 & -2 \end{vmatrix} = (-10 + 6)\vec{i} - (-6 - 2)\vec{j} + (-9 - 5)\vec{k}$$

$$= -4\vec{i} + 8\vec{j} - 14\vec{k}.$$

11. We can form the displacement vectors $\vec{a} = -\vec{i} + \vec{j} + 0\vec{k}$ from $(1,0,0)$ to $(0,1,0)$ and $\vec{b} = -\vec{i} + 0\vec{j} + \vec{k}$ from $(1,0,0)$ to $(0,0,1)$. A normal vector to the plane is $\vec{a} \times \vec{b} = \vec{i} + \vec{j} + \vec{k}$. Using the point $(1,0,0)$, the plane can be written as $(x - 1) + y + z = 0$ or $x + y + z = 1$.

12. The displacement vector from $(3, 4, 2)$ to $(-2, 1, 0)$ is:

$$\vec{a} = -5\vec{i} - 3\vec{j} - 2\vec{k}.$$

The displacement vector from $(3, 4, 2)$ to $(0, 2, 1)$ is:

$$\vec{b} = -3\vec{i} - 2\vec{j} - \vec{k}.$$

Therefore the vector normal to the plane is:

$$\vec{n} = \vec{a} \times \vec{b} = -\vec{i} + \vec{j} + \vec{k}.$$

Using the first point, the equation of the plane can be written as:

$$-(x - 3) + (y - 4) + (z - 2) = 0.$$

The equation of the plane is thus:

$$-x + y + z = 3.$$

13. (a) If we let $\overrightarrow{PQ}$ in Figure 12.33 be the vector from point P to point Q and $\overrightarrow{PR}$ be the vector from P to R, then

$$\overrightarrow{PQ} = -\vec{i} + 2\vec{k}$$
$$\overrightarrow{PR} = 2\vec{i} - \vec{k},$$

then the area of the parallelogram determined by $\overrightarrow{PQ}$ and $\overrightarrow{PR}$ is:

$$\text{Area of parallelogram} = \|\overrightarrow{PQ} \times \overrightarrow{PR}\| = \left\| \begin{vmatrix} \vec{i} & \vec{j} & \vec{k} \\ -1 & 0 & 2 \\ 2 & 0 & -1 \end{vmatrix} \right\| = \|3\vec{j}\| = 3.$$

Thus, the area of the triangle PQR is

$$\left(\begin{matrix} \text{Area of} \\ \text{triangle} \end{matrix} \right) = \frac{1}{2} \left(\begin{matrix} \text{Area of} \\ \text{parallelogram} \end{matrix} \right) = \frac{3}{2} = 1.5.$$

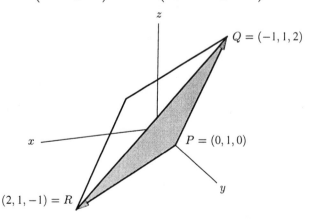

Figure 12.33

(b) Since $\vec{n} = \overrightarrow{PQ} \times \overrightarrow{PR}$ is perpendicular to the plane PQR, and from above, we have $\vec{n} = 3\vec{j}$, the equation of the plane has the form $3y = C$. At the point $(0, 1, 0)$ we get $3 = C$, therefore $3y = 3$, i.e., $y = 1$.

14. The normal vectors to the planes are $\vec{n_1} = 2\vec{i} - 3\vec{j} + 5\vec{k}$ and $\vec{n_2} = 4\vec{i} + \vec{j} - 3\vec{k}$. The line of intersection is perpendicular to both normal vectors (picture the pages in a partially open book). Hence the vector we need is $\vec{n_1} \times \vec{n_2} = 4\vec{i} + 26\vec{j} + 14\vec{k}$.

15. The vector parallel to the line of intersection is $4\vec{i} + 26\vec{j} + 14\vec{k}$ and this is normal to the desired plane. Therefore, $4x + 26y + 14z = 0$ is the equation of the plane.

16. We use the same normal vector $\vec{n} = 4\vec{i} + 26\vec{j} + 14\vec{k}$ and the point $(4, 5, 6)$ to get $4(x - 4) + 26(y - 5) + 14(z - 6) = 0$.

17. First let
$$\vec{a} = a_1\vec{i} + a_2\vec{j} + a_3\vec{k} \quad \vec{b} = b_1\vec{i} + b_2\vec{j} + b_3\vec{k} \quad \vec{c} = c_1\vec{i} + c_2\vec{j} + c_3\vec{k}$$
so $\vec{b} + \vec{c} = (b_1 + c_1)\vec{i} + (b_2 + c_2)\vec{j} + (b_3 + c_3)\vec{k}$. Now, using the general formula for cross products, we have:

$$\vec{a} \times (\vec{b} + \vec{c})$$
$$= [a_2(b_3 + c_3) - a_3(b_2 + c_2)]\vec{i} + [a_3(b_1 + c_1) - a_1(b_3 + c_3)]\vec{j} + [a_1(b_2 + c_2) - a_2(b_1 + c_1)]\vec{k}$$
$$= (a_2b_3 + a_2c_3 - a_3b_2 - a_3c_2)\vec{i} + (a_3b_1 + a_3c_1 - a_1b_3 - a_1c_3)\vec{j}$$
$$+(a_1b_2 + a_1c_2 - a_2b_1 - a_2c_1)\vec{k}$$
$$= (a_2b_3 - a_3b_2)\vec{i} + (a_2c_3 - a_3c_2)\vec{i} + (a_3b_1 - a_1b_3)\vec{j} + (a_3c_1 - a_1c_3)\vec{j}$$
$$+(a_1b_2 - a_2b_1)\vec{k} + (a_1c_2 - a_2c_1)\vec{k}$$
$$= (a_2b_3 - a_3b_2)\vec{i} + (a_3b_1 - a_1b_3)\vec{j} + (a_1b_2 - a_2b_1)\vec{k} + (a_2c_3 - a_3c_2)\vec{i} + (a_3c_1 - a_1c_3)\vec{j}$$
$$+(a_1c_2 - a_2c_1)\vec{k}$$
$$= (\vec{a} \times \vec{b}) + (\vec{a} \times \vec{c})$$

Thus, $\vec{a} \times (\vec{b} + \vec{c}) = \vec{a} \times \vec{b} + \vec{a} \times \vec{c}$.

18. Any vector $\vec{v}$ that is perpendicular to both $\vec{a}$ and $\vec{b}$ will have the property that its dot product with $\vec{a}$ and $\vec{b}$ is 0, that is

$$\vec{a} \cdot \vec{v} = a_1x + a_2y + a_3z = 0,$$
$$\vec{b} \cdot \vec{v} = b_1x + b_2y + b_3z = 0.$$

Multiply the first equation by b_1 and the second by a_1 and subtract to get

$$(b_1a_2 - a_1b_2)y + (b_1a_3 - a_1b_3)z = 0 \quad \text{or} \quad y = \frac{-(b_1a_3 - a_1b_3)z}{(b_1a_2 - a_1b_2)} \quad \text{(for } b_1a_2 \neq a_1b_2)$$

Multiply the second equation by a_2 and the first by b_2 and subtract to get

$$(b_2a_1 - a_2b_1)x + (b_2a_3 - a_2b_3)z = 0 \quad \text{or} \quad x = \frac{-(b_2a_3 - a_2b_3)z}{(b_2a_1 - a_2b_1)}.$$

So

$$\vec{v} = \frac{-(b_2a_3 - a_2b_3)z}{(b_2a_1 - a_2b_1)}\vec{i} - \frac{(b_1a_3 - a_1b_3)z}{(b_1a_2 - a_1b_2)}\vec{j} + z\vec{k}.$$

Pick $z = b_2a_1 - b_1a_2$ and multiply out, and we see that the algebraic method of finding a cross product yields the same result as our standard method.

19. (a) Figure 12.34 shows the vectors $\vec{a}$, $\vec{b}$, and $\vec{c}$ satisfying the conditions $0 < a_2 < a_1$ and $0 < b_1 < b_2$.

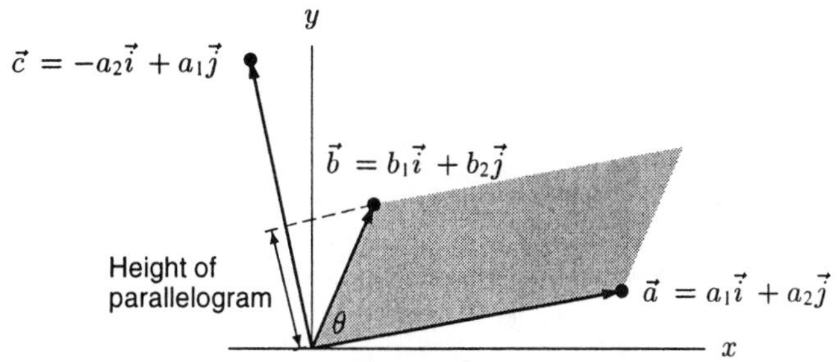

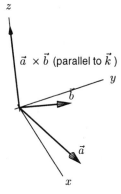

$\vec{c} = -a_2\vec{i} + a_1\vec{j}$

$\vec{b} = b_1\vec{i} + b_2\vec{j}$

Height of parallelogram

$\vec{a} = a_1\vec{i} + a_2\vec{j}$

θ

Figure 12.34

(b) $\vec{c} \cdot \vec{a} = -a_2 a_1 + a_1 a_2 = 0$ and $\vec{c} \cdot \vec{c} = a_2{}^2 + a_1{}^2 = \|\vec{a}\|^2$. Thus $\vec{c}$ is orthogonal (perpendicular) to $\vec{a}$ and has the same length as $\vec{a}$.

(c) $\vec{c} \cdot \vec{b} = -a_2 b_1 + a_1 b_2$. Since $a_1 > a_2 > 0$, and $b_2 > b_1 > 0$, we know that $\vec{c} \cdot \vec{b}$ is positive.

(d) If θ is the angle between $\vec{a}$ and $\vec{b}$ and α is the angle between $\vec{c}$ and $\vec{b}$, then $\alpha = (\frac{\pi}{2} - \theta)$. Thus $\cos\alpha = \sin\theta$, so

$$\vec{c} \cdot \vec{b} = \|\vec{c}\|\|\vec{b}\|\cos\alpha = \|\vec{a}\|\|\vec{b}\|\sin\theta.$$

Since $\|\vec{a}\| = $ Base of the parallelogram and
$\|\vec{b}\|\sin\theta = $ Height of the parallelogram, we have

$$\vec{c} \cdot \vec{b} = \text{Base} \cdot \text{Height} = \text{Area of the parallelogram formed by } \vec{a} \text{ and } \vec{b}.$$

(e) By the right-hand rule, $\vec{a} \times \vec{b}$ is in the direction of the positive z-axis. See Figure 12.35. Since we know that

$$\text{Area of the parallelogram} = \vec{c} \cdot \vec{b} = a_1 b_2 - a_2 b_1.$$

the definition of $\vec{a} \times \vec{b}$ tells us that

$$\vec{a} \times \vec{b} = (\text{Area of Parallelogram})\vec{k} = (\vec{c} \cdot \vec{b})\vec{k} = (a_1 b_2 - a_2 b_1)\vec{k}.$$

Thus,

$$\vec{a} \times \vec{b} = (a_1 b_2 - a_2 b_1)\vec{k}.$$

z

$\vec{a} \times \vec{b}$ (parallel to $\vec{k}$)

y

$\vec{b}$

$\vec{a}$

x

Figure 12.35: Cross product of two vectors in the xy-plane

20. Solve for $\vec{c}$ to get $\vec{c} = -(\vec{a} + \vec{b})$.
    So

$$
\begin{aligned}
\vec{b} \times \vec{c} &= \vec{b} \times (-(\vec{a} + \vec{b})) \\
&= -(\vec{b} \times (\vec{a} + \vec{b})) \\
&= -(\vec{b} \times \vec{a} + \vec{b} \times \vec{b}) \\
&= -(\vec{b} \times \vec{a} + \vec{0}) \\
&= -(\vec{b} \times \vec{a}) \\
&= \vec{a} \times \vec{b}.
\end{aligned}
$$

Also,

$$
\begin{aligned}
\vec{c} \times \vec{a} &= -(\vec{a} + \vec{b}) \times \vec{a} \\
&= -((\vec{a} + \vec{b}) \times \vec{a}) \\
&= -(\vec{a} \times \vec{a} + \vec{b} \times \vec{a}) \\
&= -(\vec{0} + \vec{b} \times \vec{a}) \\
&= \vec{a} \times \vec{b}.
\end{aligned}
$$

Therefore, $\vec{a} \times \vec{b} = \vec{b} \times \vec{c} = \vec{c} \times \vec{a}$.

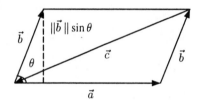

Figure 12.36

Geometrically, the magnitude of the cross product of two vectors is equal to the area of the parallelogram formed by the vectors. If $\vec{a} + \vec{b} + \vec{c} = 0$, then we can think of the vectors $\vec{a}$, $\vec{b}$, and $\vec{c}$ as forming a triangle. (See Figure 12.36.) So by showing that

$$
\vec{a} \times \vec{b} = \vec{b} \times \vec{c} = \vec{c} \times \vec{a},
$$

we are showing that the areas of the parallelograms formed by any two sides of the same triangle are equal.

21. If $\lambda = 0$, then all three cross products are $\vec{0}$, since the cross product of the zero vector with any other vector is always 0.
    If $\lambda > 0$, then $\lambda \vec{v}$ and $\vec{v}$ are in the same direction and $\vec{w}$ and $\lambda \vec{w}$ are in the same direction. Therefore the unit normal vector $\vec{n}$ is the same in all three cases. In addition, the angles between $\lambda \vec{v}$ and $\vec{w}$, and between $\vec{v}$ and $\vec{w}$, and between $\vec{v}$ and $\lambda \vec{w}$ are all θ. Thus,

$$
\begin{aligned}
(\lambda \vec{v}) \times \vec{w} &= \|\lambda \vec{v}\| \|\vec{w}\| \sin \theta \vec{n} \\
&= \lambda \|\vec{v}\| \|\vec{w}\| \sin \theta \vec{n} \\
&= \lambda (\vec{v} \times \vec{w}) \\
&= \|\vec{v}\| \|\lambda \vec{w}\| \sin \theta \vec{n} \\
&= \vec{v} \times (\lambda \vec{w})
\end{aligned}
$$

If $\lambda < 0$, then $\lambda \vec{v}$ and $\vec{v}$ are in opposite directions, as are $\vec{w}$ and $\lambda \vec{w}$ in opposite directions. Therefore if $\vec{n}$ is the normal vector in the definition of $\vec{v} \times \vec{w}$, then the right-hand rule gives $-\vec{n}$ for $(\lambda \vec{v}) \times \vec{w}$ and $\vec{v} \times (\lambda \vec{w})$. In addition, if the angle between $\vec{v}$ and $\vec{w}$ is θ, then the angle between $\lambda \vec{v}$ and $\vec{w}$ and between $\vec{v}$ and $\lambda \vec{w}$ is $(\pi - \theta)$. Since if $\lambda < 0$, we have $|\lambda| = -\lambda$, so

$$\begin{aligned}
(\lambda \vec{v}) \times \vec{w} &= \|\lambda \vec{v}\|\|\vec{w}\| \sin(\pi - \theta)(-\vec{n}) \\
&= |\lambda|\|\vec{v}\|\|\vec{w}\| \sin(\pi - \theta)(-\vec{n}) \\
&= -\lambda\|\vec{v}\|\|\vec{w}\| \sin\theta(-\vec{n}) \\
&= \lambda\|\vec{v}\|\|\vec{w}\| \sin\theta \vec{n} \\
&= \lambda(\vec{v} \times \vec{w}).
\end{aligned}$$

Similarly,

$$\begin{aligned}
\vec{v} \times (\lambda \vec{w}) &= \|\vec{v}\|\|\lambda \vec{w}\| \sin(\pi - \theta)(-\vec{n}) \\
&= -\lambda\|\vec{v}\|\|\vec{w}\| \sin\theta(-\vec{n}) \\
&= \lambda(\vec{v} \times \vec{w}).
\end{aligned}$$

22. The quantities $\left|\vec{a} \cdot (\vec{b} \times \vec{c})\right|$ and $\left|(\vec{a} \times \vec{b}) \cdot \vec{c}\right|$ both represent the volume of the same parallelepiped, namely that defined by the three vectors $\vec{a}$, $\vec{b}$, and $\vec{c}$, and therefore must be equal. Thus, the two triple products $\vec{a} \cdot (\vec{b} \times \vec{c})$ and $(\vec{a} \times \vec{b}) \cdot \vec{c}$ must be equal except perhaps for their sign. In fact, both are positive if $\vec{a}$, $\vec{b}$, $\vec{c}$ are right-handed and negative if $\vec{a}$, $\vec{b}$, $\vec{c}$ are left-handed. This can be shown by drawing a picture:

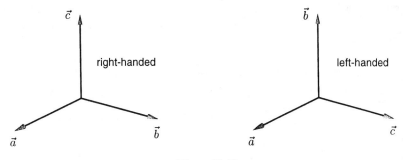

Figure 12.37

23. If θ is the angle between $\vec{a}$ and $\vec{b}$, then

$$\begin{aligned}
\|\vec{a} \times \vec{b}\|^2 &= (\|\vec{a}\|\|\vec{b}\| \sin\theta)^2 \\
&= \|\vec{a}\|^2\|\vec{b}\|^2 \sin^2\theta \\
&= \|\vec{a}\|^2\|\vec{b}\|^2(1 - \cos^2\theta) \\
&= \|\vec{a}\|^2\|\vec{b}\|^2 - \|\vec{a}\|^2\|\vec{b}\|^2 \cos^2\theta \\
&= \|\vec{a}\|^2\|\vec{b}\|^2 - (\vec{a} \cdot \vec{b})^2.
\end{aligned}$$

24. (a) Since $\vec{c}$ is perpendicular to $\vec{a} \times \vec{b}$, and since $\vec{a} \times \vec{b}$ is normal to the plane containing $\vec{a}$ and $\vec{b}$, it follows that $\vec{c}$ must be in the plane containing $\vec{a}$ and $\vec{b}$.

(b) Using the expression given in the problem for $\vec{c}$, we get

$$\vec{a} \cdot \vec{c} = \vec{a} \cdot (\vec{a} \times (\vec{b} \times \vec{a}))$$
$$= (\vec{a} \times \vec{a}) \cdot (\vec{b} \times \vec{a})$$
$$= \vec{0} \cdot (\vec{b} \times \vec{a}) = 0.$$

and

$$\vec{b} \cdot \vec{c} = \vec{b} \cdot (\vec{a} \times (\vec{b} \times \vec{a}))$$
$$= (\vec{b} \times \vec{a}) \cdot (\vec{b} \times \vec{a})$$
$$= \|\vec{b} \times \vec{a}\|^2$$
$$= \|\vec{a}\|^2 \|\vec{b}\|^2 - (\vec{a} \cdot \vec{b})^2.$$

(c) Since $\vec{c}$ lies in the plane containing $\vec{a}$ and $\vec{b}$, it is of the form $\vec{c} = x\vec{a} + y\vec{b}$ for some scalars x and y. Thus, using the fact that $\vec{a} \cdot \vec{c} = 0$ from part (b), we have

$$\vec{a} \cdot \vec{c} = \vec{a} \cdot (x\vec{a} + y\vec{b}) = x\|\vec{a}\|^2 + y(\vec{a} \cdot \vec{b}) = 0.$$

Similarly, using the fact that $\vec{b} \cdot \vec{c} = \|\vec{a}\|^2\|\vec{b}\|^2 - (\vec{a} \cdot \vec{b})^2$ from part (b), we have

$$\vec{b} \cdot \vec{c} = \vec{b} \cdot (x\vec{a} + y\vec{b}) = x(\vec{a} \cdot \vec{b}) + y\|\vec{b}\|^2 = \|\vec{a}\|^2\|\vec{b}\|^2 - (\vec{a} \cdot \vec{b})^2.$$

Solving these two linear equations in x and y, we find $x = -\vec{a} \cdot \vec{b}$ and $y = \|\vec{a}\|^2$.

25. Problem 22 tells us that $(\vec{u} \times \vec{v}) \cdot \vec{w} = \vec{u} \cdot (\vec{v} \times \vec{w})$. Using this result on the triple product of $(\vec{a} + \vec{b}) \times \vec{c}$ with any vector $\vec{d}$ together with the fact that the dot product distributes over addition gives us:

$$[(\vec{a} + \vec{b}) \times \vec{c}] \cdot \vec{d} = (\vec{a} + \vec{b}) \cdot (\vec{c} \times \vec{d})$$
$$= \vec{a} \cdot (\vec{c} \times \vec{d}) + \vec{b} \cdot (\vec{c} \times \vec{d}) \qquad \text{(dot product is distributive)}$$
$$= (\vec{a} \times \vec{c}) \cdot \vec{d} + (\vec{b} \times \vec{c}) \cdot \vec{d} \qquad \text{(using Problem 22 again)}$$
$$= [(\vec{a} \times \vec{c}) + (\vec{b} \times \vec{c})] \cdot \vec{d}. \qquad \text{(dot product is distributive)}$$

So, since $[(\vec{a} + \vec{b}) \times \vec{c}] \cdot \vec{d} = [(\vec{a} \times \vec{c}) + (\vec{b} \times \vec{c})] \cdot \vec{d}$, then

$$[(\vec{a} + \vec{b}) \times \vec{c}] \cdot \vec{d} - [(\vec{a} \times \vec{c}) + (\vec{b} \times \vec{c})] \cdot \vec{d} = 0,$$

Since the dot product is distributive, we have

$$[((\vec{a} + \vec{b}) \times \vec{c}) - (\vec{a} \times \vec{c}) - (\vec{b} \times \vec{c})] \cdot \vec{d} = 0.$$

Since this equation is true for all vectors $\vec{d}$, by letting

$$\vec{d} = ((\vec{a} + \vec{b}) \times \vec{c}) - (\vec{a} \times \vec{c}) - (\vec{b} \times \vec{c}),$$

we get

$$\|(\vec{a} + \vec{b}) \times \vec{c} - \vec{a} \times \vec{c} - \vec{b} \times \vec{c}\|^2 = 0$$

and hence

$$(\vec{a} + \vec{b}) \times \vec{c} - (\vec{a} \times \vec{c}) - (\vec{b} \times \vec{c}) = \vec{0}.$$

Thus

$$(\vec{a} + \vec{b}) \times \vec{c} = (\vec{a} \times \vec{c}) + (\vec{b} \times \vec{c}).$$

26. Write $\vec{v}$ and $\vec{w}$ in components and expand using the distributive property of the cross product.

$$\vec{v} \times \vec{w} = (v_1\vec{i} + v_2\vec{j} + v_3\vec{k}) \times (w_1\vec{i} + w_2\vec{j} + w_3\vec{k})$$
$$= v_1w_1\vec{i} \times \vec{i} + v_1w_2\vec{i} \times \vec{j} + v_1w_3\vec{i} \times \vec{k}$$
$$+ v_2w_1\vec{j} \times \vec{i} + v_2w_2\vec{j} \times \vec{j} + v_2w_3\vec{j} \times \vec{k}$$
$$+ v_3w_1\vec{k} \times \vec{i} + v_3w_2\vec{k} \times \vec{j} + v_3w_3\vec{k} \times \vec{k}$$

Now we use the fact that $\vec{i} \times \vec{i} = \vec{0}$, $\vec{i} \times \vec{j} = \vec{k}$, $\vec{i} \times \vec{k} = -\vec{j}$, $\vec{j} \times \vec{i} = -\vec{k}$, $\vec{j} \times \vec{j} = \vec{0}$, $\vec{j} \times \vec{k} = \vec{i}$, $\vec{k} \times \vec{i} = \vec{j}$, $\vec{k} \times \vec{j} = -\vec{i}$, $\vec{k} \times \vec{k} = \vec{0}$. Thus we have

$$\vec{v} \times \vec{w} = \vec{0} + v_1w_2\vec{k} + v_1w_3(-\vec{j}) + v_2w_1(-\vec{k}) + \vec{0} + v_2w_3\vec{i} + v_3w_1\vec{j} + v_3w_2(-\vec{i}) + \vec{0}$$
$$= (v_2w_3 - v_3w_2)\vec{i} + (v_3w_1 - v_1w_3)\vec{j} + (v_1w_2 - v_2w_1)\vec{k}.$$

27. (a) Let $r = \|\vec{a}\|$ and $s = \|\vec{b}\|$, and let α, β, be the angles between $\vec{a}$, $\vec{b}$, and the x-axis as shown in the figure. Using the formula

$$\sin(\beta - \alpha) = \sin\beta\cos\alpha - \cos\beta\sin\alpha,$$

and the fact that $a_1 = r\cos\alpha$, $a_2 = r\sin\alpha$, $b_1 = s\cos\beta$, and $b_2 = s\sin\beta$, we get

$$a_1b_2 - a_2b_1 = (r\cos\alpha)(s\sin\beta) - (r\sin\alpha)(s\cos\beta)$$
$$= rs(\cos\alpha\sin\beta - \sin\alpha\cos\beta)$$
$$= rs\sin(\beta - \alpha) \qquad \text{(from } \sin(\beta - \alpha) = \sin\beta\cos\alpha - \cos\beta\sin\alpha)$$
$$= \|\vec{a}\|\|\vec{b}\|\sin(\beta - \alpha)$$
$$= \text{Area of parallelogram.}$$

(b) Suppose θ is the angle between $\vec{a}$ and $\vec{b}$. We drew the figure with $\alpha < \beta$ and thus $\beta - \alpha = \theta$. If $\alpha > \beta$, then $\alpha - \beta = \theta$. In both cases we can be sure that

$$|a_1b_2 - a_2b_1| = \|\vec{a}\|\|\vec{b}\| |\sin(\beta - \alpha)|$$
$$= \|\vec{a}\|\|\vec{b}\|\sin\theta.$$

Thus, the sign of $a_1b_2 - a_2b_1$ tells us whether the rotation from $\vec{a}$ to $\vec{b}$ is counterclockwise (then $a_1b_2 - a_2b_1$ is positive) or clockwise (then $a_1b_2 - a_2b_1$ is negative).

28. The area vector for face $OAB = \frac{1}{2}\vec{b} \times \vec{a}$.
The area vector for face $OBC = \frac{1}{2}\vec{a} \times \vec{c}$.
The area vector for face $OAC = \frac{1}{2}\vec{b} \times \vec{c}$.
The area vector for face $ABC = \frac{1}{2}(\vec{b} - \vec{a}) \times (\vec{c} - \vec{a})$.

$$\frac{1}{2}\vec{b} \times \vec{a} + \frac{1}{2}\vec{c} \times \vec{b} + \frac{1}{2}\vec{a} \times \vec{c} + \frac{1}{2}(\vec{b} - \vec{a}) \times (\vec{c} - \vec{a}) =$$
$$\frac{1}{2}\vec{b} \times \vec{a} + \frac{1}{2}\vec{c} \times \vec{b} + \frac{1}{2}\vec{a} \times \vec{c} + \frac{1}{2}(\vec{b} \times \vec{c} - \vec{b} \times \vec{a} - \vec{a} \times \vec{c} - \vec{a} \times \vec{a}) = 0.$$

Solutions for Chapter 12 Review

1. (a)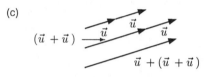

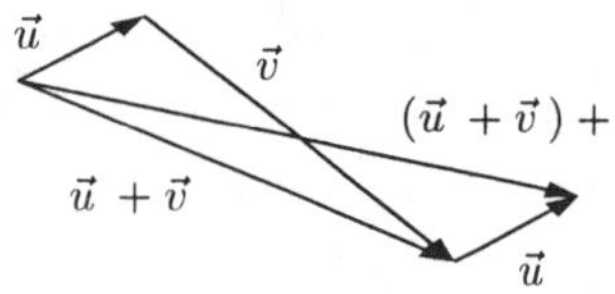

   Figure 12.38

   (b)

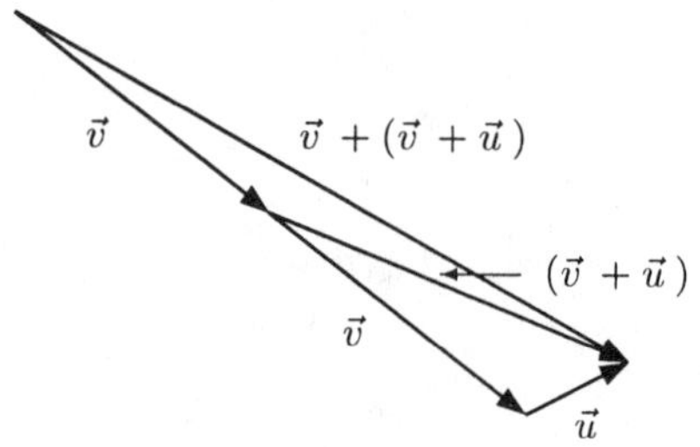

   Figure 12.39

   (c)

   Figure 12.40

2. The coordinates of the points are:

   $$A = (0,0), \quad B = (2,2), \quad C = (7,0), \quad D = (3,4), \quad E = (4,2).$$

   (a) We find $\overrightarrow{AB} = 2\vec{i} + 2\vec{j}, \overrightarrow{CD} = -4\vec{i} + 4\vec{j}$. Therefore,

   $$\vec{u} = (2.5)(2\vec{i} + 2\vec{j}) + (-0.8)(-4\vec{i} + 4\vec{j}) = 5\vec{i} + 5\vec{j} + 3.2\vec{i} - 3.2\vec{j} = 8.2\vec{i} + 1.8\vec{j},$$

   $$\vec{v} = (2.5)(-2\vec{i} - 2\vec{j}) - (-0.8)(-4\vec{i} + 4\vec{j}) = -5\vec{i} - 5\vec{j} - 3.2\vec{i} + 3.2\vec{j} = -8.2\vec{i} - 1.8\vec{j}.$$

   (b) $\vec{v} = -\vec{u}$. We know that $-\overrightarrow{AB}$ is equivalent to $\overrightarrow{BA}$. In other words,

   $$\vec{v} = -(2.5)\overrightarrow{AB} + (0.8)\overrightarrow{CD}.$$

   By factoring out a -1, we get

   $$\vec{v} = -((2.5)\overrightarrow{AB} + (-0.8)\overrightarrow{CD}) = -\vec{u}.$$

3. We find $\overrightarrow{EA}$ to be $-4\vec{i} - 2\vec{j}$. A unit vector on the $\overrightarrow{EA}$ direction is

   $$\vec{n} = \frac{-4i - 2j}{\sqrt{4^2 + 2^2}}$$

   $$= \frac{-2}{\sqrt{5}}\vec{i} - \frac{1}{\sqrt{5}}\vec{j}.$$

   So a vector of length 2 in this direction is

   $$\vec{p} = 2\vec{n} = \frac{-4}{\sqrt{5}}\vec{i} - \frac{2}{\sqrt{5}}\vec{j}.$$

   Thus, $\vec{p} = -\frac{4\sqrt{5}}{5}\vec{i} - \frac{2\sqrt{5}}{5}\vec{j}$.

4. (a) True, by the property of commutativity.
    (b) It doesn't make sense, we cannot add vectors and scalars.
    (c) True, by the property of commutativity.
    (d) This is not always true. For example, let $\vec{a} = \vec{i} + 2\vec{j}$, $\vec{b} = -2\vec{i} + \vec{j}$.

$$\|\vec{a} + \vec{b}\| = \|\vec{i} + 2\vec{j} - 2\vec{i} + \vec{j}\|$$
$$= \|-\vec{i} + 3\vec{j}\| = \sqrt{(-1)^2 + 3^2}$$
$$= \sqrt{10}$$
$$\|\vec{a}\| = \sqrt{1^2 + 2^2} = \sqrt{5}$$
$$\|\vec{b}\| = \sqrt{(-2)^2 + 1^2} = \sqrt{5}.$$

So, $\|\vec{a}\| + \|\vec{b}\| = \sqrt{5} + \sqrt{5} = 2\sqrt{5}$. Thus, $\sqrt{10} \neq 2\sqrt{5}$ and $\|\vec{a} + \vec{b}\| \neq \|\vec{a}\| + \|\vec{b}\|$.

5.

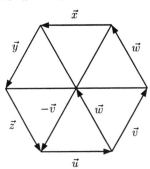

Figure 12.41

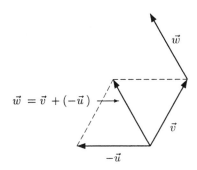

Figure 12.42

Break the hexagon up into 6 equilateral triangles, as shown in Figure 12.41.
Then $\vec{u} - \vec{v} + \vec{w} = \vec{0}$, so $\vec{w} = \vec{v} - \vec{u}$
Similarly, $\vec{x} = -\vec{u}$, $\vec{y} = -\vec{v}$, $\vec{z} = -\vec{w} = \vec{u} - \vec{v}$.

6. (a) $\vec{a} = \vec{e}$
    (b) By using a ruler, we discover $\vec{a} = \frac{1}{2.9}\vec{g}$, because $\vec{g}$ is 2.9 times as long as $\vec{a}$ and points in the same direction.
    (c) There is no such x because $\vec{b}$ and $\vec{d}$ are not parallel.
    (d) We can solve the equation as follows. Draw a line from the tail of $\vec{f}$ and parallel to $\vec{c}$, and another line through the tip of $\vec{f}$ and parallel to $\vec{d}$. Then resolve $\vec{f}$ into components $\vec{v}_1$ and $\vec{v}_2$ such that $\vec{v}_1$ and $\vec{v}_2$ are parallel to $\vec{c}$ and $\vec{d}$ respectively. So $\vec{f} = \vec{v}_1 + \vec{v}_2$ and $\vec{v}_1 = u\vec{c}$ and $\vec{v}_2 = v\vec{d}$. With a ruler, we compare the lengths of the components of $\vec{f}$ with the lengths of $\vec{c}$ and $\vec{d}$ to determine $u \approx 0.8$ and $v \approx -0.9$. Thus, we find that $\vec{f} \approx 0.8\vec{c} - 0.9\vec{d}$ shown in Figure 12.44.

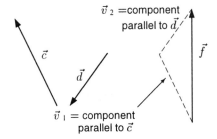

Figure 12.43

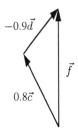

Figure 12.44

7. (a) We need $6\vec{i} + 8\vec{j} + 3\vec{k} = \lambda(2\vec{i} + (t^2 + \frac{2}{3}t + 1)\vec{j} + t\vec{k})$ for some λ. This gives

$$6 = 2\lambda$$
$$8 = (t^2 + \frac{2}{3}t + 1)\lambda$$
$$3 = t\lambda$$

From the first equation, we have $\lambda = 3$. Substituting $\lambda = 3$ into the third equation gives $t = 1$. Check the second equation, it says $8 = 8$, if $t = 1$ and $\lambda = 3$. So for $t = 1$, the two vectors are parallel to each other.

(b) Similar to part (a), we need to solve

$$2 = t\lambda$$
$$-4 = \lambda$$
$$1 = \lambda(t - 1)$$

From the first two equations we have $\lambda = -4$ and $t = -\frac{1}{2}$. Substituting this into the third equation gives $1 = 6$. Thus this system of equations has no solution, so the pair of vectors is not parallel to each other for any value of t.

(c) $2t\vec{i} + t\vec{j} + t\vec{k} = \frac{t}{3}(6\vec{i} + 3\vec{j} + 3\vec{k})$. For any t, the two vectors are parallel to each other.

8. By the definition of cross product, $2\vec{i} \times (\vec{i} + \vec{j})$ is in the direction of $\vec{k}$. The magnitude of it equals to the area of the parallelogram which is

$$\|2\vec{i}\| \cdot \|\vec{i} + \vec{j}\| \sin \frac{\pi}{4} = 2\sqrt{2} \sin \frac{\pi}{4} = 2\sqrt{2} \cdot \frac{\sqrt{2}}{2} = 2.$$

So $2\vec{i} \times (\vec{i} + \vec{j}) = 2\vec{k}$. See Figure 12.45.

9. By definition, $(\vec{i} + \vec{j}) \times (\vec{i} - \vec{j})$ is in the direction of $-\vec{k}$. The magnitude is

$$\|\vec{i} + \vec{j}\| \cdot \|\vec{i} - \vec{j}\| \sin \frac{\pi}{4} = \sqrt{2} \cdot \sqrt{2} = 2.$$

So $(\vec{i} + \vec{j}) \times (\vec{i} - \vec{j}) = -2\vec{k}$. See Figure 12.46.

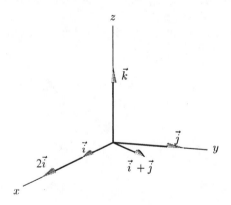

Figure 12.45

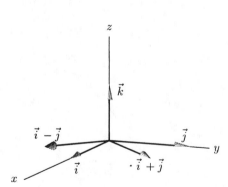

Figure 12.46

10.

$$
\begin{aligned}
[(\vec{i} + \vec{j}) \times \vec{i}] \times \vec{j} &= (\vec{i} \times \vec{i} + \vec{j} \times \vec{i}) \times \vec{j} \\
&= (\vec{0} - \vec{k}) \times \vec{j} \\
&= -\vec{k} \times \vec{j} \\
&= \vec{j} \times \vec{k} = \vec{i}.
\end{aligned}
$$

11.

$$
\begin{aligned}
(\vec{i} + \vec{j}) \times (\vec{i} \times \vec{j}) &= (\vec{i} + \vec{j}) \times \vec{k} \\
&= (\vec{i} \times \vec{k}) + (\vec{j} \times \vec{k}) \\
&= -\vec{j} + \vec{i} = \vec{i} - \vec{j}.
\end{aligned}
$$

12. True. If $\vec{a} \times \vec{b} = 0$, $\vec{a}$ and $\vec{b}$ have same direction, hence $-\vec{b} \times \vec{a} = 0$ so we have equality. If $\vec{a} \times \vec{b} \neq 0$, $\vec{a}$ and $\vec{b}$ don't have same direction, and $\vec{a} \times \vec{b}$ and $\vec{b} \times \vec{a}$ are both perpendicular to $\vec{a}$ and to $\vec{b}$ and in opposite directions. Since both $\vec{a} \times \vec{b}$ and $\vec{b} \times \vec{a}$ have the same magnitude, namely $\|\vec{a}\| \|\vec{b}\| \sin\theta$, where θ is the angle between $\vec{a}$ and $\vec{b}$, it is true that $\vec{a} \times \vec{b} = -(\vec{b} \times \vec{a})$ for all $\vec{a}$ and $\vec{b}$.

13. Let $\vec{r}_1$ be the displacement vector $\overrightarrow{PQ}$ and let $\vec{r}_2$ be the displacement vector $\overrightarrow{PR}$. Then

$$
\begin{aligned}
\vec{r}_1 &= (1 + 2)\vec{i} + (3 - 2)\vec{j} + (-1 - 0)\vec{k} = 3\vec{i} + \vec{j} - \vec{k}, \\
\vec{r}_2 &= (-4 + 2)\vec{i} + (2 - 2)\vec{j} + (1 - 0)\vec{k} = -2\vec{i} + \vec{k}, \\
\vec{r}_1 \times \vec{r}_2 &= \begin{vmatrix} \vec{i} & \vec{j} & \vec{k} \\ 3 & 1 & -1 \\ -2 & 0 & 1 \end{vmatrix} = \vec{i} - (3 - 2)\vec{j} + 2\vec{k} = \vec{i} - \vec{j} + 2\vec{k}.
\end{aligned}
$$

The area of the triangle $= \frac{1}{2}\|\vec{r}_1 \times \vec{r}_2\| = \frac{1}{2}\sqrt{1^2 + 1^2 + 2^2} = \frac{\sqrt{6}}{2}$.

14. If the planes are parallel, they have a common normal vector $\vec{n}$. Rewrite the equation of the plane as $4x - 3y - z = -8$ so that $\vec{n} = 4\vec{i} - 3\vec{j} - \vec{k}$ and the desired plane is $4(x - 0) - 3(y - 0) - (z - 0) = 0$ or $4x - 3y - z = 0$.

15. $\vec{n} = 4\vec{i} + 6\vec{k}$ (the coefficients of x, y, z are the same as the coefficients of $\vec{i}, \vec{j}$, and $\vec{k}$.)

16. The equation will be of the form $mx + ny + ez = d$, but we can divide through by d to get an equation of the form $ax + by + cz = 1$ (Note that d can not be zero, as the origin is not in the plane). Now substitute in the points: From $(0, 0, 2)$, we get $a(0) + b(0) + c(2) = 1$. From this we get $c = \frac{1}{2}$. Similarly we get $a = \frac{1}{5}$, and $b = \frac{1}{3}$. So the equation that fits these points is

$$
\frac{x}{5} + \frac{y}{3} + \frac{z}{2} = 1.
$$

17. (a) On the x-axis, $y = z = 0$, so $5x = 21$, giving $x = \frac{21}{5}$. So the only such point is $(\frac{21}{5}, 0, 0)$.
    (b) Other points are $(0, -21, 0)$, and $(0, 0, 3)$. There are many other possible answers.
    (c) $\vec{n} = 5\vec{i} - \vec{j} + 7\vec{k}$. It is the normal vector.
    (d) The vector between two points in the plane is parallel to the plane. Using the points from part (b), the vector $3\vec{k} - (-21\vec{j}) = 21\vec{j} + 3\vec{k}$ is parallel to the plane.

18.   (a)   Since

$$\overrightarrow{PQ} = (3\vec{i} + 5\vec{j} + 7\vec{k}) - (\vec{i} + 2\vec{j} + 3\vec{k}) = 2\vec{i} + 3\vec{j} + 4\vec{k},$$

and

$$\overrightarrow{PR} = (2\vec{i} + 5\vec{j} + 3\vec{k}) - (\vec{i} + 2\vec{j} + 3\vec{k}) = \vec{i} + 3\vec{j},$$

$$\overrightarrow{PQ} \times \overrightarrow{PR} = \begin{vmatrix} \vec{i} & \vec{j} & \vec{k} \\ 2 & 3 & 4 \\ 1 & 3 & 0 \end{vmatrix} = -12\vec{i} + 4\vec{j} + 3\vec{k},$$

which is a vector perpendicular to the plane containing P, Q and R. Since

$$\|\overrightarrow{PQ} \times \overrightarrow{PR}\| = \sqrt{(-12)^2 + 4^2 + 3^2} = 13,$$

the unit vectors which are perpendicular to a plane containing P, Q, and R are

$$-\frac{12}{13}\vec{i} + \frac{4}{13}\vec{j} + \frac{3}{13}\vec{k},$$

or the unit vector pointing to the opposite direction,

$$\frac{12}{13}\vec{i} - \frac{4}{13}\vec{j} - \frac{3}{13}\vec{k}.$$

(b)   The angle between PQ and PR is θ for which

$$\cos\theta = \frac{\overrightarrow{PQ} \cdot \overrightarrow{PR}}{\|\overrightarrow{PQ}\| \cdot \|\overrightarrow{PR}\|} = \frac{2 \cdot 1 + 3 \cdot 3 + 4 \cdot 0}{\sqrt{2^2 + 3^2 + 4^2} \cdot \sqrt{1^2 + 3^2 + 0^2}} = \frac{11}{\sqrt{290}},$$

so

$$\theta = \cos^{-1}\left(\frac{11}{\sqrt{290}}\right) \approx 49.76°.$$

(c)   The area of triangle $PQR = \frac{1}{2}\|\overrightarrow{PQ} \times \overrightarrow{PR}\| = \frac{13}{2}$.

(d)   Let d be the distance from R to the line through P and Q (see Figure 12.47), then

$$\frac{1}{2}d \cdot \|\overrightarrow{PQ}\| = \text{the area of } \triangle PQR = \frac{13}{2}.$$

Therefore,

$$d = \frac{13}{\|\overrightarrow{PQ}\|} = \frac{13}{\sqrt{2^2 + 3^2 + 4^2}} = \frac{13}{\sqrt{29}}.$$

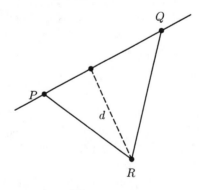

Figure 12.47

19. Let $\vec{v} = x\vec{i} + y\vec{j}$.
We want $\|\vec{v}\| = 1$ and $\|\vec{v} + \vec{i}\| = 1$, i.e.,

$$\sqrt{x^2 + y^2} = 1 \quad \text{and} \quad \sqrt{(x+1)^2 + y^2} = 1.$$

Setting these equations equal and solving for x gives:

$$\sqrt{x^2 + y^2} = \sqrt{(x+1)^2 + y^2}$$
$$x^2 + y^2 = (x+1)^2 + y^2 \quad \text{\footnotesize(after squaring both sides)}$$
$$x^2 + y^2 = x^2 + 2x + 1 + y^2$$
$$0 = 2x + 1$$
$$x = -\frac{1}{2}.$$

Since we know $x = -\frac{1}{2}$, we can use the fact that $\|\vec{v}\| = 1$ to solve for y:

$$\|\vec{v}\| = 1$$
$$\sqrt{x^2 + y^2} = 1$$
$$x^2 + y^2 = 1$$
$$y^2 = 1 - x^2$$
$$y = \pm\sqrt{1 - x^2}$$
$$= \pm\sqrt{1 - \left(\frac{1}{2}\right)^2}$$
$$= \pm\frac{\sqrt{3}}{2}.$$

Thus the vectors we are looking for are:

$$\vec{v} = -\frac{1}{2}\vec{i} + \frac{\sqrt{3}}{2} \quad \text{and} \quad \vec{v} = -\frac{1}{2}\vec{i} - \frac{\sqrt{3}}{2}\vec{j}.$$

20. Let $\vec{w} = x\vec{i} + y\vec{j} + z\vec{k}$.
We want $\|\vec{w}\| = 1$ and $\|\vec{w} + \vec{i}\|$, i.e.,

$$\sqrt{x^2 + y^2 + z^2} = 1 \quad \text{and} \quad \sqrt{(x+1)^2 + y^2 + z^2} = 1.$$

Setting these equations equal and solving for x gives:

$$\sqrt{x^2 + y^2 + z^2} = \sqrt{(x+1)^2 + y^2 + z^2}$$
$$x^2 + y^2 + z^2 = (x+1)^2 + y^2 + z^2$$
$$x^2 = (x+1)^2$$
$$x^2 = x^2 + 2x + 1$$
$$0 = 2x + 1$$
$$x = -\frac{1}{2}.$$

Using $x = -\frac{1}{2}$ and $\sqrt{x^2 + y^2 + z^2} = 1$ to find an expression for y and z gives:

$$\sqrt{x^2 + y^2 + z^2} = 1$$
$$x^2 + y^2 + z^2 = 1$$
$$y^2 + z^2 = 1 - x^2$$
$$y^2 + z^2 = 1 - \left(\frac{1}{2}\right)^2$$
$$y^2 + z^2 = \frac{3}{4}.$$

So we have $x = -\frac{1}{2}$ and $y^2 + z^2 = \frac{3}{4}$. Therefore vectors that satisfy $\|\vec{w}\| = 1$ and $\|\vec{w} + \vec{i}\| = 1$ have the form $-\frac{1}{2}\vec{i} + y\vec{j} + z\vec{k}$, where $y^2 + z^2 = \frac{3}{4}$. Geometrically, this is a circle with center $\left(-\frac{1}{2}, 0, 0\right)$ and radius $\frac{\sqrt{3}}{2}$ lying parallel to the yz-plane. Another way to visualize this set geometrically is to note that the points with $\|\vec{w}\| = 1$ lie on a sphere of radius 1 centered at the origin, and the points with $\|\vec{w} + \vec{i}\| = 1$ lie on a sphere of radius 1 centered at the point $(-1, 0, 0)$. The intersection of these spheres is a circle with center halfway between the two spheres, i.e., at the point $\left(-\frac{1}{2}, 0, 0\right)$, and lying parallel to the yz-plane.

21. Writing $\vec{P} = (P_1, P_2, \cdots, P_{50})$ where P_i is the population of the i-th state, shows that $\vec{P}$ can be thought of as a vector with 50 components.

22. Since there are 16 ounces in a pound, we multiply the vector by $1/16$ to get $0.01875\vec{i} + 0.0125\vec{j} + 0.03125\vec{k}$ in dollars per ounce.

23. The speed is a scalar which equals 30 times the circumference of the circle per minute. So it is a constant. The velocity is a vector. Since the direction of the motion changes all the time, the velocity is not constant. This implies that the acceleration is nonzero.

24. (a) Since the radius of the circle is 1 meter, the circumference is 2π meters. Thus, the object is moving at 2π meters/minute, or $\pi/30$ meters/second ≈ 0.11 meters/second.
     (b) 30 seconds after passing the point $(0, 1)$, the object is at the point $(-1, 0)$. (Since it completes 1 revolution each minute, it will move π radians in 30 seconds.) This is true regardless of whether the point is moving clockwise or counterclockwise. However, since the velocity vector, $\vec{v}$, is tangential to the curve in the direction of motion, it will have an opposite sign if the motion is in the opposite direction. So, moving clockwise $\vec{v} = 2\pi\vec{j}$, and moving counterclockwise $\vec{v} = -2\pi\vec{j}$, if the speed is measured in meters/minute.

25. (a) Target A is at the point $(30, 0, 3)$; Target B is at the point $(20, 15, 0)$; Target C is the point $(12, 30, 8)$. You fire from the point $P = (0, 0, 5)$. The vectors to each of these targets are $\overrightarrow{PA} = 30\vec{i} - 2\vec{k}$, $\overrightarrow{PB} = 20\vec{i} + 15\vec{j} - 5\vec{k}$, $\overrightarrow{PC} = 12\vec{i} + 30\vec{j} + 3\vec{k}$.
     (b) You fire from the point $Q = (0, -1, 3)$, so $\overrightarrow{QA} = 30\vec{i} + \vec{j}$, $\overrightarrow{QB} = 20\vec{i} + 16\vec{j} - 3\vec{k}$, $\overrightarrow{QC} = 12\vec{i} + 31\vec{j} + 5\vec{k}$.

26. (a) 500 km/h in the west direction, so $\vec{v} = -500\vec{i}$.
     (b) While traveling at constant altitude, the plane travels 250 km westward. Thus the coordinates of the point where the plane begins to descend are $(550, 60, 4) - (250, 0, 0) = (300, 60, 4)$.
     (c) The vector from the plane to the airport at the time it begins its descent is $(200\vec{i} + 10\vec{j}) - (300\vec{i} + 60\vec{j} + 4\vec{k}) = -100\vec{i} - 50\vec{j} - 4\vec{k}$. Velocity is a vector of length 200 km/h in the direction of $-100\vec{i} - 50\vec{j} - 4\vec{k}$. Since $\sqrt{(-100)^2 + (-50)^2 + (-4)^2} \approx 111.9$, a unit vector in the direction of

descent is $-\frac{100}{111.9}\vec{i} - \frac{50}{111.9}\vec{j} - \frac{4}{111.9}\vec{k}$. Thus

$$\text{Velocity vector} = 200\left(-\frac{100}{111.9}\vec{i} - \frac{50}{111.9}\vec{j} - \frac{4}{111.9}\vec{k}\right) = -178.7\vec{i} - 89.4\vec{j} - 7.2\vec{k}.$$

27.

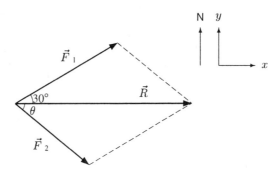

Figure 12.48

Let $\vec{R}$ be the resultant force, and let $\vec{F}_1$ and $\vec{F}_2$ be the forces exerted by the larger and smaller tugs. Then $\|\vec{F}_1\| = \frac{5}{4}\|\vec{F}_2\|$. The y components of the vectors $\vec{F}_1$ and $\vec{F}_2$ must cancel each other in order to ensure that the ship travels due east, hence

$$\|\vec{F}_1\| \sin 30° = \|\vec{F}_2\| \sin\theta,$$

so

$$\frac{5}{4}\|\vec{F}_2\| \sin 30° = \|\vec{F}_2\| \sin\theta,$$

giving $\sin\theta = \frac{5}{8}$, and hence $\theta = \sin^{-1}\frac{5}{8} = 38.7°$.

28. (a) Let x-axis be the East direction and y-axis be the North direction. From Figure 12.49,

$$\theta = \sin^{-1}(4/5) = 53.1°.$$

That is, he should steer at $53.1°$ east of south.

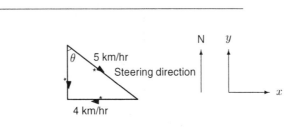

Figure 12.49

(b)

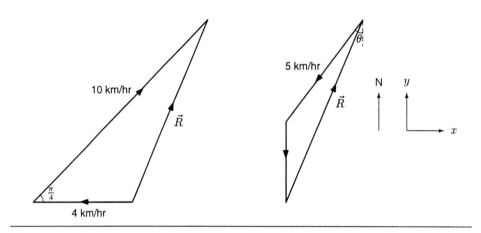

Figure 12.50

Let $\vec{R}$ be the resultant of the wind and river velocities, that is

$$\vec{R} = -4\vec{i} + (10\cos(\frac{\pi}{4})\vec{i} + 10\cos(\frac{\pi}{4})\vec{j})$$
$$= (-4 + 5\sqrt{2})\vec{i} + 5\sqrt{2}\vec{j}.$$

From Figure 12.50, we see that to get the the x-component of his rowing velocity and the x-component of $\vec{R}$ to cancel each other, we must have

$$5\sin\theta = -4 + 5\sqrt{2}$$
$$\theta = \sin^{-1}\left(\frac{-4 + 5\sqrt{2}}{5}\right)$$
$$= 37.9°.$$

However for this value of θ, the y-component of the velocity is

$$5\sqrt{2} - 5\cos(37.9°) = 3.1.$$

Since the y-component is positive, the man will not move across the river in a southward direction.

29. (a)

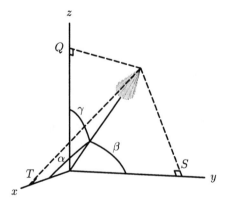

Figure 12.51

Suppose $\vec{v} = \overrightarrow{OP}$ as in Figure 12.51. The $\vec{i}$ component of $\overrightarrow{OP}$ is the projection of $\overrightarrow{OP}$ on the x-axis:

$$\overrightarrow{OT} = v \cos \alpha \vec{i}.$$

Similarly, the $\vec{j}$ and $\vec{k}$ components of $\overrightarrow{OP}$ are the projections of $\overrightarrow{OP}$ on the y-axis and the z-axis respectively. So:

$$\overrightarrow{OS} = v \cos \beta \vec{j}$$
$$\overrightarrow{OQ} = v \cos \gamma \vec{k}$$

Since $\vec{v} = \overrightarrow{OT} + \overrightarrow{OS} + \overrightarrow{OQ}$, we have

$$\vec{v} = v \cos \alpha \vec{i} + v \cos \beta \vec{j} + v \cos \gamma \vec{k}.$$

(b) Since

$$v^2 = \vec{v} \cdot \vec{v} = (v \cos \alpha \vec{i} + v \cos \beta \vec{j} + v \cos \gamma \vec{k}) \cdot$$
$$(v \cos \alpha \vec{i} + v \cos \beta \vec{j} + v \cos \gamma \vec{k})$$
$$= v^2 (\cos^2 \alpha + \cos^2 \beta + \cos^2 \gamma)$$

so

$$\cos^2 \alpha + \cos^2 \beta + \cos^2 \gamma = 1.$$

30. Let $\vec{v} = v_x \vec{i} + v_y \vec{j} + v_z \vec{k}$ be the vector. We will use the properties given in the problem to find v_x, v_y, and v_z.

(a) If $\vec{v}$ has magnitude 10, then $\|\vec{v}\| = 10$.

(b) If $\vec{v}$ makes an angle of $45°$ with the x-axis, then its x-component, v_x, is given by:

$$v_x = \vec{v} \cdot \vec{i} = \|\vec{v}\| \cos 45° = 10 \left(\frac{\sqrt{2}}{2} \right) = 7.0710.$$

(c) Similarly, if $\vec{v}$ makes a $75°$ angle with the y-axis, then its y-component, v_y, is given by:

$$v_y = \vec{v} \cdot \vec{j} = \|\vec{v}\| \cos 75° = 10(0.25882) = 2.5882.$$

(d) We now have two components of $\vec{v}$:

$$\vec{v} = 7.0710\vec{i} + 2.5882\vec{j} + v_z \vec{k}.$$

We only need to find v_z. To do this we use the fact that $\sqrt{\vec{v} \cdot \vec{v}} = \|\vec{v}\| = 10$.

$$\vec{v} \cdot \vec{v} = 100$$
$$v_x^2 + v_y^2 + v_z^2 = 100$$
$$v_z^2 = 100 - v_x^2 - v_y^2$$
$$v_z^2 = \pm\sqrt{100 - v_x^2 - v_y^2}$$
$$v_z = \pm 6.580$$

Since the problem tells us that the $\vec{k}$-component is positive, $v_z = +6.580$. Thus

$$\vec{v} = 7.0710\vec{i} + 2.5882\vec{j} + 6.580\vec{k}.$$

31. Let F be the middle point on $\overline{AB}$. Let G be the point of intersection of the perpendicular bisectors from D and E. (See Figure 12.52.) We must show that $\overrightarrow{GF}$ is perpendicular to $\overrightarrow{AB}$, that is, $\vec{g} \cdot \vec{c} = 0$. We first express $\vec{c}$ and $\vec{g}$ in terms of $\vec{a}, \vec{b}, \vec{e}, \vec{f}$. From the figure and the properties of vectors we can have:

$$\vec{c} = \vec{a} + \vec{b}$$
$$\vec{a} \cdot \vec{e} = 0$$
$$\vec{b} \cdot \vec{f} = 0$$
$$\vec{e} = \vec{a} + \vec{b} + \vec{f}.$$

Since E and F are endpoints of BC and AB, $\overrightarrow{EF}$ is parallel to $\overrightarrow{AC}$, and $\|EF\| = \frac{1}{2}\|AC\|$, i.e., $\overrightarrow{EF} = \vec{b}$, so $\vec{g} = \vec{b} - \vec{e}$. Hence

$$\begin{aligned}
\vec{g} \cdot \vec{c} &= (\vec{b} - \vec{e}) \cdot (\vec{a} + \vec{b}) \\
&= \vec{b} \cdot \vec{a} + \vec{b} \cdot \vec{b} - \vec{e} \cdot \vec{a} - \vec{e} \cdot \vec{b} \\
&= \vec{b} \cdot \vec{a} + \vec{b} \cdot \vec{b} - (\vec{a} + \vec{b} + \vec{f}) \cdot \vec{b} \quad \text{(since } \vec{e} \cdot \vec{a} = 0 \text{ and } \vec{e} = \vec{a} + \vec{b} + \vec{f}) \\
&= \vec{b} \cdot \vec{a} + \vec{b} \cdot \vec{b} - \vec{a} \cdot \vec{b} - \vec{b} \cdot \vec{b} - \vec{f} \cdot \vec{b} \\
&= -\vec{f} \cdot \vec{b} = 0.
\end{aligned}$$

Thus $\overrightarrow{GF}$ is perpendicular to $\overrightarrow{AB}$. Hence the perpendicular bisectors of a triangle intersect at a point.

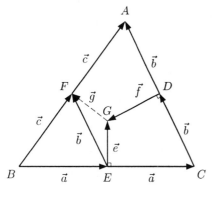

Figure 12.52

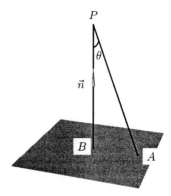

Figure 12.53

32. Find an arbitrary point on the plane $2x + 4y - z = -1$, say $A = (0, 0, 1)$. The normal $\vec{n}$ to the plane at B is $\vec{n} = 2\vec{i} + 4\vec{j} - \vec{k}$ and $\overrightarrow{PA} = -2\vec{i} + \vec{j} - 2\vec{k}$. See Figure 12.53. So the distance d from the point P to the plane is

$$\begin{aligned}
d &= \|\overrightarrow{PB}\| = \|\overrightarrow{PA}\| \cos \theta \\
&= \frac{\overrightarrow{PA} \cdot \vec{n}}{\|\vec{n}\|} \quad \text{since } \overrightarrow{PA} \cdot \vec{n} = \|\overrightarrow{PA}\|\|\vec{n}\| \cos \theta) \\
&= \frac{(-2\vec{i} + \vec{j} - 2\vec{k}) \cdot (2\vec{i} + 4\vec{j} - \vec{k})}{\sqrt{2^2 + 4^2 + (-1)^2}} \\
&= \frac{2}{\sqrt{21}}.
\end{aligned}$$

33. The displacement from $(1, 1, 1)$ to $(1, 4, 5)$ is

$$\vec{r_1} = (1 - 1)\vec{i} + (4 - 1)\vec{j} + (5 - 1)\vec{k} = 3\vec{j} + 4\vec{k}.$$

The displacement from $(-3, -2, 0)$ to $(1, 4, 5)$ is

$$\vec{r_2} = (1 + 3)\vec{i} + (4 + 2)\vec{j} + (5 - 0)\vec{k} = 4\vec{i} + 6\vec{j} + 5\vec{k}.$$

A normal vector is

$$\vec{n} = \vec{r_1} \times \vec{r_2} = \begin{vmatrix} \vec{i} & \vec{j} & \vec{k} \\ 0 & 3 & 4 \\ 4 & 6 & 5 \end{vmatrix} = (15 - 24)\vec{i} - (-16)\vec{j} + (-12)\vec{k} = -9\vec{i} + 16\vec{j} - 12\vec{k}.$$

The equation of the plane is

$$-9x + 16y - 12z = -9 \cdot 1 + 16 \cdot 1 - 12 \cdot 1 = -5$$
$$9x - 16y + 12z = 5.$$

We pick a point A on the plane, $A = (\frac{5}{9}, 0, 0)$ and let $P = (0, 0, 0)$. (See Figure 12.54.) Then $\vec{PA} = (5/9)\vec{i}$.

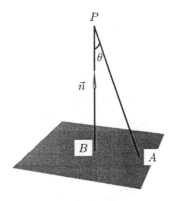

Figure 12.54

So the distance d from the point P to the plane is

$$d = \|\vec{PB}\| = \|\vec{PA}\| \cos\theta$$
$$= \frac{\vec{PA} \cdot \vec{n}}{\|\vec{n}\|} \quad \text{since } \vec{PA} \cdot \vec{n} = \|\vec{PA}\|\|\vec{n}\| \cos\theta)$$
$$= \left| \frac{(\frac{5}{9}\vec{i}) \cdot (-9\vec{i} + 16\vec{j} - 12\vec{k})}{\sqrt{9^2 + 16^2 + 12^2}} \right|$$
$$= \frac{5}{\sqrt{481}} = 0.23.$$

34. In Figure 12.55, let l_1 be a line with direction vector $\vec{v_1}$ passing through P_1. Let l_2 be a line with direction vector $\vec{v_2}$ passing through P_2. Lines l_1 and l_2 are skew if they are not parallel and do not intersect.

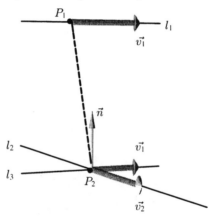

Figure 12.55

If we draw a line l_3 parallel to l_1, i.e., with a direction vector v_1, passing through P_2, as shown in Figure 12.56, then l_2 and l_3 determine a plane that is parallel to l_1:

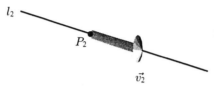

Figure 12.56

The minimum distance between l_1 and l_2 is equal to the distance of l_1 to the plane. So it is equal to the projection of $\overrightarrow{P_1P_2}$ in the direction of the normal vector of the plane. The unit normal vector is given by:

$$\vec{n} = \frac{\vec{v}_1 \times \vec{v}_2}{\|\vec{v}_1 \times \vec{v}_2\|},$$

so the component of $\overrightarrow{P_1P_2}$ in the direction of $\vec{n}$ is:

$$\overrightarrow{P_1P_2} \cdot \vec{n} = \overrightarrow{P_1P_2} \cdot \frac{\vec{v}_1 \times \vec{v}_2}{\|\vec{v}_1 \times \vec{v}_2\|}.$$

Thus the minimum distance between l_1 and l_2 is:

$$\frac{|\overrightarrow{P_1P_2} \cdot (\vec{v}_1 \times \vec{v}_2)|}{\|\vec{v}_1 \times \vec{v}_2\|}.$$

CHAPTER THIRTEEN

Solutions for Section 13.1

1. If h is small, then

$$f_x(3,2) \approx \frac{f(3+h,2) - f(3,2)}{h}.$$

With $h = 0.01$, we find

$$f_x(3,2) \approx \frac{f(3.01,2) - f(3,2)}{0.01} = \frac{\frac{3.01^2}{(2+1)} - \frac{3^2}{(2+1)}}{0.01} = 2.00333.$$

With $h = 0.0001$, we get

$$f_x(3,2) \approx \frac{f(3.0001,2) - f(3,2)}{0.0001} = \frac{\frac{3.0001^2}{(2+1)} - \frac{3^2}{(2+1)}}{0.0001} = 2.0000333.$$

Since the difference quotient seems to be approaching 2 as h gets smaller, we conclude

$$f_x(3,2) \approx 2.$$

To estimate $f_y(3,2)$, we use

$$f_y(3,2) \approx \frac{f(3,2+h) - f(3,2)}{h}.$$

With $h = 0.01$, we get

$$f_y(3,2) \approx \frac{f(3,2.01) - f(3,2)}{0.01} = \frac{\frac{3^2}{(2.01+1)} - \frac{3^2}{(2+1)}}{0.01} = -0.99668.$$

With $h = 0.0001$, we get

$$f_y(3,2) \approx \frac{f(3,2.0001) - f(3,2)}{0.0001} = \frac{\frac{3^2}{(2.0001+1)} - \frac{3^2}{(2+1)}}{0.0001} = -0.9999667.$$

Thus, it seems that the difference quotient is approaching -1, so we estimate

$$f_y(3,2) \approx -1.$$

2. Using first $\Delta x = 0.1$ and $\Delta y = 0.1$, we have the estimates:

$$f_x(1,3) \approx \frac{f(1.1,3) - f(1,3)}{0.1}$$
$$= \frac{0.0470 - 0.0519}{0.1} = -0.0493,$$

and

$$f_y(1,3) \approx \frac{f(1,3.1) - f(1,3)}{0.1}$$
$$= \frac{0.0153 - 0.0519}{0.1} = -0.3660.$$

Now, using $\Delta x = 0.01$ and $\Delta y = 0.01$, we have the estimates:

$$f_x(1,3) \approx \frac{f(1.01,3) - f(1,3)}{0.01}$$
$$= \frac{0.0514 - 0.0519}{0.01} = -0.0501,$$

and

$$f_y(1,3) \approx \frac{f(1,3.01) - f(1,3)}{0.01}$$
$$= \frac{0.0483 - 0.0519}{0.01} = -0.3629.$$

3. (a) This means you must pay a mortgage payment of $1090.08/month if you have borrowed a total of $92000 at an interest rate of 14%, on a 30-year mortgage.
    (b) This means that the rate of change of the monthly payment with respect to the interest rate is $72.82; i.e., your monthly payment will go up by approximately $72.82 for one percentage point increase in the interest rate for the $92,000 borrowed under a 30-year mortgage.
    (c) It should be *positive*, because the monthly payments will increase if the total amount borrowed is increased.
    (d) It should be *negative*, because as you increase the number of years in which to pay the mortgage, you should have to pay less each month.

4. (a) If you borrow $8000 at an interest rate of 1% per month and pay it off in 24 months, your monthly payments are $376.59.
    (b) The increase in your monthly payments for borrowing an extra dollar under the same terms as in (a) is about 4.7 cents.
    (c) If you borrow the same amount of money for the same time period as in (a), but if the interest rate increases by 1%, the increase in your monthly payments is about $44.83.

5. (a) An increase in the price of a new car will decrease the number of cars bought annually. Thus $\frac{\partial q_1}{\partial x} < 0$.
    Similarly, an increase in the price of gasoline will decrease the amount of gas sold, implying $\frac{\partial q_2}{\partial y} < 0$.
    (b) Since the demands for a car and gas complement each other, an increase in the price of gasoline will decrease the total number of cars bought. Thus $\frac{\partial q_1}{\partial y} < 0$. Similarly, we may expect $\frac{\partial q_2}{\partial x} < 0$.

6. (a) The units of $\partial c/\partial x$ are concentration/distance. (For example, (gm/cm^3)/cm.) The practical interpretation of $\partial c/\partial x$ is the rate of change of concentration with distance as you move down the blood vessel at a fixed time. We expect $\partial c/\partial x < 0$ because the further away you get from the point of injection, the less of the drug you would expect to find (at a fixed time).
    (b) The units of $\partial c/\partial t$ are concentration/time. (For example, (gm/cm^3)/sec.) The practical interpretation of $\partial c/\partial t$ is the rate of change of concentration with time, as you look at a particular point in the blood vessel. We would expect the concentration to first increase (as the drug reaches the point) and then decrease as the drug dies away. Thus, we expect $\partial c/\partial t > 0$ for small t and $\partial c/\partial t < 0$ for large t.

7. $\partial P / \partial t$: The unit is dollars per month. This is the rate at which payments change as the number of months it takes to pay off the loan changes. The sign is negative because payments decrease as the pay-off time increases.

    $\partial P / \partial r$: The unit is dollars per percentage point. This is the rate at which payments change as the interest rate changes. The sign is positive because payments increase as the interest rate increases.

8. (a) Near A, the value of z increases as x increases, so $f_x(A) > 0$.
    (b) Near A, the value of z decreases as y increases, so $f_y(A) < 0$.
    (c) $f_x(P)$ changes from positive to negative as P moves from A to B along a straight line, because after P crosses the y-axis, z decreases as x increases near P.
    $f_y(P)$ does not change sign as P moves from A to B along a straight line; it is negative along AB.

9. (a) For points near the point $(0, 5, 3)$, moving in the positive x direction, the surface is sloping down and the function is decreasing. Thus, $f_x(0, 5) < 0$.
    (b) Moving in the positive y direction near this point the surface slopes up as the function increases, so $f_y(0, 5) > 0$.

10. By using a difference quotient to approximate $f_w(10, 25)$ we get

    $$f_w(10, 25) \approx \frac{f(10 + h, 25) - f(10, 25)}{h}.$$

    Choosing $h = 5$ and reading values from Table 11.3 on page 8 of the text, we get

    $$f_w(10, 25) \approx \frac{f(15, 25) - f(10, 25)}{5} = \frac{2 - 10}{5} = -1.6$$

    This means that when the wind speed is 10 mph and the true temperature is 25°F, as the wind speed increases from 10 mph by 1 mph we feel a 1.6°F drop in temperature. This rate is negative because the temperature you feel drops as the wind speed increases.

11. Using a difference quotient with $h = 5$, we get

    $$f_T(5, 20) \approx \frac{f(5, 20 + 5) - f(5, 20)}{5} = \frac{21 - 16}{5} = 1$$

    This means that when the wind speed is 5 mph and the true temperature is 20°F, the apparent temperature increases by approximately 1°F for every increase of 1°F in the true temperature. This rate is positive because the true temperature you feel increases as true temperature increases.

12. Since the average rate of change of the temperature adjusted for wind-chill is about -2.6 (drops by 2.6°F) with every 1 mph increase in wind speed from 5 mph to 10 mph, when the true temperature stays constant at 20°F, we know that

    $$f_w(5, 20) \approx -2.6.$$

13. (a) Estimate $\partial P/\partial r$ and $\partial P/\partial L$ by using difference quotients and reading values of P from the graph:

$$\frac{\partial P}{\partial r}(8,4000) \approx \frac{P(16,4000) - P(8,4000)}{16-8}$$

$$= \frac{100-80}{8} = 2.5,$$

and

$$\frac{\partial P}{\partial L} \approx \frac{P(8,5000) - P(8,4000)}{5000-4000}$$

$$= \frac{100-80}{1000} = 0.02.$$

$P_r(8,4000) \approx 2.5$ means that at an interest rate of 8% and a loan amount of $4000 the monthly payment increases by approximately $2.50 for every one percent increase of the interest rate. $P_L(8,4000) \approx 0.02$ means the monthly payment increases by approximately $0.02 for every $1 increase in the loan amount at an 8% rate and a loan amount of $4000.

 (b) Using difference quotients and reading from the graph

$$\frac{\partial P}{\partial r}(8,6000) \approx \frac{P(14,6000) - P(8,6000)}{14-8}$$

$$= \frac{140-120}{6} = 3.33,$$

and

$$\frac{\partial P}{\partial L}(8,6000) \approx \frac{P(8,7000) - P(8,6000)}{7000-6000}$$

$$= \frac{140-120}{1000} = 0.02.$$

Again, we see that the monthly payment increases with increases in interest rate and loan amount. The interest rate is $r = 8\%$ as in part (a), but here the loan amount is $L = \$6000$. Since $P_L(8,4000) \approx P_L(8,6000)$, the increase in monthly payment per unit increase in loan amount remains the same as in part a). However, in this case, the effect of the interest rate is different: here the monthly payment increases by approximately $3.33 for every one percent increase of interest rate at $r = 8\%$ and loan amount of $6000.

 (c)

$$\frac{\partial P}{\partial r}(13,7000) \approx \frac{P(19,7000) - P(13,7000)}{19-13}$$

$$= \frac{180-160}{6} = 3.33,$$

and

$$\frac{\partial P}{\partial L}(13,7000) \approx \frac{P(13,8000) - P(13,7000)}{8000-7000}$$

$$= \frac{180-160}{1000} = 0.02.$$

The figures show that the rates of change of the monthly payment with respect to the interest rate and loan amount are roughly the same for $(r,L) = (8,6000)$ and $(r,L) = (13,7000)$.

14. Estimating from the contour diagram, using positive increments for Δx and Δy, we have, for point A,

$$\frac{\partial n}{\partial x}(A) \approx \frac{1.5 - 1}{67 - 59} = \frac{1/2}{8} = \frac{1}{16} \approx 0.06 \frac{\text{foxes/km}^2}{\text{km}}$$

$$\frac{\partial n}{\partial y}(A) \approx \frac{0.5 - 1}{60 - 51} = -\frac{1/2}{9} = -\frac{1}{18} \approx -0.06 \frac{\text{foxes/km}^2}{\text{km}}.$$

So, from point A the fox population density increases as we move eastward. The population density decreases as we move north from A.

At point B,

$$\frac{\partial n}{\partial x}(B) \approx \frac{0.75 - 1}{132 - 115} = -\frac{1/4}{17} = -\frac{1}{68} \approx -0.01 \frac{\text{foxes/km}^2}{\text{km}}$$

$$\frac{\partial n}{\partial y}(B) \approx \frac{0.5 - 1}{120 - 110} = -\frac{1/2}{10} = -\frac{1}{20} \approx -0.05 \frac{\text{foxes/km}^2}{\text{km}}.$$

So, fox population density decreases as we move both east and north of B. However, notice that the partial derivative $\partial n/\partial x$ at B is smaller in magnitude than the others. Indeed if we had taken a negative Δx we would have obtained an estimate of the opposite sign. This suggests that better estimates for B are

$$\frac{\partial n}{\partial x}(B) \approx 0 \frac{\text{foxes/km}^2}{\text{km}}$$

$$\frac{\partial n}{\partial y}(B) \approx -0.05 \frac{\text{foxes/km}^2}{\text{km}}.$$

At point C,

$$\frac{\partial n}{\partial x}(C) \approx \frac{2 - 1.5}{130 - 105} = \frac{1/2}{25} = \frac{1}{50} \approx 0.02 \frac{\text{foxes/km}^2}{\text{km}}$$

$$\frac{\partial n}{\partial y}(C) \approx \frac{2 - 1.5}{85 - 55} = \frac{1/2}{30} = \frac{1}{60} \approx 0.02 \frac{\text{foxes/km}^2}{\text{km}}.$$

So, the fox population density increases as we move east and north of C. Again, if these estimates were made using negative values for Δx and Δy we would have had estimates of the opposite sign. Thus, better estimates are

$$\frac{\partial n}{\partial x}(C) \approx 0 \frac{\text{foxes/km}^2}{\text{km}}$$

$$\frac{\partial n}{\partial y}(C) \approx 0 \frac{\text{foxes/km}^2}{\text{km}}.$$

15. The quantity $H_T(10, 0.1)$ is approximated by a difference quotient. The first partial derivative with respect to T is approximated by

$$H_T(10, 0.1) \approx \frac{H(10 + \Delta T, 0.1) - H(10, 0.1)}{\Delta T} \quad \text{for small } \Delta T.$$

We are free to choose ΔT. If we take $H(10, 0.1) = 110$ and $H(20, 0.1) = 100$, we get the approximation

$$H_T(10, 0.1) \approx \frac{H(20, 0.1) - H(10, 0.1)}{10} = \frac{100 - 110}{10} = -1.$$

(Note that you may get a different answer if you read different values from the graph.) The geometric meaning of the partial derivative $H_T(10, 0.01)$ that we just approximated is the slope of the curve in Figure 13.1 corresponding to $w = 0.1$ at the point where $T = 10$. In practical terms, we have found that for fog at $10°$ C containing 0.1 g water per m^3 of fog, a $1°$ C increase in temperature will reduce the heat requirement for dissipating the fog by about 1 calories per cubic meter of fog.

16. Reading values of H from the graph gives Table 13.1. In order to compute $H_T(T, w)$ at $T = 30$, it is useful to have values of $H(T, 2)$ for $T = 40°$ C. The column corresponding to $w = 0.4$ is not used in this problem.

TABLE 13.1   *Estimated values of $H(T, w)$ (in calories/meter^3)*

		w (gm/m^3)		
	0.1	0.2	0.3	0.4
10	110	240	330	450
20	100	180	260	350
T (°C) 30	70	150	220	300
40	65	140	200	270

The estimates for $H_T(T, w)$ in Table 13.2 are now computed using the formula

$$H_T(T, w) \approx \frac{H(T + 10, w) - H(T, w)}{10}.$$

TABLE 13.2
Estimated values of $H_T(T, w)$ (in calories/meter^3/°C)

		w (gm/m^3)	
	0.1	0.2	0.3
10	-1.0	-6.0	-7.0
T (°C) 20	-3.0	-3.0	-4.0
30	-0.5	-1.0	-2.0

17.

TABLE 13.3   *Estimated values of $H(T, w)$ (in calories/meter^3)*

		w (gm/m^3)		
	0.1	0.2	0.3	0.4
10	110	240	330	450
20	100	180	260	350
T (°C) 30	70	150	220	300
40	65	140	200	270

Values of H from the graph are given in Table 13.3. In order to compute $H_w(T, w)$ for $w = 0.3$, it is useful to have the column corresponding to $w = 0.4$. The row corresponding to $T = 40$ is not used in this problem. The partial derivative $H_w(T, w)$ can be approximated by

$$H_w(10, 0.1) \approx \frac{H(10, 0.1 + h) - H(10, 0.1)}{h} \quad \text{for small } h.$$

We choose $h = 0.1$ because we can read off a value for $H(10, 0.2)$ from the graph. If we take $H(10, 0.2) = 240$, we get the approximation

$$H_w(10, 0.1) \approx \frac{H(10, 0.2) - H(10, 0.1)}{0.1} = \frac{240 - 110}{0.1} = 1300.$$

In practical terms, we have found that for fog at $10°$ C containing 0.1 g/m^3 of water, an increase in the water content of the fog will increase the heat requirement for dissipating the fog at the rate given by $H_w(10, 0.1)$. Specifically, a 1 g/m^3 increase in the water content will increase the heat required to dissipate the fog by about 1300 calories per cubic meter of fog.

Wetter fog is harder to dissipate. Other values of $H_w(T, w)$ in Table 13.4 are computed using the formula

$$H_w(T, w) \approx \frac{H(T, w + 0.1) - H(T, w)}{0.1},$$

where we have used Table 13.3 to evaluate H.

TABLE 13.4 *Table of values of $H_w(T, w)$ (in cal/gm)*

T	w (gm/m^3)		
(°C)	0.1	0.2	0.3
10	1300	900	1200
20	800	800	900
30	800	700	800

18. (a) $\dfrac{\partial p}{\partial c} = f_c(c, s) =$ change in blood pressure as cardiac output is increased while systemic vascular resistance remains constant.

(b) Suppose that $p = kcs$. Note that c (cardiac output), a volume, s (SVR), a resistance, and p, a pressure, must all be positive. Thus k must be positive, and our level curves should be confined to the first quadrant. Several level curves are shown in Figure 13.1. Each level curve represents a different blood pressure level. Each point on a given curve is a combination of cardiac output and SVR that results in the blood pressure associated with that curve.

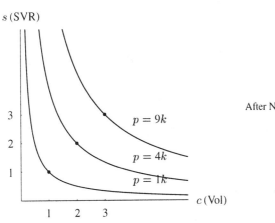

Figure 13.1

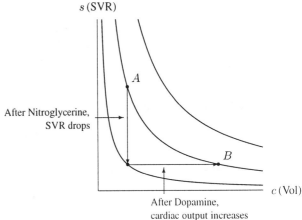

Figure 13.2

(c) Point B in Figure 13.2 shows that if the two doses are correct, the changes in pressure will cancel. The patient's cardiac output will have increased and his SVR will have decreased, but his blood pressure won't have changed.

(d) At point F in Figure 13.3, the patient's blood pressure is normalized, but his/her cardiac output has dropped and his SVR is up.

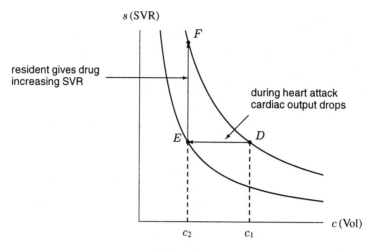

Figure 13.3

Note: c_1 and c_2 are the cardiac outputs before and after the heart attack, respectively.

Solutions for Section 13.2

1. Since $f_y(3, 2)$ equals the derivative of $f(3, y)$ at $y = 2$, we use the function

$$f(3, y) = \frac{9}{y + 1}.$$

Differentiating with respect to y, we get

$$f_y(3, y) = \frac{d}{dy}\left(\frac{9}{y+1}\right) = \frac{-9}{(y+1)^2},$$

and so

$$f_y(3, 2) = -1.$$

2. (a) The difference quotient for approximating $f_u(u, v)$ is given by

$$f_u(u, v) \approx \frac{f(u+h, v) - f(u, v)}{h}.$$

Putting $(u, v) = (1, 3)$ and $h = 0.001$, the difference quotient is

$$f_u(1, 3) \approx \frac{1.001(1.001^2 + 3^2)^{3/2} - 1(1^2 + 3^2)^{3/2}}{0.001}$$

$$\approx \frac{31.6639 - 31.6228}{0.001} \approx \frac{0.0411}{0.001} \approx 41.1$$

(b) Using the derivative formulas

$$f_u = \frac{\partial f}{\partial u} = (u^2 + v^2)^{3/2} + 3u^2(u^2 + v^2)^{1/2} = (u^2 + v^2)^{1/2}(4u^2 + v^2)$$

so

$$f_u(1,3) = (1^2 + 3^2)^{1/2} \cdot (4 \cdot 1^2 + 3^2) \approx 41.11$$

We see that the approximation in part (a) was reasonable.

3. $z_x = 2xy + 10x^4y$

4. $z_x = \cos(5x^3y - 3xy^2) \cdot (15x^2y - 3y^2) = (15x^2y - 3y^2)\cos(5x^3y - 3xy^2)$

5. Differentiating with respect to x gives

$$g_x(x,y) = \frac{\partial}{\partial x} \ln(ye^{xy}) = (ye^{xy})^{-1} \frac{\partial}{\partial x}(ye^{xy}) = (ye^{xy})^{-1} \cdot y\frac{\partial}{\partial x}(e^{xy})$$

$$= (ye^{xy})^{-1} \cdot y \cdot y \cdot e^{xy}$$

$$= y$$

6. $F_m = g$

7. $\dfrac{\partial}{\partial x}(a\sqrt{x}) = a \cdot \dfrac{1}{2}x^{-1/2} = \dfrac{a}{2\sqrt{x}}$

8. $\dfrac{\partial}{\partial x}(xe^{\sqrt{xy}}) = e^{\sqrt{xy}} + xe^{\sqrt{xy}} \cdot \dfrac{1}{2}(xy)^{-1/2}y = e^{\sqrt{xy}}(1 + \dfrac{xy}{2\sqrt{xy}}) = e^{\sqrt{xy}}(1 + \dfrac{\sqrt{xy}}{2})$

9. $\dfrac{\partial}{\partial y}(3x^5y^7 - 32x^4y^3 + 5xy) = 21x^5y^6 - 96x^4y^2 + 5x$

10.

$$z_y = \frac{(15xy - 8)(21x^2y^6 - 2y) - 15x(3x^2y^7 - y^2)}{(15xy - 8)^2}$$

$$= \frac{315x^3y^7 - 168x^2y^6 - 30xy^2 + 16y - 45x^3y^7 + 15xy^2}{(15xy - 8)^2}$$

$$= \frac{270x^3y^7 - 168x^2y^6 - 15xy^2 + 16y}{(15xy - 8)^2}$$

11. $\dfrac{\partial A}{\partial h} = \dfrac{1}{2}(a + b)$

12. $\dfrac{\partial}{\partial m}\left(\dfrac{1}{2}mv^2\right) = \dfrac{1}{2}v^2$

13. $\dfrac{\partial}{\partial B}\left(\dfrac{1}{u_0}B^2\right) = \dfrac{2B}{u_0}$

14. $\dfrac{\partial}{\partial r}\left(\dfrac{2\pi r}{v}\right) = \dfrac{2\pi}{v}$

15. $F_v = \dfrac{2mv}{r}$

16. $\dfrac{\partial}{\partial v_0}(v_0 + at) = 1 + 0 = 1$

17. $\dfrac{\partial F}{\partial m_2} = \dfrac{Gm_1}{r^2}$

18. $a_v = \dfrac{2v}{r}$

19. $\dfrac{\partial}{\partial T}\left(\dfrac{2\pi r}{T}\right) = -\dfrac{2\pi r}{T^2}$

20. $\dfrac{\partial}{\partial t}(v_0 t + \frac{1}{2}at^2) = v_0 + \frac{1}{2}\cdot 2at = v_0 + at$

21. $u_E = \frac{1}{2}\epsilon_0 \cdot 2E + 0 = \epsilon_0 E$

22. $\dfrac{\partial f_0}{\partial L} = \dfrac{1}{2\pi}(-\frac{1}{2})(LC)^{-3/2}C = -\dfrac{1}{4\pi}\dfrac{C}{LC\sqrt{LC}} = -\dfrac{1}{4\pi L\sqrt{LC}}$

23. $\dfrac{\partial y}{\partial t} = \cos(ct - 5x)c = c\cos(ct - 5x)$

24. $\dfrac{\partial}{\partial M}\left(\dfrac{2\pi r^{3/2}}{\sqrt{GM}}\right) = 2\pi r^{3/2}(-\frac{1}{2})(GM)^{-3/2}(G) = -\pi r^{3/2}\cdot\dfrac{G}{GM\sqrt{GM}} = -\dfrac{\pi r^{3/2}}{M\sqrt{GM}}$

25. $z_x = \dfrac{1}{2ay}(-2)\dfrac{1}{x^3} + \dfrac{15x^4 abc}{y} = -\dfrac{1}{ax^3 y} + \dfrac{15abcx^4}{y} = \dfrac{15a^2 bcx^7 - 1}{ax^3 y}$

26.
$$\dfrac{\partial\alpha}{\partial\beta} = \dfrac{(2y\beta + 5)e^{x\beta-3}x - 2ye^{x\beta-3}}{(2y\beta + 5)^2} = \dfrac{[(2y\beta + 5)x - 2y]e^{x\beta-3}}{(2y\beta + 5)^2} = \dfrac{(5x - 2y + 2xy\beta)e^{x\beta-3}}{(2y\beta + 5)^2}$$

27.
$$\dfrac{\partial}{\partial\lambda}\left(\dfrac{x^2 y\lambda - 3\lambda^5}{\sqrt{\lambda^2 - 3\lambda + 5}}\right)$$
$$= \left[(x^2 y - 15\lambda^4)\sqrt{\lambda^2 - 3\lambda + 5} - \dfrac{(2\lambda - 3)(x^2 y\lambda - 3\lambda^5)}{2\sqrt{\lambda^2 - 3\lambda + 5}}\right]\dfrac{1}{\lambda^2 - 3\lambda + 5}$$
$$= \dfrac{(x^2 y - 15\lambda^4)\cdot 2(\lambda^2 - 3\lambda + 5) - (2\lambda - 3)(x^2 y\lambda - 3\lambda^5)}{2(\lambda^2 - 3\lambda + 5)\sqrt{\lambda^2 - 3\lambda + 5}}$$
$$= \dfrac{x^2 y[2(\lambda^2 - 3\lambda + 5) - (2\lambda - 3)\lambda] - 15\lambda^4 \cdot 2(\lambda^2 - 3\lambda + 5) + (2\lambda - 3)\cdot 3\lambda^5}{2(\lambda^2 - 3\lambda + 5)\sqrt{\lambda^2 - 3\lambda + 5}}$$
$$= \dfrac{x^2 y(-3\lambda + 10) - 3\lambda^4(8\lambda^2 - 27\lambda + 50)}{2(\lambda^2 - 3\lambda + 5)\sqrt{\lambda^2 - 3\lambda + 5}}$$

28.
$$\dfrac{\partial m}{\partial v} = m_0 \dfrac{\partial}{\partial v}((1 - v^2/c^2)^{-1/2}) = -\dfrac{m_0}{2}(1 - v^2/c^2)^{-3/2}\left(\dfrac{-2v}{c^2}\right) = \dfrac{m_0 v/c^2}{(1 - v^2/c^2)^{3/2}}.$$

29.
$$\dfrac{\partial}{\partial w}(\sqrt{2\pi xyw - 13x^7 y^3 v}) = \dfrac{1}{2}(2\pi xyw - 13x^7 y^3 v)^{-1/2}(2\pi xy - 0)$$
$$= \dfrac{\pi xy}{\sqrt{2\pi xyw - 13x^7 y^3 v}}$$

30.

$$\frac{\partial}{\partial w}\left(\frac{x^2yw - xy^3w^7}{w-1}\right)^{-7/2}$$

$$= -\frac{7}{2}\left(\frac{x^2yw - xy^3w^7}{w-1}\right)^{-9/2}\left(\frac{(w-1)(x^2y - 7xy^3w^6) - (x^2yw - xy^3w^7)(1)}{(w-1)^2}\right)$$

$$= -\frac{7}{2}\left(\frac{w-1}{x^2yw - xy^3w^7}\right)^{9/2}\left(\frac{(w-1)(x^2y - 7xy^3w^6) - (x^2yw - xy^3w^7)}{(w-1)^2}\right)$$

$$= \frac{7}{2}\left(\frac{w-1}{x^2yw - xy^3w^7}\right)^{-9/2}\cdot\frac{x^2y + 6xy^3w^7 - 7xy^3w^6}{(w-1)^2}$$

31. $z_x = 7x^6 + yx^{y-1}$, and $z_y = 2^y \ln 2 + x^y \ln x$

32. $z_x = -\sin x$, $z_x(2,3) = -\sin 2 \approx -0.9$

33. We regard x as constant and differentiate with respect to y using the product rule:

$$\frac{\partial z}{\partial y} = 2e^{x+2y}\sin y + e^{x+2y}\cos y$$

Substituting $x = 1, y = 0.5$ gives

$$\left.\frac{\partial z}{\partial y}\right|_{(1,0.5)} = 2e^2\sin(0.5) + e^2\cos(0.5) = 13.6.$$

34.

$$\frac{\partial f}{\partial x} = \ln(y\cos x) + x\frac{1}{y\cos x}(-y\sin x) = \ln(y\cos x) - x\tan x$$

$$\left.\frac{\partial f}{\partial x}\right|_{(\pi/3,1)} = \ln(\cos\pi/3) - \pi/3\tan\pi/3 \approx -2.51$$

35. (a) From contour diagram,

$$f_x(2,1) \approx \frac{f(2.3,1) - f(2,1)}{2.3 - 2}$$
$$= \frac{6-5}{0.3} = 3.3,$$
$$f_y(2,1) \approx \frac{f(2,1.4) - f(2,1)}{1.4 - 1}$$
$$= \frac{6-5}{0.4} = 2.5.$$

(b) A table of values for f is given in Table 13.5

TABLE 13.5

		y	
	0.9	1.0	1.1
1.9	4.42	4.61	4.82
x 2	4.81	5.00	5.21
2.1	5.22	5.41	5.62

From Table 13.5 we estimate $f_x(2,1)$ and $f_y(2,1)$ using difference quotients

$$f_x(2,1) \approx \frac{5.41 - 5.00}{2.1 - 2} = 4.1$$

$$f_y(2,1) \approx \frac{5.21 - 5.00}{1.1 - 1} = 2.1.$$

We obtain better estimates by finer data in the table.

(c) $f_x = 2x$, $f_y = 2y$. So the true values are $f_x(2,1) = 4$, $f_y(2,1) = 2$.

36.   We compute the partial derivatives:

$$\frac{\partial Q}{\partial K} = b\alpha K^{\alpha-1}L^{1-\alpha} \quad \text{so} \quad K\frac{\partial Q}{\partial K} = b\alpha K^\alpha L^{1-\alpha}$$

$$\frac{\partial Q}{\partial L} = b(1-\alpha)K^\alpha L^{-\alpha} \quad \text{so} \quad L\frac{\partial Q}{\partial L} = b(1-\alpha)K^\alpha L^{1-\alpha}$$

Adding these two results, we have:

$$K\frac{\partial Q}{\partial K} + L\frac{\partial Q}{\partial L} = b(\alpha + 1 - \alpha)K^\alpha L^{1-\alpha}$$
$$= Q.$$

37.   (a)   To calculate $\partial B/\partial t$, we hold P constant and differentiate B with respect to t:

$$\frac{\partial B}{\partial t} = \frac{\partial}{\partial t}(Pe^{rt}) = Pre^{rt}.$$

In financial terms, $\partial B/\partial t$ represents the change in the amount of money in the bank as one unit of time passes by.

(b)   To calculate $\partial B/\partial P$, we hold t constant and differentiate B with respect to P:

$$\frac{\partial B}{\partial P} = \frac{\partial}{\partial P}(Pe^{rt}) = e^{rt}.$$

In financial terms, $\partial B/\partial P$ represents the change in the amount of money in the bank at time t as you increase the amount of money that was initially deposited by one unit.

38.   (a)   Substituting $t = 0$ and $t = 1$ into the formula for H gives:

$$H(x,0) = 100\sin(\pi x)$$
$$H(x,1) = 100e^{-0.1}\sin(\pi x) = 90.5\sin(\pi x).$$

The graphs of $H(x,0)$ and $H(x,1)$ are shown in Figure 13.4.

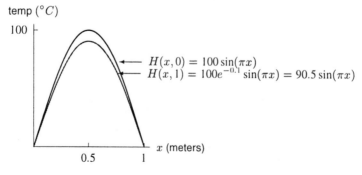

Figure 13.4

(b) To calculate $H_x(x,t)$, we hold t constant and differentiate with respect to x:

$$H_x(x,t) = \frac{\partial}{\partial x} H(x,t) = \frac{\partial}{\partial x}(100e^{-0.1t}\sin(\pi x)) = 100\pi e^{-0.1t}\cos(\pi x)$$
$$H_x(0.2,t) = 100\pi e^{-0.1t}\cos(0.2\pi) = 254.2e^{-0.1t}\,°\text{C/meter}$$
$$H_x(0.8,t) = 100\pi e^{-0.1t}\cos(0.8\pi) = -254.2e^{-0.1t}\,°\text{C/meter}.$$

The practical interpretation of these partial derivatives is the rate of change in temperature at $x = 0.2$ and $x = 0.8$ as we increase the distance from the end $x = 0$. Notice that $e^{-0.1t}$ is positive for all t. Given the formula for $H(x,t)$, we see that the closer the position to the center of the rod, the hotter the temperature. The partial derivative $H_x(0.2,t)$ has a positive sign because, at $x = 0.2$ as we increase x, we get closer to the center of the rod which is hottest. The partial derivative $H_x(0.8,t)$ has a negative sign because, at $x = 0.8$ as we increase x, we get further away from the center of the rod which is hottest.

(c) To calculate $H_t(x,t)$, we hold x constant and differentiate with respect to t:

$$\frac{\partial H}{\partial t} = \frac{\partial}{\partial t}(100e^{-0.1t}\sin(\pi x)) = -10e^{-0.1t}\sin(\pi x)\,°\text{C/second}.$$

For all t, and for $0 < x < 1$ (that is, for all t and all x inside the rod), the partial derivative $H_t(x,t)$ is negative. In terms of heat, $H_t(x,t)$ represents the rate at which the temperature of the rod is changing as time passes at position x and time t. Thus, the temperature inside the rod is always decreasing.

39. The function in $h(x,t)$ tells us the height of the head of the spectator in seat x at time t seconds. Thus, $h_x(2,5)$ is in feet per seat and $h_t(2,5)$ is in feet per second. So

$$h_x(x,t) = -0.5\sin(0.5x - t)$$
$$h_x(2,5) = -0.5\sin(0.5(2) - 5) \approx -0.38 \text{ ft/seat}.$$
$$h_t(x,t) = \sin(0.5x - t)$$
$$h_t(2,5) = 0.76 \text{ ft/second}.$$

The value of $h_x(2,5)$ is the rate of change of height of heads as you move along the row of seats. The value of $h_t(2,5)$ is the vertical velocity of the head of the person in seat $x = 2$ at time $t = 5$.

40. Since $f_x(x,y) = 4x^3y^2 - 3y^4$, we could have

$$f(x,y) = x^4y^2 - 3xy^4.$$

In that case,

$$f_y(x,y) = \frac{\partial}{\partial y}(x^4y^2 - 3xy^4) = 2x^4y - 12xy^3$$

as expected. More generally, we could have $f(x,y) = x^4y^2 - 3xy^4 + C$, where C is any constant.

Solutions for Section 13.3

1. We have
$$z = e^y + x + x^2 + 6.$$

The partial derivatives are

$$\left.\frac{\partial z}{\partial x}\right|_{(x,y)=(1,0)} = (2x+1)\Big|_{(x,y)=(1,0)} = 3$$

$$\left.\frac{\partial z}{\partial y}\right|_{(x,y)=(1,0)} = e^y\Big|_{(x,y)=(1,0)} = 1.$$

So the equation of the tangent plane is

$$z = 9 + 3(x-1) + y = 6 + 3x + y.$$

2. The partial derivatives are
$$z_x = e^{x/y} \quad \text{and} \quad z_y = \frac{-x}{y}e^{x/y} + e^{x/y},$$

so

$$z_x(1,1) = e \quad \text{and} \quad z_y(1,1) = -e^1 + e^1 = 0.$$

The tangent plane to $z = ye^{x/y}$ at $(x,y) = (1,1)$ has equation

$$z = z(1,1) + z_x(1,1)(x-1) + z_y(1,1)(y-1)$$
$$= e + e(x-1) + 0(y-1)$$
$$= ex.$$

3. The partial derivatives are
$$z_x = x \quad \text{and} \quad z_y = 4y,$$

so

$$z(2,1) = 4, \quad z_x(2,1) = 2 \quad \text{and} \quad z_y(2,1) = 4.$$

The tangent plane to $z = \frac{1}{2}(x^2 + 4y^2)$ at $(x,y) = (2,1)$ has equation

$$z = z(2,1) + z_x(2,1)(x-2) + z_y(2,1)(y-1)$$
$$= 4 + 2(x-2) + 4(y-1)$$
$$= -4 + 2x + 4y.$$

4. (a) Since the equation of a tangent plane should be linear, this answer is wrong.
   (b) The student didn't substitute the values $x = 2$, $y = 3$ into the formulas for the partial derivatives used in the formula of a tangent plane.
   (c) Let $f(x,y) = z = x^3 - y^2$. Since $f_x(x,y) = 3x^2$ and $f_y(x,y) = -2y$, substituting $x = 2$, $y = 3$ gives $f_x(2,3) = 12$ and $f_y(2,3) = -6$. Then the equation of the tangent plane is

$$z = 12(x-2) - 6(y-3) - 1, \quad \text{or} \quad z = 12x - 6y - 7.$$

5. We have

$$f_x(3, 1) = \left.\frac{\partial f}{\partial x}\right|_{(3,1)} = 2xy|_{(3,1)} = 6,$$

and

$$f_y(3, 1) = \left.\frac{\partial f}{\partial y}\right|_{(3,1)} = x^2|_{(3,1)} = 9.$$

Also $f(3, 1) = 9$. So the local linearization is,

$$z = 9 + 6(x - 3) + 9(y - 1).$$

6. (a) The two tables of values are Table 13.6 and 13.7.

TABLE 13.6

		y		
		1.9	2.0	2.1
	0.9	0.3847	0.3697	0.3510
x	1.0	0.3481	0.3345	0.3176
	1.1	0.3150	0.3027	0.2873

TABLE 13.7

		y		
		1.99	2.00	2.01
	0.99	0.3394	0.3379	0.3363
x	1.00	0.3360	0.3345	0.3330
	1.01	0.3327	0.3312	0.3297

Both tables are nearly linear. To check this, observe that the increments in each row (column) are equal, or nearly so. Table 13.7 is more linear due to finer data.

(b) Table 13.7 shows $f(1, 2) \approx 0.3345$. Also

$$f_x(1, 2) \approx \frac{f(1.01, 2) - f(1, 2)}{1.01 - 1}$$

$$= \frac{0.3312 - 0.3345}{0.01} = -0.3300,$$

$$f_y(1, 2) \approx \frac{f(1, 2.01) - f(1, 2)}{2.01 - 2}$$

$$= \frac{0.3330 - 0.3345}{0.01} = -0.1500.$$

Using the estimates for the partial derivatives that we just made from the Table 13.7, we get that the local linearization of f around $(1, 2)$ is

$$f(x, y) \approx f(1, 2) + f_x(1, 2)(x - 1) + f_y(1, 2)(y - 2)$$
$$= 0.3345 - 0.33(x - 1) - 0.15(y - 2).$$

Now we use $f_x = -e^{-x} \sin(y)$ and $f_y = e^{-x} \cos(y)$, giving

$$f_x(1, 2) = -0.3345,$$
$$f_y(1, 2) = -0.1531.$$

These values of the partial derivatives tell us that the local linearization of f around $(1, 2)$ is

$$f(x, y) \approx 0.3345 - 0.3345(x - 1) - 0.1531(y - 2).$$

Notice that the two linearizations agree up to two decimal places.

7. Making use of the values of P_r and P_L from the solution to Problem 13 on page 104, we have the local linearizations:

For $(r, L) = (8, 4000)$,

$$P(r, L) \approx 80 + 2.5(r - 8) + 0.02(L - 4000),$$

For $(r, L) = (8, 6000)$,

$$P(r, L) \approx 120 + 3.33(r - 8) + 0.02(L - 6000),$$

For $(r, L) = (13, 7000)$,

$$P(r, L) \approx 160 + 3.33(r - 13) + 0.02(L - 7000).$$

8. The linear approximation near $(480, 20)$ is given by

$$f(T, p) \approx f(480, 20) + f_T(480, 20)(T - 480) + f_p(480, 20)(p - 20).$$

Directly from the table on page 113, we have

$$f_T(480, 20) \approx \frac{f(480, 20) - f(500, 20)}{-20} = \frac{27.85 - 28.46}{-20} = 0.0305$$

$$f_p(480, 20) \approx \frac{f(480, 20) - f(480, 22)}{-2} = \frac{27.85 - 25.31}{-2} = -1.27$$

This yields the linear approximation near $(480, 20)$

$$f(T, p) \approx 27.85 + 0.0305(T - 480) - 1.27(p - 20)$$
$$= 0.0305T - 1.27p + 38.61.$$

9. (a) The linear approximation gives

$$f(520, 24) \approx 24.20, \quad f(480, 24) \approx 23.18,$$
$$f(500, 22) \approx 25.52, \quad f(500, 26) \approx 21.86.$$

The approximations for $f(520, 24)$ and $f(500, 26)$ agree exactly with the values in the table; the other two do not. The reason for this is that the partial derivatives were estimated using difference quotients with these values.

(b) We could get a more balanced estimate by using a difference quotient that uses the values on both sides. Thus, we could estimate the partial derivatives as follows:

$$f_T(500, 24) \approx \frac{f(520, 24) - f(480, 24)}{40}$$
$$= \frac{(24.20 - 23.19)}{40} = 0.02525,$$

and

$$f_p(500, 24) \approx \frac{f(500, 26) - f(500, 22)}{4}$$
$$= \frac{(21.86 - 25.86)}{4} = -1.$$

This yields the linear approximation

$$V = f(T, p) \approx 23.69 + 0.02525(T - 500) - (p - 24) \text{ ft}^3.$$

This approximation yields values

$$f(520, 24) \approx 24.195, \quad f(480, 24) \approx 23.185,$$
$$f(500, 22) \approx 25.69, \quad f(500, 26) \approx 21.69.$$

Although none of these predictions are accurate, the error in the predictions that were wrong before has been reduced. This new linearization is a better all-round approximation for values near $(500, 24)$.

10. We want an approximation of the form:

$$i_c = f(i_b, v_c) \approx r + p(i_b - (-300)) + q(v_c - (-8))$$

where $r = f(-300, -8)$, and $p = f_{i_b}(-300, -8)$, and $q = f_{v_c}(-300, -8)$. (The coefficients p and q are known to electrical engineers as *hybrid parameters*, or *h-parameters* for short.)

From Figure 13.22, it can be seen that $r = f(-300, -8) = -30$ mA.

We can approximate p with difference quotients:

$$p = f_{i_b}(-300, -8) \approx \frac{f(-300 + h, -8) - f(-300, -8)}{h}$$

We must choose h as small as possible but so that a value for $f(-300 + h, -8)$ can be read off from Figure 13.22. There are two logical choices, $h = 100$ and $h = -100$, which give the two approximations

$$p \approx \frac{f(-200, -8) - f(-300, -8)}{100} = \frac{-21 - (-30) \text{mA}}{100 \mu \text{A}} = 0.09 \text{ mA}/\mu \text{ A}$$

$$p \approx \frac{f(-400, -8) - f(-300, -8)}{-100} = \frac{-38 - (-30) \text{ mA}}{-100 \mu \text{ A}} = 0.08 \text{ mA}/\mu \text{ A}$$

We use the average value, $p \approx (0.09 + 0.08)/2 = 0.085$ mA $/ \mu$ A.

The other partial derivative, $f_{v_c}(-300, -8)$, is the slope of the $i_b = -300$ curve at $v_c = -8$. Since it is relatively straight around that point, we can approximate the slope by choosing two points far apart. At $v_c = -6$ we have $i_c = -30$, and at $v_c = -11$ we have $i_c = -32$, so the slope is

$$\frac{-32 - (-30) \text{mA}}{-11 - (-6) \text{ V}} = 0.4 \text{ mA/V}$$

In conclusion, for i_b near -300μA and v_c near -8 V,

$$i_c = f(i_b, v_c) \approx -30 + 0.085(i_b + 300) + 0.4(v_c + 8) \text{ mA}$$

11. $df = y \cos(xy) \, dx + x \cos(xy) \, dy$

12. Since $z_x = -e^{-x} \cos(y)$ and $z_y = -e^{-x} \sin(y)$, we have

$$dz = -e^{-x} \cos(y) dx - e^{-x} \sin(y) dy.$$

13. Since $g_u = 2u + v$ and $g_v = u$, we have

$$dg = (2u + v) \, du + u \, dv$$

14. $dh = e^{-3t} \cos(x + 5t) \, dx + (-3e^{-3t} \sin(x + 5t) + 5e^{-3t} \cos(x + 5t)) \, dt$

15. We have $df = f_x\,dx + f_y\,dy$. Finding the partial derivatives, we have $f_x = e^{-y}$ so $f_x(1,0) = e^{-0} = 1$, and $f_y = -xe^{-y}$ so $f_y(1,0) = -1e^{-0} = -1$. Thus, $df = dx - dy$.

16. We have $dg = g_x\,dx + g_t\,dt$. Finding the partial derivatives, we have $g_x = 2x\sin(2t)$ so $g_x(2,\frac{\pi}{4}) = 4\sin(\pi/2) = 4$, and $g_t = 2x^2\cos(2t)$ so $g_t(2,\frac{\pi}{4}) = 8\cos(\frac{\pi}{2}) = 0$. Thus $dg = 4\,dx$.

17. We have $dP = P_L\,dL + P_K\,dK$.
    Now
    $$P_K = (1.01)(0.75)K^{-0.25}L^{0.25}$$
    $$P_K(100,1) \approx 2.395,$$
    and
    $$P_L = (1.01)(0.25)K^{0.75}L^{-0.75}$$
    $$P_L(100,1) \approx 0.008$$
    Thus
    $$dP \approx 2.395\,dK + 0.008\,dL$$

18. We have $dF = F_m\,dm + F_r\,dr$.
    $$F_m = \frac{G}{r^2}, F_m(100,10) = \frac{G}{(10)^2} = \frac{G}{100} = 0.01G$$
    $$F_r = \frac{-2Gm}{r^3}, F_r(100,10) = \frac{-2G100}{(10)^3} = \frac{-G}{5} = -0.2G.$$
    Thus,
    $$dF = 0.01G\,dm - 0.2G\,dr.$$

19. Since $f_x(x,y) = \frac{x}{\sqrt{x^2+y^3}}$ and $f_y(x,y) = \frac{3y^2}{2\sqrt{x^2+y^3}}$,
    $f_x(1,2) = \frac{1}{\sqrt{1^2+2^3}} = \frac{1}{3}$ and $f_y(1,2) = \frac{3\cdot2^2}{2\sqrt{1^2+2^3}} = 2$.
    Thus the differential at the point $(1,2)$ is
    $$df = df(1,2) = f_x(1,2)dx + f_y(1,2)dy = \frac{1}{3}dx + 2dy.$$

    Using the differential at the point $(1,2)$, we can estimate $f(1.04,1.98)$. Since
    $$\triangle f \approx f_x(1,2)\triangle x + f_y(1,2)\triangle y$$
    where $\triangle f = f(1.04,1.98) - f(1,2)$ and $\triangle x = 1.04 - 1$ and $\triangle y = 1.98 - 2$, we have
    $$f(1.04,1.98) \approx f(1,2) + f_x(1,2)(1.04-1) + f_y(1,2)(1.98-2)$$
    $$= \sqrt{1^2+2^3} + \frac{0.04}{3} - 2(0.02) \approx 2.973.$$

20. Local linearization gives us the approximation
    $$T(x,y) \approx T(2,1) + T_x(2,1)(x-2) + T_y(2,1)(y-1)$$
    $$T(x,y) \approx 135 + 16(x-2) - 15(y-1).$$
    Thus,
    $$T(2.04,0.97) \approx 135 + 16(2.04-2) - 17.5(0.97-1) = 136.09°C.$$

21. (a) If the volume is held constant, $\Delta V = 0$, so $\Delta U \approx 27.32 \Delta T$. Thus the energy increases if the temperature increases.

(b) If the temperature is held constant, then $\Delta T = 0$, so $\Delta U \approx 840 \Delta V$. Thus the energy increases if the volume increases (yes, it sounds bizarre, but remember the temperature is being held constant).

(c) First, we convert 100 cm^3 to 0.0001 m^3. Now, using the differential approximation,

$$\begin{aligned} \Delta U &\approx 840 \, \Delta V + 27.32 \, \Delta T \\ &= (840)(-0.0001) + (27.32)(2) \\ &= -0.084 + 54.64 \approx 55 \text{ joules.} \end{aligned}$$

22. (a) For a mass m of liquid, we have $\rho = m/V$, so

$$d\rho = \frac{-m}{V^2} \, dV = \frac{-m}{V^2} \beta V \, dT = -\beta \frac{m}{V} \, dT = -\beta \rho \, dT.$$

(b) From part (a), we have $\Delta \rho \approx -\beta \rho \Delta T$, so

$$\beta \approx -\frac{1}{\rho} \cdot \frac{\Delta \rho}{\Delta T}.$$

Thus, in the limit as ΔT and $\Delta \rho$ become very small, we have

$$\beta = -\frac{1}{\rho} \frac{d\rho}{dT} = -\frac{1}{\rho} \left(\begin{array}{c} \text{Slope of tangent line} \\ \text{in Figure 13.5} \end{array} \right).$$

We use Figure 13.5 to estimate ρ, $\Delta \rho$, and ΔT. We use these values to approximate β.

From Figure 13.5 we see that $\rho \approx 997$ when $T = 20$. In addition, we see that $\rho \approx 1000$ when $T = 0$. Between these points, the temperature change is $\Delta T = 20 - 0 = 20$, and the density change is $\Delta \rho = 997 - 1000 = -3$. Thus, $\Delta \rho \approx -\beta \rho \Delta T$

$$\beta \approx -\frac{1}{\rho} \cdot \frac{\Delta \rho}{\Delta T} = -\frac{1}{997} \cdot \frac{(-3)}{20} \approx 0.00015.$$

At $T = 80$ we have $\rho \approx 973$ and at $T = 60$, we have $\rho \approx 983$. Thus $\Delta T = 80 - 60 = 20$, and $\Delta \rho = 973 - 983 = -10$, so when $T = 80$ we have,

$$\beta \approx -\frac{1}{\rho} \cdot \frac{\Delta \rho}{\Delta T} = -\frac{1}{973} \cdot \frac{(-10)}{20} \approx 0.0005.$$

As you can see from Figure 13.5, using $\Delta T = 20$ may not give a very good approximation. To get a better approximation, use a smaller value of ΔT.

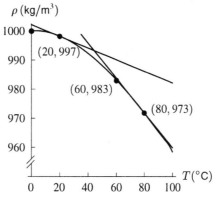

Figure 13.5

23.   In order to compute the error in l, we find dl:

$$dl = \frac{\partial l}{\partial s}ds + \frac{\partial l}{\partial k}dk = \left(1 - \frac{k^2}{s^2}\right)ds + \frac{2k}{s}dk.$$

Now a 1% error in s and k means $|\Delta s| = 0.01s$, and $|\Delta k| = 0.01k$. The maximum error in l is then

$$\Delta l \approx \left(1 - \frac{k^2}{s^2}\right)|\Delta s| + \left(\frac{2k}{s}\right)|\Delta k|.$$

Here we take the absolute value of Δs and Δk to get the worst possible error; otherwise Δs and Δk could have opposite sign and their effects could partly cancel out. Notice that $(1 - k^2/s^2) > 0$ and $2k/s > 0$ since $0 < k < s$.

Therefore,

$$\Delta l \approx \left(1 - \frac{k^2}{s^2}\right)0.01s + \frac{2k}{s}0.01k = 0.01\left(s - \frac{k^2}{s} + \frac{2k^2}{s}\right)$$

$$= 0.01\left(s + \frac{k^2}{s}\right) = 0.01l.$$

So the maximum error in l is 1%.

Next, we want to compute the error in g. Thus we find dg:

$$dg = \frac{\partial g}{\partial l}dl + \frac{\partial g}{\partial T}dT = \frac{4\pi^2}{T^2}dl - \frac{8\pi^2 l}{T^3}dT,$$

Notice that the maximum error occurs when Δl and ΔT have opposite signs. Then the maximum error in g is

$$\Delta g \approx \frac{4\pi^2}{T^2}|\Delta l| + \frac{8\pi^2 l}{T^3}|\Delta T|$$

$$= \frac{4\pi^2}{T^2}0.01l + \frac{8\pi^2 l}{T^3}0.005T$$

$$= 0.02\frac{4\pi^2 l}{T^2} = 0.02g.$$

So the maximum error in the computed value of g is 2%.

24.   At temperature t_0, the length is l_0 and the period is

$$T_0 = 2\pi\sqrt{\frac{l_0}{g}}.$$

Now suppose the temperature changes by Δt, causing a change in the length, Δl. Then

$$\Delta l \approx \frac{dl}{dt}\Delta t = l_0\alpha\Delta t,$$

The change in period, ΔT, caused by this change in length is given by

$$\Delta T \approx \frac{dT}{dl}\Delta l = \frac{2\pi}{\sqrt{g}} \cdot \frac{1}{2}l^{-1/2}\Delta l = \frac{\pi}{\sqrt{gl}}\Delta l.$$

Substituting for Δl we have

$$\Delta T \approx \frac{\pi l_0 \alpha}{\sqrt{gl}}\Delta t.$$

Now we have to figure out how many seconds a day the clock loses or gains as a result of this change in period. We assume the units of l and g are chosen to given T in seconds. When the period is T, the pendulum executes N oscillations per day, where

$$N = \frac{\text{Number of seconds in a day}}{\text{Period in seconds}} = \frac{24 \cdot 60 \cdot 60}{T} = \frac{86400}{T}.$$

Thus, when the period changes by ΔT, we have

$$\Delta N \approx \frac{dN}{dT} \Delta T = -\frac{86400}{T^2} \Delta T.$$

Substituting for ΔT gives

$$\Delta N \approx -\frac{86400 \pi l_0 \alpha}{T^2 \sqrt{gl}} \Delta t.$$

The clock records T_0 seconds as having passed for each oscillation executed. Therefore the number of seconds lost or gained per day when N changes by ΔN is given by

$$\text{Number of seconds lost/gained per day} = T_0 \Delta N.$$

Substituting for ΔN we have

$$\text{Number of seconds lost/gained per day} = T_0 \left(\frac{-86400 \pi l_0 \alpha}{T^2 \sqrt{gl}} \Delta t \right)$$

Substituting for T_0 and T gives

$$\text{Number of seconds lost/gained per day} = 2\pi \sqrt{\frac{l_0}{g}} \left(\frac{-86400 \pi l_0 \alpha}{(4\pi^2 l/g)\sqrt{gl}} \Delta t \right)$$

$$= -43200 \frac{l_0^{3/2}}{l^{3/2}} \alpha \Delta t$$

Since $l_0^{3/2}/l^{3/2} = 1/(1 + \alpha t)^{3/2}$, the value of l_0 cancels and we have:

$$\text{Number of seconds lost/gained per day} = \frac{-43200\alpha}{(1 + \alpha t)^{3/2}} \Delta t$$

This answer is independent of l_0.

25. (a) The area of a circle of radius r is given by

$$A = \pi r^2$$

and the perimeter is

$$L = 2\pi r.$$

Thus we get

$$r = \frac{L}{2\pi}$$

and

$$A = \pi \left(\frac{L}{2\pi} \right)^2 = \frac{L^2 \pi}{4\pi^2} = \frac{L^2}{4\pi}.$$

Thus we get

$$\pi = \frac{L^2}{4A}.$$

(b) We will treat π as a function of L and A.

$$d\pi = \frac{\partial \pi}{\partial L}dL + \frac{\partial \pi}{\partial A}dA = \frac{2L}{4A}dL - \frac{L^2}{4A^2}dA.$$

If L is in error by a factor λ, then $\Delta L = \lambda L$, and if A is in error by a factor μ, then $\Delta A = \mu A$. Therefore,

$$\Delta \pi \approx \frac{2L}{4A}\Delta L - \frac{L^2}{4A^2}\Delta A$$
$$= \frac{2L}{4A}\lambda L - \frac{L^2}{4A^2}\mu A$$
$$= \frac{2\lambda L^2}{4A} - \frac{\mu L^2}{4A} = (2\lambda - \mu)\frac{L^2}{4A} = (2\lambda - \mu)\pi,$$

so π is in error by a factor of $2\lambda - \mu$.

Solutions for Section 13.4

1. (a) We use the definition of the directional derivative

$$f_{\vec{u}}(a,b) = \lim_{h \to 0} \frac{f(a + hu_1, b + hu_2) - f(a,b)}{h}$$

where $\vec{u} = u_1\vec{i} + u_2\vec{j}$ is a unit vector. In this case the vector $\vec{u}$ is the direction of the point $(3,5)$ and so is parallel to $(3-1)\vec{i} + (5-4)\vec{j}$. Thus the unit vector

$$\vec{u} = \frac{2}{\sqrt{5}}\vec{i} + \frac{1}{\sqrt{5}}\vec{j} \approx 0.894\vec{i} + 0.447\vec{j}.$$

We approximate by taking a small value of h, giving

$$f_{\vec{u}}(1,4) \approx \frac{f(1 + 0.894h, 4 + 0.447h) - f(1,4)}{h}$$

We choose a value of h say $h = 0.01$

$$f_{\vec{u}}(1,4) \approx \frac{f(1.00894, 4.00447) - f(1,4)}{0.01}$$
$$= \frac{1.00984 + \ln(4.00447) - 1 - \ln 4}{0.01}$$
$$= 1.01$$

(b) Here $\vec{u}$ is the same as in part (a), and we take $h = 0.01$, giving

$$f_{\vec{u}}(3,5) \approx \frac{f(3,5) - f(3 - 0.894h, 5 - 0.447h)}{h}$$
$$= \frac{f(3,5) - f(2.99106, 4.99553)}{h}$$
$$= \frac{3 + \ln 5 - 2.99106 - \ln(4.99553)}{0.01}$$
$$= 0.98$$

2. The rate of change of f at the point $(2, 1)$ in the direction of $\vec{u}$ is the directional derivative of f at the point $(2, 1)$ in the direction of $\vec{u}$. Thus we want to calculate $f_{\vec{u}}(2, 1)$. Since $\vec{u}$ is a unit vector,

$$
\begin{aligned}
f_{\vec{u}}(2,1) &= \lim_{h\to 0} \frac{f(2+\frac{h}{\sqrt{2}}, 1+\frac{h}{\sqrt{2}}) - f(2,1)}{h} \\
&= \lim_{h\to 0} \frac{(2(2+\frac{h}{\sqrt{2}})^2 + (1+\frac{h}{\sqrt{2}})^2) - (2(2^2)+1^2)}{h} \\
&= \lim_{h\to 0} \frac{8 + 4\sqrt{2}h + h^2 + 1 + \sqrt{2}h + \frac{h^2}{2} - 8 - 1}{h} \\
&= \lim_{h\to 0}(5\sqrt{2} + \frac{3h}{2}) = 5\sqrt{2} \approx 7.07.
\end{aligned}
$$

3. $f_{\vec{i}}(3,1)$ means the rate of change of f in the x direction at $(3,1)$. Thus,

$$
f_{\vec{i}}(3,1) \approx \frac{f(4,1) - f(3,1)}{1} = \frac{2-1}{1} = 1.
$$

4. $f_{\vec{j}}(3,1)$ means the rate of change of f in the y direction at $(3,1)$. If we guess that $f(3,2) \approx 0.5$, we have

$$
f_{\vec{j}}(3,1) \approx \frac{f(3,2) - f(3,1)}{1} \approx \frac{0.5-1}{1} = -0.5.
$$

5. Since $\vec{u} = (\vec{i} - \vec{j})/\sqrt{2}$, we head away from the point $(3,1)$ toward the point $(4,0)$.
   From the graph, we see that $f(3,1) = 1$ and $f(4,0) = 4$. Since the points $(3,1)$ and $(4,0)$ are a distance $\sqrt{2}$ apart, we have

$$
f_{\vec{u}}(3,1) \approx \frac{f(4,0) - f(3,1)}{\sqrt{2}} = \frac{4-1}{\sqrt{2}} = 2.12.
$$

6. Since $\vec{u} = (-\vec{i} + \vec{j})/\sqrt{2}$, we head away from the point $(3,1)$ toward the point $(2,2)$. From the graph we estimate that $f(3,1) = 1$ and $f(2,2) = 0$. These points are a distance of $\sqrt{2}$ apart, so

$$
f_{\vec{u}} \approx \frac{f(2,2) - f(3,1)}{\sqrt{2}} = \frac{0-1}{\sqrt{2}} = -0.71.
$$

7. Since $f_{\vec{i}} = f_x$, the partial derivative in the x-direction, we are looking for the part of the rectangular region where f increases as we move in the increasing x-direction. (i.e. horizontally to the right). This is the region where $x > 2$.

8. Since $f_{\vec{j}} = f_y$, the partial derivative in the y-direction, we are looking for the part of the rectangular region where f decreases as we move in the increasing y-direction. (i.e. up). This is the region where $y < 2$.

9. (a) In the $\vec{i} - \vec{j}$ direction the function is decreasing, so the value of $g_{\vec{u}}(2,5)$ is negative.
   (b) In the $\vec{i} + \vec{j}$ direction the function is decreasing, so the value of $g_{\vec{u}}(2,5)$ negative as well.

10. (a) $\operatorname{grad} f = \frac{\partial f}{\partial x}\vec{i} + \frac{\partial f}{\partial y}\vec{j} = 2x\vec{i} + \frac{1}{y}\vec{j}$

    (b) $\operatorname{grad} f$ at $(4,1)$ is $(2\cdot4)\vec{i} + \frac{1}{1}\vec{j} = 8\vec{i} + \vec{j}$.

11. Since the partial derivatives are

$$z_x = \left(\frac{1}{y}\right) \cos\left(\frac{x}{y}\right), \quad z_y = \left(\frac{-x}{y^2}\right) \cos\left(\frac{x}{y}\right)$$

we have

$$\nabla z = \frac{1}{y} \cos\left(\frac{x}{y}\right)\vec{i} - \frac{x}{y^2} \cos\left(\frac{x}{y}\right)\vec{j}.$$

12. Since the partial derivatives are

$$z_x = e^y, \quad \text{and} \quad z_y = xe^y$$

we have

$$\nabla z = e^y\vec{i} + xe^y\vec{j}.$$

13. Since the partial derivatives are

$$z_x = e^y, \quad \text{and} z_y = xe^y + e^y + ye^y,$$

we have

$$\nabla z = e^y\vec{i} + e^y(1 + x + y)\vec{j}.$$

14. Since the partial derivatives are

$$z_x = \frac{1}{1 + \left(\frac{x}{y}\right)^2}\left(\frac{1}{y}\right) = \frac{y}{y^2 + x^2}, \quad \text{and} \quad z_y = \frac{1}{1 + \left(\frac{x}{y}\right)^2}\left(-\frac{x}{y^2}\right) = -\frac{x}{y^2 + x^2},$$

we have

$$\nabla z = \frac{y}{y^2 + x^2}\vec{i} - \frac{x}{y^2 + x^2}\vec{j}.$$

15. Since the partial derivatives are

$$z_x = (2x) \cos(x^2 + y^2), \quad \text{and} \quad z_y = (2y) \cos(x^2 + y^2),$$

$$\nabla z = 2x \cos(x^2 + y^2)\vec{i} + 2y \cos(x^2 + y^2)\vec{j}.$$

16. Since the partial derivatives are

$$z_x = \frac{e^y(x + y) - xe^y}{(x + y)^2} = \frac{ye^y}{(x + y)^2}$$

$$z_y = \frac{xe^y(x + y) - xe^y}{(x + y)^2} = \frac{e^y(x^2 + xy - x)}{(x + y)^2}$$

we have

$$\nabla z = \frac{ye^y}{(x + y)^2}\vec{i} + \frac{e^y(x^2 + xy - x)}{(x + y)^2}\vec{j}$$

17. Since the partial derivatives are

$$\frac{\partial f}{\partial m} = 2m + 0 = 2m$$

$$\frac{\partial f}{\partial n} = 0 + 2n = 2n$$

we have

$$\text{grad } f = \frac{\partial f}{\partial m}\vec{i} + \frac{\partial f}{\partial n}\vec{j} = 2m\vec{i} + 2n\vec{j}.$$

18. Since the partial derivatives are

$$\frac{\partial f}{\partial x} = \frac{15}{2}x^4 - 0 = \frac{15}{2}x^4$$

$$\frac{\partial f}{\partial y} = 0 - \frac{24}{7}y^5 = -\frac{24}{7}y^5$$

we have

$$\text{grad } f = \frac{\partial f}{\partial x}\vec{i} + \frac{\partial f}{\partial y}\vec{j} = \left(\frac{15}{2}x^4\right)\vec{i} - \left(\frac{24}{7}y^5\right)\vec{j}.$$

19. Since the partial derivatives are

$$\frac{\partial f}{\partial s} = (t^2 - 2t + 4)(-\frac{1}{2})s^{-3/2} = -\frac{(t^2 - 2t + 4)}{2s\sqrt{s}}$$

$$\frac{\partial f}{\partial t} = \frac{1}{\sqrt{s}}(2t - 2)$$

we have

$$\text{grad } f = \frac{\partial f}{\partial s}\vec{i} + \frac{\partial f}{\partial t}\vec{j} = \left(-\frac{(t^2 - 2t + 4)}{2s\sqrt{s}}\right)\vec{i} + \left(\frac{1}{\sqrt{s}}(2t - 2)\right)\vec{j}.$$

20. Since the partial derivatives are

$$\frac{\partial f}{\partial \alpha} = \frac{(2\alpha - 3\beta)(2 + 0) - (2 - 0)(2\alpha + 3\beta)}{(2\alpha - 3\beta)^2}$$

$$= \frac{4\alpha - 6\beta - (4\alpha + 6\beta)}{(2\alpha - 3\beta)^2}$$

$$= -\frac{12\beta}{(2\alpha - 3\beta)^2}$$

$$\frac{\partial f}{\partial \beta} = \frac{(2\alpha - 3\beta)(0 + 3) - (0 - 3)(2\alpha + 3\beta)}{(2\alpha - 3\beta)^2}$$

$$= \frac{(6\alpha - 9\beta) + (6\alpha + 9\beta)}{(2\alpha - 3\beta)^2}$$

$$= \frac{12\alpha}{(2\alpha - 3\beta)^2}$$

we have

$$\text{grad } f = \frac{\partial f}{\partial \alpha}\vec{i} + \frac{\partial f}{\partial \beta}\vec{j} = \left(-\frac{12\beta}{(2\alpha - 3\beta)^2}\right)\vec{i} + \left(\frac{12\alpha}{(2\alpha - 3\beta)^2}\right)\vec{j}.$$

21. Since the partial derivatives are

$$\frac{\partial f}{\partial \alpha} = \frac{1}{2}(5\alpha^2 + \beta)^{-1/2}(10\alpha + 0) = \frac{5\alpha}{\sqrt{5\alpha^2 + \beta}}$$

$$\frac{\partial f}{\partial \beta} = \frac{1}{2}(5\alpha^2 + \beta)^{-1/2}(0 + 1) = \frac{1}{2\sqrt{5\alpha^2 + \beta}},$$

we have

$$\operatorname{grad} f = \frac{\partial f}{\partial \alpha}\vec{i} + \frac{\partial f}{\partial \beta}\vec{j} = \left(\frac{5\alpha}{\sqrt{5\alpha^2 + \beta}}\right)\vec{i} + \left(\frac{1}{2\sqrt{5\alpha^2 + \beta}}\right)\vec{j}.$$

22. Since the partial derivatives are

$$\frac{\partial f}{\partial x} = \cos(xy) \cdot (y) - \sin(xy) \cdot (y) = y[\cos(xy) - \sin(xy)]$$

$$\frac{\partial f}{\partial y} = \cos(xy) \cdot (x) - \sin(xy) \cdot (x) = x[\cos(xy) - \sin(xy)]$$

we have

$$\operatorname{grad} f = \frac{\partial f}{\partial x}\vec{i} + \frac{\partial f}{\partial y}\vec{j}$$

$$= y[\cos(xy) - \sin(xy)]\vec{i} + x[\cos(xy) - \sin(xy)]\vec{j}.$$

23. Since the partial derivatives are

$$\frac{\partial f}{\partial m} = 10m + 0 = 10m$$

$$\frac{\partial f}{\partial n} = 0 + 12n^3 = 12n^3$$

we have

$$\operatorname{grad} f\Big|_{(5,2)} = \frac{\partial f}{\partial m}\vec{i} + \frac{\partial f}{\partial n}\vec{j}\Big|_{(5,2)}$$

$$= 10m\vec{i} + 12n^3\vec{j}\Big|_{(5,2)}$$

$$= 10(5)\vec{i} + 12(2)^3\vec{j}$$

$$= 50\vec{i} + 96\vec{j}$$

24. Since the partial derivatives are

$$\frac{\partial f}{\partial x} = 2xy + 7y^3$$

$$\frac{\partial f}{\partial y} = x^2 + 21xy^2$$

we have

$$\left.\text{grad } f\right|_{(1,2)} = \left.\left(\frac{\partial f}{\partial x}\vec{i} + \frac{\partial f}{\partial y}\vec{j}\right)\right|_{(1,2)}$$

$$= \left.\left((2xy + 7y^3)\vec{i} + (x^2 + 21xy^2)\vec{j}\right)\right|_{(1,2)}$$

$$= \left(2(1)(2) + 7(2)^3\right)\vec{i} + \left((1)^2 + 21(1)(2)^2\right)\vec{j}$$

$$= (4 + 56)\vec{i} + (1 + 84)\vec{j}$$

$$= 60\vec{i} + 85\vec{j}$$

25. Since the partial derivatives are

$$\frac{\partial f}{\partial x} = \frac{1}{2}(\tan x + y)^{-1/2}\left(\frac{1}{\cos^2 x} + 0\right) = \frac{1}{2\cos^2 x\sqrt{\tan x + y}},$$

and

$$\frac{\partial f}{\partial y} = \frac{1}{2}(\tan x + y)^{-1/2}(0 + 1) = \frac{1}{2\sqrt{\tan x + y}},$$

then

$$\text{grad } f = \frac{\partial f}{\partial x}\vec{i} + \frac{\partial f}{\partial y}\vec{j} = \left(\frac{1}{2\cos^2 x\sqrt{\tan x + y}}\right)\vec{i} + \left(\frac{1}{2\sqrt{\tan x + y}}\right)\vec{j}.$$

Hence we have

$$\left.\text{grad } f\right|_{(0,1)} = \left(\frac{1}{2(\cos(0))^2\sqrt{\tan(0) + 1}}\right)\vec{i} + \left(\frac{1}{2\sqrt{\tan(0) + 1}}\right)\vec{j}$$

$$= \left(\frac{1}{2(1)^2\sqrt{0 + 1}}\right)\vec{i} + \left(\frac{1}{2\sqrt{0 + 1}}\right)\vec{j}$$

$$= \frac{1}{2}\vec{i} + \frac{1}{2}\vec{j}$$

26. Since the partial derivatives are

$$\frac{\partial f}{\partial x} = \cos\left(x^2\right) \cdot (2x) + 0 = 2x\cos\left(x^2\right)$$

$$\frac{\partial f}{\partial y} = 0 - \sin y = -\sin y$$

we have

$$\left.\text{grad } f\right|_{\left(\frac{\sqrt{\pi}}{2},0\right)} = \left.\left(\frac{\partial f}{\partial x}\vec{i} + \frac{\partial f}{\partial y}\vec{j}\right)\right|_{\left(\frac{\sqrt{\pi}}{2},0\right)}$$

$$= \left.\left((2x\cos(x^2))\vec{i} + (-\sin y)\vec{j}\right)\right|_{\left(\frac{\sqrt{\pi}}{2},0\right)} = \left(2(\frac{\sqrt{\pi}}{2})\cos(\frac{\pi}{4})\right)\vec{i} + (-\sin 0)\vec{j}$$

$$= \left(\sqrt{\pi}(\frac{1}{\sqrt{2}})\right)\vec{i} + 0$$

$$= \left(\sqrt{\frac{\pi}{2}}\right)\vec{i}$$

27.   (a)   First we will find a unit vector in the same direction as the vector $\vec{v} = 3\vec{i} - 2\vec{j}$. Since this vector has magnitude $\sqrt{13}$, a unit vector is

$$\vec{u}_1 = \frac{1}{\|\vec{v}\|}\vec{v} = \frac{3}{\sqrt{13}}\vec{i} - \frac{2}{\sqrt{13}}\vec{j}.$$

The partial derivatives are

$$f_x(x,y) = \frac{(1+x^2) - (x+y)\cdot 2x}{(1+x^2)^2} = \frac{1 - x^2 - 2xy}{(1+x^2)^2},$$

$$\text{and} \quad f_y(x,y) = \frac{1}{1+x^2},$$

then, at the point P, we have

$$f_x(P) = f_x(1,-2) = \frac{1 - 1^2 - 2\cdot 1\cdot(-2)}{(1+1^2)^2} = 1,$$

$$f_y(P) = f_y(1,-2) = f_y(1,-2) = \frac{1}{1+1^2} = \frac{1}{2}.$$

Thus

$$f_{\vec{u}_1}(P) = \operatorname{grad} f(P)\cdot\vec{u}_1$$
$$= (\vec{i} + \frac{1}{2}\vec{j})\cdot(\frac{3}{\sqrt{13}}\vec{i} - \frac{2}{\sqrt{13}}\vec{j})$$
$$= \frac{3}{\sqrt{13}} - \frac{1}{\sqrt{13}} = \frac{2}{\sqrt{13}}.$$

(b)   The unit vector in the same direction as the vector $\vec{v} = -\vec{i} + 4\vec{j}$ is

$$\vec{u}_2 = \frac{1}{\|\vec{v}\|}\vec{v} = \frac{1}{\sqrt{(-1)^2 + 4^2}}(-\vec{i} + 4\vec{j})$$
$$= -\frac{1}{\sqrt{17}}\vec{i} + \frac{4}{\sqrt{17}}\vec{j}.$$

Since we have calculated from part (a) that $f_x(P) = 1$ and $f_y(P) = 1/2$,

$$f_{\vec{u}_2}(P) = \operatorname{grad} f(P)\cdot\vec{u}_2$$
$$= (\vec{i} + \frac{1}{2}\vec{j})\cdot(-\frac{1}{\sqrt{17}}\vec{i} + \frac{4}{\sqrt{17}}\vec{j})$$
$$= -\frac{1}{\sqrt{17}} + \frac{2}{\sqrt{17}} = \frac{1}{\sqrt{17}}.$$

(c)   The direction of greatest increase is $\operatorname{grad} f$ at P. By part (a) we have found that

$$f_x(P) = 1 \quad\text{and}\quad f_y(P) = \frac{1}{2}.$$

Therefore the direction of greatest increase is

$$\operatorname{grad} f(P) = \vec{i} + \frac{1}{2}\vec{j}.$$

28. (a) The directional derivative should be a number, not a vector.

    (b) Since the partial derivatives are

    $$f_x(x,y) = 2xe^y, \quad f_y(x,y) = x^2e^y, \quad f_x(1,0) = 2, \quad f_y(1,0) = 1.$$

    we have

    $$\nabla f(1,0) = 2\vec{i} + \vec{j}.$$

    The unit vector $\vec{u}$ in the direction of $\vec{v}$ is $\frac{4}{5}\vec{i} + \frac{3}{5}\vec{j}$. Thus, the correct answer is

    $$f_{\vec{u}}(1,0) = \nabla f(1,0) \cdot \vec{u} = 2 \cdot \frac{4}{5} + 1 \cdot \frac{3}{5} = \frac{8}{5} + \frac{3}{5} = \frac{11}{5}.$$

29. (a) The partial derivatives are given by

    $$f_x = e^x(\tan y) + 4xy, \quad f_y = e^x(\sec^2 y) + 2x^2.$$

    Thus

    $$f_x(0, \frac{\pi}{4}) = 1 \quad \text{and} \quad f_y(0, \frac{\pi}{4}) = 2,$$

    and so

    $$\text{grad } f(0, \frac{\pi}{4}) = \vec{i} + 2\vec{j}.$$

    The unit vector $\vec{u_1}$ in the direction of $\vec{i} - \vec{j}$ is $\frac{1}{\sqrt{2}}(\vec{i} - \vec{j})$. Then the directional derivative of f at $(0, \frac{\pi}{4})$ in the direction of $\vec{i} - \vec{j}$ is

    $$f_{\vec{u_1}}(0, \frac{\pi}{4}) = \text{grad } f(0, \frac{\pi}{4}) \cdot \vec{u_1}$$
    $$= (\vec{i} + 2\vec{j}) \cdot (\frac{1}{\sqrt{2}}\vec{i} - \frac{1}{\sqrt{2}}\vec{j})$$
    $$= \frac{1}{\sqrt{2}} - \sqrt{2} = -\frac{\sqrt{2}}{2}.$$

    (b) The unit vector $\vec{u_2}$ in the direction of $\vec{i} + \sqrt{3}\vec{j}$ is $\vec{u_2} = \frac{1}{2}(\vec{i} + \sqrt{3}\vec{j})$. From part (a),

    $$\text{grad } f(0, \frac{\pi}{4}) = \vec{i} + 2\vec{j}.$$

    Then the directional derivative of f at $(0, \frac{\pi}{4})$ in the direction of $\vec{i} + \sqrt{3}\vec{j}$ is

    $$f_{\vec{u_2}}(0, \frac{\pi}{4}) = \text{grad } f(0, \frac{\pi}{4}) \cdot \vec{u_2} = (\vec{i} + 2\vec{j}) \cdot (\frac{1}{2}\vec{i} + \frac{\sqrt{3}}{2}\vec{j})$$
    $$= \frac{1}{2} + \sqrt{3}.$$

30. (a) The gradient vector at the point $x = 1, y = 2$ is

    $$\nabla z = \nabla(x^2 y) = 2xy\vec{i} + x^2\vec{j} = 4\vec{i} + \vec{j}.$$

    The unit vector making an angle of $5\pi/4$ with the x-axis is

    $$\vec{u} = \cos\frac{5\pi}{4}\vec{i} + \sin\frac{5\pi}{4}\vec{j} = -\frac{\sqrt{2}}{2}\vec{i} - \frac{\sqrt{2}}{2}\vec{j}.$$

    The directional derivative in this direction is

    $$f_{\vec{u}}(1,2) = \nabla z \cdot \vec{u} = (4\vec{i} + \vec{j}) \cdot \frac{\sqrt{2}}{2}(-\vec{i} - \vec{j}) = -\frac{\sqrt{2}}{2}(4 + 1) = -\frac{5\sqrt{2}}{2}.$$

    (b) The directional derivative is a maximum in the direction of the gradient vector $\nabla z = 4\vec{i} + \vec{j}$.

31.   We want the directional derivative in the direction of $\vec{u}$ at $(1,2)$, so we want to calculate $f_{\vec{u}}(1,2)$. Since $f_x = 2x$ and $f_y = 2y$, at the point $(1,2)$, we have grad $f(1,2) = 2\vec{i} + 4\vec{j}$. Since $\vec{u}$ is a unit vector, we obtain

$$
\begin{aligned}
f_{\vec{u}}(1,2) &= \text{grad } f(1,2) \cdot \vec{u} \\
&= (2\vec{i} + 4\vec{j}) \cdot (0.6\vec{i} + 0.8\vec{j}) \\
&= 2(0.6) + 4(0.8) \\
&= 4.4.
\end{aligned}
$$

32.   The vector from $(2,1)$ to $(1,3)$ is $\vec{v_1} = (1-2)\vec{i} + (3-1)\vec{j} = -\vec{i} + 2\vec{j}$. A unit vector in this direction is $\vec{u_1} = -\frac{1}{\sqrt{5}}\vec{i} + \frac{2}{\sqrt{5}}\vec{j}$.
A vector from $(2,1)$ to $(5,5)$ is $\vec{v_2} = (5-2)\vec{i} + (5-1)\vec{j} = 3\vec{i} + 4\vec{j}$. A unit vector in this direction is $\vec{u_2} = \frac{3}{5}\vec{i} + \frac{4}{5}\vec{j}$.
The directional derivative along $\vec{u_1}$ is

$$
z_{\vec{u_1}}(1,2) = \nabla z \cdot \left(-\frac{1}{\sqrt{5}}\vec{i} + \frac{2}{\sqrt{5}}\vec{j}\right) = -\frac{1}{\sqrt{5}}\frac{\partial z}{\partial x} + \frac{2}{\sqrt{5}}\frac{\partial z}{\partial y}.
$$

So

$$
-\frac{1}{\sqrt{5}}\frac{\partial z}{\partial x} + \frac{2}{\sqrt{5}}\frac{\partial z}{\partial y} = -\frac{2}{\sqrt{5}},
$$

that is,
(1)

$$
-\frac{\partial z}{\partial x} + 2\frac{\partial z}{\partial y} = -2.
$$

The directional derivative along $\vec{u_2}$ is

$$
z_{\vec{u_2}}(1,2) = \nabla z \cdot \left(\frac{3}{5}\vec{i} + \frac{4}{5}\vec{j}\right) = \frac{3}{5}\frac{\partial z}{\partial x} + \frac{4}{5}\frac{\partial z}{\partial y},
$$

so

$$
\frac{3}{5}\frac{\partial z}{\partial x} + \frac{4}{5}\frac{\partial z}{\partial y} = 1,
$$

that is,
(2)

$$
3\frac{\partial z}{\partial x} + 4\frac{\partial z}{\partial y} = 5.
$$

Now we solve the system of equations (1) and (2). Multiplying equation (1) by 3 gives
(3)

$$
-3\frac{\partial z}{\partial x} + 6\frac{\partial}{\partial y} = -6.
$$

Adding (2) and (3) we get:

$$
10\frac{\partial z}{\partial y} = -1.
$$

So

$$
\frac{\partial z}{\partial y} = -0.1
$$

and from equation (1)

$$
\frac{\partial z}{\partial x} = 2\frac{\partial z}{\partial y} + 2 = 2\left(-\frac{1}{10}\right) + 2 = 1.8.
$$

33. Directional derivative $= \nabla f \cdot \vec{u}$, where $\vec{u} = $ unit vector. If we move from $(4, 5)$ to $(5, 6)$, we move in the direction $\vec{i} + \vec{j}$ so $\vec{u} = \frac{1}{\sqrt{2}}\vec{i} + \frac{1}{\sqrt{2}}\vec{j}$. So,

$$\nabla f \cdot \vec{u} = f_x\left(\frac{1}{\sqrt{2}}\right) + f_y\left(\frac{1}{\sqrt{2}}\right) = 2.$$

Similarly, if we move from $(4, 5)$ to $(6, 6)$, the direction is $2\vec{i} + \vec{j}$ so $\vec{u} = \frac{2}{\sqrt{5}}\vec{i} + \frac{1}{\sqrt{5}}\vec{j}$. So

$$\nabla f \cdot \vec{u} = f_x\left(\frac{2}{\sqrt{5}}\right) + f_y\left(\frac{1}{\sqrt{5}}\right) = 3.$$

Solving the system of equations for f_x and f_y

$$f_x + f_y = 2\sqrt{2}$$
$$2f_x + f_y = 3\sqrt{5}$$

gives

$$f_x = 3\sqrt{5} - 2\sqrt{2}$$
$$f_y = 4\sqrt{2} - 3\sqrt{5}.$$

Thus at $(4, 5)$,

$$\nabla f = (3\sqrt{5} - 2\sqrt{2})\vec{i} + (4\sqrt{2} - 3\sqrt{5})\vec{j}.$$

34. (a)   We have

$$T(x, y) = \frac{100}{x^2 + y^2 + 1}$$

The level curves with $T = c$ (constant) have

$$T(x, y) = \frac{100}{x^2 + y^2 + 1} = c$$

giving

$$x^2 + y^2 + 1 = \frac{100}{c}$$

which is the equation of a circle:

$$x^2 + y^2 = \frac{100}{c} - 1$$

if $0 < c < 100$ or of a point (the origin) if $c = 100$. Thus the level curves are circles centered at the origin. See Figure 13.6 where $r = $ radius.

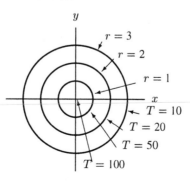

Figure 13.6

$$\text{At } r = 1, \ x^2 + y^2 = r^2 = 1, \text{ and } T = \frac{100}{(1)^2 + 1} = \frac{100}{2} = 50.$$

$$\text{At } r = 2, \text{ we have } \quad T = \frac{100}{(2)^2 + 1} = \frac{100}{5} = 20$$

$$\text{At } r = 3, \text{ we have } \quad T = \frac{100}{(3)^2 + 1} = \frac{100}{10} = 10$$

$$\text{At } r = 0, \text{ we have } \quad T = \frac{100}{(0)^2 + 1} = \frac{100}{1} = 100$$

(b)   Since $r = 0$ minimizes the denominator, the maximum temperature occurs at the origin $(0,0)$ and has value

$$T = \frac{100}{1 + (0)^2} = 100.$$

Note that T tends to zero as r tends to infinity.

(c)   The direction of the greatest increase in temperature is towards the origin. At a given point, this direction is in the opposite direction to the position vector. For the point $(3, 2)$, this is $-3\vec{i} - 2\vec{j}$. We can verify this by checking the gradient, since ∇T gives the direction of maximum increase:

$$\nabla T = \frac{\partial T}{\partial x}\vec{i} + \frac{\partial T}{\partial y}\vec{j}$$

$$\frac{\partial T}{\partial x} = \frac{\partial}{\partial x}[100(x^2 + y^2 + 1)^{-1}] = (-1)(100)(x^2 + y^2 + 1)^{-2}(2x) = -\frac{200x}{(1 + x^2 + y^2)^2}$$

$$\frac{\partial T}{\partial y} = \frac{\partial}{\partial y}[100(x^2 + y^2 + 1)^{-1}] = (-1)(100)(x^2 + y^2 + 1)^{-2}(2y) = -\frac{200y}{(1 + x^2 + y^2)^2} \quad \text{At } (3, 2) \text{ we}$$

have

$$\nabla T = \left(\frac{(-200)(3)}{(1 + 3^2 + 2^2)^2}\right)\vec{i} + \left(\frac{(-200)(2)}{(1 + 3^2 + 2^2)^2}\right)\vec{j}$$

$$= -\left(\frac{600}{(14)^2}\right)\vec{i} - \left(\frac{400}{(14)^2}\right)\vec{j} = \frac{200}{196}(-3\vec{i} - 2\vec{j})$$

The magnitude of this rate of change is

$$\|\nabla T\| = \sqrt{\left(\frac{200}{(14)^2}\right)^2 (9 + 4)} = \frac{200\sqrt{13}}{14^2} = \frac{50\sqrt{13}}{49}.$$

Note that the gradient is a scalar multiple of $-3\vec{i} - 2\vec{j}$, the direction vector we obtained earlier. It points towards the origin.

(d)   The greatest decrease is in the direction opposite to the gradient, that is in the direction of the vector $200(3\vec{i} + 2\vec{j})/196$.

(e)   The direction in which the temperature does nor increase or decrease is perpendicular to the gradient. Let $\vec{u} = u_1\vec{i} + u_2\vec{j}$ be normal to ∇T; then $\vec{u} \cdot \nabla T = 0$. Since at $(3, 2)$, ∇T is parallel to $-3\vec{i} - 2\vec{j}$, we have

$$\vec{u} \cdot (-3\vec{i} - 2\vec{j}) = (u_1\vec{i} + u_2\vec{j}) \cdot (-3\vec{i} - 2\vec{j}) = -3u_1 - 2u_2 = 0$$

$$u_1 = -\frac{2}{3}u_2.$$

Any vector whose components satisfy this ratio will do. So any vector of the form $-\frac{2}{3}u_2\vec{i} + u_2\vec{j}$ is perpendicular to ∇T and thus also points along the level curve. Any multiple of $-2\vec{i} + 3\vec{j}$ is such a vector.

35. The vector grad $f(4, 1) = 2\vec{i} - \vec{j}$ is perpendicular to the level curve at the point $(4, 1)$. Thus, the vector $\vec{i} + 2\vec{j}$ is tangent to the curve at this point. This tells us that the slope of the tangent line is $2/1 = 2$. Since it goes through the point $(4, 1)$, the equation of the tangent is

$$y - 1 = 2(x - 4)$$
$$y = 2x - 7.$$

36. (a) Negative. ∇f is perpendicular to the level curve at the point P, so its x-component which is $\nabla f \cdot \vec{i}$ is negative.

   (b) Positive. The y-component of ∇f is in the same direction as $\vec{j}$ at P and hence the dot product will be positive.

   (c) Positive. The partial derivative with respect to x at Q is positive because the value of f is increasing in the positive x direction at Q. (Note that Q lies between the level curves with values 3 and 4 and that the one with value 4 is further in the positive x direction from Q.)

   (d) Negative. Again, Q lies between the level curves with values 3 and 4 and the one with value 3 is further from Q in the positive y direction, so the partial derivative with respect to y at Q is negative.

37. $\|\nabla f\|$ at P is larger because the level curves are closer there.

38. The gradient vectors are perpendicular to the level curves. To determine the length of the gradient vector, we estimate the rate of change of the function from the contour diagram. At $(1, 1)$, the value of f changes from 1 to 2 in a distance of $\sqrt{2}$ (as it moves from $(1, 1)$ to $(2, 2)$), so the length of grad f is $\frac{1}{\sqrt{2}} \approx 0.7$. At $(1, 4)$, the value of f changes slightly faster (the lines are closer together), so grad f is slightly longer here. In fact, the value of f changes from 2 to 3 as we move from $(1, 4)$ to $(2.2, 4.2)$ a distance of about 1.2. So the new length is $1/1.2 = 0.8$.

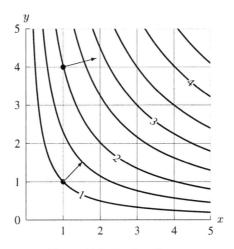

Figure 13.7: Gradient Vectors

39. (a) P corresponds to greatest rate of increase of f and Q corresponds to greatest rate of decrease of f. See Figure 13.8.

   (b) The points are marked in Figure 13.9.

   (c) Amplitude is $\|\text{grad } f\|$. The equation is

$$f_{\vec{u}} = \|\text{grad } f\| \cos \theta.$$

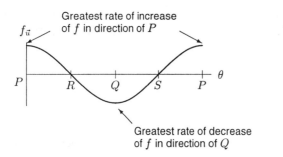

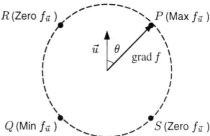

Figure 13.8

Figure 13.9

40. (a) To estimate the change in f, we use the gradient vector to estimate the change in f in moving from P to Q. Because the contours are approximately parallel, moving from P to Q takes you to the same contour as moving from P to R. (See Figure 13.10.) If θ is the angle between $\vec{u}$ and grad $f(a, b)$, then

$$\begin{array}{c} \text{Change in } f \\ \text{between } P \text{ and } Q \end{array} = \text{Change in } f$$

$$= \left(\begin{array}{c} \text{Rate of change} \\ \text{in direction } PR \end{array} \right) \left(\begin{array}{c} \text{Distance traveled} \\ \text{between } P \text{ and } R \end{array} \right)$$

$$\approx \| \text{grad } f \| (h \cos \theta).$$

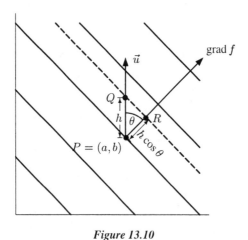

Figure 13.10

(b) Since $\vec{u}$ is a unit vector, we use the definition of $f_{\vec{u}}(a, b)$ to estimate

$$f_{\vec{u}}(a, b) \approx \frac{\text{Change in } f}{h} \approx \frac{\| \text{grad } f(a, b) \| h \cos \theta}{h}$$
$$= \| \text{grad } f(a, b) \| \cos \theta = \| \text{grad } f \| \| \vec{u} \| \cos \theta = \text{grad } f(a, b) \cdot \vec{u}.$$

This approximation gets better as we choose h smaller and smaller, and in the limit we get the formula:

$$f_{\vec{u}}(a, b) = \text{grad } f(a, b) \cdot \vec{u}.$$

Solutions for Section 13.5

1. The unit vector $\vec{u}$ in the direction of $\vec{v} = 2\vec{i} + \vec{j} - 2\vec{k}$ is

$$\vec{u} = \frac{\vec{v}}{\|\vec{v}\|} = \left(\frac{2}{3}\right)\vec{i} + \left(\frac{1}{3}\right)\vec{j} - \left(\frac{2}{3}\right)\vec{k}.$$

The partial derivatives are

$$f_x(x, y, z) = 2x + 3y,$$
$$f_y(x, y, z) = 3x,$$
$$f_z(x, y, z) = 2.$$

Hence,

$$f_{\vec{u}}(2, 0, -1) = f_x(2, 0, -1)\left(\frac{2}{3}\right) + f_y(2, 0, -1)\left(\frac{1}{3}\right) + f_z(2, 0, -1)\left(-\frac{2}{3}\right)$$

$$= 4\left(\frac{2}{3}\right) + 6\left(\frac{1}{3}\right) + 2\left(-\frac{2}{3}\right)$$

$$= \frac{10}{3}$$

2. (a) The unit vector $\vec{u}_1$ in the direction of $\vec{v}_1 = \vec{i} - \vec{k}$ is $\vec{u}_1 = \frac{1}{\sqrt{2}}\vec{i} - \frac{1}{\sqrt{2}}\vec{k}$. We have

$$f_x(x, y, z) = 6xy^2, \qquad \text{and } f_x(-1, 0, 4) = 0$$
$$f_y(x, y, z) = 6x^2y + 2z, \qquad \text{and } f_y(-1, 0, 4) = 8$$
$$f_z(x, y, z) = 2y, \qquad \text{and } f_z(-1, 0, 4) = 0.$$

So,

$$f_{\vec{u}_1}(-1, 0, 4) = f_x(-1, 0, 4)\left(\frac{1}{\sqrt{2}}\right) + f_y(-1, 0, 4)(0) + f_z(-1, 0, 4)\left(-\frac{1}{\sqrt{2}}\right)$$

$$= 0\left(\frac{1}{\sqrt{2}}\right) + 8(0) + 0\left(-\frac{1}{\sqrt{2}}\right)$$

$$= 0.$$

(b) The unit vector $\vec{u}_2 = -\frac{1}{\sqrt{19}}\vec{i} + \frac{3}{\sqrt{19}}\vec{j} + \frac{3}{\sqrt{19}}\vec{k}$ is in the direction of $\vec{v}_2 = -\vec{i} + 3\vec{j} + 3\vec{k}$. Using the partial derivatives from part (a),

$$f_{\vec{u}_2}(-1, 0, 4) = 0\left(-\frac{1}{\sqrt{19}}\right) + 8\left(\frac{3}{\sqrt{19}}\right) + 0\left(\frac{3}{\sqrt{19}}\right) = \frac{24}{\sqrt{19}}. \qquad \approx 5.506$$

3. $z_x = \frac{-x}{\sqrt{17 - x^2 - y^2}}$ and $z_y = \frac{-y}{\sqrt{17 - x^2 - y^2}}$, so

$z(3, 2) = \sqrt{17 - 9 - 4} = 2$, $z_x(3, 2) = -\frac{3}{2}$, and $z_y(3, 2) = \frac{-2}{2} = -1$.

The tangent plane to $z = \sqrt{17 - x^2 - y^2}$ at $(x, y) = (3, 2)$ is

$$z = z(3, 2) + z_x(3, 2)(x - 3) + z_y(3, 2)(y - 2)$$

$$= 2 + (\frac{-3}{2})(x - 3) + (-1)(y - 2)$$

$$= \frac{17}{2} - \frac{3}{2}x - y,$$

or $2z + 3x + 2y = 17$.

4. $z_x = -8/yx^2$ and $z_y = -8/xy^2$, so
    $z(1, 2) = 8/(1)(2) = 4$, $z_x(1, 2) = -8/(2)(1)^2 = -4$ and $z_y(1, 2) = -8/(1)(2)^2 = -2$.
    The tangent plane to $z = 8/xy$ at $(x, y) = (1, 2)$ is

    $$\begin{aligned}
    z &= z(1, 2) + z_x(1, 2)(x - 1) + z_y(1, 2)(y - 2) \\
    &= 4 - 4(x - 1) - 2(y - 2) \\
    &= 12 - 4x - 2y.
    \end{aligned}$$

5. The surface is given by $F(x, y, z) = 0$ where $F(x, y, z) = x - y^3 z^7$. The normal direction is $\nabla F = \dfrac{\partial F}{\partial x}\vec{i} + \dfrac{\partial F}{\partial y}\vec{j} + \dfrac{\partial F}{\partial z}\vec{k} = \vec{i} - 3y^2 z^7 \vec{j} - 7y^3 z^6 \vec{k}$.

    Thus, at $(1, -1, -1)$ a normal vector is $\vec{i} + 3\vec{j} + 7\vec{k}$. The tangent plane has the equation
    $\left[(x\vec{i} + y\vec{j} + z\vec{k}) - (\vec{i} - \vec{j} - \vec{k}) \right] \cdot (\vec{i} + 3\vec{j} + 7\vec{k}) = 0$, that is,

    $$x + 3y + 7z = -9.$$

6. (a) To get a normal vector to the surface $z = \cos x \sin y$ at the point $(0, \pi/2, 1)$, we first represent the surface S by the equation $F(x, y, z) = z - \cos x \sin y = 0$. Then we calculate the gradient of F which is normal to S
    $$\nabla F = \sin x \sin y \vec{i} - \cos x \cos y \vec{j} + \vec{k}.$$

    At the point $(0, \pi/2, 1)$,
    $$\nabla F(0, \pi/2, 1) = \vec{k}.$$

    (b) The plane with normal $\vec{k}$ and through the point $(0, \pi/2, 1)$ is
    $$z = 1.$$

    So $z = 1$ is the equation of the tangent plane at the point $(0, \pi/2, 1)$

7. (a) The vector $\operatorname{grad} f(x, y)$ is perpendicular to the level curve of f through (x, y):
    $$\operatorname{grad} f(x, y) = (e^x - 1) \cos y \vec{i} - (e^x - x) \sin y \vec{j}$$

    Thus, at the point $(2, 3)$,
    $$\operatorname{grad} f(2, 3) = (e^2 - 1) \cos 3 \vec{i} - (e^2 - 2) \sin 3 \vec{j}$$

    The vector $\operatorname{grad} f$ points in the direction of greatest increase in f, so the vector we want is
    $$\begin{aligned}
    -\operatorname{grad} f(2, 3) &= -(e^2 - 1) \cos 3 \vec{i} + (e^2 - 2) \sin 3 \vec{j} \\
    &= 6.33\vec{i} + 0.76\vec{j}
    \end{aligned}$$

    (b) To find a vector normal to the surface, we write the surface in the form
    $$F(x, y, z) = (e^x - x) \cos y - z = 0.$$

    Then
    $$\operatorname{grad} F = (e^x - 1) \cos y \vec{i} - (e^x - x) \sin y \vec{j} - \vec{k}.$$

So, at P,

$$\text{grad } F = (e^2 - 1)\cos 3\vec{i} - (e^2 - 2)\sin y\vec{j} - \vec{k}$$
$$= -6.33\vec{i} - 0.76\vec{j} - \vec{k}.$$

The vector $\vec{v}$ is perpendicular to grad F, so

$$\vec{v} \cdot \text{grad } F = (5\vec{i} + 4\vec{j} + a\vec{k}) \cdot (-6.33\vec{i} - 0.76\vec{j} - \vec{k}) = 0.$$

This gives

$$-5(6.33) - 4(0.76) - a = 0$$

so

$$a = -34.69$$

8. (a) The graph of $z = y^2$ is in Figure 13.11.

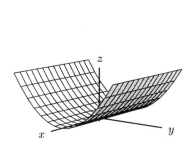

$y = 4$	$c = 16$
$y = 3$	$c = 9$
$y = 2$	$c = 4$
$y = 1$	$c = 1$
$y = 0$	$c = 0$
$y = -1$	$c = 1$
$y = -2$	$c = 4$

Figure 13.11   **Figure 13.12**

(b) $z = c$ gives $y^2 = c$, so the level curves are $y = \pm\sqrt{c}$. See Figure 13.12.
(c) The level curves in part (b) show that the direction of the greatest increase is in the y direction if the point is in the upper half xy-plane (where $y > 0$). Since the point $(2, 3, 9)$ is in the upper half xy-plane, we climb fastest in the direction $\vec{j}$, parallel to the y-axis.

9. (a)

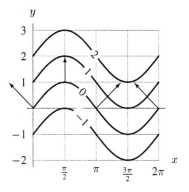

Figure 13.13

(b) The bug is walking parallel to the y-axis. Looking to the right or left, the bug sees higher contours — thus it is in a valley.
(c) See Figure 13.13.

10. (a) In the direction of ∇F:

$$\nabla F\bigg|_{(-1,1,1)} = \left((2x + 2xz^2)\vec{i} + (4y^3)\vec{j} + (2x^2z)\vec{k}\right)\bigg|_{(-1,1,1)} = -4\vec{i} + 4\vec{j} + 2\vec{k}.$$

(b) The rate of change in the direction of ∇F with respect to distance $= \|\nabla F\| = \sqrt{16 + 16 + 4} = 6$. Now we want rate of change with respect to time. If we move at 4 units/sec:

$$\text{rate of change} \left(\frac{\text{conc}}{\text{time}}\right) = \left(\text{rate of change}\left(\frac{\text{conc}}{\text{dist}}\right)\right) \times \left(\text{rate of change}\left(\frac{\text{dist}}{\text{time}}\right)\right)$$

$$= 6 \times 4 = 24(\text{in units of concentration/sec}).$$

11. (a) A normal vector at (x, y, z) is given by

$$\nabla F(x, y, z) = \frac{\partial F}{\partial x}\vec{i} + \frac{\partial F}{\partial y}\vec{j} + \frac{\partial F}{\partial z}\vec{k}$$

$$= 2x\vec{i} - \frac{1}{z^2}\vec{j} + \frac{2y}{z^3}\vec{k}$$

This is parallel to the xy-plane when it is perpendicular to $\vec{k}$, i.e. when

$$\left(2x\vec{i} - \frac{1}{z^2}\vec{j} + \frac{2y}{z^3}\vec{k}\right) \cdot \vec{k} = 0,$$

that is,

$$\frac{2y}{z^3} = 0 \quad \text{so} \quad y = 0.$$

Points that meet these requirements are those (x, y, z) such that

$$x^2 - \frac{y}{z^2} = 0 \quad \text{and} \quad y = 0,$$

i.e., points such that $x = y = 0$ and $z \neq 0$, that is, the z-axis minus the origin. (Observe that we must exclude the points where $z = 0$ because the surface is not defined there: the expression $x^2 - (y/z^2)$ is undefined when $z = 0$.)

(b) A normal vector at $(0, 0, 1)$ is $(2 \cdot 0)\vec{i} - \frac{1}{1^2}\vec{j} + \frac{2 \cdot 0}{1^3}\vec{k} = -\vec{j}$
The equation of the tangent plane at the point $(0, 0, 1)$ is

$$((x - 0)\vec{i} + (y - 0)\vec{j} + (z - 1)\vec{k}) \cdot (-\vec{j}) = 0,$$

i.e.,

$$-y = 0, \quad \text{so} \quad y = 0.$$

A normal vector at $(1, 1, 1)$ is $(2 \cdot 1)\vec{i} - (\frac{1}{1^2})\vec{j} + (\frac{2 \cdot 1}{1^3})\vec{k} = 2\vec{i} - \vec{j} + 2\vec{k}$.
The equation of the tangent plane at $(1, 1, 1)$ is

$$((x - 1)\vec{i} + (y - 1)\vec{j} + (z - 1)\vec{k}) \cdot (2\vec{i} - \vec{j} + 2\vec{k}) = 0,$$

i.e.,

$$2x - 2 - y + 1 + 2z - 2 = 0,$$

so

$$2x - y + 2z = 3.$$

(c) The direction of maximum increase of F is ∇F. At $(0, 0, 1)$ this is $-\vec{j}$, which is a unit vector. So $\vec{u_1} = -\vec{j}$.
At $(1, 1, 1)$ this is $2\vec{i} - \vec{j} + 2\vec{k}$, so the unit vector $\vec{u_2} = \frac{2\vec{i} - \vec{j} + 2\vec{k}}{\sqrt{2^2 + (-1)^2 + 2^2}} = \frac{2}{3}\vec{i} - \frac{1}{3}\vec{j} + \frac{2}{3}\vec{k}$.

12. (a) The plane $z = 5$ is horizontal. The surface $z = 1 + x^2 + y^2$ is bowl-shaped, with its lowest point at $(0, 0, 1)$. At this point its tangent plane is horizontal and therefore parallel to the plane $z = 5$.

(b) The tangent plane to the surface $z = f(x, y)$ at the point where $(x, y) = (a, b)$ has equation

$$z = f(a, b) + f_x(a, b)(x - a) + f_y(a, b)(y - b).$$

Thus, for $z = 1 + x^2 + y^2$, the tangent plane is

$$z = (1 + a^2 + b^2) + 2a(x - a) + 2b(y - b)$$
$$= (1 - a^2 - b^2) + 2ax + 2by.$$

This is parallel to $z = 5 + 6x - 10y$ when $6 = 2a$, and $-10 = 2b$ so $a = 3$, $b = -5$. Then $z = f(3, -5) = 1 + 3^2 + (-5)^2 = 35$, so the point on the surface whose tangent plane is parallel to $z = 5 + 6x - 10y$ is $(3, -5, 35)$.

13. The point $(4, 1, 3)$ lies on the surface. The surface is the level surface of the function

$$F(x, y, z) = f(x, y) - z = 0.$$

The normal to the surface at the point $(4, 1, 3)$ is

$$\operatorname{grad} F(4, 1, 3) = f_x(4, 1)\vec{i} + f_y(4, 1)\vec{j} - \vec{k} = 2\vec{i} - \vec{j} - \vec{k}.$$

Thus the equation of the tangent plane is

$$2x - y - z = 2(4) - 1 - 3 = 4$$
$$2x - y - z = 4$$

14. (a) The vector $\operatorname{grad} f = 2\vec{i} - 5\vec{j}$ is perpendicular to the level curve at the point $(1, 3)$. Now the vector $5\vec{i} + 2\vec{j}$ is perpendicular to the vector $2\vec{i} - 5\vec{j}$. Thus, the vector $5\vec{i} + 2\vec{j}$ is tangent to curve. (There are many other vectors with this property, such as $-5\vec{i} - 2\vec{j}$, $10\vec{i} + 4\vec{j}$, etc.) The slope of the tangent line is therefore $2/5$. Since the tangent line goes through the point $(1, 3)$, its equation is

$$y - 3 = \frac{2}{5}(x - 1)$$

or

$$y = \frac{2}{5}x + \frac{13}{5}.$$

(b) The surface $z = f(x, y)$ can be written in the form

$$F(x, y, z) = f(x, y) - z = 0.$$

The normal the this surface is

$$\operatorname{grad} F = f_x\vec{i} + f_y\vec{j} - \vec{k}.$$

Thus, at the point $(1, 3, 7)$, the normal is

$$\operatorname{grad} F(1, 3, 7) = 2\vec{i} - 5\vec{j} - \vec{k}.$$

Thus, the equation of the tangent plane is

$$2x - 5y - z = 2(1) - 5(3) - 7 = -20$$
$$2x - 5y - z + 20 = 0.$$

15.   The tangent plane to $z = \sqrt{2x^2 + 2y^2 - 25}$ at $(x, y) = (4, 3)$ is

$$z = z(4, 3) + z_x(4, 3)(x - 4) + z_y(4, 3)(y - 3)$$

$$= \sqrt{2(4)^2 + 2(3)^2 - 25} + \frac{2(4)}{\sqrt{2(4)^2 + 2(3)^2 - 25}}(x - 4) + \frac{2(3)}{\sqrt{2(4)^2 + 2(3)^2 - 25}}(y - 3)$$

$$= 5 + \frac{8}{5}(x - 4) + \frac{6}{5}(y - 3).$$

The tangent plane to $z = \frac{1}{5}(x^2 + y^2)$ at $(x, y) = (4, 3)$ is

$$z = z(4, 3) + z_x(4, 3)(x - 4) + z_y(4, 3)(y - 3)$$

$$= \frac{1}{5}(4^2 + 3^2) + \frac{2}{5}(4)(x - 4) + \frac{2}{5}(3)(y - 3)$$

$$= 5 + \frac{8}{5}(x - 4) + \frac{6}{5}(y - 3).$$

Thus the two surfaces are tangential at the point $(4, 3, 5)$.

16.   The points of intersection are $(x, y) = (a, b)$ such that

$$\frac{1}{2}(a^2 + b^2 - 1) = \frac{1}{2}(1 - a^2 - b^2)$$

$$a^2 + b^2 - 1 = 1 - (a^2 + b^2)$$

$$2(a^2 + b^2) = 2$$

$$a^2 + b^2 = 1,$$

which are points on the unit circle.

The tangent plane to $z = \frac{1}{2}(x^2 + y^2 - 1)$ at $(x, y) = (a, b)$ such that $a^2 + b^2 = 1$ is

$$z = z(a, b) + z_x(a, b)(x - a) + z_y(a, b)(y - b)$$

$$= 0 + a(x - a) + b(y - b)$$

$$= ax + by - (a^2 + b^2)$$

$$= ax + by - 1, \text{ or } ax + by - z = 1.$$

Similarly, the tangent to $z = \frac{1}{2}(1 - x^2 - y^2)$ at $(x, y) = (a, b)$ such that $a^2 + b^2 = 1$ is $z = -ax - by + 1$ or $ax + by + z = 1$. The normal vector to the plane $ax + by - z = 1$ is $\vec{n_1} = a\vec{i} + b\vec{j} - \vec{k}$ and the normal vector to the plane $ax + by + z = 1$ is $\vec{n_2} = a\vec{i} + b\vec{j} + \vec{k}$.

Since $\vec{n_1} \cdot \vec{n_2} = a^2 + b^2 - 1 = 1 - 1 = 0$, $\vec{n_1}$ and $\vec{n_2}$ are perpendicular, hence the two surfaces are orthogonal at all points of intersection.

17.   If write $\vec{r} = x\vec{i} + y\vec{j} + z\vec{k}$, then we know

$$\text{grad } f(x, y, z) = g(x, y, z)(x\vec{i} + y\vec{j} + z\vec{k}) = g(x, y, z)\vec{r}$$

so grad f is everywhere radially outward, and therefore perpendicular to a sphere centered at the origin. If f were not constant on such a sphere, then grad f would have a component tangent to the sphere. Thus, f must be constant on any sphere centered at the origin.

18.

$$\text{grad}(\vec{\mu} \cdot \vec{r}) = \text{grad}(\mu_1 x + \mu_2 y + \mu_3 z)$$

$$= \mu_1\vec{i} + \mu_2\vec{j} + \mu_3\vec{k} = \vec{\mu}.$$

19.

$$\begin{aligned}
\text{grad}(\|\vec{r}\|^a) &= \text{grad}((x^2 + y^2 + z^2)^{a/2}) \\
&= \frac{a}{2}(x^2 + y^2 + z^2)^{(a/2)-1}(2x)\vec{i} + \frac{a}{2}(x^2 + y^2 + z^2)^{(a/2)-1}(2y)\vec{j} \\
&\quad + \frac{a}{2}(x^2 + y^2 + z^2)^{(a/2)-1}(2z)\vec{k} \\
&= a(x^2 + y^2 + z^2)^{(a-2)/2}(x\vec{i} + y\vec{j} + z\vec{k}) \\
&= a\|\vec{r}\|^{a-2}\vec{r}.
\end{aligned}$$

20. We must calculate the gradient of φ.

$$\begin{aligned}
\text{grad}\,\varphi(x, y, z) &= \text{grad}\,\frac{GmM}{\|\vec{r}\|} \\
&= GMm\,\text{grad}\,\frac{1}{\sqrt{x^2 + y^2 + z^2}}
\end{aligned}$$

Now

$$\frac{\partial}{\partial x}\left(\frac{1}{\sqrt{x^2 + y^2 + z^2}}\right) = \frac{\partial}{\partial x}\left((x^2 + y^2 + z^2)^{-1/2}\right) = -\frac{1}{2}(x^2 + y^2 + z^2)^{-3/2}2x = \frac{-x}{(x^2 + y^2 + z^2)^{3/2}}.$$

The partial derivatives with respect to y and z are similar, so

$$\text{grad}\,\frac{1}{\sqrt{x^2 + y^2 + z^2}} = -\frac{x\vec{i} + y\vec{j} + z\vec{k}}{(x^2 + y^2 + z^2)^{3/2}}.$$

Thus,

$$\begin{aligned}
\text{grad}\,\varphi &= -GMm\frac{x\vec{i} + y\vec{j} + z\vec{k}}{(x^2 + y^2 + z^2)^{3/2}} \\
&= -GMm\frac{\vec{r}}{\|\vec{r}\|^3} \\
&= \vec{F}
\end{aligned}$$

Solutions for Section 13.6

1. Using the chain rule we see:

$$\begin{aligned}
\frac{dz}{dt} &= \frac{\partial z}{\partial x}\frac{dx}{dt} + \frac{\partial z}{\partial y}\frac{dy}{dt} \\
&= -y^2 e^{-t} + 2xy\cos t \\
&= -(\sin t)^2 e^{-t} + 2e^{-t}\sin t\cos t \\
&= \sin(t)e^{-t}(2\cos t - \sin t)
\end{aligned}$$

We can also solve the problem using one variable methods:

$$z = e^{-t}(\sin t)^2$$

$$\frac{dz}{dt} = \frac{d}{dt}(e^{-t}(\sin t)^2)$$

$$= \frac{de^{-t}}{dt}(\sin t)^2 + e^{-t}\frac{d(\sin t)^2}{dt}$$

$$= -e^{-t}(\sin t)^2 + 2e^{-t}\sin t\cos t$$

$$= e^{-t}\sin t(2\cos t - \sin t)$$

2.   Using the chain rule we see:

$$\frac{dz}{dt} = \frac{\partial z}{\partial x}\frac{dx}{dt} + \frac{\partial z}{\partial y}\frac{dy}{dt}$$

$$= 2t(\sin y + y\cos x) + \frac{1}{t}(x\cos y + \sin x)$$

$$= 2t\sin(\ln t) + 2t\ln(t)\cos(t^2) + t\cos(\ln t) + \frac{\sin t^2}{t}$$

This problem can also be solved using one variable methods. Attempting to solve the problem that way will demonstrate the advantage of using the chain rule.

3.   This is a case where substituting is easier:

$$z = \ln(t^{-2} + t)$$

$$\frac{dz}{dt} = \frac{1 - 2t^{-3}}{t^{-2} + t}$$

$$= \frac{t^3 - 2}{t + t^4}$$

If you use the chain rule the solution is:

$$\frac{dz}{dt} = \frac{\partial z}{\partial x}\frac{dx}{dt} + \frac{\partial z}{\partial y}\frac{dy}{dt}$$

$$= \frac{-2x}{t^2(x^2 + y^2)} + \frac{y}{\sqrt{t}(x^2 + y^2)}$$

$$= \frac{-2}{t^3((1/t)^2 + t)} + \frac{1}{(1/t)^2 + t}$$

$$= \frac{-2}{t + t^4} + \frac{t^2}{1 + t^3}$$

$$= \frac{t^3 - 2}{t + t^4}$$

4.   Substituting into the chain rule gives

$$\frac{dz}{dt} = \frac{\partial z}{\partial x}\frac{dx}{dt} + \frac{\partial z}{\partial y}\frac{dy}{dt} = \cos\left(\frac{x}{y}\right)\left(\frac{1}{y}\right)(2) + \cos\left(\frac{x}{y}\right)\left(\frac{-x}{y^2}\right)(-2t)$$

$$= \cos\left(\frac{x}{y}\right)\left(\frac{2y + 2xt}{y^2}\right) = 2\left(\cos\left(\frac{2t}{1 - t^2}\right)\right)\frac{1 + t^2}{(1 - t^2)^2}$$

5. Substituting into the chain rule gives

$$\frac{dz}{dt} = \frac{\partial z}{\partial x}\frac{dx}{dt} + \frac{\partial z}{\partial y}\frac{dy}{dt} = e^y(2) + xe^y(-2t)$$

$$= 2e^y(1 - xt) = 2e^{1-t^2}(1 - 2t^2).$$

6. Substituting into the chain rule gives

$$\frac{dz}{dt} = \frac{\partial z}{\partial x}\frac{dx}{dt} + \frac{\partial z}{\partial y}\frac{dy}{dt} = e^y(2) + (xe^y + e^y + ye^y)(-2t)$$

$$= 2e^y(1 - xt - t - yt) = 2e^{1-t^2}(1 - 2t^2 - 2t + t^3).$$

7. Since z is a function of two variables x and y which are functions of two variables u and v, the two chain rule identities which apply are:

$$\frac{\partial z}{\partial u} = \frac{\partial z}{\partial x}\frac{\partial x}{\partial u} + \frac{\partial z}{\partial y}\frac{\partial y}{\partial u}$$

$$\frac{\partial z}{\partial v} = \frac{\partial z}{\partial x}\frac{\partial x}{\partial v} + \frac{\partial z}{\partial y}\frac{\partial y}{\partial v}$$

First to find $\partial z/\partial u$

$$\frac{\partial z}{\partial u} = (e^{-y} - ye^{-x})\sin v + (-xe^{-y} + e^{-x})(-v\sin u)$$

$$= (e^{-v\cos u} - v\cos(u)e^{-u\sin v})\sin v - (-u\sin(v)e^{-v\cos u} + e^{-u\sin v})v\sin u$$

Now we find $\partial z/\partial v$ using the same method.

$$\frac{\partial z}{\partial v} = (e^{-y} - ye^{-x})u\cos v + (-xe^{-y} + e^{-x})\cos u$$

$$= (e^{-v\cos u} - v\cos(u)e^{-u\sin v})u\cos v + (-u\sin(v)e^{-v\cos u} + e^{-u\sin v})\cos u$$

8. Since z is a function of two variables x and y which are functions of two variables u and v, the two chain rule identities which apply are:

$$\frac{\partial z}{\partial u} = \frac{\partial z}{\partial x}\frac{\partial x}{\partial u} + \frac{\partial z}{\partial y}\frac{\partial y}{\partial u}$$

$$\frac{\partial z}{\partial v} = \frac{\partial z}{\partial x}\frac{\partial x}{\partial v} + \frac{\partial z}{\partial y}\frac{\partial y}{\partial v}$$

This problem is most easily solved by substitution:

$$z = \cos(u^2((\cos v)^2 + (\sin v)^2))$$

$$= \cos u^2$$

$$\frac{\partial z}{\partial u} = -2u\sin u^2$$

$$\frac{\partial z}{\partial v} = 0$$

This problem can also be solved using the chain rule but it is more difficult.

9. Since z is a function of two variables x and y which are functions of two variables u and v, the two chain rule identities which apply are:

$$\frac{\partial z}{\partial u} = \frac{\partial z}{\partial x}\frac{\partial x}{\partial u} + \frac{\partial z}{\partial y}\frac{\partial y}{\partial u} = e^y(\frac{1}{u}) + xe^y \cdot 0 = \frac{e^v}{u}.$$

$$\frac{\partial z}{\partial v} = \frac{\partial z}{\partial x}\frac{\partial x}{\partial v} + \frac{\partial z}{\partial y}\frac{\partial y}{\partial v} = e^y(0) + xe^y \cdot 1 = e^v \ln u.$$

10. Since z is a function of two variables x and y which are functions of two variables u and v, the two chain rule identities which apply are:

$$\frac{\partial z}{\partial u} = \frac{\partial z}{\partial x}\frac{\partial x}{\partial u} + \frac{\partial z}{\partial y}\frac{\partial y}{\partial u} = e^y(\frac{1}{u}) + e^y(1+x+y) \cdot 0 = \frac{e^v}{u}.$$

$$\frac{\partial z}{\partial v} = \frac{\partial z}{\partial x}\frac{\partial x}{\partial v} + \frac{\partial z}{\partial y}\frac{\partial y}{\partial v} = e^y(0) + e^y(1+x+y) \cdot 1 = (1 + \ln u + v)e^v.$$

11. Since z is a function of two variables x and y which are functions of two variables u and v, the two chain rule identities which apply are:

$$\frac{\partial z}{\partial u} = \frac{\partial z}{\partial x}\frac{\partial x}{\partial u} + \frac{\partial z}{\partial y}\frac{\partial y}{\partial u} = e^y(2u) + xe^y(2u)$$
$$= 2ue^y(1+x) = 2ue^{(u^2-v^2)}(1+u^2+v^2).$$
$$\frac{\partial z}{\partial v} = \frac{\partial z}{\partial x}\frac{\partial x}{\partial v} + \frac{\partial z}{\partial y}\frac{\partial y}{\partial v} = e^y(2v) + xe^y(-2v)$$
$$= 2ve^y(1-x) = 2ve^{(u^2-v^2)}(1-u^2-v^2).$$

12. Since z is a function of two variables x and y which are functions of two variables u and v, the two chain rule identities which apply are:

$$\frac{\partial z}{\partial u} = \frac{\partial z}{\partial x}\frac{\partial x}{\partial u} + \frac{\partial z}{\partial y}\frac{\partial y}{\partial u} = e^y(2u) + (xe^y + e^y + ye^y)(2u)$$
$$= 2ue^y(1+x+1+y) = 2ue^y(x+y+2)$$
$$= 2ue^{(u^2-v^2)}(u^2+v^2+u^2-v^2+2) = 2ue^{(u^2-v^2)}(2u^2+2)$$
$$\frac{\partial z}{\partial v} = \frac{\partial z}{\partial x}\frac{\partial x}{\partial v} + \frac{\partial z}{\partial y}\frac{\partial y}{\partial v} = e^y(2v) + (xe^y + e^y + ye^y)(-2v)$$
$$= 2ve^y(1-x-1-y) = -2ve^y(x+y)$$
$$= -2ve^{(u^2-v^2)}(u^2+v^2+u^2-v^2) = -4u^2ve^{(u^2-v^2)}.$$

13. Since z is a function of two variables x and y which are functions of two variables u and v, the two chain rule identities which apply are:

$$\frac{\partial z}{\partial u} = \frac{\partial z}{\partial x}\frac{\partial x}{\partial u} + \frac{\partial z}{\partial y}\frac{\partial y}{\partial u} = \left(\cos\left(\frac{x}{y}\right)\right)\left(\frac{1}{y}\right)\frac{1}{u} + \left(\cos\left(\frac{x}{y}\right)\right)\left(\frac{-x}{y^2}\right) \cdot 0$$
$$= \frac{1}{vu}\cos\left(\frac{\ln u}{v}\right).$$
$$\frac{\partial z}{\partial v} = \frac{\partial z}{\partial x}\frac{\partial x}{\partial v} + \frac{\partial z}{\partial y}\frac{\partial y}{\partial v} = \left(\cos\left(\frac{x}{y}\right)\right)\left(\frac{1}{y}\right) \cdot 0 + \left(\cos\left(\frac{x}{y}\right)\right)\left(\frac{-x}{y^2}\right) \cdot 1 = -\frac{\ln u}{v^2}\cos\left(\frac{\ln u}{v}\right).$$

14. Since z is a function of two variables x and y which are functions of two variables u and v, the two chain rule identities which apply are:

$$\frac{\partial z}{\partial v} = \frac{\partial z}{\partial x}\frac{\partial x}{\partial u} + \frac{\partial z}{\partial y}\frac{\partial y}{\partial u} = \frac{1}{1+(\frac{x}{y})^2}(\frac{1}{y})(2u) + \frac{1}{1+(\frac{x}{y})^2}(\frac{-x}{y^2})(2u)$$

$$= 2u(\frac{y-x}{y^2+x^2}) = \frac{-2uv^2}{u^4+v^4}$$

$$\frac{\partial z}{\partial v} = \frac{\partial z}{\partial x}\frac{\partial x}{\partial v} + \frac{\partial z}{\partial y}\frac{\partial y}{\partial v} = \frac{1}{1+(\frac{x}{y})^2}(\frac{1}{y})(2v) + \frac{1}{1+(\frac{x}{y})^2}(\frac{-x}{y^2})(-2v)$$

$$= 2v(\frac{y+x}{y^2+x^2}) = \frac{2vu^2}{u^4+v^4}.$$

15. The tree diagram in Figure 13.14 tells us that

$$\frac{\partial w}{\partial u} = \frac{\partial w}{\partial x}\frac{\partial x}{\partial u} + \frac{\partial w}{\partial y}\frac{\partial y}{\partial u} + \frac{\partial w}{\partial z}\frac{\partial z}{\partial u},$$

$$\frac{\partial w}{\partial v} = \frac{\partial w}{\partial x}\frac{\partial x}{\partial v} + \frac{\partial w}{\partial y}\frac{\partial y}{\partial v} + \frac{\partial w}{\partial z}\frac{\partial z}{\partial v}.$$

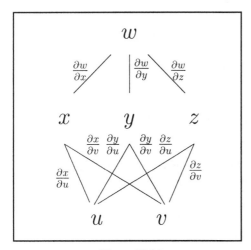

Figure 13.14

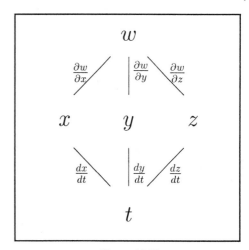

Figure 13.15

16. From the tree diagram in Figure 13.15, we get

$$\frac{dw}{dt} = \frac{\partial w}{\partial x}\frac{dx}{dt} + \frac{\partial w}{\partial y}\frac{dy}{dt} + \frac{\partial w}{\partial z}\frac{dz}{dt}.$$

17.

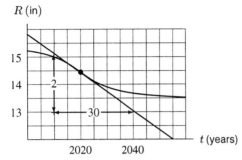

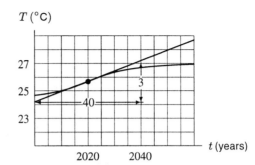

Figure 13.16: Global warming predictions:
Rainfall as a function of time

Figure 13.17: Global warming predictions:
Temperature as a function of time

We know that, as long as the temperature and rainfall stay close to their current values of $R = 15$ inches and $T = 30°C$, a change, ΔR, in rainfall and a change, ΔT, in temperature produces a change, ΔC, in corn production given by

$$\Delta C \approx 3.3\Delta R - 5\Delta T.$$

Now both R and T are functions of time t (in years), and we want to find the effect of a small change in time, Δt, on R and T. Figure 13.16 shows that the slope of the graph for R versus t is about $-2/30 \approx -0.07$ in/year when $t = 2020$. Similarly, Figure 13.17 shows the slope of the graph of T versus t is about $3/40 \approx 0.08°C$/year when $t = 2020$. Thus, around the year 2020,

$$\Delta R \approx -0.07\Delta t \quad \text{and} \quad \Delta T \approx 0.08\Delta t.$$

Substituting these into the equation for ΔC, we get

$$\Delta C \approx (3.3)(-0.07)\Delta t - (5)(0.08)\Delta t \approx -0.6\Delta t.$$

Since at present $C = 100$, corn production will decline by about 0.6 % between the years 2020 and 2021. Now $\Delta C \approx -0.6\Delta t$ tells us that when $t = 2020$,

$$\frac{\Delta C}{\Delta t} \approx -0.6, \quad \text{and therefore, that} \quad \frac{dC}{dt} \approx -0.6.$$

18.   The voltage at any time t is given by $V = IR$ where R is the resistance for the whole circuit. (In this case $R = R_1 R_2/(R_1 + R_2)$.) So the rate at which the voltage is changing is

$$\begin{aligned}
\frac{dV}{dt} &= \frac{dI}{dt}R + I\frac{dR}{dt} \\
&= \frac{dI}{dt}R + I\left(\frac{\partial R}{\partial R_1}\frac{dR_1}{dt} + \frac{\partial R}{\partial R_2}\frac{dR_2}{dt}\right) \\
&= \frac{dI}{dt}R + I\left(\frac{R_2^2}{(R_1 + R^2)^2}\frac{dR_1}{dt} + \frac{R_1^2}{(R_1 + R_2)^2}\frac{dR_2}{dt}\right) \\
&= 0.01\left(\frac{15}{8}\right) + 2\left(\frac{25}{64}(0.5) + \frac{9}{64}(-0.1)\right) \\
&= 0.3812.
\end{aligned}$$

So the voltage is increasing by 0.3812 volts/sec.

19. Let $g(t) = f(tx, ty)$. We use the chain rule, with $u = tx$ and $v = ty$ as our variables. Then we have

$$g'(t) = \frac{\partial f(u,v)}{\partial u}\frac{du}{dt} + \frac{\partial f(u,v)}{\partial v}\frac{dv}{dt}$$
$$= f_u(u,v)\, x + f_v(u,v)\, y.$$

At $t = 1$, we have $u = x$ and $v = y$. So

$$g'(1) = x\, f_x(x,y) + y\, f_y(x,y).$$

On the other hand, since $f(x,y)$ is homogeneous of degree p we also have $g(t) = t^p f(x,y)$. Thus we have $g'(t) = p\, t^{p-1} f(x,y)$ and

$$g'(1) = p\, f(x,y).$$

Thus,

$$x\, f_x(x,y) + y\, f_y(x,y) = p\, f(x,y).$$

20. Since $\left(\dfrac{\partial U}{\partial P}\right)_V$ involves the variables P and V, we are viewing U as a function of these two variables, so $U = U_3(P, V)$. Then

$$\left(\frac{\partial U}{\partial P}\right)_V = \frac{\partial U_3(P,V)}{\partial P}.$$

21. We will use analysis similar to that in Example 6. Since V is a function of P and T, we have

$$dV = \left(\frac{\partial V}{\partial T}\right)_P dT + \left(\frac{\partial V}{\partial P}\right)_T dP$$

We are interested in $\left(\frac{\partial U}{\partial V}\right)_T$ so we use the formula for dU corresponding to U_2. Substituting g for dV into this formula for dU gives

$$dU = \left(\frac{\partial U}{\partial T}\right)_V dT + \left(\frac{\partial U}{\partial V}\right)_T \left(\left(\frac{\partial V}{\partial T}\right)_P dT + \left(\frac{\partial V}{\partial P}\right)_T dP\right)$$
$$= \left(\left(\frac{\partial U}{\partial T}\right)_V + \left(\frac{\partial U}{\partial V}\right)_T \left(\frac{\partial V}{\partial T}\right)_P\right) dT + \left(\frac{\partial U}{\partial V}\right)_T \left(\frac{\partial V}{\partial P}\right)_T dP$$

But we are also interested in $\left(\dfrac{\partial U}{\partial P}\right)_T$ so we compare with the formula for dU corresponding to U_1.

$$dU = \left(\frac{\partial U}{\partial T}\right)_P dT + \left(\frac{\partial U}{\partial P}\right)_T dP.$$

Since the coefficients of dP must be identical, we get

$$\left(\frac{\partial U}{\partial P}\right)_T = \left(\frac{\partial U}{\partial V}\right)_T \left(\frac{\partial V}{\partial P}\right)_T.$$

22. From Example 6, we know that

$$\left(\frac{\partial U}{\partial T}\right)_{P} = \left(\frac{\partial U}{\partial T}\right)_{V} + \left(\frac{\partial U}{\partial V}\right)_{T} \left(\frac{\partial V}{\partial T}\right)_{P}.$$

From Problem 21, we know that

$$\left(\frac{\partial U}{\partial T}\right)_{P} = \left(\frac{\partial U}{\partial V}\right)_{P} \left(\frac{\partial V}{\partial T}\right)_{P}.$$

Thus, equating these expressions for $\left(\frac{\partial U}{\partial T}\right)_{P}$ we have

$$\left(\frac{\partial U}{\partial V}\right)_{P} \left(\frac{\partial V}{\partial T}\right)_{P} = \left(\frac{\partial U}{\partial T}\right)_{V} + \left(\frac{\partial U}{\partial V}\right)_{T} \left(\frac{\partial V}{\partial T}\right)_{P}$$

Dividing by $\left(\frac{\partial V}{\partial T}\right)_{P}$ gives

$$\left(\frac{\partial U}{\partial V}\right)_{P} = \frac{\left(\frac{\partial U}{\partial T}\right)_{V}}{\left(\frac{\partial V}{\partial T}\right)_{P}} + \left(\frac{\partial U}{\partial V}\right)_{T}$$

23. (a)   We will use the chain rule identities,

$$\frac{\partial z}{\partial r} = \frac{\partial z}{\partial x}\frac{\partial x}{\partial r} + \frac{\partial z}{\partial y}\frac{\partial y}{\partial r} \quad \text{and} \quad \frac{\partial z}{\partial \theta} = \frac{\partial z}{\partial x}\frac{\partial x}{\partial \theta} + \frac{\partial z}{\partial y}\frac{\partial y}{\partial \theta}.$$

These equations are to be in terms of $\partial z/\partial x$ and $\partial z/\partial y$, so we may calculate the other terms, switching from Cartesian to polar coordinates. Recall polar coordinates :

$$x = r\cos\theta, \quad y = r\sin\theta$$

Thus we have

$$\frac{\partial x}{\partial r} = \frac{\partial(r\cos\theta)}{\partial r} = \cos\theta$$

$$\frac{\partial y}{\partial r} = \frac{\partial(r\sin\theta)}{\partial r} = \sin\theta$$

$$\frac{\partial x}{\partial \theta} = \frac{\partial(r\cos\theta)}{\partial \theta} = -r\sin\theta$$

$$\frac{\partial y}{\partial \theta} = \frac{\partial(r\sin\theta)}{\partial \theta} = r\cos\theta$$

Now, substituting into the equations for $\partial z/\partial r$ and $\partial z/\partial \theta$, we get

$$(1) \qquad \frac{\partial z}{\partial r} = \cos\theta\frac{\partial z}{\partial x} + \sin\theta\frac{\partial z}{\partial y}$$

$$(2) \qquad \frac{\partial z}{\partial \theta} = -r\sin\theta\frac{\partial z}{\partial x} + r\cos\theta\frac{\partial z}{\partial y}.$$

We will call these equations (1) and (2).

(b)   Now we solve for $\partial z/\partial x$ and $\partial z/\partial y$. From (2) we get:

$$(3) \quad \frac{\partial z}{\partial x} = \left(\frac{\partial z}{\partial \theta} - r \cos \theta \frac{\partial z}{\partial y} \right) \left(\frac{-1}{r \sin \theta} \right),$$

Now substitute (3) into (1):

$$\frac{\partial z}{\partial r} = \cos \theta \left(\frac{\partial z}{\partial \theta} - r \cos \theta \frac{\partial z}{\partial y} \right) \left(\frac{-1}{r \sin \theta} \right) + \sin \theta \frac{\partial z}{\partial y}$$

$$= -\frac{\cos \theta}{r \sin \theta} \frac{\partial z}{\partial \theta} + \frac{\cos^2 \theta}{\sin \theta} \frac{\partial z}{\partial y} + \sin \theta \frac{\partial z}{\partial y}$$

Now solve for $\partial z/\partial y$:

$$\frac{\partial z}{\partial y} \left(\frac{\cos^2 \theta}{\sin \theta} + \frac{\sin^2 \theta}{\sin \theta} \right) = \frac{\partial z}{\partial r} + \frac{\cos \theta}{r \sin \theta} \frac{\partial z}{\partial \theta}$$

$$\frac{\partial z}{\partial y} \left(\frac{1}{\sin \theta} \right) = \frac{\partial z}{\partial r} + \frac{\cos \theta}{r \sin \theta} \frac{\partial z}{\partial \theta}$$

$$\frac{\partial z}{\partial y} = \sin \theta \frac{\partial z}{\partial r} + \frac{\cos \theta}{r} \frac{\partial z}{\partial \theta}.$$

Now, substitute $\partial z/\partial y$ into equation (3) and solve for $\partial z/\partial x$.

$$\frac{\partial z}{\partial r} = \left(\frac{\partial z}{\partial \theta} - r \cos \theta \frac{\partial z}{\partial y} \right) \frac{-1}{r \sin \theta}$$

$$= \frac{-1}{r \sin \theta} \frac{\partial z}{\partial \theta} + \frac{\cos \theta}{\sin \theta} \left(\sin \theta \frac{\partial z}{\partial r} + \frac{\cos \theta}{r} \frac{\partial z}{\partial \theta} \right)$$

$$= \cos \theta \frac{\partial z}{\partial r} + \frac{\cos^2 \theta - 1}{r \sin \theta} \frac{\partial z}{\partial \theta}$$

$$= \cos \theta \frac{\partial z}{\partial r} - \frac{\sin^2 \theta}{r \sin \theta} \frac{\partial z}{\partial \theta}$$

$$= \cos \theta \frac{\partial z}{\partial r} - \frac{\sin \theta}{r} \frac{\partial z}{\partial \theta}.$$

(c)   Now we use the chain rule to get $\partial z/\partial x$ and $\partial z/\partial y$.

$$(4) \quad \frac{\partial z}{\partial y} = \frac{\partial z}{\partial r} \frac{\partial r}{\partial y} + \frac{\partial z}{\partial \theta} \frac{\partial \theta}{\partial y}, \qquad \frac{\partial z}{\partial x} = \frac{\partial z}{\partial r} \frac{\partial r}{\partial x} + \frac{\partial z}{\partial \theta} \frac{\partial \theta}{\partial x}$$

We will call this equation (4).

As before, we will calculate some of these partials using $r = \sqrt{x^2 + y^2}$ and $\theta = \arctan(y/x)$

$$\frac{\partial r}{\partial y} = \frac{\partial \sqrt{x^2 + y^2}}{\partial y} = \frac{y}{\sqrt{x^2 + y^2}} = \sin \theta$$

$$\frac{\partial \theta}{\partial y} = \frac{\partial \arctan(y/x)}{\partial y} = \frac{1}{1 + (\frac{y}{x})^2} (x^{-1}) = \frac{x}{(x^2 + y^2)} = \frac{\cos \theta}{r}$$

$$\frac{\partial r}{\partial x} = \frac{x}{\sqrt{x^2 + y^2}} = \cos \theta$$

$$\frac{\partial \theta}{\partial x} = \frac{1}{1 + (\frac{y}{x})^2} \left(-\frac{y}{x^2} \right) = -\frac{y}{(x^2 + y^2)} = -\frac{\sin \theta}{r}$$

Now, substituting these into (4), we get:

$$\frac{\partial z}{\partial y} = \sin\theta \frac{\partial z}{\partial r} + \frac{\cos\theta}{r}\frac{\partial z}{\partial \theta}$$

$$\frac{\partial z}{\partial x} = \cos\theta \frac{\partial z}{\partial r} - \frac{\sin\theta}{r}\frac{\partial z}{\partial \theta}$$

Note that these equations match with those found in part (b).

24. Using $x = r\cos\theta$ and $y = r\sin\theta$ we compute $\partial z/\partial r$ and $\partial z/\partial\theta$ in terms of $\partial z/\partial x$ and $\partial z/\partial y$:

$$\frac{\partial z}{\partial r} = \frac{\partial z}{\partial x}\frac{\partial x}{\partial r} + \frac{\partial z}{\partial y}\frac{\partial y}{\partial r} = \frac{\partial z}{\partial x}\cos\theta + \frac{\partial z}{\partial y}\sin\theta$$

$$\frac{\partial z}{\partial \theta} = \frac{\partial z}{\partial x}\frac{\partial x}{\partial \theta} + \frac{\partial z}{\partial y}\frac{\partial y}{\partial \theta} = \frac{\partial z}{\partial x}(-r\sin\theta) + \frac{\partial z}{\partial y}r\cos\theta$$

So we have

$$\left(\frac{\partial z}{\partial r}\right)^2 = \left(\frac{\partial z}{\partial x}\right)^2 \cos^2\theta + 2\frac{\partial z}{\partial x}\frac{\partial z}{\partial y}\cos\theta\sin\theta + \left(\frac{\partial z}{\partial y}\right)^2 \sin^2\theta$$

In addition we have,

$$\frac{1}{r}\frac{\partial z}{\partial \theta} = \frac{\partial z}{\partial x}(-\sin\theta) + \frac{\partial z}{\partial y}\cos\theta$$

thus,

$$\frac{1}{r^2}\left(\frac{\partial z}{\partial \theta}\right)^2 = \left(\frac{\partial z}{\partial x}\right)^2 \sin^2\theta - 2\frac{\partial z}{\partial x}\frac{\partial z}{\partial y}\sin\theta\cos\theta + \left(\frac{\partial z}{\partial y}\right)^2 \cos^2\theta$$

Adding we get

$$\left(\frac{\partial z}{\partial r}\right)^2 + \frac{1}{r^2}\left(\frac{\partial z}{\partial \theta}\right)^2 = \left(\frac{\partial z}{\partial x}\right)^2 + \left(\frac{\partial z}{\partial y}\right)^2$$

25. Use chain rule for the equation $0 = F(x, y, f(x, y))$. Differentiating both sides with respect to x, remembering $z = f(x, y)$ and regarding y as a constant gives:

$$0 = \frac{\partial F}{\partial x}\frac{dx}{dx} + \frac{\partial F}{\partial z}\frac{dz}{dx}.$$

Since $dx/dx = 1$, we get

$$-\frac{\partial F}{\partial x} = \frac{\partial F}{\partial z}\frac{\partial z}{\partial x},$$

so

$$\frac{\partial z}{\partial x} = -\frac{\partial F/\partial x}{\partial F/\partial z}.$$

Similarly, differentiating both sides of the equation $0 = F(x, y, f(x, y))$ with respect to y gives:

$$0 = \frac{\partial F}{\partial y}\frac{dy}{dy} + \frac{\partial F}{\partial z}\frac{dz}{dy}.$$

Since $dy/dy = 1$, we get

$$-\frac{\partial F}{\partial y} = \frac{\partial F}{\partial z}\frac{\partial z}{\partial y},$$

so

$$\frac{\partial z}{\partial y} = -\frac{\partial F/\partial y}{\partial F/\partial z}.$$

Solutions for Section 13.7

1. Calculating the partial derivatives:

$$\frac{\partial f}{\partial x} = 2(x+y), \qquad \frac{\partial^2 f}{\partial x^2} = 2.$$

Therefore, we get

$$\frac{\partial f}{\partial y} = 2(x+y), \qquad \frac{\partial^2 f}{\partial y^2} = 2, \qquad \frac{\partial^2 f}{\partial y \partial x} = 2, \qquad \frac{\partial^2 f}{\partial x \partial y} = 2.$$

2. Calculating the partial derivatives:

$$\frac{\partial f}{\partial x} = 3(x+y)^2, \quad \frac{\partial^2 f}{\partial x^2} = 6(x+y).$$

$$\frac{\partial f}{\partial y} = 3(x+y)^2, \quad \frac{\partial^2 f}{\partial y^2} = 6(x+y).$$

Consequently, we get

$$\frac{\partial^2 f}{\partial x \partial y} = 6(x+y), \quad \frac{\partial^2 f}{\partial y \partial x} = 6(x+y).$$

3. Since $f(x,y) = xe^y$, the partial derivatives are

$$f_x = e^y, \quad f_y = xe^y$$
$$f_{xx} = 0, \quad f_{xy} = e^y = f_{yx}, \quad f_{yy} = xe^y.$$

4. Since $f = (x+y)e^y$, the partial derivatives are

$$f_x = e^y, \quad f_y = e^y(x+1+y)$$
$$f_{xx} = 0, \quad f_{yx} = e^y = f_{xy}$$
$$f_{yy} = xe^y + e^y + e^y + ye^y = e^y(x+2+y).$$

5. Since $f(x,y) = \sin(x^2 + y^2)$, we have

$$f_x = (\cos(x^2+y^2))2x \quad , f_y = (\cos(x^2+y^2))2y$$
$$f_{xx} = -(\sin(x^2+y^2))4x^2 + 2\cos(x^2+y^2)$$
$$f_{xy} = -(\sin(x^2+y^2))4xy = f_{yx}$$
$$f_{yy} = -(\sin(x^2+y^2))4y^2 + 2\cos(x^2+y^2).$$

6.   Since $f(x, y) = \sqrt{x^2 + y^2}$, we have

$$f_x = \frac{x}{\sqrt{x^2 + y^2}}, \quad f_y = \frac{y}{\sqrt{x^2 + y^2}}$$

$$f_{xx} = \frac{\sqrt{x^2 + y^2} - x\left(\frac{x}{\sqrt{x^2+y^2}}\right)}{x^2 + y^2} = \frac{y^2}{(x^2 + y^2)^{3/2}}$$

$$f_{xy} = -\frac{1}{2}\frac{x(2y)}{(x^2 + y^2)^{3/2}} = \frac{-xy}{(x^2 + y^2)^{3/2}} = f_{yx}$$

$$f_{yy} = \frac{\sqrt{x^2 + y^2} - y\frac{y}{\sqrt{x^2+y^2}}}{(x^2 + y^2)} = \frac{x^2}{(x^2 + y^2)^{3/2}}.$$

7.   Since $f(x, y) = \sin(x/y)$, the first partial derivatives are:

$$f_x = (\cos(\frac{x}{y}))\frac{1}{y}, \quad f_y = (\cos(\frac{x}{y}))(\frac{-x}{y^2}).$$

Thus, the second partial derivatives are

$$f_{xx} = -(\sin(\frac{x}{y}))(\frac{1}{y^2})$$

$$f_{xy} = -(\sin(\frac{x}{y}))(\frac{-x}{y^2})(\frac{1}{y}) + (\cos(\frac{x}{y}))(\frac{-1}{y^2}) = f_{yx}$$

$$f_{yy} = -(\sin(\frac{x}{y}))(\frac{-x}{y^2})^2 + (\cos(\frac{x}{y}))(\frac{2x}{y^3}).$$

8.   Since $f(x, y) = \tan^{-1}(x + y)$, we have

$$f_x = \frac{1}{1 + (x + y)^2} = f_y$$

$$f_{xx} = \frac{-1}{(1 + (x + y)^2)^2}2(x + y) = \frac{-2(x + y)}{(1 + (x + y)^2)^2} = f_{yy} = f_{xy} = f_{yx}.$$

9.   Since $z_y = g(x), z_{yy} = 0$, because g is a function of x only.

10. (a) $z_{yx} = z_{xy} = 4y$
     (b) $z_{xyx} = \frac{\partial}{\partial x}(z_{xy}) = \frac{\partial}{\partial x}(4y) = 0$
     (c) $z_{xyy} = z_{yxy} = \frac{\partial}{\partial y}(4y) = 4$

11. (a) $f_x(P) > 0$ because f increases as you go to the right.
     (b) $f_y(P) = 0$ because f does not change as you go up.
     (c) $f_{xx}(P) > 0$ because f_x increases as you go to the right. (The rate of change of f is larger when you go to the right since the level curves are closer together. Thus f_x changes from a small positive to a large positive number.)
     (d) $f_{yy}(P) = 0$ because f_y does not change as you go up.
     (e) $f_{xy}(P) = 0$ because f_x does not change as you go up.

12. (a) $f_x(P) > 0$ because f increases as you go to the right.
     (b) $f_y(P) = 0$ because f does not change as you go up.

(c) $f_{xx}(P) < 0$ because f_x decreases as you go to the right. (Since the level curves are further apart as you go to the right, the rate of change of f decreases. Thus, f_x changes from a large positive to a small positive number.)

(d) $f_{yy}(P) = 0$ because f_y does not change as you go up.

(e) $f_{xy}(P) = 0$ because f_x does not change as you go up.

13. (a) $f_x(P) < 0$ because f decreases as you go to the right.

(b) $f_y(P) = 0$ because f does not change as you go up.

(c) $f_{xx}(P) < 0$ because f_x decreases as you go to the right (f_x changes from a small negative number to a large negative number).

(d) $f_{yy}(P) = 0$ because f_y does not change as you go up.

(e) $f_{xy}(P) = 0$ because f_x does not change as you go up.

14. (a) $f_x(P) < 0$ because f decreases as you go to the right.

(b) $f_y(P) = 0$ because f does not change as you go up.

(c) $f_{xx}(P) > 0$ because f_x increases as you go to the right (f_x changes from a large negative number to a small negative number).

(d) $f_{yy}(P) = 0$ because f_y does not change as you go up.

(e) $f_{xy}(P) = 0$ because f_x does not change as you go up.

15. (a) $f_x(P) = 0$ because f does not change as you go to the right.

(b) $f_y(P) < 0$ because f decreases as you go up.

(c) $f_{xx}(P) = 0$ because f_x does not change as you go to the right.

(d) $f_{yy}(P) < 0$ because f_y decreases as you go up. (f_y changes from a negative number with smaller magnitude to a negative number with larger magnitude.)

(e) $f_{xy}(P) = 0$ because f_x does not change as you go up.

16. (a) $f_x(P) = 0$ because f does not change as you go to the right.

(b) $f_y(P) > 0$ because f increases as you go up.

(c) $f_{xx}(P) = 0$ because f_x does not change as you go to the right.

(d) $f_{yy}(P) < 0$ because f_y decreases as you go up (Since the level curves are further apart as you move up, the rate of change of f is slower, that is, f_y decreases as you move up.)

(e) $f_{xy}(P) = 0$ because f_x does not change as you go up.

17. (a) $f_x(P) < 0$ because f decreases as you go to the right.

(b) $f_y(P) < 0$ because f decreases as you go up.

(c) $f_{xx}(P) = 0$ because f_x does not change as you go to the right. (Notice that the level curves are equidistant and parallel, so the partial derivatives of f do not change if you move horizontally or vertically.)

(d) $f_{yy}(P) = 0$ because f_y does not change as you go up.

(e) $f_{xy}(P) = 0$ because f_x does not change as you go up.

18. (a) $f_x(P) > 0$ because f increases as you go to the right.

(b) $f_y(P) < 0$ because f decreases as you go up.

(c) $f_{xx}(P) < 0$ because the level curves are further apart as you go to the right, so the rate of increase of f is slower as you move to the right. Therefore, f_x decreases as you go to the right.

(d) $f_{yy}(P) < 0$ because f_y decreases as you move up (f_y changes from a negative number with smaller magnitude to a negative number with larger magnitude).

(e) $f_{xy}(P) > 0$ because the rate of change of f with respect to x at P is lower than at points above P. Therefore f_x increases as you move up.

19. (a) $f_x(P) < 0$ because f decreases as you go to the right.

(b) $f_y(P) > 0$ because f increases as you go up.

(c) $f_{xx}(P) > 0$ because f_x increases as you move right (f_x changes from negative numbers with larger magnitude to negative numbers with smaller magnitude).

(d) $f_{yy}(P) > 0$ because the level curves are closer together as you move up, so f_y increases as you go up.

(e) $f_{xy}(P) < 0$ because the rate of change of f with respect to x is a negative number at P and a negative number with larger magnitude higher up. Therefore f_x decreases as the point moves up.

20. (a)   The definition of f_x is:

$$f_x(a, b) = \lim_{h \to 0} \frac{f(a + h, b) - f(a, b)}{h}.$$

(b)   We define $f_{xy} = (f_x)_y$ as follows:

$$f_{xy}(a, b) = (f_x)_y(a, b) = \lim_{k \to 0} \frac{f_x(a, b + k) - f_x(a, b)}{k}.$$

(c)   Substituting the expression for f_x into the definition f_{xy}:

$$\begin{aligned} f_{xy}(a, b) &= \lim_{k \to 0} \frac{f_x(a, b + k) - f_x(a, b)}{k} \\ &= \lim_{k \to 0} \frac{1}{k} \left(\lim_{h \to 0} \frac{f(a + h, b + k) - f(a, b + k)}{h} - \lim_{h \to 0} \frac{f(a + h, b) - f(a, b)}{h} \right) \\ &= \lim_{k \to 0} \lim_{h \to 0} \frac{f(a + h, b + k) - f(a, b + k) - f(a + h, b) + f(a, b)}{hk}. \end{aligned}$$

(d)   Similarly,

$$f_{yx}(a, b) = \lim_{h \to 0} \lim_{k \to 0} \frac{f(a + h, b + k) - f(a + h, b) - f(a, b + k) + f(a, b)}{hk}.$$

(e)   The numerators in the two expressions in part (c) and (d) are the same (just swap the middle terms), so the only difference between them is the order in which the limits are taken. To be sure f_{xy} and f_{yx} are equal, we have to assume we can swap the order of the limits. Swapping limits can be a tricky business, but it can be done in this case if f_{xy} and f_{yx} are continuous.

Solutions for Section 13.8

1. (a)   We must first find the estimated rate of change from $t = 0$ to $t = 1$, with $u_t(4, 0)$ and $u_t(8, 0)$. From the heat equation, we know that we must first approximate $u_{xx}(4, 0)$ and $u_{xx}(8, 0)$. As in example 3, we will use the fact that u_{xx} is approximately a difference quotient of u_x, with the estimated slopes of two nearby points.

First for $u(4, 1)$:

$$u_x(3, 0) \approx \frac{(u(4, 0) - u(2, 0))}{(4 - 2)} = \frac{(56 - 52)}{2} = 2$$

$$u_x(5, 0) \approx \frac{(u(6, 0) - u(4, 0))}{(6 - 4)} = \frac{(62 - 56)}{2} = 3$$

$$u_{xx}(4, 0) \approx \frac{(u_x(5, 0) - u_x(3, 0))}{(5 - 3)} = \frac{(3 - 2)}{2} = 0.5$$

$$u_t(4, 0) = 0.1 u_{xx}(4, 0) \approx 0.1(0.5) = 0.05$$

Now for $u(8, 1)$:

$$u_x(7,0) \approx \frac{(u(8,0) - u(6,0))}{(8 - 6)} = \frac{(70 - 62)}{2} = 4$$

$$u_x(9,0) \approx \frac{(u(10,0) - u(8,0))}{(10 - 8)} = \frac{(80 - 70)}{2} = 5$$

$$u_{xx}(8,0) \approx \frac{(u_x(9,0) - u_x(7,0))}{(9 - 7)} = \frac{(5 - 4)}{2} = 0.5$$

$$u_t(8,0) = 0.1u_{xx}(8,0) \approx 0.1(0.5) = 0.05$$

So, using the local linear approximation we have:

$$u(4,1) \approx u(4,0) + u_t(4,0)(1) \approx 56 + 0.05(1) = 56.05°C$$

$$u(8,1) \approx u(8,0) + u_t(8,0)(1) \approx 70 + 0.05(1) = 70.05°C.$$

(b)   Now, we will follow the same process for $u(6,2)$.

$$u_x(5,1) \approx \frac{(u(6,1) - u(4,1))}{(6 - 4)} = \frac{(62.05 - 56.05)}{2} = 3$$

$$u_x(7,1) \approx \frac{(u(8,1) - u(6,1))}{(8 - 6)} = \frac{(70.05 - 62.05)}{2} = 4$$

$$u_{xx}(6,1) \approx \frac{(u_x(7,1) - u_x(5,1))}{(7 - 5)} = \frac{(4 - 3)}{2} = 0.5$$

$$u_t(6,1) = 0.1u_{xx}(6,1) \approx 0.1(0.5) = 0.05$$

So, we have

$$u(6,2) \approx u(6,1) + u_t(6,1)(1) \approx 62.05 + (.05)(1) = 62.1°C.$$

2.

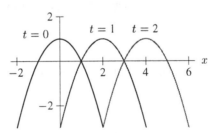

Figure 13.18

Graphs of $u(x,t) = 1 - (x - 2t)^2$ for $t = 0, 1, 2$ are in Figure 13.18. The graphs show a wave which appears to be traveling to the right. Its speed in that direction is distance divided by time(Δt). Between the first wave and the second, we have $\Delta t = 1$ and distance $= 2$ units, so the speed of the wave is 2 to the right.

3.   Substitute partial derivatives of I into equation and get

$$ce^{ax+by+ct} = D\left(a^2 e^{ax+by+ct} + b^2 e^{ax+by+ct}\right).$$

Cancelling $e^{ax+by+ct}$, we obtain

$$c = D(a^2 + b^2).$$

So this equation must be satisfied for the diffusion equation to hold.

4. Taking partial derivatives of u, we get:

$$\frac{\partial u}{\partial x} = xae^{ax+by} + e^{ax+by} + aye^{ax+by}$$

$$\frac{\partial u}{\partial y} = bxe^{ax+by} + bye^{ax+by} + e^{ax+by}$$

$$\frac{\partial^2 u}{\partial x \partial y} = abxe^{ax+by} + be^{ax+by} + abye^{ax+by} + ae^{ax+by}$$

Substituting into the differential equation and simplifying, we solve for the constants:

$$\frac{\partial^2 u}{\partial x \partial y} - \frac{\partial u}{\partial x} - \frac{\partial u}{\partial y} + u = 0$$

$$(abxe^{ax+by} + be^{ax+by} + abye^{ax+by} + ae^{ax+by})$$
$$-(xae^{ax+by} + e^{ax+by} + aye^{ax+by})$$
$$-(bxe^{ax+by} + bye^{ax+by} + e^{ax+by})$$
$$+(x+y)e^{ax+by} = 0$$

$$(ab - a - b + 1)xe^{ax+by} + (ab - a - b + 1)ye^{ax+by} + (b + a - 1 - 1)e^{ax+by} = 0$$
$$((ab - a - b + 1)(x + y) + (a + b - 2))e^{ax+by} = 0$$
$$((a - 1)(b - 1)(x + y) + a - (2 - b))e^{ax+by} = 0$$

Dividing by the non-zero factor e^{ax+by} gives us

$$(a - 1)(b - 1)(x + y) + (a - (2 - b)) = 0$$

In order for this equation to be true for all values of x and y, both $(a - 1)(b - 1)$ and $(a - (2 - b))$ must be equal to zero. (Why?) Thus we solve the system of two equations by setting $a = 2 - b$ to get:

$$((2 - b) - 1)(b - 1) = 0$$
$$-(b - 1)^2 = 0$$

This has one solution: $b = 1$. Thus our only values for a and b that satisfy the given differential equation are $a = 1, b = 1$.

5. Differentiating, we get

$$F_x = -e^{-x} \sin y, \; F_y = e^{-x} \cos y, \; F_{xx} = e^{-x} \sin y, \; F_{yy} = -e^{-x} \sin y = -F_{xx}.$$

Thus, $F_{xx} + F_{yy} = 0$.

6. Differentiating, we get

$$F_x = \frac{1}{1 + \left(\frac{y}{x}\right)}\left(-\frac{y}{x^2}\right) = \frac{-y}{x^2 + y^2}$$

$$F_y = \frac{1}{1 + \left(\frac{y}{x}\right)^2}\left(\frac{1}{x}\right) = \frac{x}{x^2 + y^2}$$

$$F_{xx} = (-y)\frac{-1}{(x^2 + y^2)^2}(2x) = \frac{2xy}{(x^2 + y^2)^2}$$

$$F_{yy} = \frac{-x}{(x^2 + y^2)^2}(2y) = \frac{-2xy}{(x^2 + y^2)^2} = -F_{xx}$$

Thus, $F_{xx} + F_{yy} = 0$.

7. Differentiating, we get

$$F_x = e^x \sin y + e^y \cos x \quad F_y = e^x \cos y + e^y \sin x$$
$$F_{xx} = e^x \sin y - e^y \sin x \quad F_{yy} = -e^x \sin y + e^y \sin x = -F_{xx}$$

Thus, $F_{xx} + F_{yy} = 0$.

8. To show it is a solution to $f_{tt} = c^2 f_{xx}$, we must take the x and t partial derivatives:

$$f_x = 0.003(\pi) \sin(2765t) \cos(\pi x)$$
$$f_{xx} = -0.003(\pi)^2 \sin(2765t) \sin(\pi x)$$
$$f_t = 0.003(2765) \cos(2765t) \sin(\pi x)$$
$$f_{tt} = -0.003(2765)^2 \sin(2765t) \sin(\pi x)$$

Now, substituting these in to the wave equation we have

$$-22935.675 \sin(2765t) \sin(\pi x) = c^2(-.02961 \sin(2765t) \sin(\pi x))$$

So, $f(x,t)$ solves the wave equation if we have $c^2 = \frac{22935.675}{0.02961}$. Thus, if $c \approx 880$ then $f(x,t)$ is a solution to the wave equation.

9. Write $V = f(u)$ where $u = x + ct$, then using the chain rule

$$\frac{\partial V}{\partial x} = \frac{df}{du} \cdot \frac{\partial u}{\partial x} = f'(u)(1).$$

Similarly,

$$\frac{\partial V}{\partial t} = \frac{df}{du} \cdot \frac{\partial u}{\partial t} = f'(u)(c) = cf'(u).$$

Thus

$$\frac{\partial V}{\partial t} = cf'(u) = c\frac{\partial V}{\partial x}.$$

10. (a) The fact that the ends are fixed means that the displacement is always zero at $x = 0$ and $x = L$. Therefore $y(0,t) = y(L,t) = 0$ for all t.
    (b) The initial shape is obtained by setting $t = 0$, so $y(x,0) = f(x)$ for all x.
    (c) The velocity at time t is given by $y_t(x,t)$. Thus, the initial velocity is $y_t(x,0)$, so $y_t(x,0) = g(x)$ for all x.

11. The graph of $T = u(x,t)$ is shown in Figure 13.19

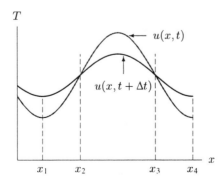

Figure 13.19

If $u(x, t)$ satisfies the heat equation $u_t = u_{xx}$, then when $u_{xx} > 0$ we have $u_t > 0$, so the temperature would be increasing with time. Similarly when $u_{xx} < 0$, the temperature is decreasing. Thus, since the graph is concave down on $x_2 \le x \le x_3$, we have $u_{xx} < 0$ and $u_t < 0$, on the same interval. So for $x_2 \le x \le x_3$ the temperature, meaning the height of the graph, will decrease between time t and $t + \Delta t$. On the other hand, for $0 \le x \le x_2$ and $x_3 \le x \le x_4$ the graph is concave up, so $u_{xx} > 0$ and $u_t > 0$, and the temperature increases in these intervals.

12. Since $V = m/r$, where r is the distance from (x, y, z) to (x_0, y_0, z_0), we have

$$V = \frac{m}{\sqrt{(x - x_0)^2 + (y - y_0)^2 + (z - z_0)^2}},$$

$$\frac{\partial V}{\partial x} = \frac{-m(x - x_0)}{\left((x - x_0)^2 + (y - y_0)^2 + (z - z_0)^2\right)^{\frac{3}{2}}},$$

$$\frac{\partial^2 V}{\partial x^2} = \frac{3m(x - x_0)^2}{\left((x - x_0)^2 + (y - y_0)^2 + (z - z_0)^2\right)^{\frac{5}{2}}} - \frac{m}{\left((x - x_0)^2 + (y - y_0)^2 + (z - z_0)^2\right)^{\frac{3}{2}}}$$

Similarly,

$$\frac{\partial^2 V}{\partial y^2} = \frac{3m(y - y_0)^2}{\left((x - x_0)^2 + (y - y_0)^2 + (z - z_0)^2\right)^{\frac{5}{2}}} - \frac{m}{\left((x - x_0)^2 + (y - y_0)^2 + (z - z_0)^2\right)^{\frac{3}{2}}}$$

and,

$$\frac{\partial^2 V}{\partial z^2} = \frac{3m(z - z_0)^2}{\left((x - x_0)^2 + (y - y_0)^2 + (z - z_0)^2\right)^{\frac{5}{2}}} - \frac{m}{\left((x - x_0)^2 + (y - y_0)^2 + (z - z_0)^2\right)^{\frac{3}{2}}},$$

So

$$\nabla^2 V = \frac{\partial^2 V}{\partial x^2} + \frac{\partial^2 V}{\partial y^2} + \frac{\partial^2 V}{\partial z^2}$$

$$= \frac{3m\left((x - x_0)^2 + (y - y_0)^2 + (z - z_0)^2\right)}{\left((x - x_0)^2 + (y - y_0)^2 + (z - z_0)^2\right)^{\frac{5}{2}}} - 3\frac{m}{\left((x - x_0)^2 + (y - y_0)^2 + (z - z_0)^2\right)^{\frac{3}{2}}}$$

$$= 0$$

13. First let us take the partials.

$$u_t = ae^{at} \sin(bx)$$
$$u_x = be^{at} \cos(bx)$$
$$u_{xx} = -b^2 e^{at} \sin(bx)$$

Now, plugging them in the equation, we have

$$ae^{at} \sin(bx) = u_t = u_{xx} = -b^2 e^{at} \sin(bx)$$

So, we have $a = -b^2$.

14. From the initial condition $y(x, 0) = 0$ we have that

$$F(x + 0) + G(x - 0) = 0,$$

for $0 < x < 5$. This gives

$$F(x) = -G(x).$$

From $\frac{\partial y}{\partial t}\big|_{t=0} = 5\sin(\pi x)$, we have

$$2\left[F'(x+2t) - G'(x-2t)\right]\Big|_{t=0} = 5\sin(\pi x)$$

that is, for $0 < x < 5$,

$$F'(x) - G'(x) = \frac{5}{2}\sin(\pi x)$$

Using $F(x) = -G(x)$, we get that

$$F'(x) + F'(x) = \frac{5}{2}\sin(\pi x).$$

Thus

$$F'(x) = \frac{5}{4}\sin(\pi x),$$

so

$$F(x) = \frac{-5}{4\pi}\cos(\pi x) + C.$$

Hence

$$y = F(x+2t) + G(x-2t) = F(x+2t) - F(x-2t)$$
$$= \left(-\frac{5}{4\pi}\cos(\pi x + 2\pi t) + C\right) + \left(\frac{5}{4\pi}\cos(\pi x - 2\pi t) - C\right)$$
$$= \frac{5}{4\pi}(\cos(\pi x - 2\pi t) - \cos(\pi x + 2\pi t))$$

is a solution to the wave equation. We see that it satisfies $y(0,t) = y(5,t) = 0$.

15. (a) Since $T = u(x,t)$ and the rod must be maintained at $0°$ C at both ends, the temperature at $x = 0$ and $x = 1$ must be zero at all times. So, the boundary conditions are

$$u(0,t) = 0 \qquad u(1,t) = 0$$

(b) Now we will take the partial derivatives and substitute them into the heat equation $u_t = u_{xx}$:

$$u_t = ae^{at}\sin(bx)$$
$$u_x = be^{at}\cos(bx)$$
$$u_{xx} = -b^2e^{at}\sin(bx)$$

Substituting these in the equation, we have

$$ae^{at}\sin(bx) = u_t = u_{xx} = -b^2e^{at}\sin(bx)$$

So, from these equations we have $a = -b^2$. Now from the boundary conditions, we have:

$$u(0,t) = e^{at}\sin(0) = 0 \quad \text{and} \quad u(1,t) = e^{at}\sin(b) = 0$$

The first condition gives us no new information about a and b, but the second does. Since

$$\sin b = 0$$
$$b = \arcsin 0 + \pi k \qquad \text{where } k = \text{any integer}$$

So, we found, $a = -b^2 = -(\pi k)^2$ for $k = 0, \pm 1, \pm 2, \pm 3...$

160   CHAPTER THIRTEEN /SOLUTIONS

16.   (a)   Taking partial derivatives of u, we get

$$u_t = -\frac{\pi}{4}(\pi t)^{-\frac{3}{2}}e^{-x^2/(4t)} + \frac{x^2}{4t^2}e^{-x^2/(4t)}\frac{1}{2\sqrt{\pi t}}$$

$$= -\frac{\pi}{4(\pi t)^{\frac{3}{2}}}e^{-x^2/(4t)} + \frac{x^2}{8t^2\sqrt{\pi t}}e^{-x^2/(4t)}$$

$$= -\frac{1}{4t\sqrt{\pi t}}e^{-x^2/(4t)} + \frac{x^2}{8t^2\sqrt{\pi t}}e^{-x^2/(4t)}$$

$$u_x = \frac{1}{2\sqrt{\pi t}}\left(\frac{-2x}{4t}\right)e^{-x^2/(4t)}$$

$$= -\frac{x}{4t\sqrt{\pi t}}e^{-x^2/(4t)}$$

$$u_{xx} = -\frac{1}{4t\sqrt{\pi t}}e^{-x^2/(4t)} - \frac{x}{4t\sqrt{\pi t}}\left(\frac{-2x}{4t}\right)e^{-x^2/(4t)}$$

$$= -\frac{1}{4t\sqrt{\pi t}}e^{-x^2/(4t)} + \frac{x^2}{8t^2\sqrt{\pi t}}e^{-x^2/(4t)}$$

So we have

$$u_t = u_{xx},$$

showing that u satisfies the heat equation.

   (b)

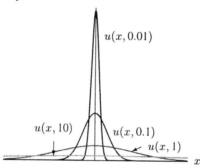

$u(x, 0.01)$

$u(x, 10)$ $u(x, 0.1)$

$u(x, 1)$

x

Figure 13.20

Note that as time progresses the heat at the origin decreases and flows out towards the ends of the rod, until at $t = 10$ the temperature appears to be leveling out towards being constant throughout the rod.

17.   We must first find the partial derivatives:

$$u_x = be^{-at}\cos(bx)\sin(cy)$$
$$u_{xx} = -b^2e^{-at}\sin(bx)\sin(cy)$$
$$u_y = ce^{-at}\sin(bx)\cos(cy)$$
$$u_{yy} = -c^2e^{-at}\sin(bx)\sin(cy)$$
$$u_t = -ae^{-at}\sin(bx)\sin(cy)$$

Now, substituting them in to the two-dimensional heat equation we have

$$-ae^{-at}\sin(bx)\sin(cy) = A(-b^2e^{-at}\sin(bx)\sin(cy) - c^2e^{-at}\sin(bx)\sin(cy))$$
$$= -A(b^2 + c^2)e^{-at}\sin(bx)\sin(cy)$$

Since this relationship holds for all values of x and y, we must have

$$A = \frac{a}{(b^2 + c^2)}.$$

Notice that $a/(b^2 + c^2)$ is a constant. Since $A > 0$, we must have $a > 0$.

18. (a)   Taking partial derivatives of u, we have

$$u_x = a \cos(ax) \sin(at)$$
$$u_{xx} = -a^2 \sin(ax) \sin(at)$$
$$u_t = a \sin(ax) \cos(at)$$
$$u_{tt} = -a^2 \sin(ax) \sin(at)$$

Thus, $u_{xx} = u_{tt}$.

   (b)   We have just showed that $u(x, t)$ satisfies the PDE, so now we must look at the boundary conditions. These equations must be true:

$$u(0, t) = \sin[a(0)] \sin(at) = 0$$
$$u(1, t) = \sin[a(1)] \sin(at) = 0.$$

The first condition is always true, for all t and a since $\sin 0 = 0$. The second condition is true for all t only when $\sin[a(1)] = 0$, which means we have $a = k\pi$ where $k = 0, \pm 1, \pm 2, \pm 3 \dots$. But we wanted $a > 0$, so $a = k\pi$, where $k = 1, 2, 3 \dots$.

19. (a)   First we must take the partial derivatives of $u(x, t)$.

$$u_x = f'(x - ct)$$
$$u_{xx} = f''(x - ct)$$
$$u_t = -cf'(x - ct)$$
$$u_{tt} = c^2 f''(x - ct)$$

So, $u_{tt} = c^2 f''(x - ct) = c^2 u_{xx}$.

   (b)   Again we must take the partial derivatives of $u(x, t)$.

$$u_x = g'(x + ct)$$
$$u_{xx} = g''(x + ct)$$
$$u_t = cg'(x + ct)$$
$$u_{tt} = c^2 g''(x + ct)$$

So, $u_{tt} = c^2 g''(x + ct) = c^2 u_{xx}$.

   (c)   We know $u_1 = f(x - ct)$ and $u_2 = g(x + ct)$ are both solutions, that is,

$$(u_1)_{tt} = c^2 (u_1)_{xx}$$
$$(u_2)_{tt} = c^2 (u_2)_{xx}$$

Since we have

$$(u_1 + u_2)_{tt} = (u_1)_{tt} + (u_2)_{tt}$$

and similarly

$$(u_1 + u_2)_{xx} = (u_1)_{xx} + (u_2)_{xx},$$

we can add the two equations, giving

$$(u_1 + u_2)_{tt} = c^2 (u_1 + u_2)_{xx}.$$

Thus $u(x, t) = f(x - ct) + g(x + ct)$ is a solution to the wave equation.

20. Taking partial derivatives, we have

$$u_x = a \cos{(ax)} \sin{(by)} \sin{(kt)}$$
$$u_{xx} = -a^2 \sin{(ax)} \sin{(by)} \sin{(kt)}$$
$$u_y = b \sin{(ax)} \cos{(by)} \sin{(kt)}$$
$$u_{yy} = -b^2 \sin{(ax)} \sin{(by)} \sin{(kt)}$$
$$u_t = k \sin{(ax)} \sin{(by)} \cos{(kt)}$$
$$u_{tt} = -k^2 \sin{(ax)} \sin{(by)} \sin{(kt)}$$

So, substituting in to the wave equation we have

$$-k^2 \sin{(ax)} \sin{(by)} \sin{(kt)} = c^2[-a^2 \sin{(ax)} \sin{(by)} \sin{(kt)} - b^2 \sin{(ax)} \sin{(by)} \sin{(kt)}]$$
$$k^2 = c^2(a^2 + b^2).$$

Thus, the condition upon a, b and k is that they must satisfy:

$$c^2 = \frac{k^2}{(a^2 + b^2)}.$$

Solutions for Section 13.9

1. The quadratic Taylor expansion about $(0,0)$ is given by

$$f(x,y) \approx Q(x,y) = f(0,0) + f_x(0,0)x + f_y(0,0)y + \frac{1}{2}f_{xx}(0,0)x^2 + f_{xy}(0,0)xy + \frac{1}{2}f_{yy}(0,0)y^2.$$

First we find all the relevant derivatives

$$f(x,y) = e^{-2x^2-y^2}$$
$$f_x(x,y) = -4xe^{-2x^2-y^2}$$
$$f_y(x,y) = -2ye^{-2x^2-y^2}$$
$$f_{xx}(x,y) = -4e^{-2x^2-y^2} + 16x^2e^{-2x^2-y^2}$$
$$f_{yy}(x,y) = -2e^{-2x^2-y^2} + 4y^2e^{-2x^2-y^2}$$
$$f_{xy}(x,y) = 8xye^{-2x^2-y^2}$$

Now we evaluate each of these derivatives at $(0,0)$ and substitute into the formula to get as our final answer:

$$Q(x,y) = 1 - 2x^2 - y^2$$

2. The quadratic Taylor expansion about $(0,0)$ is given by

$$f(x,y) \approx Q(x,y) = f(0,0) + f_x(0,0)x + f_y(0,0)y + \frac{1}{2}f_{xx}(0,0)x^2 + f_{xy}(0,0)xy + \frac{1}{2}f_{yy}(0,0)y^2$$

So first we find all the relevant derivatives

$$f(x, y) = \sin 2x + \cos y$$
$$f_x(x, y) = 2 \cos 2x$$
$$f_y(x, y) = -\sin y$$
$$f_{xx}(x, y) = -4 \sin 2x$$
$$f_{yy}(x, y) = -\cos y$$
$$f_{xy}(x, y) = 0$$

We substitute into the formula to get for our answer:

$$Q(x, y) = 1 + 2x - \frac{1}{2}y^2$$

3. The quadratic Taylor expansion about $(0, 0)$ is given by

$$f(x, y) \approx Q(x, y) = f(0, 0) + f_x(0, 0)x + f_y(0, 0)y + \frac{1}{2}f_{xx}(0, 0)x^2 + f_{xy}(0, 0)xy + \frac{1}{2}f_{yy}(0, 0)y^2.$$

So first we find all the relevant derivatives:

$$f(x, y) = \ln(1 + x^2 - y)$$
$$f_x(x, y) = \frac{2x}{1 + x^2 - y}$$
$$f_y(x, y) = \frac{-1}{1 + x^2 - y}$$
$$f_{xx}(x, y) = \frac{2(1 + x^2 - y) - 4x^2}{(1 + x^2 - y)^2}$$
$$f_{yy}(x, y) = \frac{-1}{(1 + x^2 - y)^2}$$
$$f_{xy}(x, y) = \frac{2x}{(1 + x^2 - y)^2}$$

Substituting into the formula we get as our answer:

$$Q(x, y) = -y + x^2 - \frac{y^2}{2}$$

4. We have $z(1, 1) = e$ and the relevant derivatives are:

$$z_x(x, y) = e^y \quad \text{so} \quad z_x(1, 1) = e$$
$$z_y(x, y) = xe^y \quad \text{so} \quad z_y(1, 1) = e$$
$$z_{xx}(x, y) = 0 \quad \text{so} \quad z_{xx}(1, 1) = 0$$
$$z_{xy}(x, y) = e^y \quad \text{so} \quad z_{xy}(1, 1) = e$$
$$z_{yy}(x, y) = xe^y \quad \text{so} \quad z_{yy}(1, 1) = e.$$

Thus the linear approximation, $L(x, y)$ to $z(x, y)$ at $(1, 1)$, is given by:

$$z(x, y) \approx L(x, y) = z(1, 1) + z_x(1, 1)(x - 1) + z_y(1, 1)(y - 1)$$
$$= e + e(x - 1) + e(y - 1).$$

The quadratic approximation, $Q(x, y)$ to $z(x, y)$ near $(1, 1)$, is given by:

$$z(x, y) \approx Q(x, y) = z(1, 1) + z_x(1, 1)(x - 1) + z_y(1, 1)(y - 1) + \frac{1}{2}z_{xx}(1, 1)(x - 1)^2$$

$$+ z_{xy}(1, 1)(x - 1)(y - 1) + \frac{1}{2}z_{yy}(1, 1)(y - 1)^2$$

$$= e + e(x - 1) + e(y - 1) + e(x - 1)(y - 1) + \frac{1}{2}e(y - 1)^2 .$$

Now

$$L(1.1, 1.1) = e + e \cdot 0.1 + e \cdot 0.1 = e \cdot 1.2 = 3.262$$

$$Q(1.1, 1.1) = L(1.1, 1.1) + e \cdot 0.1 \cdot 0.1 + \frac{1}{2}e \cdot (0.1)^2 = 3.303$$

and

$$z(1.1, 1.1) = 1.1e^{1.1} = 3.304 .$$

5.   We have $z(1, 1) = 2e$ and the relevant derivatives are:

$$z_x(x, y) = e^y \quad \text{so} \quad z_x(1, 1) = e$$
$$z_y(x, y) = e^y(x + 1 + y) \quad \text{so} \quad z_y(1, 1) = 3e$$
$$z_{xx}(x, y) = 0 \quad \text{so} \quad z_{xx}(1, 1) = 0$$
$$z_{xy}(x, y) = e^y \quad \text{so} \quad z_{xy}(1, 1) = e$$
$$z_{yy}(x, y) = e^y(x + 2 + y) \quad \text{so} \quad z_{yy}(1, 1) = 4e .$$

Thus the linear approximation, $L(x, y)$ to $z(x, y)$ at $(1, 1)$, is given by:

$$z(x, y) \approx L(x, y) = z(1, 1) + z_x(1, 1)(x - 1) + z_y(1, 1)(y - 1)$$
$$= 2e + e(x - 1) + 3e(y - 1) .$$

The quadratic approximation, $Q(x, y)$ to $z(x, y)$ at $(1, 1)$, is given by:

$$z(x, y) \approx Q(x, y) = z(1, 1) + z_x(1, 1)(x - 1) + z_y(1, 1)(y - 1)$$
$$+ \frac{1}{2}z_{xx}(1, 1)(x - 1)^2 + z_{xy}(1, 1)(x - 1)(y - 1) + \frac{1}{2} + z_{yy}(1, 1)(y - 1)^2$$
$$= 2e + e(x - 1) + 3e(y - 1) + e(x - 1)(y - 1) + 2e(y - 1)^2 .$$

Now

$$L(1.1, 1.1) = 6.524, \quad Q(1.1, 1.1) = 6.605, \quad z(1.1, 1.1) = 6.609 .$$

6.   We have $z(1, 1) = \sin 2$ and the relevant derivatives are:

$$z_x(x, y) = 2x \cos(x^2 + y^2) \quad \text{so} \quad z_x(1, 1) = 2 \cos 2$$
$$z_y(x, y) = 2y \cos(x^2 + y^2) \quad \text{so} \quad z_y(1, 1) = 2 \cos 2$$
$$z_{xx}(x, y) = -4x^2 \sin(x^2 + y^2) + 2 \cos(x^2 + y^2) \quad \text{so} \quad z_{xx}(1, 1) = -4 \sin 2 + 2 \cos 2$$
$$z_{xy}(x, y) = -4xy \sin(x^2 + y^2) \quad \text{so} \quad z_{xy}(1, 1) = -4 \sin 2$$
$$z_{yy}(x, y) = -4y^2 \sin(x^2 + y^2) + 2 \cos(x^2 + y^2) \quad \text{so} \quad z_{yy}(1, 1) = -4 \sin 2 + 2 \cos 2 .$$

Thus the linear approximation, $L(x, y)$ to $z(x, y)$ at $(1, 1)$, is given by:

$$z(x, y) \approx L(x, y) = z(1, 1) + z_x(1, 1)(x - 1) + z_y(1, 1)(y - 1)$$
$$= \sin 2 + 2\cos 2(x - 1) + 2\cos 2(y - 1) \,.$$

The quadratic approximation, $Q(x, y)$ to $z(x, y)$ at $(1, 1)$, is given by:

$$z(x, y) \approx Q(x, y) = z(1, 1) + z_x(1, 1)(x - 1) + z_y(1, 1)(y - 1)$$
$$+ \frac{1}{2}z_{xx}(1, 1)(x - 1)^2 + z_{xy}(1, 1)(x - 1)(y - 1) + \frac{1}{2}z_{yy}(1, 1)(y - 1)^2$$
$$= \sin 2 + 2\cos 2(x - 1) + 2\cos 2(y - 1) + (-2\sin 2 + \cos 2)(x - 1)^2$$
$$- 4\sin 2(x - 1)(y - 1) + (-2\sin 2 + \cos 2)(y - 1)^2 \,.$$

We have

$$L(1.1, 1.1) = 0.743$$
$$Q(1.1, 1.1) = 0.662$$
$$z(1.1, 1.1) = 0.661$$

7. We have $z(1, 1) = \sqrt{2}$ and the relevant derivatives are:

$$z_x(x, y) = \frac{x}{\sqrt{x^2 + y^2}} \quad \text{so} \quad z_x(1, 1) = \frac{1}{\sqrt{2}}$$

$$z_y(x, y) = \frac{y}{\sqrt{x^2 + y^2}} \quad \text{so} \quad z_y(1, 1) = \frac{1}{\sqrt{2}}$$

$$z_{xx}(x, y) = \frac{y^2}{(x^2 + y^2)^{3/2}} \quad \text{so} \quad z_{xx}(1, 1) = \frac{1}{2\sqrt{2}}$$

$$z_{xy}(x, y) = -\frac{xy}{(x^2 + y^2)^{3/2}} \quad \text{so} \quad z_{xy}(1, 1) = -\frac{1}{2\sqrt{2}}$$

$$z_{yy}(x, y) = \frac{x^2}{(x^2 + y^2)^{3/2}} \quad \text{so} \quad z_{yy}(1, 1) = \frac{1}{2\sqrt{2}} \,.$$

Thus the linear approximation, $L(x, y)$ to $z(x, y)$ at $(1, 1)$, is given by:

$$z(x, y) \approx L(x, y) = z(1, 1) + z_x(1, 1)(x - 1) + z_y(1, 1)(y - 1)$$
$$= \sqrt{2} + \frac{1}{\sqrt{2}}(x - 1) + \frac{1}{\sqrt{2}}(y - 1) \,.$$

The quadratic approximation, $Q(x, y)$ to $z(x, y)$ at $(1, 1)$, is given by:

$$z(x, y) \approx Q(x, y) = z(1, 1) + z_x(1, 1)(x - 1) + z_y(1, 1)(y - 1)$$
$$+ \frac{1}{2}z_{xx}(1, 1)(x - 1)^2 + z_{xy}(1, 1)(x - 1)(y - 1) + \frac{1}{2}z_{yy}(1, 1)(y - 1)^2$$
$$= \sqrt{2} + \frac{1}{\sqrt{2}}(x - 1) + \frac{1}{\sqrt{2}}(y - 1) + \frac{1}{4\sqrt{2}}(x - 1)^2 - \frac{1}{2\sqrt{2}}(x - 1)(y - 1)$$
$$+ \frac{1}{4\sqrt{2}}(y - 1)^2 \,.$$

We have

$$L(1.1, 1.1) = 1.556, \quad Q(1.1, 1.1) = 1.556, \quad z(1.1, 1.1) = 1.556 \,.$$

8. We have $z(1, 1) = \arctan 2$ and the relevant derivatives are

$$z_x(x, y) = z_y(x, y) = \frac{1}{1 + (x + y)^2}$$

so

$$z_x(1, 1) = z_y(1, 1) = \frac{1}{5}$$

and

$$z_{xx}(x, y) = z_{xy}(x, y) = z_{yy}(x, y) = -\frac{2(x + y)}{(1 + (x + y)^2)^2}$$

so

$$z_{xx}(1, 1) = z_{xy}(1, 1) = z_{yy}(1, 1) = -\frac{4}{25}.$$

Thus, the linear approximation, $L(x, y)$ to $z(x, y)$ at $(1, 1)$, is given by:

$$z(x, y) \approx L(x, y) = z(1, 1) + z_x(1, 1)(x - 1) + zy(1, 1)(y - 1)$$

$$= \arctan 2 + \frac{1}{5}(x - 1) + \frac{1}{5}(y - 1).$$

The quadratic approximation, $Q(x, y)$ to $z(x, y)$ at $(1, 1)$, is given by:

$$z(x, y) \approx Q(x, y) = z(1, 1) + z_x(1, 1)(x - 1) + z_y(1, 1)(y - 1)$$

$$+ \frac{1}{2}z_{xx}(1, 1)(x - 1)^2 + z_{xy}(1, 1)(x - 1)(y - 1) + \frac{1}{2}z_{yy}(1, 1)(y - 1)^2$$

$$= \arctan 2 + \frac{1}{5}(x - 1) + \frac{1}{5}(y - 1) - \frac{2}{25}(x - 1)^2 - \frac{4}{25}(x - 1)(y - 1)$$

$$- \frac{2}{25}(y - 1)^2.$$

We have:

$$L(1.1, 1.1) = 1.167, \quad Q(1.1, 1.1) = 1.144, \quad z(1.1, 1.1) = 1.144.$$

9. The partial derivatives of $z = \dfrac{xe^y}{x + y}$ are:

$$z_x = \frac{e^y(x + y) - xe^y}{(x + y)^2} = \frac{ye^y}{(x + y)^2}$$

$$z_y = \frac{xe^y(x + y) - xe^y}{(x + y)^2} = \frac{e^y(x^2 + xy - x)}{(x + y)^2}$$

$$z_{xx} = -\frac{2ye^y}{(x + y)^3}$$

$$z_{xy} = \frac{e^y(y + 1)(x + y)^2 - ye^y2(x + y)}{(x + y)^4} = \frac{e^y[(y + 1)(x + y) - 2y]}{(x + y)^3}$$

$$= \frac{e^y(y^2 + xy + x - y)}{(x + y)^3} = z_{yx}$$

$$z_{yy} = \frac{e^y(x^2 + xy - x + x)(x + y)^2 - e^y(x^2 + xy - x)2(x + y)}{(x + y)^4}$$

$$= \frac{e^y[(x^2 + xy)(x + y) - 2(x^2 + xy - x)]}{(x + y)^3}$$

$$= \frac{e^y(x^3 + 2x^2y + xy^2 - 2x^2 - 2xy + 2x)}{(x + y)^3}$$

The values of the partial derivatives at the point $(1, 1)$ are:

$$z(1, 1) = \frac{e}{2}$$

$$z_x(1, 1) = \frac{e}{4} = z_y(1, 1)$$

$$z_{xx}(1, 1) = \frac{-2}{8}e = \frac{-e}{4}$$

$$z_{xy}(1, 1) = \frac{2}{8}e = \frac{e}{4}$$

$$z_{yy}(1, 1) = \frac{2}{8}e = \frac{e}{4}.$$

Thus the linear approximation, $L(x, y)$ to $z(x, y)$, is given by:

$$z(x, y) \approx L(x, y) = z(1, 1) + z_x(1, 1)(x - 1) + z_y(1, 1)(y - 1)$$
$$= \frac{e}{2} + \frac{e}{4}(x - 1) + \frac{e}{4}(y - 1).$$

The quadratic approximation, $Q(x, y)$ to $z(x, y)$ at $(1, 1)$, is given by:

$$z(x, y) \approx Q(x, y) = z(1, 1) + z_x(1, 1)(x - 1) + z_y(1, 1)(y - 1)$$
$$+ \frac{1}{2}z_{xx}(1, 1)(x - 1)^2 + z_{xy}(1, 1)(x - 1)(y - 1) + \frac{1}{2}z_{yy}(1, 1)(y - 1)^2$$
$$= \frac{e}{2} + \frac{e}{4}(x - 1) + \frac{e}{4}(y - 1) - \frac{e}{8}(x - 1)^2 + \frac{e}{4}(x - 1)(y - 1) + \frac{e}{8}(y - 1)^2.$$

We have:

$$L(1.1, 1.1) = 1.495, \quad Q(1.1, 1.1) = 1.502, \quad z(1.1, 1.1) = 1.502 .$$

10. We have $z(1, 1) = \sin 1$ and the relevant derivatives are:

$$z_x(x, y) = \frac{1}{y}\cos\frac{x}{y} \quad \text{so} \quad z_x(1, 1) = \cos 1$$

$$z_y(x, y) = -\frac{x}{y^2}\cos\frac{x}{y} \quad \text{so} \quad z_y(1, 1) = -\cos 1$$

$$z_{xx}(x, y) = -\frac{1}{y^2}\sin\frac{x}{y} \quad \text{so} \quad z_{xx}(1, 1) = -\sin 1$$

$$z_{xy}(x, y) = -\frac{x}{y^3}\sin\frac{x}{y} - \frac{1}{y^2}\cos\frac{x}{y} \quad \text{so} \quad z_{xy}(1, 1) = \sin 1 - \cos 1$$

$$z_{yy}(x, y) = -\frac{x^2}{y^4}\sin\frac{x}{y} + \frac{2x}{y^3}\cos\frac{x}{y} \quad \text{so} \quad z_{yy}(1, 1) = -\sin 1 + 2\cos 1.$$

Thus the linear approximation, $L(x, y)$ to $z(x, y)$ at $(1, 1)$, is given by:

$$z(x, y) \approx L(x, y) = z(1, 1) + z_x(1, 1)(x - 1) + z_y(1, 1)(y - 1)$$
$$= \sin 1 + \cos 1(x - 1) - \cos 1(y - 1).$$

The quadratic approximation, $Q(x, y)$ to $z(x, y)$ at $(1, 1)$, is given by:

$$z(x, y) \approx Q(x, y) = z(1, 1) + z_x(1, 1)(x - 1) + z_y(1, 1)(y - 1)$$

$$+\frac{1}{2}z_{xx}(1,1)(x-1)^2 + z_{xy}(1,1)(x-1)(y-1) + \frac{1}{2}z_{yy}(1,1)(y-1)^2$$

$$= \sin 1 + \cos 1(x-1) - \cos 1(y-1) - \frac{1}{2}\sin 1(x-1)^2$$

$$+(\sin 1 - \cos 1)(x-1)(y-1) + \frac{1}{2}(-\sin 1 + 2\cos 1)(y-1)^2.$$

We have:

$$L(1.1, 1.1) = 0.841, \quad Q(1.1, 1.1) = 0.859, \quad z(1.1, 1.1) = 0.841 .$$

Notice that in this case the linear approximation is better than the quadratic one. This happens because $L(1.1, 1.1) = z(1.1, 1.1)$.

11. We have $z(1,1) = \dfrac{\pi}{4}$ and the relevant derivatives are:

$$z_x(x,y) = \frac{y}{x^2 + y^2} \quad \text{so} \quad z_x(1,1) = \frac{1}{2}$$

$$z_y(x,y) = -\frac{x}{x^2 + y^2} \quad \text{so} \quad z_y(1,1) = -\frac{1}{2}$$

$$z_{xx}(x,y) = -\frac{2xy}{(x^2 + y^2)^2} \quad \text{so} \quad z_{xx}(1,1) = -\frac{1}{2}$$

$$z_{xy}(x,y) = -\frac{x^2 - y^2}{(x^2 + y^2)^2} \quad \text{so} \quad z_{xy}(1,1) = 0$$

$$z_{yy}(x,y) = -\frac{2xy}{(x^2 + y^2)^2} \quad \text{so} \quad z_{yy}(1,1) = \frac{1}{2}.$$

Thus the linear approximation, $L(x,y)$ to $z(x,y)$ at $(1,1)$, is given by:

$$z(x,y) \approx L(x,y) = z(1,1) + z_x(1,1)(x-1) + z_y(1,1)(y-1)$$

$$= \frac{\pi}{4} + \frac{1}{2}(x-1) - \frac{1}{2}(y-1).$$

The quadratic approximation, $Q(x,y)$ to $z(x,y)$ at $(1,1)$, is given by:

$$z(x,y) \approx Q(x,y) = z(1,1) + z_x(1,1)(x-1) + z_y(1,1)(y-1)$$

$$+\frac{1}{2}z_{xx}(1,1)(x-1)^2 + z_{xy}(1,1)(x-1)(y-1) + \frac{1}{2}z_{yy}(1,1)(y-1)^2$$

$$= \frac{\pi}{4} + \frac{1}{2}(x-1) - \frac{1}{2}(y-1) - \frac{1}{4}(x-1)^2 + \frac{1}{4}(y-1)^2.$$

We have:

$$L(1.1, 1.1) = \frac{\pi}{4} \approx 0.785, \quad Q(1.1, 1.1) = \frac{\pi}{4} \approx 0.785, \quad z(1.1, 1.1) = \frac{\pi}{4} \approx 0.785 .$$

12. Let us first calculate the values of all the partial derivatives at $(0,0)$ that we need:

$$\begin{aligned}
f(x,y) &= (x + 2y + 1)^{1/2}, & f(0,0) &= 1, \\
f_x(x,y) &= \tfrac{1}{2}(x + 2y + 1)^{-1/2}, & f_x(0,0) &= 1/2, \\
f_y(x,y) &= (x + 2y + 1)^{-1/2}, & f_y(0,0) &= 1, \\
f_{xx}(x,y) &= -\tfrac{1}{4}(x + 2y + 1)^{-3/2}, & f_{xx}(0,0) &= -1/4, \\
f_{xy}(x,y) &= -\tfrac{1}{2}(x + 2y + 1)^{-3/2}, & f_{xy}(0,0) &= -1/2, \\
f_{yy}(x,y) &= -(x + 2y + 1)^{-3/2}, & f_{yy}(0,0) &= -1.
\end{aligned}$$

(a) The local linearization $L(x,y)$ of f at $(0,0)$ is given by

$$f(x,y) \approx L(x,y) = f(0,0) + f_x(0,0)x + f_y(0,0)y = 1 + \frac{1}{2}x + y.$$

(b) The second-order Taylor polynomial, $Q(x,y)$, for f at $(0,0)$ is given by

$$f(x,y) \approx Q(x,y)$$

$$= f(0,0) + f_x(0,0)x + f_y(0,0)y + \frac{f_{xx}(0,0)}{2}x^2 + f_{xy}(0,0)xy + \frac{f_{yy}(0,0)}{2}y^2$$

$$= 1 + \frac{1}{2}x + y - \frac{1}{8}x^2 - \frac{1}{2}xy - \frac{1}{2}y^2.$$

Notice that the local linearization of f is the same as the linear part of the Taylor polynomial of degree 2 for f. The extra terms in the Taylor polynomial of degree 2 can be thought of as "correction terms" to the linear approximation.

(c) Table 13.8 records the values of $f(x,y)$, $L(x,y)$, and $Q(x,y)$. Observe that the quadratic approximations $Q(x,y)$ are closer to the true values $f(x,y)$ than are the linear approximations $L(x,y)$. Of course both approximations are exact at $(0,0)$.

TABLE 13.8 *Linear and quadratic approximations to f near $(0,0)$*

Point	Linear	Quadratic	True
(x,y)	$L(x,y)$	$Q(x,y)$	$f(x,y)$
$(0,0)$	1	1	1
$(0.1, 0.1)$	1.15	1.13875	1.140175
$(-0.1, 0.1)$	1.05	1.04875	1.048809
$(0.1, -0.1)$	0.95	0.94875	0.948683
$(-0.1, -0.1)$	0.85	0.83875	0.836660

13. The contour diagrams in Figures 13.21–13.26 use the fact that

$$f(x,y) = \sqrt{x + 2y + 1},$$

$$L(x,y) = 1 + \frac{1}{2}x + y,$$

$$Q(x,y) = 1 + \frac{1}{2}x + y - \frac{1}{8}x^2 - \frac{1}{2}xy - \frac{1}{2}y^2.$$

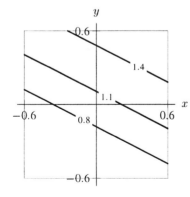

Figure 13.21: $f(x,y)$

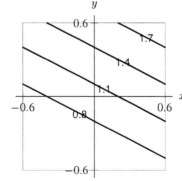

Figure 13.22: $L(x,y)$

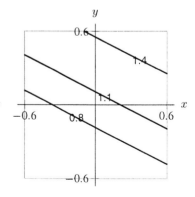

Figure 13.23: $Q(x,y)$

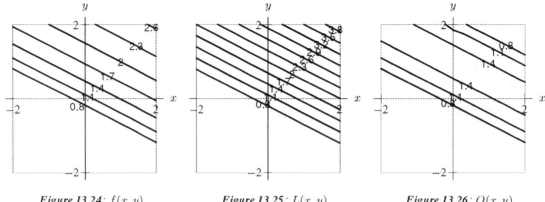

Figure 13.24: $f(x, y)$ Figure 13.25: $L(x, y)$ Figure 13.26: $Q(x, y)$

The contours for $f(x, y)$ and $L(x, y)$ are straight lines; those for $L(x, y)$ are equally spaced because $L(x, y)$ is a linear function. The contours for $f(x, y)$ are straight lines because if we set

$$f(x, y) = \sqrt{x + 2y + 1} = \text{constant}$$

then

$$x + 2y + 1 = \text{constant}.$$

However, the contours of $f(x, y)$ are not equally spaced because $f(x, y)$ is not linear.

In the "close up" diagram $[-0.6, 0.6] \times [-0.6, 0.6]$, the contours of $Q(x, y)$ look like lines (though they aren't). The contour diagram of $Q(x, y)$ is more similar to the contour diagram of $f(x, y)$ than is $L(x, y)$. This is because $Q(x, y)$ is a better approximation to $f(x, y)$ than is $L(x, y)$.

In the $[-2, 2] \times [-2, 2]$ diagram, the values on the level curves of $L(x, y)$ and $Q(x, y)$ show that neither of them is a good approximation to $f(x, y)$ away from the origin.

14. (a) Since P and Q lie on the same level curve, we have $a = k$.
     (b) We have $b = f_x$ and $c = f_y$. Since the gradient of f at P (respectively Q) points towards M or away from M, from the figure, we see $f_x(P)$ and $f_y(P)$ have opposite signs, while $f_x(Q)$ and $f_y(Q)$ have the same signs. Thus Q is the point (x_1, y_1), so P is (x_2, y_2).
     (c) Since $b = f_x(Q) > 0$ and $c = f_y(Q) > 0$, the value of f must increase as we go away from M. Thus, M must be a minimum (the surface is a valley).
     (d) Since M is a minimum, $m = f_x(P) < 0$ and $n = f_y(P) > 0$.

15. (a) Calculate the partial derivatives:

$$
\begin{aligned}
&f(x, y) = \sin x \sin y &\quad& f(0, 0) = 0 &\quad& f(\tfrac{\pi}{2}, \tfrac{\pi}{2}) = 1 \\
&f_x(x, y) = \cos x \sin y &\quad& f_x(0, 0) = 0 &\quad& f_x(\tfrac{\pi}{2}, \tfrac{\pi}{2}) = 0 \\
&f_y(x, y) = \sin x \cos y &\quad& f_y(0, 0) = 0 &\quad& f_y(\tfrac{\pi}{2}, \tfrac{\pi}{2}) = 0 \\
&f_{xx}(x, y) = -\sin x \sin y &\quad& f_{xx}(0, 0) = 0 &\quad& f_{xx}(\tfrac{\pi}{2}, \tfrac{\pi}{2}) = -1 \\
&f_{xy}(x, y) = \cos x \cos y &\quad& f_{xy}(0, 0) = 1 &\quad& f_{xy}(\tfrac{\pi}{2}, \tfrac{\pi}{2}) = 0 \\
&f_{yy}(x, y) = -\sin x \sin y &\quad& f_{yy}(0, 0) = 0 &\quad& f_{yy}(\tfrac{\pi}{2}, \tfrac{\pi}{2}) = -1
\end{aligned}
$$

Thus, the Taylor polynomial about $(0, 0)$ is

$$f(x, y) \approx Q_1(x, y) = xy.$$

The Taylor polynomial about $\left(\frac{\pi}{2}, \frac{\pi}{2}\right)$ is

$$f(x,y) \approx Q_2(x,y) = 1 - \frac{1}{2}\left(x - \frac{\pi}{2}\right)^2 - \frac{1}{2}\left(y - \frac{\pi}{2}\right)^2.$$

(b)

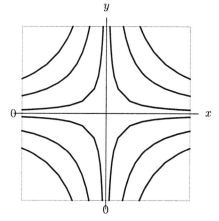

Figure 13.27: $f(x,y) \approx xy$: Quadratic approximation about $(0,0)$

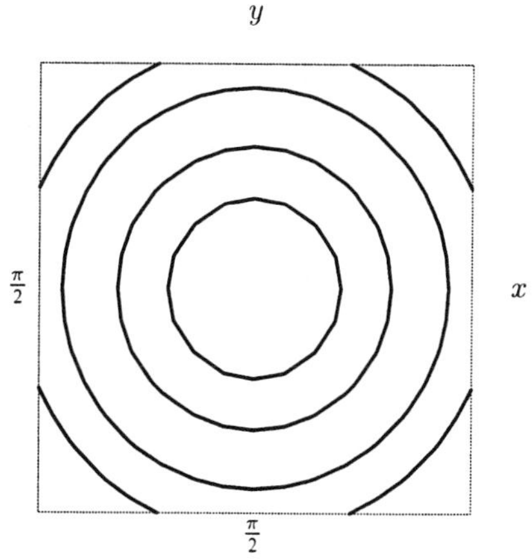

Figure 13.28: $f(x,y) \approx$ $1 - \frac{1}{2}(x - \frac{\pi}{2})^2 - \frac{1}{2}(y - \frac{\pi}{2})^2$: Quadratic approximation about $\left(\frac{\pi}{2}, \frac{\pi}{2}\right)$

16. We have $f(0,0) = 1$ and the relevant derivatives are:

$$\begin{aligned}
f_x(x,y) &= -\sin x \cos y & \text{so} \quad f_x(0,0) &= 0 \\
f_y(x,y) &= -\cos x \sin y & \text{so} \quad f_y(0,0) &= 0 \\
f_{xx}(x,y) &= -\cos x \cos y & \text{so} \quad f_{xx}(0,0) &= -1 \\
f_{xy}(x,y) &= \sin x \sin y & \text{so} \quad f_{xy}(0,0) &= 0 \\
f_{yy}(x,y) &= -\cos x \cos y & \text{so} \quad f_{yy}(0,0) &= -1 \\
f_{xxx}(x,y) &= \sin x \cos y \\
f_{xxy}(x,y) &= \cos x \sin y \\
f_{xyy}(x,y) &= \sin x \cos y \\
f_{yyy}(x,y) &= \cos x \sin y \, .
\end{aligned}$$

(a) Thus:

$$L(x,y) = f(0,0) + f_x(0,0) \cdot x + f_y(0,0) \cdot y = 1.$$

Therefore

$$|E_L(x,y)| = |f(x,y) - L(x,y)| \leq 2 \cdot M_L(x^2 + y^2) \leq M_L \cdot 0.04$$

for $|x| \leq 0.1$ and $|y| \leq 0.1$, where

$$M_L \geq |f_{xx}|, |f_{xy}|, |f_{yy}|$$

for

$$d(x,y) = (x^2 + y^2)^{\frac{1}{2}} \leq d_0.$$

In this case

$$d_0 = (0.1^2 + 0.1^2)^{\frac{1}{2}} = 0.14.$$

Looking at the formulas for f_{xx}, f_{xy}, f_{yy}, we see we can take $M_L = 1$. Thus,

$$|E_L(x,y)| \leq 2(x^2 + y^2) \leq 0.04 \quad \text{for} \quad |x| \leq 0.1 \quad \text{and} \quad |y| \leq 0.1.$$

(b)

$$Q(x,y) = f(0,0) + f_x(0,0) \cdot x + f_y(0,0) \cdot y + \frac{1}{2}f_{xx}(0,0) \cdot x^2 + f_{xy}(0,0)xy$$

$$+\frac{1}{2}f_{yy}(0,0) \cdot y^2 = 1 - \frac{1}{2}x^2 - \frac{1}{2}y^2.$$

$$|E_Q(x,y)| = |f(x,y) - Q(x,y)| \leq \frac{4}{3}M_Q d(x,y)^3$$

$$= \frac{4}{3}M_Q(x^2 + y^2)^{\frac{3}{2}} \leq 0.004 \cdot M_Q \quad \text{for} \quad |x| \leq 0.1 \quad \text{and} \quad |y| \leq 0.1$$

where

$$M_Q \geq |f_{xxx}|, |f_{xyy}|, |f_{xxy}|, |f_{yyy}|$$

for (x, y) such that

$$d(x,y) = (x^2 + y^2)^{\frac{1}{2}} \leq d_0 = 0.14.$$

Since each of these third derivatives has one factor which is a sine, we can take

$$M_Q = \sin 0.14 = 0.14.$$

Thus

$$|E_Q(x,y)| \leq 0.19(x^2 + y^2)^{\frac{3}{2}} \leq 0.00053.$$

(c)

$$E_L(0.1, 0.1) = f(0.1, 0.1) - L(0.1, 0.1) = 0.99 - 1 = 0.001$$
$$E_Q(0.1, 0.1) = f(0.1, 0.1) - Q(0.1, 0.1) = 0.99003 - 0.99 = 0.00003$$

Note: although we want to bound E_L (respectively E_Q) on the square $|x| \leq 0.1$ and $|y| \leq 0.1$, we need to compute M_L (respectively M_Q) on the disk containing the above square (i.e., the disk $d(x, y) \leq d_0 = 0.14$).

17.   We have $f(0,0) = 1$ and the relevant derivatives are:

$$f_x(x,y) = (e^x - 1)\cos y \quad \text{so} \quad f_x(0,0) = 0$$
$$f_y(x,y) = -(e^x - x)\sin y \quad \text{so} \quad f_y(0,0) = 0$$
$$f_{xx}(x,y) = e^x \cos y \quad \text{so} \quad f_{xx}(0,0) = 1$$
$$f_{xy}(x,y) = -(e^x - 1)\sin y \quad \text{so} \quad f_{xy}(0,0) = 0$$
$$f_{yy}(x,y) = -(e^x - x)\cos y \quad \text{so} \quad f_{yy}(0,0) = -1$$
$$f_{xxx}(x,y) = e^x \cos y$$
$$f_{xxy}(x,y) = -e^x \sin y$$
$$f_{xyy}(x,y) = -(e^x - 1)\cos y$$
$$f_{yyy}(x,y) = (e^x - x)\sin y \ .$$

(a)   Thus,

$$L(x,y) = f(0,0) + f_x(0,0) \cdot x + f_y(0,0) \cdot y = 1$$

and

$$|E_L(x,y)| = |f(x,y) - L(x,y)| \leq 2 \cdot M_L(x^2 + y^2) \leq 0.04 M_L$$

for $|x| \leq 0.1$ and $|y| \leq 0.1$, where $M_L \geq |f_{xx}|, |f_{xy}|, |f_{yy}|$ for (x,y) such that $d(x,y) = (x^2+y^2)^{\frac{1}{2}} \leq d_0 = (0.1^2 + 0.1^2)^{\frac{1}{2}} = 0.14$.

As $f_{xx}(x,y) = e^x \cos y$, the maximum of $|f_{xx}(x,y)|$ on the disk $d(x,y) \leq 0.14$ is $e^{0.14} \cdot \cos 0 \leq 1.151$.

Similarly, $f_{xy}(x,y) = -(e^x - 1) \sin y$, and $|f_{xy}(x,y)| \leq |\sin y| \leq 1$.

Also, $f_{yy}(x,y) = -(e^x - x)\cos y$ so $|f_{yy}(x,y)| \leq |e^x - x| \leq 1.011$. So we take $M_L = 1.151$ and get

$$|E_L(x,y)| \leq 0.047 \quad \text{for} \quad |x| \leq 0.1 \quad \text{and} \quad |y| \leq 0.1.$$

(b)

$$Q(x,y) = f(0,0) + f_x(0,0) \cdot x + f_y(0,0) \cdot y + \frac{1}{2} f_{xx}(0,0) \cdot x^2$$

$$+ f_{xy}(0,0) \cdot xy + \frac{1}{2} f_{yy}(0,0) \cdot y^2 = 1 + \frac{x^2}{2} - \frac{y^2}{2}.$$

and

$$|E_Q(x,y)| = |f(x,y) - Q(x,y)| \leq \frac{4}{3} M_Q(x^2 + y^2)^{\frac{3}{2}} \leq \frac{4}{3} M_Q(0.02)^{\frac{3}{2}} = 0.004 M_Q$$

where $M_Q \geq |f_{xxx}|, |f_{xxy}|, |f_{xyy}|, |f_{yyy}|$ for (x,y) such that $d(x,y) \leq 0.14$. Reasoning as in part (a), we take $M_Q = 1.151$ and so we get

$$|E_Q(x,y)| \leq 0.0047.$$

(c)

$$E_L(0.1, 0.1) = f(0.1, 0.1) - L(0.1, 0.1) = 1.0002 - 1 = 0.0002$$
$$E_Q(0.1, 0.1) = f(0.1, 0.1) - L(0.1, 0.1) = 1.0002 - 1 = 0.0002$$

Note: Although we want to bound E_L (respectively E_Q) on the square $|x| \leq 0.1$ and $|y| \leq 0.1$, we need to compute M_L (respectively M_Q) on the disk containing the above square (i.e., the disk $d(x,y) \leq d_0 = 0.14$).

18. We have $f(0,0) = 1$ and the relevant derivatives are:

$$f(x,y) = f_y(x,y) = f_{xx}(x,y) = f_{xy}(x,y) = f_{yy}(x,y)$$
$$= f_{xxx}(x,y) = f_{xxy}(x,y) = f_{xyy}(x,y) = f_{yyy}(x,y) = e^{x+y}$$

so

$$f_x(0,0) = f_y(0,0) = f_{xx}(0,0) = f_{xy}(0,0) = f_{yy}(0,0) = 1.$$

(a) Thus

$$L(x,y) = f(0,0) + f_x(0,0) \cdot x + f_y(0,0) \cdot y = 1 + x + y$$

and

$$|E_L(x,y)| = |f(x,y) - L(x,y)| \leq 2M_L \cdot (x^2 + y^2) \leq 0.04 \cdot M_L$$

for $|x| \leq 0$ and $|y| \leq 0.1$, where $M_L \geq |f_{xx}|, |f_{xy}|, |f_{yy}|$ for (x,y) such that $d(x,y) = (x^2+y^2)^{\frac{1}{2}} \leq d_0 = (0.1^2 + 0.1^2)^{\frac{1}{2}} = 0.14$.

Given the partial derivatives we see that we can take $M_L \geq e^{0.28} = 1.33$ and

$$|E_L(x,y)| \leq 0.053.$$

(b)

$$Q(x,y) = f(0,0) + f_x(0,0) \cdot x + f_y(0,0) \cdot y + \frac{1}{2}f_{xx}(0,0) \cdot x^2$$

$$+ f_{xy}(0,0) \cdot xy + \frac{1}{2}f_{yy}(0,0) \cdot y^2 = 1 + x + y + \frac{1}{2}x^2 + xy + \frac{1}{2}y^2$$

and

$$|E_Q(x,y)| = |f(x,y) - Q(x,y)| \le \frac{4}{3}M_Q(x^2 + y^2)^{\frac{3}{2}} \le \frac{4}{3}M_Q(0.02)^{\frac{3}{2}} = 0.004M_Q$$

for $|x| \le 0.1$ and $|y| \le 0.1$, where $M_Q \ge |f_{xxx}|, f_{xyy}|, |f_{xxy}|, |f_{yyy}|$ for (x,y) such that $d(x,y) = (x^2 + y^2)^{\frac{1}{2}} \le d_0 = (0.1^2 + 0.1^2)^{\frac{1}{2}} = 0.14$. So $M_Q = 1.33$ and

$$|E_Q(x,y)| \le 0.0053.$$

(c)

$$E_L(0.1, 0.1) = f(0.1, 0.1) - L(0.1, 0.1) = 1.2214 - 1.2 = 0.0214$$
$$E_Q(0.1, 0.1) = f(0.1, 0.1) - Q(0.1, 0.1) = 1.2214 - 1.22 = 0.0014$$

Note: Although we want to bound E_L (respectively E_Q) on the square $|x| \le 0.1$ and $|y| \le 0.1$, we need to compute M_L (respectively M_Q) on the disk containing the above square (i.e., the disk $d(x,y) \le d_0 = 0.14$).

19. We have $f(0,0) = 0$ and the relevant derivatives are:

$$f_x(x,y) = (x^2 + y^2 + 2x)e^{x+y} \quad \text{so} \quad f_x(0,0) = 0$$
$$f_y(x,y) = (x^2 + y^2 + 2y)e^{x+y} \quad \text{so} \quad f_y(0,0) = 0$$
$$f_{xx}(x,y) = (x^2 + y^2 + 4x + 2)e^{x+y} \quad \text{so} \quad f_{xx}(0,0) = 2$$
$$f_{xy}(x,y) = (x^2 + y^2 + 2x + 2y)e^{x+y} \quad \text{so} \quad f_{xx}(0,0) = 0$$
$$f_{yy}(x,y) = (x^2 + y^2 + 4y + 2)e^{x+y} \quad \text{so} \quad f_{yy}(0,0) = 2$$
$$f_{xxx}(x,y) = (x^2 + y^2 + 6x + 6)e^{x+y}$$
$$f_{xxy}(x,y) = (x^2 + y^2 + 4x + 2y + 2)e^{x+y}$$
$$f_{xyy}(x,y) = (x^2 + y^2 + 2x + 4y + 2)e^{x+y}$$
$$f_{yyy}(x,y) = (x^2 + y^2 + 6y + 6)e^{x+y}$$

(a) Thus,

$$L(x,y) = f(0,0) + f_x(0,0)x + f_y(0,0)y = 0$$

and

$$|E_L(x,y)| = |f(x,y) - L(x,y)| \le 2M_L(x^2 + y^2) \le 0.04M_L$$

for $|x| \le 0.1$ and $|y| \le 0.1$, where $M_L \ge |f_{xx}|, |f_{xy}|, |f_{yy}|$ for (x,y) such that $d(x,y) = (x^2 + y^2)^{\frac{1}{2}} \le d_0 = (0.1^2 + 0.1^2)^{\frac{1}{2}} = 0.14$.
    We have

- $|f_{xx}(x,y)| \le [(0.14)^2 + (0.14)^2 + 4 \cdot (0.14) + 2] \cdot e^{0.28} = 3.439$
- $|f_{xy}(x,y)| \le [(0.14)^2 + (0.14)^2 + 2 \cdot (0.14) + 2 \cdot (0.14)] \cdot e^{0.28} = 0.793$
- $|f_{yy}(x,y)| \le [(0.14)^2 + (0.14)^2 + 4 \cdot (0.14) + 2] \cdot e^{0.28} = 3.439$

for (x,y) such that $d(x,y) \le 0.14$, therefore we can take M_L to equal 3.44 and so we have:

$$|E_L(x,y)| \le 0.04 \cdot M_L = 0.14.$$

(b)

$$Q(x, y) = f(0,0) + f_x(0,0)x + f_y(0,0)y + \frac{1}{2}f_{xx}(0,0)x^2 + f_{xy}(0,0)xy + \frac{1}{2}f_{yy}(0,0)y^2$$
$$= x^2 + y^2,$$

and

$$|E_Q(x,y)| = |f(x,y) - Q(x,y)| \le \frac{4}{3}M_Q(x^2+y^2)^{\frac{3}{2}} \le \frac{4}{3}M_Q(0.02)^{\frac{3}{2}} = 0.004M_Q,$$

for $|x| \le 0.1$ and $|y| \le 0.1$, where $M_Q \ge |f_{xxx}|, |f_{xxy}|, |f_{xyy}|, |f_{yyy}|$ for $d(x,y) = (x^2+y^2)^{\frac{1}{2}} \le d_0 = (0.1^2 + 0.1^2)^{\frac{1}{2}} = 0.14$.

We have

- $|f_{xxx}(x,y)| \le f_{xxx}(0.14, 0.14) = (0.14^2 + 0.14^2 + 6 \cdot 0.14 + 6)e^{0.28} = 9.1$
- $|f_{xxy}(x,y)| \le f_{xxy}(0.14, 0.14) = (0.14^2 + 0.14^2 + 4 \cdot 0.14 + 2 \cdot 0.14 + 2)e^{0.28} = 3.81$
- $|f_{xyy}(x,y)| \le f_{xyy}(0.14, 0.14) = (0.14^2 + 0.14^2 + 2 \cdot 0.14 + 4 \cdot 0.14 + 2)e^{0.28} = 3.81$
- $|f_{yyy}(x,y)| \le f_{xyy}(0.14, 0.14) = (0.14^2 + 0.14^2 + 6 \cdot 0.14 + 6)e^{0.28} = 9.1$

So we can take $M_Q = 9.1$ and we have

$$|E_Q(x,y) \le 0.004 \cdot M_Q = 0.036.$$

(c)

$$E_L(0.1, 0.1) = f(0.1, 0.1) - L(0.1, 0.1) = f(0.1, 0.1) = 0.024428$$
$$E_Q(0.1, 0.1) = f(0.1, 0.1) - Q(0.1, 0.1) = 0.024428 - 0.01 = 0.014428$$

Note: Although we want to bound E_L (respectively E_Q) on the square $|x| \le 0.1$ and $|y| \le 0.1$, we need to compute M_L (respectively M_Q) on the disk containing the above square (i.e., the disk $d(x,y) \le d_0 = 0.14$).

20.   (a)   We think of h and k as constants. So if

$$g(t) = f(ht, kt),$$

then g is a function of t. Thus, by the chain rule

$$g'(t) = \frac{\partial f}{\partial x}\frac{dx}{dt} + \frac{\partial f}{\partial y}\frac{dy}{dt} = (f_x)h + (f_y)k.$$

Using the fact that $g''(t)$ can also be calculated by the chain rule, we have

$$g''(t) = \frac{d}{dt}(g'(t)) = \frac{\partial}{\partial x}(hf_x + kf_y) \cdot \frac{dx}{dt} + \frac{\partial}{\partial y}(hf_x + kf_y) \cdot \frac{dy}{dt}$$
$$= (hf_{xx} + kf_{yx})h + (hf_{xy} + kf_{yy})k$$
$$= h^2 f_{xx} + 2hk f_{xy} + k^2 f_{yy}.$$

Notice that each term in $g'(t)$ gives rise to two terms in $g''(t)$: one with an extra partial derivative with respect to x and an extra factor of h, the other with an extra partial derivative with respect to y and an extra factor of k. Exactly the same pattern happens to terms in the binomial expansion of $(a+b)^n$ when we multiply by $(a+b)$ to get $(a+b)^{n+1}$. This leads us to expect that $g'''(t)$ has the same form as $(a+b)^3$:

$$g'''(t) = h^3 f_{xxx} + 3h^2 k f_{xxy} + 3hk^2 f_{xyy} + k^3 f_{yyy}.$$

(b)   Since $L(x, y)$ is the linear approximation to $f(x, y)$, we have

$$L(x, y) = f(0, 0) + f_x(0, 0)x + f_y(0, 0)y.$$

Substituting $x = ht$, $y = kt$ gives

$$L(ht, kt) = f(0, 0) + f_x(0, 0)ht + f_y(0, 0)kt.$$

Since $P_1(t)$ is the linear approximation to $g(t)$, we have

$$\begin{aligned} P_1(t) &= g(0) + g'(0)t \\ &= f(0, 0) + \big(f_x(0, 0)h + f_y(0, 0)k\big)\, t \end{aligned}$$

Thus,

$$L(ht, kt) = P_1(t).$$

Similarly, the quadratic approximation, $Q(x, y)$, is given by

$$Q(x, y) = f(0, 0) + f_x(0, 0)x + f_y(0, 0)y + \frac{f_{xx}(0, 0)}{2}x^2 + f_{xy}(0, 0)xy + \frac{f_{yy}(0, 0)}{2}y^2.$$

Substituting $x = ht$, $y = kt$, we see that $Q(ht, kt)$ is given by

$$\begin{aligned} &Q(ht, kt) \\ &= f(0, 0) + f_x(0, 0)ht + f_y(0, 0)kt + \frac{f_{xx}(0, 0)}{2}(ht)^2 + f_{xy}(0, 0)hkt^2 + \frac{f_{yy}(0, 0)}{2}(kt)^2 \\ &= f(0, 0) + \big(f_x(0, 0)h + f_y(0, 0)k\big)\, t + \frac{1}{2}\big(f_{xx}(0, 0)h^2 + 2f_{xy}(0, 0)hk + f_{yy}(0, 0)k^2\big)\, t^2 \\ &= g(0) + g'(0)t + \frac{g''(0)}{2}t^2. \end{aligned}$$

Thus,

$$Q(ht, kt) = P_2(t).$$

(c)   Since $E_L(x, y)$ and $E_Q(x, y)$ are the errors in the linear and quadratic approximation to $f(x, y)$ and E_1 and E_2 are the errors in the corresponding approximation to $g(t)$, we have

$$E_L(ht, kt) = f(ht, kt) - L(ht, kt) = g(t) - P_1(t) = E_1(t),$$

so

$$E_L(ht, kt) = E_1(t).$$

By a similar argument,

$$E_Q(ht, kt) = E_2(t).$$

(d)   We want to show that there are bounds on $|E_L|$ and $|E_Q|$. We assume that the second-order partial derivatives of f are bounded; that is, that

$$|f_{xx}|, |f_{xy}|, |f_{yy}| \le M_L \quad \text{for } d(x, y) \le d_0.$$

We have shown that

$$|g''(t)| = |h^2 f_{xx} + 2hk f_{xy} + k^2 f_{yy}|.$$

Now, the largest that $|g''(t)|$ could be occurs when all the terms on the right add, rather than cancel, so

$$|g''(t)| \le |h|^2 |f_{xx}| + 2|h||k||f_{xy}| + |k|^2 |f_{yy}|.$$

When $|h| \leq d, |k| \leq d$, we have

$$|g''(t)| \leq d^2 M_L + 2d^2 M_L + d^2 M_L = 4d^2 M_L \quad \text{for } d \leq d_0.$$

Since we now know that $|g''(t)|$ is bounded we can apply the one-variable result, giving

$$|E_L(ht, kt)| = |E_1(t)| \leq \frac{4M_L d^2}{2}.$$

Substituting $t = 1$ gives

$$|E_L(h, k)| \leq 2M_L d^2 \quad \text{for} \quad d = \sqrt{h^2 + k^2} \leq d_0.$$

To estimate E_Q, we assume that the third-order partial derivatives are bounded, that is

$$|f_{xxx}|, |f_{xxy}|, |f_{xyy}|, |f_{yyy}| \leq M_Q \quad \text{for } d(x, y) \leq d_0.$$

Then, by a similar argument we can show that

$$|g'''(t)| \leq |h|^3 |f_{xxx}| + 3|h|^2 |k| |f_{xxy}| + 3|h| |k|^2 |f_{xyy}| + |k|^3 |f_{yyy}|$$

so

$$|g''(t)| \leq 8d^3 M_Q,$$

giving

$$|E_Q(ht, kt)| = |E_2(t)| \leq \frac{8M_Q d^3}{6} t^3.$$

Thus, for $t = 1$,

$$|E_Q(h, k)| \leq \frac{4d^3 M_Q}{3} \quad \text{for} \quad d = \sqrt{h^2 + k^2} \leq d_0.$$

Solutions for Section 13.10

1. (a) The contour diagram for $f(x, y) = \frac{x}{y} + \frac{y}{x}$ is shown in Figure 13.29
    (b) If $x \neq 0$ and $y \neq 0$ then f is differentiable at (x, y). Now we need to look at points of the form $(x, 0)$, where $x \neq 0$ and $(0, y)$, where $y \neq 0$. The function f is not differentiable at these points as it is not continuous.
    (c) For $x \neq 0$ and $y \neq 0$,

$$f_x(x, y) = \frac{1}{y} - \frac{y}{x^2}.$$

So f_x exists for $x \neq 0$, $y \neq 0$, and it is continuous.

For all points $(x_0, 0)$ on the x-axis we have:

$$f_x(x_0, 0) = \lim_{x \to x_0} \frac{f(x, 0) - f(x_0, 0)}{x - x_0} = 0.$$

Thus, f_x exists but is not continuous at these points.

For points $(0, y_0)$ on the y-axis we have:

$$\lim_{x \to 0} \frac{f(x, y_0) - f(0, y_0)}{x} = \lim_{x \to 0} \left(\frac{1}{y_0} + \frac{y_0}{x^2} \right).$$

This limit doesn't exist, so the partial derivative $f_x(0, y_0)$ doesn't exist.

Similarly, for $x \neq 0$ and $y \neq 0$,

$$f_y(x, y) = -\frac{x}{y^2} + \frac{1}{x}.$$

For points $(0, y_0)$ on the y-axis we have $f_y(0, y_0) = 0$, while $f_y(x_0, 0)$ doesn't exist for $x_0 \neq 0$.

Both $f_x(x, y)$ and $f_y(x, y)$ are continuous at (x, y) only for $x \neq 0$ and $y \neq 0$.

(d) We claim f is not continuous at $(0, 0)$. Let $x = t$ and $y = t$, where $t \to 0$, $t \neq 0$. Then

$$f(x, y) = f(t, t) = 2, \qquad \text{for} \quad t \neq 0.$$

So,

$$\lim_{t \to 0} f(t, t) = 2 \neq f(0, 0) = 0,$$

and therefore

$$\lim_{(x,y) \to (0,0)} f(x, y) \neq f(0, 0).$$

Thus, f is not differentiable at $(0, 0)$ since f is not continuous at $(0, 0)$.

(e) From part (c) we have $f_x(0, 0) = 0$ and $f_y(0, 0) = 0$. The functions f_x and f_y are not continuous at $(0, 0)$.

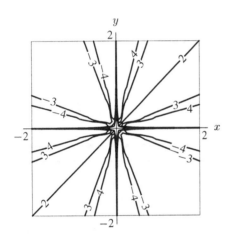

Figure 13.29

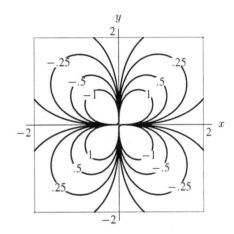

Figure 13.30

2. (a) The contour diagram for $f(x, y) = 2xy/(x^2 + y^2)^2$ is shown in Figure 13.30.

(b) The function f is differentiable at all points $(x, y) \neq (0, 0)$, as it is a rational fraction with denominator $(x^2 + y^2)^2 = 0$ only when $(x, y) = (0, 0)$.

(c) The partial derivatives of f at points $(x, y) \neq (0, 0)$ are given by

$$f_x(x, y) = \frac{2y(y^2 - 3x^2)}{(x^2 + y^2)^3},$$

$$f_y(x, y) = \frac{2x(x^2 - 3y^2)}{(x^2 + y^2)^3}.$$

Both f_x and f_y are continuous for $(x, y) \neq (0, 0)$.

(d) The function f is not continuous at $(0,0)$. To see this, let $x = y = t$ for $t \neq 0$. Then,

$$f(x,y) = f(t,t) = \frac{2t^2}{4t^4} = \frac{1}{2t^2},$$

and so $\lim_{t \to 0} f(t,t)$ does not exist. Hence, f is not differentiable at $(0,0)$.

(e) At $(0,0)$, the partial derivatives of f are given by

$$f_x(0,0) = \lim_{x \to 0} \frac{f(x,0) - f(0,0)}{x} = \lim_{x \to 0} \frac{0-0}{x} = 0,$$

$$f_y(0,0) = \lim_{y \to 0} \frac{f(0,y) - f(0,0)}{y} = \lim_{y \to 0} \frac{0-0}{y} = 0.$$

We claim that $\lim_{(x,y) \to (0,0)} f_x(x,y)$ doesn't exist. To see this, let $x = y = t$ for $t \neq 0$. Then,

$$f_x(x,y) = f_x(t,t) = \frac{2t(t^2 - 3t^2)}{(2t^2)^3} = \frac{-4t^3}{8t^6} = -\frac{1}{2t^3}$$

and so the limit

$$\lim_{t \to 0} f_x(t,t) = \lim_{t \to 0} \frac{-1}{2t^3}$$

does not exist. Hence, f_x is not continuous at $(0,0)$. Similarly, f_y is not continuous at $(0,0)$.

3. (a) The contour diagram for $f(x,y) = xy/\sqrt{x^2 + y^2}$ is shown in Figure 13.31.

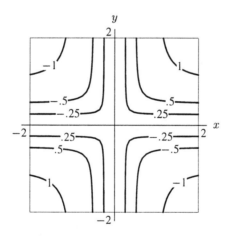

Figure 13.31

(b) By the chain rule, f is differentiable at all points (x,y) where $x^2 + y^2 \neq 0$, and so at all points $(x,y) \neq (0,0)$.

(c) The partial derivatives of f are given by

$$f_x(x,y) = \frac{y^3}{(x^2 + y^2)^{3/2}}, \qquad \text{for} \quad (x,y) \neq (0,0),$$

and

$$f_y(x,y) = \frac{x^3}{(x^2 + y^2)^{3/2}}, \qquad \text{for} \quad (x,y) \neq (0,0).$$

Both f_x and f_y are continuous at $(x,y) \neq (0,0)$.

(d)   If f were differentiable at $(0,0)$, the chain rule would imply that the function

$$g(t) = \begin{cases} f(t,t), & t \neq 0 \\ 0, & t = 0 \end{cases}$$

would be differentiable at $t = 0$. But

$$g(t) = \frac{t^2}{\sqrt{2t^2}} = \frac{1}{\sqrt{2}} \cdot \frac{t^2}{|t|} = \frac{1}{\sqrt{2}} \cdot |t|,$$

which is not differentiable at $t = 0$. Hence, f is not differentiable at $(0,0)$.

(e)   The partial derivatives of f at $(0,0)$ are given by

$$f_x(0,0) = \lim_{x \to 0} \frac{f(x,0) - f(0,0)}{x} = \lim_{x \to 0} \frac{\frac{x \cdot 0}{\sqrt{x^2 + 0^2}} - 0}{x} = \lim_{x \to 0} \frac{0 - 0}{x} = 0,$$

$$f_y(0,0) = \lim_{y \to 0} \frac{f(0,y) - f(0,0)}{y} = \lim_{y \to 0} \frac{\frac{0 \cdot y}{\sqrt{0^2 + y^2}} - 0}{y} = \lim_{y \to 0} \frac{0 - 0}{y} = 0.$$

The limit $\lim_{(x,y) \to (0,0)} f_x(x,y)$ doesn't exist since if we choose $x = y = t, t \neq 0$, then

$$f_x(x,y) = f_x(t,t) = \frac{t^3}{(2t^2)^{3/2}} = \frac{t^3}{2\sqrt{2} \cdot |t|^3} = \begin{cases} \frac{1}{2\sqrt{2}}, & t > 0, \\ -\frac{1}{2\sqrt{2}}, & t < 0. \end{cases}$$

Thus, f_x is not continuous at $(0,0)$. Similarly, f_y is not continuous at $(0,0)$.

4.   (a)   The contour diagram for $f(x,y) = x^2 y/(x^4 + y^2)$ is shown in Figure 13.32.

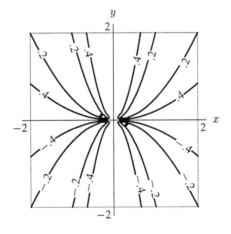

Figure 13.32

(b)   The function f is differentiable at all $(x,y) \neq (0,0)$ as it is a rational fraction with denominator which is zero only when $(x,y) = (0,0)$.

(c)   The partial derivatives of f are given by

$$f_x(x,y) = \frac{2xy(y^2 - x^4)}{(x^4 + y^2)^2}, \qquad \text{for} \quad (x,y) \neq (0,0),$$

$$f_y(x,y) = \frac{x^2(x^4 - y^2)}{(x^4 + y^2)^2}, \qquad \text{for} \quad (x,y) \neq (0,0).$$

Both f_x and f_y are continuous at $(x,y) \neq (0,0)$.

(d) We use the definition of differentiability. If f were differentiable at $(0,0)$, then the linear approximation of f at $(0,0)$ would be $L(x,y) = mx + ny$, where $m = f_x(0,0)$ and $n = f_y(0,0)$. We have

$$f_x(0,0) = \lim_{x \to 0} \frac{f(x,0) - f(0,0)}{x} = 0,$$

$$f_y(0,0) = \lim_{y \to 0} \frac{f(0,y) - f(0,0)}{y} = 0.$$

So, we need to compute the limit:

$$\lim_{(x,y) \to (0,0)} \frac{f(x,y) - L(x,y)}{\sqrt{x^2 + y^2}} = \lim_{(x,y) \to (0,0)} \frac{x^2 y}{(x^4 + y^2)\sqrt{x^2 + y^2}}.$$

This limit is not zero since if we choose $x = y = t, t > 0$, we have

$$\frac{x^2 y}{(x^4 + y^2)\sqrt{x^2 + y^2}} = \frac{t}{\sqrt{2} \cdot |t| \cdot (t^2 + 1)} = \frac{1}{\sqrt{2}(t^2 + 1)},$$

which converges to $1/\sqrt{2} \neq 0$, as $t \to 0, t > 0$. Hence, f is not differentiable at $(0,0)$.

(e) The partial derivative f_x is not continuous at $(0,0)$ since if we choose $x = y = t \neq 0$, we have

$$f_x(x,y) = f_x(t,t) = \frac{2t^2(t^2 - t^4)}{(t^4 + t^2)^2} = \frac{2(1 - t^2)}{(t^2 + 1)^2},$$

and so,

$$\lim_{t \to 0} f_x(t,t) = 2 \neq f_x(0,0) = 0.$$

Similarly f_y is not continuous at $(0,0)$.

5. (a)

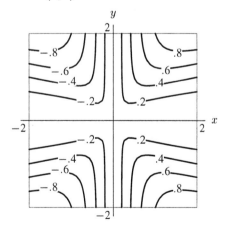

Figure 13.33

(b) f is differentiable at all $(x,y) \neq (0,0)$ as it is a rational function with nonvanishing denominator.

(c)

$$f_x(0,0) = \lim_{x \to 0} \frac{f(x,0) - f(0,0)}{x} = 0$$

$$f_y(0,0) = \lim_{y \to 0} \frac{f(0,y) - f(0,0)}{y} = 0$$

(d)   Let us use the definition. If f were differentiable, the linear approximation of f would be $L(x, y) = mx + ny$, where $m = f_x(0,0) = 0$ and $n = f_y(0,0) = 0$. So let's compute

$$\lim_{(x,y)\to(0,0)} \frac{f(x,y) - L(x,y)}{\sqrt{x^2 + y^2}} = \lim_{(x,y)\to(0,0)} \frac{xy^2}{(x^2 + y^2)^{3/2}}.$$

This limit is not zero as, for $x = y = t \to 0, t > 0$,

$$\frac{xy^2}{(x^2 + y^2)^{3/2}} = \frac{t^3}{2\sqrt{2}|t|^3} \xrightarrow[\substack{t \to 0 \\ t > 0}]{} \frac{1}{2\sqrt{2}}.$$

Hence f is not differentiable at $(0,0)$.

(e)

$$g(t) = f(x(t), y(t)) = \frac{ab^2t^3}{(a^2 + b^2)t^2} = \frac{ab^2}{a^2 + b^2}t$$

So

$$g'(0) = \frac{ab^2}{a^2 + b^2}.$$

(f) $f_x(0,0) \cdot x'(0) + f_y(0,0) \cdot y'(0) = 0$, as $f_x(0,0) = f_y(0,0) = 0$. Suppose the chain rule holds, then

$$g'(t) = f_x(x(t), y(t)) \cdot x'(t) + f_x(x(t), y(t))y'(t).$$

But $g'(0) = \frac{ab^2}{a^2+b^2}$ from part (e) and $g'(0) \neq 0$ since $a \neq 0$ and $b \neq 0$. Hence the chain rule doesn't hold. This happens because f was not differentiable at $(0,0)$.

(g)   If $\vec{u} = a\vec{i} + b\vec{j}$, then $a^2 + b^2 = 1$ as $\vec{u}$ is a unit vector. Thus,

$$f_{\vec{u}}(0,0) = \lim_{t\to 0} \frac{f(at, bt)}{t} = \lim_{t\to 0} \frac{g(t)}{t} = g'(0) = \frac{a^2b}{a^2 + b^2} = a^2b.$$

6.   (a)   The contour diagram of $f(x, y) = xy^2/(x^2 + y^4)$ is shown in Figure 13.34.

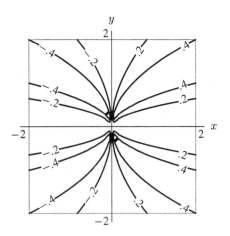

Figure 13.34

(b) Let $\vec{u} = a\vec{i} + b\vec{j}$ be the unit vector. Then

$$f_{\vec{u}}(0,0) = \lim_{t \to 0} \frac{f(at, bt)}{t} = \lim_{t \to 0} \frac{ab^2 t^3}{t(a^2 t^2 + b^4 t^4)} = \frac{ab^2}{a^2} = \frac{b^2}{a} \quad \text{if} \quad a \neq 0$$

and

$$f_{\vec{u}}(0,0) = 0 \qquad \text{if} \quad a = 0.$$

(c) f is not continuous at $(0,0)$. To see this let $x = t^2, y = t, t \to 0, t \neq 0$. Then

$$f(x,y) = \frac{t^4}{t^4 + t^4} = \frac{1}{2} \xrightarrow{t \to 0} \frac{1}{2} \neq f(0,0) = 0.$$

So f is not differentiable at $(0,0)$ either.

7. (a) The contour diagram of $f(x,y) = \sqrt{|xy|}$ is shown in Figure 13.35.

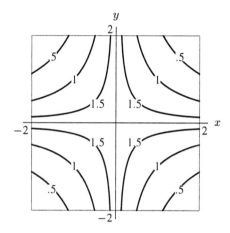

Figure 13.35

(b) The graph of $f(x,y) = \sqrt{|xy|}$ is shown in Figure 13.36.

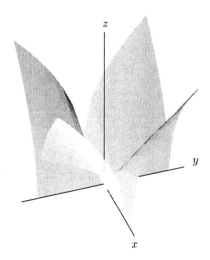

Figure 13.36

(c) f is clearly differentiable at (x,y) where $x \neq 0$ and $y \neq 0$. So we need to look at points $(x_0,0)$, $x_0 \neq 0$ and $(0,y_0)$, $y_0 \neq 0$. At $(x_0,0)$:

$$f_x(x_0,0) = \lim_{x \to x_0} \frac{f(x,0) - f(x_0,0)}{x - x_0} = 0$$

$$f_y(x_0,0) = \lim_{y \to 0} \frac{f(x_0,y) - f(x_0,0)}{y} = \lim_{y \to 0} \frac{\sqrt{|x_0 y|}}{y}$$

which doesn't exist. So f is not differentiable at the points $(x_0,0)$, $x_0 \neq 0$. Similarly, f is not differentiable at the points $(0,y_0)$, $y_0 \neq 0$.

(d)

$$f_x(0,0) = \lim_{x \to 0} \frac{f(x,0) - f(0,0)}{x} = 0$$

$$f_y(0,0) = \lim_{y \to 0} \frac{f(0,y) - f(0,0)}{y} = 0$$

(e) Let $\vec{u} = (\vec{i} + \vec{j})/\sqrt{2}$:

$$f_{\vec{u}}(0,0) = \lim_{t \to 0^+} \frac{f(\frac{t}{\sqrt{2}}, \frac{t}{\sqrt{2}}) - f(0,0)}{t} = \lim_{t \to 0^+} \frac{\sqrt{\frac{t^2}{2}}}{t} = \frac{1}{\sqrt{2}}.$$

We know that $\nabla f(0,0) = \vec{0}$ because both partial derivatives are 0. But if f were differentiable, $f_{\vec{u}}(0,0) = \nabla f(0,0) \cdot \vec{u} = f_x(0,0) \cdot \frac{1}{\sqrt{2}} + f_y(0,0) \cdot \frac{1}{\sqrt{2}} = 0$. But since, in fact, $f_{\vec{u}}(0,0) = 1/\sqrt{2}$, we conclude that f is not differentiable.

8. The fact that f is differentiable says that its graph is well approximated by a plane near (a,b). Since a plane is a graph of continuous function, it is reasonable to expect f to be continuous too. To prove this, we have to show that

$$\lim_{(x,y) \to (a,b)} f(x,y) = f(a,b) \quad \text{or equivalently} \quad \lim_{\substack{h \to 0 \\ k \to 0}} f(a+h, b+k) = f(a,b).$$

Suppose f is differentiable at (a,b). Then there is a linear function

$$L(x,y) = f(a,b) + m(x-a) + n(y-b)$$

such that

$$\lim_{\substack{h \to 0 \\ k \to 0}} \frac{f(a+h,b+k) - L(a+h,b+k)}{\sqrt{h^2 + k^2}} = 0.$$

If this limit is 0, then we also have

$$\lim_{\substack{h \to 0 \\ k \to 0}} f(a+h,b+k) - L(a+h,b+k) = 0.$$

Substituting for $L(a+h,b+k)$, this gives

$$\lim_{\substack{h \to 0 \\ k \to 0}} (f(a+h,b+k) - f(a,b) - mh - nk) = 0.$$

So,

$$\lim_{\substack{h \to 0 \\ k \to 0}} f(a+h,b+k) - f(a,b) = 0.$$

Therefore f is continuous at (a,b).

9. (a) If f were differentiable at $(0,0)$, then

$$f_{\vec{u}}(0,0) = \operatorname{grad} f(0,0) \cdot \vec{u} = f_x(0,0) \cdot \frac{1}{\sqrt{2}} + f_y(0,0) \cdot \frac{1}{\sqrt{2}} = 0$$

which contradict the information that $f_{\vec{u}}(0,0) = 3$.

(b) Let

$$f(x,y) = \begin{cases} \dfrac{3}{\sqrt{2}} \left(\dfrac{x^2}{y} + \dfrac{y^2}{x} \right), & x \neq 0 \text{ and } y \neq 0, \\ 0, & x = 0 \text{ or } y = 0. \end{cases}$$

Then $f_x(0,0) = 0$, $f_y(0,0) = 0$:

$$f_{\vec{u}}(0,0) = \lim_{t \to 0} \frac{f\left(\frac{t}{\sqrt{2}}, \frac{t}{\sqrt{2}}\right) - 0}{t} = \lim_{t \to 0} \frac{3}{\sqrt{2}t} \left(\frac{t^2}{2} \cdot \frac{\sqrt{2}}{t} + \frac{t^2}{2} \cdot \frac{\sqrt{2}}{t} \right) = 3.$$

10. (a) Differentiating gives

$$f_x(x,y) = \frac{(3x^2y^3 - y^3)(x^2+y^2) - 2x(x^3y - xy^3)}{(x^2+y^2)^2} = \frac{x^4y + 4x^2y^3 - y^5}{(x^2+y^2)^2}$$

similarly,

$$f_y(x,y) = \frac{x^5 - 4x^3y^2 - xy^4}{(x^2+y^2)^2} \qquad \text{for} \quad (x,y) \neq (0,0).$$

(b) We find the partial derivatives at the origin by using the limit definition:

$$f_x(0,0) = \lim_{x \to 0} \frac{f(x,0) - f(0,0)}{x} = 0$$

$$f_y(0,0) = \lim_{y \to 0} \frac{f(0,y) - f(0,0)}{y} = 0$$

(c) Let's compute:

$$\lim_{(x,y) \to (0,0)} f_x(x,y) = \lim_{(x,y) \to (0,0)} \frac{x^4y + 4x^2y^3 - y^5}{(x^2+y^2)^2}.$$

Let's switch to polar coordinates: $x = r\cos\theta$, $y = r\sin\theta$. Then $(x,y) \to (0,0)$ is equivalent to $r \to 0$. Therefore:

$$\lim_{(x,y) \to (0,0)} f_x(x,y) = \lim_{r \to 0} \frac{r^5(\cos^4\theta \sin\theta + 4\cos^4\sin^3\theta - \sin^5\theta)}{r^4} = 0 = f_x(0,0).$$

Similarly, $\lim_{(x,y) \to (0,0)} f_y(x,y) = 0 = f_y(0,0)$. So f_x and f_y are continuous.

(d) From part (c) f is differentiable at $(0,0)$.

Solutions for Chapter 13 Review

1. $\dfrac{\partial z}{\partial x} = \dfrac{\partial}{\partial x}\left[(x^2+x-y)^7\right] = 7(x^2+x-y)^6(2x+1) = (14x+7)(x^2+x-y)^6.$

$\dfrac{\partial z}{\partial y} = \dfrac{\partial}{\partial y}\left[(x^2+x-y)^7\right] = -7(x^2+x-y)^6.$

2. $\dfrac{\partial F}{\partial L} = \dfrac{\partial}{\partial L}\left[3\sqrt{LK}\right] = 3\dfrac{\partial}{\partial L}\left[(LK)^{1/2}\right] = \dfrac{3K}{2\sqrt{LK}} = \dfrac{3}{2}\sqrt{\dfrac{K}{L}}.$

3. $\dfrac{\partial f}{\partial p} = \dfrac{\partial}{\partial p}\left[e^{p/q}\right] = \dfrac{1}{q}e^{p/q}.$

   $\dfrac{\partial f}{\partial q} = \dfrac{\partial}{\partial q}\left[e^{p/q}\right] = \dfrac{\partial}{\partial q}\left[e^{pq^{-1}}\right] = (-pq^{-2})e^{pq^{-1}} = -\dfrac{p}{q^2}e^{p/q}.$

4. $\dfrac{\partial f}{\partial x} = \dfrac{\partial}{\partial x}\left[e^{xy}(\ln y)\right] = \dfrac{\partial e^{xy}}{\partial x}\ln y = ye^{xy}(\ln y).$

5.
$$\dfrac{\partial z}{\partial x} = 4x^3 - 7x^6y^3 + 5y^2, \quad \dfrac{\partial z}{\partial y} = -3x^7y^2 + 10xy.$$

6.
$$\dfrac{\partial z}{\partial \theta} = \dfrac{\sec^2\theta}{r}, \quad \dfrac{\partial z}{\partial r} = -\dfrac{\tan\theta}{r^2}.$$

7.
$$\dfrac{\partial w}{\partial s} = \ln(s+t) + \dfrac{s}{s+t}, \quad \dfrac{\partial w}{\partial t} = \dfrac{s}{s+t}.$$

8.
$$\dfrac{\partial w}{\partial u} = \dfrac{1}{1 + (ue^{-v})^2} \cdot (e^{-v}) = \dfrac{e^{-v}}{1 + u^2e^{-2v}},$$
$$\dfrac{\partial w}{\partial v} = \dfrac{1}{1 + (ue^{-v})^2} \cdot (-ue^{-v}) = \dfrac{-ue^{-v}}{1 + u^2e^{-2v}}.$$

9. Since $\|\vec{v}\| = 5$, we see that $\vec{v}$ is not a unit vector. The unit vector $\vec{u}$ in the direction of $\vec{v}$ is

$$\vec{u} = \dfrac{\vec{v}}{\|\vec{v}\|} = \dfrac{4}{5}\vec{i} - \dfrac{3}{5}\vec{j}.$$

The partial derivatives are $f_x(x, y) = 2xy$ and $f_y(x, y) = x^2$. So

$$f_{\vec{u}}(2, 6) = f_x(2, 6) \cdot \left(\dfrac{4}{5}\right) + f_y(2, 6) \cdot \left(-\dfrac{3}{5}\right) = 24\left(\dfrac{4}{5}\right) + 4\left(-\dfrac{3}{5}\right) = \dfrac{84}{5}.$$

10. True. It is the rate of change of f in the direction of $\vec{u}$ at the point (x_0, y_0).

11. False. $f_{\vec{u}}(a, b) = \|\nabla f(a, b)\|\cos\theta$, where θ is the angle between grad f and $\vec{u}$.

12. Must be true, because at any point grad f is perpendicular to level curves through that point.

13. True. Take the direction perpendicular to grad f at that point. If grad $f = 0$, any direction will do.

14. Is never true. If $\|\operatorname{grad} f\| = 0$, then grad $f = 0$, so grad $f \cdot \vec{u} = 0$ for any unit vector $\vec{u}$. Thus the directional derivative must be zero.

15. Is never true, because $f_{\vec{i}} = \|\operatorname{grad} f\|\cos\theta \leq \|\operatorname{grad} f\|$ at any point.

16. Might be true. For example for $f(x, y) = \sqrt{7}x - 3y$, we have grad $f = \sqrt{7}\vec{i} - 3\vec{j}$ then $\|\operatorname{grad} f\| = \sqrt{7 + 9} = 4$ and $f_{\vec{j}} = f_y = -3$.

17. (a) The difference quotient for evaluating $f_w(2,2)$ is

$$f_w(2,2) \approx \frac{f(2+0.01,2) - f(2,2)}{h} = \frac{e^{(2.01)\ln 2} - e^{2\ln 2}}{0.01} = \frac{e^{\ln(2^{2.01})} - e^{\ln(2^2)}}{0.01}$$

$$= \frac{2^{(2.01)} - 2^2}{0.01} \approx 2.78$$

The difference quotient for evaluating $f_z(2,2)$ is

$$f_z(2,2) \approx \frac{f(2,2+0.01) - f(2,2)}{h}$$

$$= \frac{e^{2\ln(2.01)} - e^{2\ln 2}}{0.01} = \frac{(2.01)^2 - 2^2}{0.01} = 4.01$$

(b) Using the derivative formulas we get

$$f_w = \frac{\partial f}{\partial w} = \ln z \cdot e^{w\ln z} = z^w \cdot \ln z$$

$$f_z = \frac{\partial f}{\partial z} = e^{w\ln z} \cdot \frac{w}{z} = w \cdot z^{w-1}$$

so

$$f_w(2,2) = 2^2 \cdot \ln 2 \approx 2.773$$
$$f_z(2,2) = 2 \cdot 2^{2-1} = 4.$$

18. (a) Estimating $T(x,t)$ from the Figure 13.10, at $x = 15, t = 20$ gives

$$\frac{\partial T}{\partial x}(15,20) \approx \frac{T(23,20) - T(15,20)}{23 - 15} = \frac{20 - 23}{8} = -\frac{3}{8}{}^{\circ}\text{C/m},$$

$$\frac{\partial T}{\partial t}(15,20) \approx \frac{T(15,25) - T(15,20)}{25 - 20} = \frac{25 - 23}{5} = \frac{2}{5}{}^{\circ}\text{C/min}.$$

At 15 m from heater at time $t = 20$ min, the room temperature decreases by approximately $3/8{}^{\circ}$C per meter and increases by approximately $2/5{}^{\circ}$C per minute.

(b) We have the estimates,

$$\frac{\partial T}{\partial x}(5,12) \approx \frac{T(7,12) - T(5,12)}{7 - 5} = \frac{25 - 26}{2} = -0.5{}^{\circ}\text{C/m},$$

$$\frac{\partial T}{\partial t}(5,12) \approx \frac{T(5,40) - T(5,12)}{40 - 12} = \frac{30 - 26}{28} = \frac{4}{28} = \frac{1}{7}{}^{\circ}\text{C/min}.$$

At $x = 5, t = 12$ the temperature decreases by approximately $1/2{}^{\circ}$C per meter and increases by approximately $1/7{}^{\circ}$C per minute.

19. (a) $f_t(\frac{\pi}{2}, \frac{\pi}{2})$ is negative since we move from zero contour of f to negative contours as t slightly increases and $x = \frac{\pi}{2}$.

$f_t(\frac{\pi}{2}, \pi)$ is positive since f increases with time t from $t = \pi$ and $x = \frac{\pi}{2}$.

When $f_t(\frac{\pi}{2}, b)$ is positive, the point on the string at $x = \frac{\pi}{2}$ is moving upwards at $t = b$. It moves downward for $f_t(\frac{\pi}{2}, b)$ negative at $t = b$.

(b) $f_t(\frac{\pi}{2}, t)$ is positive for $\pi < t < 2\pi$. For these t, f moves to larger and larger contours at $x = \frac{\pi}{2}$.

(c) For any fixed values of t between 0 and $\frac{\pi}{2}$, f increases with x for x between 0 and $\frac{3\pi}{2}$. Also for any fixed t between $\frac{3\pi}{2}$ and $\frac{5\pi}{2}$, f increases with x for x between 0 and $\frac{3\pi}{2}$. Hence, $f_x(x, t)$ is positive for $0 < x < 3\pi/2$ and t either satisfying $0 < t < \pi/2$ or $3\pi/2 < t < 5\pi/2$.

20. The partial derivative, $\partial Q/\partial b$ is the rate of change of the quantity of beef purchased with respect to the price of beef, when the price of chicken stays constant. If the price of beef increases and the price of chicken stays the same, we expect consumers to buy less beef and more chicken. Thus when b increases, we expect Q to decrease, so $\partial Q/\partial b < 0$.

   On the other hand, $\partial Q/\partial c$ is the rate of change of the quantity of beef purchased with respect to the price of chicken, when the price of beef stays constant. An increase in the price of chicken is likely to cause consumers to buy less chicken and more beef. Thus when c increases, we expect Q to increase, so $\partial Q/\partial c > 0$.

21. The derivative $\partial c/\partial x = b$ is the rate of change of the cost of producing one unit of the product with respect to the amount of labor used (in man hours) when the amount of raw material used stays the same. Thus $\partial c/\partial x = b$ represents the hourly wage.

22. The sign of $\partial f/\partial P_1$ tells you whether f (the number of people who ride the bus) increases or decreases when P_1 is increased. Since P_1 is the price of taking the bus, as it increases, f should decrease. This is because fewer people will be willing to pay the higher price, and more people will choose to ride the train.

   On the other hand, the sign of $\dfrac{\partial f}{\partial P_2}$ tells you the change in f as P_2 increases. Since P_2 is the cost of riding the train, as it increases, f should increase. This is because fewer people will be willing to pay the higher fares for the train, and more people will choose to ride the bus.

   Therefore, $\dfrac{\partial f}{\partial P_1} < 0$ and $\dfrac{\partial f}{\partial P_2} > 0$.

23. (a) $\dfrac{\partial g}{\partial m} = \dfrac{G}{r^2}$ and $\dfrac{\partial g}{\partial r} = -\dfrac{2Gm}{r^3}$

   (b) For constant r, the graph of g against m is a straight line through the origin with slope $\dfrac{\partial g}{\partial m} = \dfrac{G}{r^2}$. Thus g increases as m increases, while r is constant. See Figure 13.37.

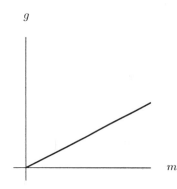

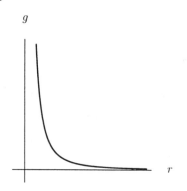

Figure 13.37 *Figure 13.38*

constant m, the graph of g against r has the shape shown in Figure 13.38, (the same shape as the graph of $y = \frac{1}{x^2}$). The slope of the graph in Figure 13.38 is $\dfrac{\partial g}{\partial r} = -\dfrac{2Gm}{r^3}$. So as r increases, g decreases, since the slope is negative.

24. (a) $\dfrac{\partial f}{\partial K} = 20K^{-\frac{2}{3}}L^{\frac{2}{3}}$ and $\dfrac{\partial f}{\partial L} = 40K^{\frac{1}{3}}L^{-\frac{1}{3}}$.

Hence, $\dfrac{\partial f}{\partial K} = \dfrac{\partial f}{\partial L}$ when $20K^{-\frac{2}{3}}L^{\frac{2}{3}} = 40K^{\frac{1}{3}}L^{\frac{-1}{3}}$ or $L = 2K$.

(b)   As before: $\dfrac{\partial f}{\partial K} = \dfrac{\partial f}{\partial L}$ so $acK^{a-1}L^b = bcK^aL^{b-1}$. Thus $aL = bK$.

25.   Average productivity increases as x_1 increases if $\frac{\partial}{\partial x_1}$ (average productivity) > 0. Now

$$\frac{\partial}{\partial x_1}(\text{average productivity}) = \frac{\partial}{\partial x_1}\left(\frac{P}{x_1}\right)$$

$$= \frac{1}{x_1}\frac{\partial P}{\partial x_1} + P\frac{\partial}{\partial x_1}\left(\frac{1}{x_1}\right)$$

$$= \frac{1}{x_1}\frac{\partial P}{\partial x_1} - \frac{P}{x_1^2}$$

$$= \frac{1}{x_1}\left(\frac{\partial P}{\partial x_1} - \frac{P}{x_1}\right)$$

So $\frac{\partial}{\partial x_1}$ (average productivity) > 0 means that $\left(\dfrac{\partial P}{\partial x_1} - \dfrac{P}{x_1}\right) > 0$, i.e.,

$$\frac{\partial P}{\partial x_1} > \frac{P}{x_1}.$$

26.   The differential is

$$dP = \frac{\partial P}{\partial L}\,dL + \frac{\partial P}{\partial K}\,dK = 10L^{-0.75}K^{0.75}\,dL + 30L^{0.25}K^{-0.25}dK.$$

When $L = 2$ and $K = 16$, this is

$$dP \approx 47.6\,dL + 17.8\,dK.$$

27. $\dfrac{\partial S}{\partial C} = \dfrac{1}{2}ab\cos C$, so with a, b held constant,

$$\Delta S \approx \frac{1}{2}ab(\cos C)\Delta C.$$

Now $\Delta C = \frac{\pi}{1080}$, so the error in S is

$$\Delta S \approx \frac{1}{2}ab(\cos C)\frac{\pi}{1080} = \frac{1}{2}ab(\sin C)\left(\frac{\cos C}{\sin C}\right)\frac{\pi}{1080} = \frac{\pi S}{1080\tan C}.$$

28.   (a)   When $V = 25$ and $P = 1$, we have $T = 305.7$. The differential dT is

$$dT = \frac{\partial T}{\partial V}\,dV + \frac{\partial T}{\partial P}\,dP = \left(-16.574\frac{1}{V^2} + 1.06\frac{1}{V^3} + 12.187P\right)dV + (0.3879 + 12.187V)\,dP.$$

When $V = 25$ and $P = 1$ this is

$$dT = 12.16\,dV + 305.06\,dP.$$

(b)   If $\Delta P = 0.1$ and $\Delta T = 0$, then

$$0 \approx (12.16)\Delta V + (305.06)(0.1),$$

so

$$\Delta V \approx -\frac{30.506}{12.16} \approx -2.5.$$

Thus the volume would have to decrease by about 2.5 dm^3, or about 10%.

29. $f_{\vec{u}} = \nabla f \cdot (\frac{1}{\sqrt{5}}\vec{i} + \frac{2}{\sqrt{5}}\vec{j})$, where $\vec{u}$ = unit vector $\frac{1}{\sqrt{5}}\vec{i} + \frac{2}{\sqrt{5}}\vec{j}$.

Now $\nabla f = \nabla(xe^y) = e^y\vec{i} + xe^y\vec{j}$.

Thus $\nabla f \cdot (\frac{1}{\sqrt{5}}\vec{i} + \frac{2}{\sqrt{5}}\vec{j}) = (e^y\vec{i} + xe^y\vec{j}) \cdot (\frac{1}{\sqrt{5}}\vec{i} + \frac{2}{\sqrt{5}}\vec{j}) = \frac{(e^y + 2xe^y)}{\sqrt{5}} = \frac{e^y}{\sqrt{5}}(2x+1)$.

At the point $(1,1)$, we have $f_{\vec{u}} = \frac{e^1}{\sqrt{5}}[2(1)+1] = \frac{3e}{\sqrt{5}}$.

30. $f_x = 2x$, $f_y = -2y$, so grad $f(3,-1) = 6\vec{i} + 2\vec{j}$. For the direction $\theta = \pi/4$, the direction is $\vec{u} = \frac{1}{\sqrt{2}}\vec{i} + \frac{1}{\sqrt{2}}\vec{j}$,

so $f_{\vec{u}}(3,-1) = (6\vec{i} + 2\vec{j}) \cdot (\frac{1}{\sqrt{2}}\vec{i} + \frac{1}{\sqrt{2}}\vec{j}) = \frac{8}{\sqrt{2}} = 4\sqrt{2}$.

The directional derivative is largest in the direction of the gradient vector grad $f(3,-1) = 6\vec{i} + 2\vec{j}$.

31. One way to do this is to estimate the gradient vector and then find grad $f(x,y) \cdot \vec{u}$. This is a useful approach since it is easier to estimate grad f than to estimate $f_{\vec{u}}$ directly. Since grad $f(x,y) = (f_x(x,y), f_y(x,y))$ we can simply estimate the x- and y-derivatives of f at $(3,1)$ to find grad f at that point. In the x-direction we see that the function is increasing. This implies that the x-derivative is positive. To estimate its value we estimate the slope in the x-direction. Applying the same reasoning to find f_y, we get:

$$f_x(3,1) \approx \frac{f(3+1,1) - f(3,1)}{1} = 2 - 1 = 1,$$

$$f_y(3,1) \approx \frac{f(3,1) - f(3,1-0.6)}{0.6} \approx \frac{1-2}{0.6} \approx -1.67.$$

This gives us our estimated value for grad $f(3,1) \approx \vec{i} - 1.67\vec{j}$. Now, if $\vec{u} = (-2\vec{i} + \vec{j})/\sqrt{5}$,

$$f_{\vec{u}}(3,1) = \text{grad } f(3,1) \cdot (-2\vec{i} + \vec{j})/\sqrt{5}$$
$$\approx (\vec{i} - 1.67\vec{j}) \cdot (-2\vec{i} + \vec{j})/\sqrt{5}$$
$$\approx -1.64$$

32. Looking at the contour diagram, we see that the contours are almost straight lines that are reasonably evenly spaced, so treating the function as linear is not a bad approximation. The monthly payment m is a function of two variables, P, the dollars you borrow, and r, the interest rate. We are going to use the formula

$$m(P,r) \approx m(P_0, r_0) + m_p(P_0, r_0)(P - P_0) + m_r(P_0, r_0)(r - r_0)$$

to approximate the payment $m(P,r)$. The amount of $6000 and the interest rate 11% would probably be reasonable choices for our P_0 and r_0.

Directly from the Figure 10 on page 31, we have

$$m(6000, 0.11) = \$130$$

Next we approximate $m_p(6000, 0.11)$ by a difference quotient.

$$m_p(6000, 0.11) \approx \frac{m(6000 + h, 0.11) - m(6000, 0.11)}{h}$$

Here we choose $h = 500$, and we get $m(6500, 0.11) = 140$ from the figure, so

$$m_p(6000, 0.11) \approx \frac{140 - 130}{500} = 0.02$$

Next we approximate $m_r(6000, 0.11)$ by taking $h = 0.03$ and $m(6000, 0.11 + 0.03) = m(6000, 0.14) = 140$ from the figure,

$$m_r(6000, 0.11) \approx \frac{m(6000, 0.11 + h) - m(6000, 0.11)}{h} = \frac{140 - 130}{0.03} \approx 333.33$$

Thus we have:

$$m(P, r) \approx 130 + 0.02(P - 6000) + 333.33(r - 0.11)$$
$$= -26.67 + 0.02P + 333.33r \quad \text{in dollars}$$

The constants in the answer tell us several useful things. The slope of 0.02 along the P axis tells us how much our monthly payment will increase if we decide to borrow more money; we can expect to pay about 2 cents on each extra dollar every month. The slope in the r direction tells us how much our payment will change if the interest rate changes. We can expect to pay an extra \$333.33 each month for each point the interest rate goes up. The constant c is a gauge of the non-linearity of this function. We know that if we borrow no money at zero percent interest, we should expect to not have a monthly payment ($m = 0$). The constant we calculated is negative which implies that the bank will pay us \$26.67 each month if we do not borrow any money! So, we should bear in mind that the function we have calculated is only useful close to the point at which we made the approximation, in this case $P_0 = 6000$ and $r_0 = 11\%$.

33. In other words, find points on the level surface $F(x, y, z) = x^2 + y^2 + z^2 = 8$ (which happens to describe a sphere of radius $2\sqrt{2}$ centered on the origin) where the normal to the sphere is parallel to $\vec{i} - \vec{j} + 3\vec{k}$, which is the normal vector of the plane $x - y + 3z = 0$. The normal to the surface is given by $\nabla F = 2x\vec{i} + 2y\vec{j} + 2z\vec{k}$. To find points (x, y, z) where the vector $2x\vec{i} + 2y\vec{j} + 2z\vec{k}$ is parallel to (i.e. a scalar multiple of) $\vec{i} - \vec{j} + 3\vec{k}$, solve $2x\vec{i} + 2y\vec{j} + 2z\vec{k} = \lambda(\vec{i} - \vec{j} + 3\vec{k})$ for λ. This equation implies that $x = \lambda/2$, $y = -\lambda/2$, and $z = 3\lambda/2$. We have

$$x^2 + y^2 + z^2 = 8$$
$$\left(\frac{\lambda}{2}\right)^2 + \left(\frac{-\lambda}{2}\right)^2 + \left(\frac{3\lambda}{2}\right)^2 = 8$$
$$\frac{\lambda^2 + \lambda^2 + 9\lambda^2}{4} = 8$$
$$11\lambda^2 = 32$$
$$\lambda = \pm\sqrt{\frac{32}{11}} = \pm\frac{4\sqrt{2}}{\sqrt{11}}$$

This gives us

$$(x, y, z) = \pm 4\sqrt{\frac{2}{11}}\left(\frac{1}{2}, -\frac{1}{2}, \frac{3}{2}\right)$$

34. The temperature increases fastest in the direction of the gradient vector, namely $\text{grad } T = -2x\vec{i} - 2y\vec{j}$. Figure 13.39 shows that the gradient vector points radially inward.

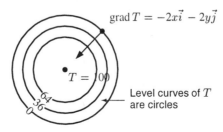

Figure 13.39

35. (a) Fix $y = 3$. When x changes from 2.00 to 2.01, $f(x, 3)$ decreases from 7.56 to 7.42. So

$$\left.\frac{\partial f}{\partial x}\right|_{(2,3)} \approx \left.\frac{\Delta f}{\Delta x}\right|_{(2,3)} = \frac{7.42 - 7.56}{2.01 - 2.00} = \frac{-0.14}{0.01} = -14.$$

Fix $x = 2$, when y changes from 3.00 to 3.02, $f(2, y)$ increases from 7.56 to 7.61. So

$$\left.\frac{\partial f}{\partial y}\right|_{(2,3)} \approx \left.\frac{\Delta f}{\Delta y}\right|_{(2,3)} = \frac{7.61 - 7.56}{3.02 - 3.00} = \frac{0.05}{0.02} = 2.5.$$

(b) Since the unit vector $\vec{u}$ of the direction $\vec{i} + 3\vec{j}$ is

$$\vec{u} = \frac{\vec{i} + 3\vec{j}}{\|\vec{i} + 3\vec{j}\|} = \frac{1}{\sqrt{10}}\vec{i} + \frac{3}{\sqrt{10}}\vec{j},$$

$$f_{\vec{u}}(2,3) = \text{grad } f(2,3) \cdot \vec{u} \approx \left(\left.\frac{\Delta f}{\Delta x}\right|_{(2,3)}\vec{i} + \left.\frac{\Delta f}{\Delta y}\right|_{(2,3)}\vec{j}\right) \cdot \vec{u}$$

$$= (-14\vec{i} + 2.5\vec{j}) \cdot \left(\frac{1}{\sqrt{10}}\vec{i} + \frac{3}{\sqrt{10}}\vec{j}\right) = -\frac{6.5}{\sqrt{10}} \approx -2.055.$$

(c) Maximum rate equals $\|\text{grad } f\| \approx \sqrt{(-14)^2 + (2.5)^2} \approx 14.221$ in the direction of the gradient which is approximately equal to $-14\vec{i} + 2.5\vec{j}$.

(d) The equation of the level curve is

$$f(x, y) = f(2, 3) = 7.56.$$

(e) The vector must be perpendicular to grad f, so $\vec{v} = 2.5\vec{i} + 14\vec{j}$ is a possible answer. (There are many others.).

(f) The differential at the point $(2, 3)$ is

$$df = -14\,dx + 2.5\,dy.$$

If $dx = 0.03$, $dy = 0.04$, we get

$$df = -14(0.03) + 2.5(0.04) = -0.32.$$

The df approximates the change in f when (x, y) changes from $(2, 3)$ to $(2.03, 3.04)$.

36. (a) The equation of the level curve is

$$g(x, y) = \text{constant} = g(1, 3)$$

so

$$g(x, y) = 4.$$

(b) The point has coordinates $(1, 3, 4)$.

(c) The surface is the level surface of the function

$$G(x, y, z) = g(x, y) - z = 0.$$

The normal to this surface at the point $(1, 3, 4)$ is

$$\text{grad } G(1, 3, 4) = g_x(1, 3)\vec{i} + g_y(1, 3)\vec{j} - \vec{k} = -\vec{i} + 2\vec{j} - \vec{k}.$$

Thus the equation of the tangent plane is

$$-x + 2y - z = -(1) + 2(3) - 4 = 1$$
$$-x + 2y - z = 1.$$

37. (a) $\dfrac{\partial w}{\partial u} = \dfrac{\partial w}{\partial x} \cdot \dfrac{\partial x}{\partial u} + \dfrac{\partial w}{\partial y} \cdot \dfrac{\partial y}{\partial u} + \dfrac{\partial w}{\partial z} \cdot \dfrac{\partial z}{\partial u} = 3y \cdot \dfrac{1}{u} + (3x + z) \cdot \sin v + y \cdot v.$

At $(u, v) = (1, \pi)$ we have $y = 1, x = -1, z = \pi$.

Thus, $\dfrac{\partial w}{\partial u}\bigg|_{(1,\pi)} = 3 + (-3 + \pi) \cdot 0 + \pi = 3 + \pi.$

$\dfrac{\partial w}{\partial v} = \dfrac{\partial w}{\partial x} \cdot \dfrac{\partial x}{\partial v} + \dfrac{\partial w}{\partial y} \cdot \dfrac{\partial y}{\partial v} + \dfrac{\partial w}{\partial z} \cdot \dfrac{\partial z}{\partial v} = 3y(- \sin v) + (3x + z) \cdot u \cos v + yu.$

Thus, $\dfrac{\partial w}{\partial v}\bigg|_{(1,\pi)} = 3(1)(0) + (-3 + \pi)1(-1) + 1(1) = 4 - \pi.$

(b) $\dfrac{dw}{dt} = \dfrac{\partial w}{\partial u} \cdot \dfrac{du}{dt} + \dfrac{\partial w}{\partial v} \cdot \dfrac{dv}{dt}.$

$\dfrac{du}{dt} = \pi \cos \pi t$, so $\dfrac{du}{dt}\bigg|_{t=1} = -\pi$ and $\dfrac{dv}{dt} = 2\pi t$, so $\dfrac{dv}{dt}\bigg|_{t=1} = 2\pi.$

Thus $\dfrac{dw}{dt}\bigg|_{t=1} = (3 + \pi)(-\pi) + (4 - \pi)2\pi = 5\pi - 3\pi^2.$

38. The total population is given by $N = \text{Density} \cdot \text{Area} = \rho \pi r^2$ and the density in 1997 was

$$\rho = \frac{3{,}000{,}000}{\pi\, 25^2} = \frac{4800}{\pi}.$$

So the rate of growth for the total population is

$$\frac{dN}{dt} = \frac{\partial N}{\partial \rho}\frac{d\rho}{dt} + \frac{\partial N}{\partial r}\frac{dr}{dt}$$
$$= \pi r^2 \frac{d\rho}{dt} + 2\rho \pi r \frac{dr}{dt}$$
$$= \pi(25^2)(200) + \frac{4800}{\pi}\pi(2)(25)(0.1)$$
$$= 24000 + 125000\,\pi$$
$$= 416{,}699 \text{ people/year.}$$

39.

$$\frac{\partial V}{\partial x} = xF'(2x + y)\frac{\partial}{\partial x}(2x + y) + 1 \cdot F(2x + y)$$
$$= 2xF'(2x + y) + F(2x + y)$$
$$\frac{\partial V}{\partial y} = xF'(2x + y)\frac{\partial}{\partial y}(2x + y) = xF'(2x + y)$$
$$x\frac{\partial V}{\partial x} - 2x\frac{\partial V}{\partial y} = 2x^2 F'(2x + y) + xF(2x + y) - 2x^2 F'(2x + y)$$
$$= xF(2x + y) = V$$

40. We know from Problem 39 that $V(x, y) = xF(2x + y)$ will solve the equation for any function F, so it is enough to find an F such that $xF(2x + y) = y^2$ when $x = 1$.

In other words, $1 \cdot F(2 + y) = y^2$, i.e. $F(y) = (y - 2)^2$ for all y. So one solution is $V(x, y) = xF(2x + y) = x(2x + y - 2)^2 = 4x^3 + xy^2 + 4x + 4x^2y - 8x^2 - 4xy.$

41. Let us first collect the computations that we will need.

$$f(x, y) = \cos{(x + 2y)} \sin{(x - y)},$$
$$f_x(x, y) = \cos{(x + 2y)} \cos{(x - y)} - \sin{(x - y)} \sin{(x + 2y)}$$
$$= \cos{(x + 2y + x - y)} = \cos{(2x + y)},$$
$$f_y(x, y) = -\cos{(x + 2y)} \cos{(x - y)} - 2 \sin{(x - y)} \sin{(x + 2y)}$$
$$= -\cos{(x + 2y - (x - y))} - \sin{(x + 2y)} \sin{(x - y)}$$
$$= -\cos{(3y)} + \frac{1}{2} \left[\cos{(2x + y)} - \cos{(3y)} \right]$$
$$= \frac{1}{2} \cos{(2x + y)} - \frac{3}{2} \cos{(3y)},$$
$$f_{xx}(x, y) = -2 \sin{(2x + y)},$$
$$f_{xy}(x, y) = -\sin{(2x + y)},$$
$$f_{yy}(x, y) = -\frac{1}{2} \sin{(2x + y)} + \frac{9}{2} \sin{(3y)}.$$

Then

$$f(0, 0) = 0,$$
$$f_x(0, 0) = 1,$$
$$f_y(0, 0) = -1,$$
$$f_{xx}(0, 0) = 0,$$
$$f_{xy}(0, 0) = 0,$$
$$f_{yy}(0, 0) = 0.$$

Hence the quadratic Taylor polynomial $P(x, y)$ of $f(x, y)$ at $(0, 0)$ is

$$P(x, y) = f(0, 0) + f_x(0, 0)x + f_y(0, 0)y$$
$$+ \frac{1}{2} f_{xx}(0, 0)x^2 + f_{xy}(0, 0)xy + \frac{1}{2} f_{yy}(0, 0)y^2$$
$$= x - y.$$

42. (a) The first-order Taylor polynomial of a function f about a point (a, b) is equal to

$$f(a, b) + f_x(a, b)(x - a) + f_y(a, b)(y - b).$$

Computing the partial derivatives, we get:

$$f_x = 2(x - 1)e^{(x-1)^2 + (y-3)^2}$$
$$f_y = 2(y - 3)e^{(x-1)^2 + (y-3)^2}$$
$$f_x(0, 0) = 2(-1)e^{(-1)^2 + (-3)^2}$$
$$= -2e^{10}$$
$$f_y(0, 0) = 2(-3)e^{(-1)^2 + (-3)^2}$$
$$= -6e^{10}$$

Thus,

$$f(x, y) \approx e^{10} - 2e^{10}x - 6e^{10}y$$

(b) The second-order Taylor polynomial of a function f about the point $(1, 3)$ is given by

$$f(1, 3) + f_x(1, 3)(x - 1) + f_y(1, 3)(y - 3)$$
$$+ \frac{1}{2}f_{xx}(1, 3)(x - 1)^2 + f_{xy}(1, 3)(x - 1)(y - 3) + \frac{1}{2}f_{yy}(1, 3)(y - 3)^2.$$

Computing the partial derivatives, we get:

$$f_x = 2(x - 1)e^{(x-1)^2 + (y-3)^2}$$
$$f_y = 2(y - 3)e^{(x-1)^2 + (y-3)^2}$$
$$f_{xx} = (4(x - 1)^2 + 2)e^{(x-1)^2 + (y-3)^2}$$
$$f_{xy} = 4(x - 1)(y - 3)e^{(x-1)^2 + (y-3)^2}$$
$$f_{yy} = (4(y - 3)^2 + 2)e^{(x-1)^2 + (y-3)^2}$$

Substituting in the point $(1, 3)$ to these partial derivatives, we get:

$$f_x(1, 3) = 0$$
$$f_y(1, 3) = 0$$
$$f_{xy}(1, 3) = 0$$
$$f_{xx}(1, 3) = (4(0)^2 + 2)e^{0^2 + 0^2} = 2$$
$$f_{yy}(1, 3) = (4(0)^2 + 2)e^{0^2 + 0^2} = 2$$

Thus,

$$f(x, y) \approx e^0 + 0(x - 1) + 0(y - 3) + \frac{2}{2}(x - 2)^2 + 0(x - 1)(y - 3) + \frac{2}{2}(y - 3)^2$$

$$f(x, y) \approx 1 + (x - 1)^2 + (y - 3)^2.$$

(c) A vector perpendicular to the level curve is grad f. At the point $(0, 0)$, we have

$$\text{grad } f = f_x(0, 0)\vec{i} + f_y(0, 0)\vec{j}$$

Computing partial derivatives, we have

$$f_x = 2(x - 1)e^{(x-1)^2 + (y-3)^2}$$
$$f_y = 2(y - 3)e^{(x-1)^2 + (y-3)^2}$$
$$f_x(0, 0) = 2(-1)e^{(-1)^2 + (-3)^2}$$
$$= -2e^{10}$$
$$f_y(0, 0) = 2(-3)e^{(-1)^2 + (-3)^2}$$
$$= -6e^{10}$$

Therefore, a perpendicular vector is grad $f = -2e^{10}\vec{i} - 6e^{10}\vec{j}$. Any multiple of grad f, say $-2\vec{i} - 6\vec{j}$, will do.

(d)   Since the surface can be represented by the level surface

$$F(x, y, z) = f(x, y) - z = 0,$$

a vector perpendicular to the surface at $(0, 0)$ is given by

$$\text{grad } F = f_x(0,0)\vec{i} + f_y(0,0)\vec{j} - \vec{k} = -2e^{10}\vec{i} - 6e^{10}\vec{j} - \vec{k}$$

43.   (a)   Calculating the necessary partial derivatives:

$$T(x, y, z, t) = \frac{1}{(4\pi Kt)^{3/2}} e^{-(x^2+y^2+z^2)/4Kt}$$

$$\frac{\partial T}{\partial t} = \frac{1}{(4\pi Kt)^{3/2}} e^{-(x^2+y^2+z^2)/4Kt} \left[\frac{x^2+y^2+z^2}{4Kt^2} - \frac{3}{2}\frac{1}{t} \right]$$

$$\frac{\partial^2 T}{\partial x^2} = \frac{1}{(4\pi Kt)^{3/2}} e^{-(x^2+y^2+z^2)/4Kt} \left[\frac{x^2}{4K^2t^2} - \frac{1}{2Kt} \right]$$

Thus,

$$K(T_{xx} + T_{yy} + T_{zz}) = \frac{1}{(4\pi Kt)^{3/2}} e^{-(x^2+y^2+z^2)/4Kt} \left[\frac{x^2+y^2+z^2}{4Kt^2} - \frac{3}{2t} \right] = T_t$$

(b)   For $t = $ const., the level surfaces are given by

$$x^2 + y^2 + z^2 = c^2, \qquad c \quad \text{constant}$$

which is an equation of a sphere centered at origin.

(c)   With $t = $ constant, $\text{grad } T = T_x\vec{i} + T_y\vec{j} + T_z\vec{k}$. Since

$$\frac{\partial T}{\partial x} = -\frac{2x}{4Kt}T = -\frac{x}{2Kt}T$$

and similarly for T_y and T_z. Thus we have

$$\text{grad } T(x, y, z) = -(x\vec{i} + y\vec{j} + z\vec{k})\frac{1}{2Kt}\frac{1}{(4\pi Kt)^{3/2}} e^{-(x^2+y^2+z^2)/4Kt}$$

$$= (-\vec{r})\frac{1}{2Kt}\frac{1}{(4\pi Kt)^{3/2}} e^{-r^2/4Kt}, \quad \text{where} \quad \vec{r} = x\vec{i} + y\vec{j} + z\vec{k}$$

The heat flows toward lower temperatures, that is in the direction of $-\text{grad } T$. Since $\text{grad } T$ is in the direction of $-\vec{r}$, heat is flowing outward and with exponentially decreasing magnitude.

44.   (a)   To do this you need to imagine the surfaces with the normal vector $\vec{n}$ at P.

For (I),	For (II),	For (III),	For (IV),
$\vec{n}$ is $(-,-,+)$	$\vec{n}$ is $(+,+,+)$	$\vec{n}$ is $(+,-,+)$	$\vec{n}$ is $(-,+,+)$
or $(+,+,-)$	or $(-,-,-)$	or $(-,+,-)$	or $(+,-,-)$
so (E)	so (F)	so (G)	so (H)

(b)   Since the equation of the tangent plane is of the form

$$n_1 x + n_2 y + n_3 z = k,$$

the coefficients of x, y, z in the plane must have the same sign as the components of the normal. Hence (I)-(E)-(L); (II)-(F)-(J); (III)-(G)-(M); (IV)-(H)-(K).

CHAPTER FOURTEEN

Solutions for Section 14.1

1. The point A is not a critical point and the contour lines look like parallel lines. The point B is a critical point and is a local maximum; the point C is a saddle point.

2. (a) P is a local maximum.
    (b) Q is a saddle point.
    (c) R is a local minimum.
    (d) S is none of these.

3. Figure 14.1 shows the gradient vectors around P and Q pointing perpendicular to the contours and in the direction of increasing values of the function.

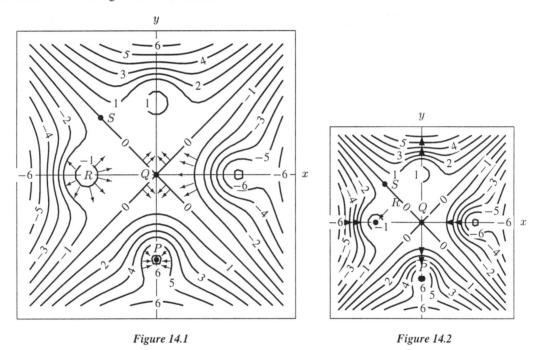

Figure 14.1   Figure 14.2

4. Figure 14.2 shows the direction of ∇f at the points where $\|\nabla f\|$ is largest, since at those points the contours are closest together.

5. The partial derivatives are $f_x(x, y) = 3x^2 - 3$ which vanishes for $x = \pm 1$ and $f_y(x, y) = 3y^2 - 3$ which vanishes for $y = \pm 1$. The points $(1, 1), (1, -1), (-1, 1), (-1, -1)$ where both partials vanish are the critical points. To determine the nature of these critical points we calculate their discriminant and use the second derivative test. The discriminant is

$$D = f_{xx}(x, y)f_{yy}(x, y) - f_{xy}^2(x, y) = (6x)(6y) - 0 = 36xy.$$

At $(1, 1)$ and $(-1, -1)$ the discriminant is positive. Since $f_{xx}(1, 1) = 6$ is positive, $(1, 1)$ is a local minimum. And since $f_{xx}(-1, -1) = -6$ is negative, $(-1, -1)$ is a local maximum. The remaining two points, $(1, -1)$ and $(-1, 1)$ are saddle points since the discriminant is negative.

6. First, we identify the critical points. The partials are $f_x(x, y) = 3x^2$ and $f_y(x, y) = -2ye^{-y^2}$. These will vanish simultaneously when $x = 0$ and $y = 0$, so our only critical point is $(0, 0)$. The discriminant is

$$D = f_{xx}(x, y)f_{yy}(x, y) - f_{xy}^2(x, y) = (6x)(4y^2e^{-y^2} - 2e^{-y^2}) - 0 = 12xe^{-y^2}(2y^2 - 1).$$

Unfortunately, the discriminant is zero at the origin so the second derivative test can tell us nothing about our critical point. We can, however, see that we are at a saddle point by looking at the behavior of $f(x, y)$ along the line $y = 0$. Here we have $f(x, 0) = x^3 + 1$, so for positive x, we have $f(x, 0) > 1 = f(0, 0)$ and for negative x, we have $f(x, 0) < 1 = f(0, 0)$. So $f(x, y)$ has neither a maximum nor a minimum at $(0, 0)$.

7. Find the critical point(s) by setting

$$f_x = (xy + 1) + (x + y) \cdot y = y^2 + 2xy + 1 = 0,$$
$$f_y = (xy + 1) + (x + y) \cdot x = x^2 + 2xy + 1 = 0,$$

then we get $x^2 = y^2$, and so $x = y$ or $x = -y$.

If $x = y$, then $x^2 + 2x^2 + 1 = 0$, that is, $3x^2 = -1$, and there is no real solution. If $x = -y$, then $x^2 - 2x^2 + 1 = 0$, which gives $x^2 = 1$. Solving it we get $x = 1$ or $x = -1$, then $y = -1$ or $y = 1$, respectively. Hence, $(1, -1)$ and $(-1, 1)$ are critical points.
Since

$$f_{xx}(x, y) = 2y,$$
$$f_{xy}(x, y) = 2y + 2x \quad \text{and}$$
$$f_{yy}(x, y) = 2x,$$

the discriminant is

$$D(x, y) = f_{xx}f_{yy} - f_{xy}^2$$
$$= 2y \cdot 2x - (2y + 2x)^2$$
$$= -4(x^2 + xy + y^2).$$

thus

$$D(1, -1) = -4(1^2 + 1 \cdot (-1) + (-1)^2) = -4 < 0,$$
$$D(-1, 1) = -4((-1)^2 + (-1) \cdot 1 + 1^2) = -4 < 0.$$

Therefore $(1, -1)$ and $(-1, 1)$ are saddle points.

8. At a critical point, $f_x = 0$, $f_y = 0$.

$$f_x = 8y - (x + y)^3 = 0, \text{ we know } 8y = (x + y)^3.$$

$$f_y = 8x - (x + y)^3 = 0, \text{ we know } 8x = (x + y)^3.$$

Therefore we must have $x = y$. Since $(x + y)^3 = (2y)^3 = 8y^3$, this tells us that $8y - 8y^3 = 0$. Solving gives $y = 0, \pm 1$.
Thus the critical points are $(0, 0), (1, 1), (-1, -1)$.
$f_{yy} = f_{xx} = -3(x + y)^2$, and $f_{xy} = 8 - 3(x + y)^2$.
The discriminant is

$$D(x, y) = f_{xx}f_{yy} - f_{xy}^2$$
$$= 9(x + y)^4 - \left(64 - 48(x + y)^2 + 9(x + y)^4\right)$$
$$= -64 + 48(x + y)^2.$$

$D(0,0) = -64 < 0$, so $(0,0)$ is a saddle point.
$D(1,1) = -64 + 192 > 0$ and $f_{xx}(1,1) = -12 < 0$, so $(1,1)$ is a local maximum.
$D(-1,-1) = -64 + 192 > 0$ and $f_{xx}(-1,-1) = -12 < 0$, so $(-1,-1)$ is a local maximum.

9. To find critical points, set partial derivatives equal to zero:

$$E_x = \sin x = 0 \quad \text{when} \quad x = 0, \pm\pi, \pm2\pi, \cdots$$

$$E_y = y = 0 \quad \text{when} \quad y = 0.$$

The critical points are

$$\cdots (-2\pi, 0), (-\pi, 0), (0, 0), (\pi, 0), (2\pi, 0), (3\pi, 0) \cdots$$

To classify, calculate $D = E_{xx}E_{yy} - (E_{xy})^2 = \cos x$.
At the points $(0,0), (\pm2\pi, 0), (\pm4\pi, 0), (\pm6\pi, 0), \cdots$

$$D = (1) > 0 \quad \text{and} \quad E_{xx} > 0 \quad (\text{Since} E_{xx}(0, 2k\pi) = \cos(2k\pi) = 1).$$

Therefore $(0,0), (\pm2\pi, 0), (\pm4\pi, 0), (\pm6\pi, 0), \cdots$ are local minima.
At the points $(\pm\pi, 0), (\pm3\pi, 0), (\pm5\pi, 0), (\pm7\pi, 0), \cdots$, we have $\cos(2k+1)\pi = -1$, so

$$D = (-1) < 0.$$

Therefore $(\pm\pi, 0), (\pm3\pi, 0), (\pm5\pi, 0), (\pm7\pi, 0), \cdots$ are saddle points.

10. At a critical point,

$$f_x = \cos x \sin y = 0 \quad \text{so} \quad \cos x = 0 \text{ or } \sin y = 0;$$

and

$$f_y = \sin x \cos y = 0 \quad \text{so} \quad \sin x = 0 \text{ or } \cos y = 0.$$

Case 1: Assume $\cos x = 0$. This gives

$$x = \cdots - \frac{3\pi}{2}, -\frac{\pi}{2}, \frac{\pi}{2}, \frac{3\pi}{2}, \cdots$$

(This can be written more compactly as: $x = k\pi + \pi/2$, for $k = 0, \pm1, \pm2, \cdots$.)
If $\cos x = 0$, then $\sin x = \pm1 \neq 0$. Thus in order to have $f_y = 0$ we need $\cos y = 0$, giving

$$y = \cdots - \frac{3\pi}{2}, -\frac{\pi}{2}, \frac{\pi}{2}, \frac{3\pi}{2}, \cdots$$

(More compactly, $y = l\pi + \pi/2$, for $l = 0, \pm1, \pm2, \cdots$)
Case 2: Assume $\sin y = 0$. This gives

$$y = \cdots - 2\pi, -\pi, 0, \pi, 2\pi, \cdots$$

(More compactly, $y = l\pi$, for $l = 0, \pm1, \pm2, \cdots$)
If $\sin y = 0$, then $\cos y = \pm1 \neq 0$, so to get $f_y = 0$ we need $\sin x = 0$, giving

$$x = \cdots, -2\pi, -\pi, 0, \pi, 2\pi, \cdots$$

(More compactly, $x = k\pi$ for $k = 0, \pm0, \pm1, \pm2, \cdots$)
Hence we get all the critical points of $f(x, y)$. Those from Case 1 are as follows:

$$\cdots \left(-\frac{\pi}{2}, -\frac{\pi}{2}\right), \left(-\frac{\pi}{2}, \frac{\pi}{2}\right), \left(-\frac{\pi}{2}, \frac{3\pi}{2}\right) \cdots$$

$$\cdots \left(\frac{\pi}{2}, -\frac{\pi}{2}\right), \left(\frac{\pi}{2}, \frac{\pi}{2}\right), \left(\frac{\pi}{2}, \frac{3\pi}{2}\right) \cdots$$

$$\cdots \left(\frac{3\pi}{2}, -\frac{\pi}{2}\right), \left(\frac{3\pi}{2}, \frac{\pi}{2}\right), \left(\frac{3\pi}{2}, \frac{3\pi}{2}\right) \cdots$$

Those from Case 2 are as follows:

$$\cdots (-\pi, -\pi), (-\pi, 0), (-\pi, \pi), (-\pi, 2\pi) \cdots$$

$$\cdots (0, -\pi), (0, 0), (0, \pi), (0, 2\pi) \cdots$$

$$\cdots (\pi, -\pi), (\pi, 0), (\pi, \pi), (\pi, 2\pi) \cdots$$

More compactly these points can be written as,

$$(k\pi, l\pi), \text{ for } k = 0, \pm1, \pm2, \cdots, l = 0, \pm1, \pm2, \cdots$$
$$\text{and } (k\pi + \frac{\pi}{2}, l\pi + \frac{\pi}{2}), \text{ for } k = 0, \pm1, \pm2, \cdots, l = 0, \pm1, \pm2, \cdots$$

To classify the critical points, we find the discriminant. We have

$$f_{xx} = -\sin x \sin y, \qquad f_{yy} = -\sin x \sin y, \qquad \text{and} \qquad f_{xy} = \cos x \cos y.$$

Thus the discriminant is

$$D(x, y) = f_{xx}f_{yy} - f_{xy}^2$$
$$= (-\sin x \sin y)(-\sin x \sin y) - (\cos x \cos y)^2$$
$$= \sin^2 x \sin^2 y - \cos^2 x \cos^2 y$$
$$= \sin^2 y - \cos^2 x. \text{ (Use: } \sin^2 x = 1 - \cos^2 x \text{ and factor.)}$$

At points of the form $(k\pi, l\pi)$ where $k = 0, \pm1, \pm2, \cdots; l = 0, \pm1, \pm2, \cdots$, we have
$D(x, y) = -1 < 0$ so $(k\pi, l\pi)$ are saddle points.
At points of the form $(k\pi + \frac{\pi}{2}, l\pi + \frac{\pi}{2})$ where $k = 0, \pm1, \pm2, \cdots; l = 0, \pm1, \pm2, \cdots$
$D(k\pi + \frac{\pi}{2}, l\pi + \frac{\pi}{2}) = 1 > 0$, so we have two cases:
If k and l are both even or k and l are both odd, then
$f_{xx} = -\sin x \sin y = -1 < 0$, so $(k\pi + \frac{\pi}{2}, l\pi + \frac{\pi}{2})$ are local maximum points.
If k is even but l is odd or k is odd but l is even, then
$f_{xx} = 1 > 0$ so $(k\pi + \frac{\pi}{2}, l\pi + \frac{\pi}{2})$ are local minimum points.

11. The partial derivatives are

$$P_x = -6x - 4 + 2y \quad \text{and} \quad P_y = 2x - 10y + 48.$$

Set $P_x = 0$ and $P_y = 0$ to find the critical point, thus

$$2y - 6x = 4 \quad \text{and} \quad 10y - 2x = 48.$$

Now, solve these equations simultaneously to obtain $x = 1$ and $y = 5$.
    Since $P_{xx} = -6$, $P_{yy} = -10$ and $P_{xy} = 2$ for all (x, y), at $(1, 5)$ the discriminant $D = (-6)(-10) -$
$(2)^2 = 56 > 0$ and $P_{xx} < 0$. Thus $P(x, y)$ has a local maximum value at $(1, 5)$.

12. At a local maximum value of f,

$$\frac{\partial f}{\partial x} = -2x - B = 0.$$

We are told that this is satisfied by $x = -2$. So $-2(-2) - B = 0$ and $B = 4$. In addition,

$$\frac{\partial f}{\partial y} = -2y - C = 0$$

and we know this holds for $y = 1$, so $-2(1) - C = 0$, giving $C = -2$. We are also told that the value of f is 15 at the point $(2, 1)$, so

$$15 = f(-2, 1) = A - ((-2)^2 + 4(-2) + 1^2 - 2(1)) = A - (-5), \text{ so } A = 10.$$

Now we check that these values of A, B, and C give $f(x, y)$ a local maximum at the point $(-2, 1)$. Since

$$f_{xx}(-2, 1) = -2,$$
$$f_{yy}(-2, 1) = -2$$

and

$$f_{xy}(-2, 1) = 0,$$

we have that the discriminant $D = (-2)(-2) - 0 > 0$ and $f_{xx}(-2, 1) < 0$. Thus, f has a local maximum value 15 at $(-2, 1)$.

13. At the origin $f(0, 0) = 0$. Since $x^6 \geq 0$ and $y^6 \geq 0$, the point $(0, 0)$ is a local (and global) minimum. The second derivative test does not tell you anything since $D = 0$.

14. At the origin $g(0, 0) = 0$. Since $y^3 \geq 0$ for $y > 0$ and $y^3 < 0$ for $y < 0$, the function g takes on both positive and negative values near the origin, which must therefore be a saddle point. The second derivative test does not tell you anything since $D = 0$.

15. At the origin $h(0, 0) = 1$. Since $\cos x$ and $\cos y$ are never above 1, the origin must be a local (and global) maximum. The second derivative test

$$D = h_{xx}h_{yy} - (h_{xy})^2 = \left((-\cos x \cos y)(-\cos x \cos y) - (\sin x \sin y)^2\right)\Big|_{x=0, y=0}$$
$$= \left(\cos^2 x \cos^2 y - \sin^2 x \sin^2 y\right)\Big|_{x=0, y=0}$$
$$= 1 > 0$$

and $h_{xx}(0, 0) < 0$, so $(0, 0)$ is a local maximum.

16. (a) To find the critical points, we must solve the equations

$$\frac{\partial f}{\partial x} = e^x(1 - \cos y) = 0$$
$$\frac{\partial f}{\partial y} = e^x(\sin y) = 0.$$

The first equation has solution

$$y = 0, \pm 2\pi, \pm 4\pi, \ldots.$$

The second equation has solution

$$y = 0, \pm \pi, \pm 2\pi, \pm 3\pi, \ldots.$$

Since x can be anything, the lines

$$y = 0, \pm 2\pi, \pm 4\pi, \ldots$$

are lines of critical points.

(b) We calculate

$$D = (f_{xx})(f_{yy})^2 - (f_{xy})^2 = e^x(1 - \cos y)e^x \cos y - (e^x \sin y)^2$$
$$= e^{2x}(\cos y - \cos^2 y - \sin^2 y)$$
$$= e^{2x}(\cos y - 1)$$

At any critical point on one of the lines $y = 0$, $y = \pm2\pi$, $y = \pm4\pi$, ...,

$$D = e^{2x}(1 - 1) = 0.$$

Thus, D tells us nothing. However, all along these critical lines, the value of the function, f, is zero. Since the function f is never negative, the critical points are all both local and global minima.

17. (a) $(1, 3)$ is a critical point. Since $f_{xx} > 0$ and the discriminant

$$D = f_{xx}f_{yy} - f_{xy}^2 = f_{xx}f_{yy} - 0^2 = f_{xx}f_{yy} > 0,$$

the point $(1, 3)$ is a minimum.

(b)

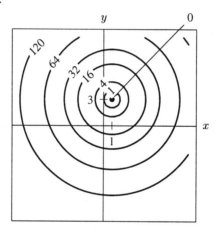

Figure 14.3

18. (a) (a, b) is a critical point. Since the discriminant $D = f_{xx}f_{yy} - f_{xy}^2 = -f_{xy}^2 < 0$, (a, b) is a saddle point.
(b)

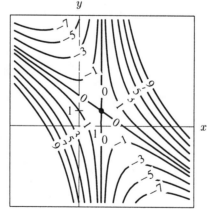

Figure 14.4

19. The partial derivatives are

$$f_x(x,y) = 3x^2 - 3y^2 \qquad \text{and} \qquad f_y(x,y) = -6xy.$$

Now $f_x(x,y)$ will vanish if $x = \pm y$ and $f_y(x,y)$ will vanish if either $x = 0$ or $y = 0$. Since the partial derivatives are defined everywhere, the only critical points are where $f_x(x,y)$ and $f_y(x,y)$ vanish simultaneously. $(0,0)$ is the only critical point.

To find the contour for $f(x,y) = 0$, we solve the equation $x^3 - 3xy^2 = 0$. This can be factored into

$$f(x,y) = x(x - \sqrt{3}\,y)(x + \sqrt{3}\,y) = 0$$

whose roots are $x = 0$, $x = \sqrt{3}\,y$ and $x = -\sqrt{3}\,y$. Each of these roots describes a line through the origin; the three of them divide the plane into six regions. Crossing any one of these lines will change the sign of only one of the three factors of $f(x,y)$, which will change the sign of $f(x,y)$.

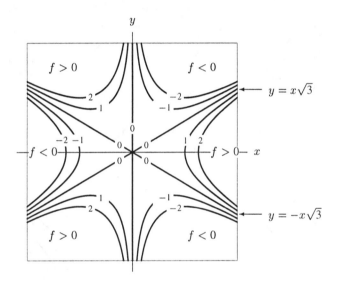

Figure 14.5

20. The first order partial derivatives are

$$f_x(x,y) = 2kx - 2y \qquad \text{and} \qquad f_y(x,y) = 2ky - 2x.$$

And the second order partial derivatives are

$$f_{xx}(x,y) = 2k \qquad f_{xy}(x,y) = -2 \qquad f_{yy}(x,y) = 2k$$

Since $f_x(0,0) = f_y(0,0) = 0$, the point $(0,0)$ is a critical point. The discriminant is

$$D = (2k)(2k) - 4 = 4(k^2 - 1).$$

For $k = \pm 2$, the discriminant is positive, $D = 12$. When $k = 2$, $f_{xx}(0,0) = 4$ which is positive so we have a local minimum at the origin. When $k = -2$, $f_{xx}(0,0) = -4$ so we have a local maximum at the origin. In the case $k = 0$, $D = -4$ so the origin is a saddle point.

Lastly, when $k = \pm 1$ the discriminant is zero, so the second derivative test can tell us nothing. Luckily, we can factor $f(x, y)$ when $k = \pm 1$. When $k = 1$,

$$f(x, y) = x^2 - 2xy + y^2 = (x - y)^2.$$

This is always greater than or equal to zero. So $f(0, 0) = 0$ is a minimum and the surface is a trough-shaped parabolic cylinder with its base along the line $x = y$.

When $k = -1$,

$$f(x, y) = -x^2 - 2xy - y^2 = -(x + y)^2.$$

This is always less than or equal to zero. So $f(0, 0) = 0$ is a maximum. The surface is a parabolic cylinder, with its top ridge along the line $x = -y$.

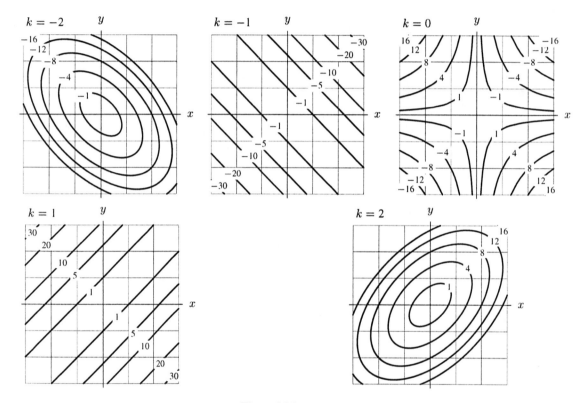

Figure 14.6

Solutions for Section 14.2

1. Mississippi lies entirely within a region designated as "80s" so we expect both the high and low daily temperatures within the state to be in the 80s. The South-Western most corner of the state is close to a region designated as "90s" so, we would expect the temperature here to be in the high 80s, say 87-88. The northern most portion of the state is relatively central to the "80s" region. We might expect the temperature there to be between 83-87.

Alabama also lies completely within a region designated as "80s" so both the high and low daily temperatures within the state are in the 80s. The south-eastern tip of the state is close to a "90s" region so we would expect the temperature here to be ≈ 88-89 degrees. The northern most part of the state is relatively central to the "80s" region so the temperature there is ≈ 83-87 degrees.

Pennsylvania is also in the "80s" region, but it is touched by the boundary line between the "80s" and a "70s" region. Thus we expect the low daily temperature to occur there and be about 70 degrees. The state is also touched by a boundary line which contains a "90s" region so the high will occur there and be 89-90 degrees.

New York is split by a boundary between an "80s" and a "70s" region, the northern portion of the state is apt to be about 74-76 while the southern portion is likely to be in the low 80s, maybe 81-84 or so.

California contains many different zones. The northern coastal areas will probably have the state daily low at 65-68, although without another contour on that side, it is difficult to judge how quickly the temperature is dropping off to the west. The tip of Southern California is in a 100s region, so there we expect the state daily high to be 100-101.

Arizona will have a low around 85-87 in the northwest corner and a high in the 100s, perhaps 102-107 in its southern regions.

Massachusetts will probably have a high around 81-84 and a low at 70.

2. Since $f(x, y) \leq 0$ for all x, y and since $f(0, 0) = 0$, the function has a global maximum (it is 0) and no global minimum.

3. Suppose x is fixed. Then for large values of y the sign of f is determined by the highest power of y, namely y^3. Thus,

$$f(x, y) \to \infty \quad \text{as} \quad y \to \infty$$
$$f(x, y) \to -\infty \quad \text{as} \quad y \to -\infty.$$

So f does not have a global maximum or minimum.

4. The function f has no global maximum or global minimum; g has a global minimum (it is 0) but no global maximum; h has no global maximum or minimum.

5. To maximize $z = x^2 + y^2$, it suffices to maximize x^2 and y^2. We can maximize both of these at the same time by taking the point $(1, 1)$, where $z = 2$. It occurs on the boundary of the square. (Note: We also have maxima at the points $(-1, -1), (-1, 1)$ and $(1, -1)$ which are on the boundary of the square.)

To minimize $z = x^2 + y^2$, we choose the point $(0, 0)$, where $z = 0$. It does not occur on the boundary of the square.

6. To maximize this function, it suffices to maximize x^2 and minimize y^2. We can do this by choosing the point $(1, 0)$, or $(-1, 0)$ where $z = 1$. These occur on the boundary of the square.

To minimize $z = x^2 - y^2$, it suffices to maximize y^2 and minimize x^2. We can do this by taking the point $(0, 1)$, or $(0, -1)$ where $z = -1$. These occur on the boundary of the square.

7. To maximize $z = -x^2 - y^2$ it suffices to minimize x^2 and y^2. Thus, the maximum is at $(0, 0)$, where $z = 0$. It doesn't occur on the boundary of the square.

To minimize $z = -x^2 - y^2$, it suffices to maximize x^2 and y^2. Do this by taking the point $(1, 1)$, $(-1, -1), (-1, 1)$, or $(1, -1)$ where $z = -2$. These occur on the boundary of the square.

8. (a) This tells us that an increase in the price of either product causes a decrease in the quantity demanded of both products. An example of products with this relationship is tennis rackets and tennis balls. An increase in the price of either product is likely to lead to a decrease in the quantity demanded of both products as they are used together. In economics, it is rare for the quantity demanded of a product to increase if its price increases, so for q_1, the coefficient of p_1 is negative as expected. The coefficient of p_2 in the expression could be either negative or positive. In this case, it is negative showing that the two products are complementary in use. If it was positive, however, it would indicate that the two products

are competitive in use, for example Coke and Pepsi.

(b)   The revenue for A would be $q_1 p_1 = 150 p_1 - 2p_1^2 - p_1 p_2$, and the revenue for B would be $q_2 p_2 = 200 p_2 - p_1 p_2 - 3p_2^2$. The total sales revenue of both products, R, would be

$$R(p_1, p_2) = 150 p_1 + 200 p_2 - 2p_1 p_2 - 2p_1^2 - 3p_2^2.$$

Note that R is a function of p_1 and p_2. To find the critical points of R, set $\nabla R = 0$, i.e.,

$$\frac{\partial R}{\partial p_1} = \frac{\partial R}{\partial p_2} = 0.$$

This gives

$$\frac{\partial R}{\partial p_1} = 150 - 2p_2 - 4p_1 = 0$$

and

$$\frac{\partial R}{\partial p_2} = 200 - 2p_1 - 6p_2 = 0$$

Solving simultaneously, we have $p_1 = 25$ and by substituting in, we get $p_2 = 25$. Therefore the point $(25, 25)$ is a critical point for R. Further,

$$\frac{\partial^2 R}{\partial p_1^2} = -4, \frac{\partial^2 R}{\partial p_2^2} = -6, \frac{\partial^2 R}{\partial p_1 \partial p_2} = -2,$$

so the discriminant at this critical point is

$$D = (-4)(-6) - (-2)^2 = 20.$$

Since $D > 0$ and $\frac{\partial^2 R}{\partial p_1^2} < 0$, this critical point is a local maximum. Since R is quadratic in p_1 and p_2, this is a global maximum. Therefore the maximum possible revenue is

$$\begin{aligned} R &= 150(25) + 200(25) - 2(25)(25) - 2(25)^2 - 3(25)^2 \\ &= (6)(25)^2 + 8(25)^2 - 7(25)^2 \\ &= 4375. \end{aligned}$$

This can be obtained when $p_1 = p_2 = 25$. Note that at these prices, $q_1 = 75$ units, and $q_2 = 100$ units.

9.   The total revenue is

$$R = pq = (60 - 0.04q)q = 60q - 0.04q^2,$$

and as $q = q_1 + q_2$, this gives

$$R = 60q_1 + 60q_2 - 0.04q_1^2 - 0.08q_1 q_2 - 0.04q_2^2.$$

Therefore, the profit is

$$\begin{aligned} P(q_1, q_2) &= R - C_1 - C_2 \\ &= -13.7 + 60q_1 + 60q_2 - 0.07q_1^2 - 0.08q_2^2 - 0.08q_1 q_2. \end{aligned}$$

At a local maximum point, we have grad $P = \vec{0}$:

$$\frac{\partial P}{\partial q_1} = 60 - 0.14q_1 - 0.08q_2 = 0,$$

$$\frac{\partial P}{\partial q_2} = 60 - 0.16q_2 - 0.08q_1 = 0.$$

Solving these equations, we find that

$$q_1 = 300 \quad \text{and} \quad q_2 = 225.$$

To see whether or not we have found a local maximum, we compute the second-order partial derivatives:

$$\frac{\partial^2 P}{\partial q_1^2} = -0.14, \quad \frac{\partial^2 P}{\partial q_2^2} = -0.16, \quad \frac{\partial^2 P}{\partial q_1 \partial q_2} = -0.08.$$

Therefore,

$$D = \frac{\partial^2 P}{\partial q_1^2}\frac{\partial^2 P}{\partial q_2^2} - \frac{\partial^2 P}{\partial q_1 \partial q_2} = (-0.14)(-0.16) - (-0.08)^2 = 0.016,$$

and so we have found a local maximum point. The graph of $P(q_1, q_2)$ has the shape of an upside down paraboloid since P is quadratic in q_1 and q_2, hence $(300, 225)$ is a global maximum point.

10. (a) The revenue $R = p_1 q_1 + p_2 q_2$. Profit $= P = R - C = p_1 q_1 + p_2 q_2 - 2q_1^2 - 2q_2^2 - 10$.

$$\frac{\partial P}{\partial q_1} = p_1 - 4q_1 = 0 \quad \text{gives } q_1 = \frac{p_1}{4}$$

$$\frac{\partial P}{\partial q_2} = p_2 - 4q_2 = 0 \quad \text{gives } q_2 = \frac{p_2}{4}$$

Since $\frac{\partial^2 P}{\partial q_1^2} = -4$, $\frac{\partial^2 P}{\partial q_2^2} = -4$ and $\frac{\partial^2 P}{\partial q_1 \partial q_2} = 0$, at $(p_1/4, p_2/4)$ we have that the discriminant, $D = (-4)(-4) > 0$ and $\frac{\partial^2 P}{\partial q_1^2} < 0$, thus P has a local maximum value at $(q_1, q_2) = (p_1/4, p_2/4)$. Since P is quadratic in q_1 and q_2, this is a global maximum. So $P = \frac{p_1^2}{4} + \frac{p_2^2}{4} - 2\frac{p_1^2}{16} - 2\frac{p_2^2}{16} - 10 = \frac{p_1^2}{8} + \frac{p_2^2}{8} - 10$ is the maximum profit.

(b) The rate of change of the maximum profit as p_1 increases is

$$\frac{\partial}{\partial p_1}(\text{max } P) = \frac{2p_1}{8} = \frac{p_1}{4}.$$

11. We calculate the partial derivatives and set them to zero.

$$\frac{\partial (\text{range})}{\partial t} = -10t - 6h + 400 = 0$$

$$\frac{\partial (\text{range})}{\partial h} = -6t - 6h + 300 = 0.$$

$$10t + 6h = 400$$
$$6t + 6h = 300$$

solving we obtain

$$4t = 100$$

so

$$t = 25$$

Solving for h, we obtain $6h = 150$, yielding $h = 25$. Since the range is quadratic in h and t, the second derivative test tells us this is a local and global maximum. So the optimal conditions are $h = 25\%$ humidity and $t = 25°C$.

12. We want to maximize the theater's profit, P, as a function of the two variables (prices) p_c and p_a. As always, $P = R - C$, where R is the revenue, $R = q_c p_c + q_a p_a$, and C is the cost, which is of the form $C = k(q_c + q_a)$ for some constant k. Thus,

$$P(p_c, p_a) = q_c p_c + q_a p_a - k(q_c + q_a)$$
$$= rp_c^{-3} - krp_c^{-4} + sp_a^{-1} - ksp_a^{-2}$$

To find the critical points, solve

$$\frac{\partial P}{\partial p_c} = -3rp_c^{-4} + 4krp_c^{-5} = 0$$

$$\frac{\partial P}{\partial p_a} = -sp_a^{-2} + 2ksp_a^{-3} = 0.$$

We get $p_c = 4k/3$ and $p_a = 2k$.

This critical point is a global maximum by the following useful, general argument. Suppose that $F(x, y) = f(x) + g(y)$, where f has a global maximum at $x = b$ and g has a global maximum at $y = d$. Then for all x, y:

$$F(x, y) = f(x) + g(y) \leq f(b) + g(d) = F(b, d),$$

so F has global maximum at $x = b$, $y = d$.

The profit function in this problem has the form

$$P(p_c, p_a) = f(p_c) + g(p_a),$$

and the usual single-variable calculus argument using f' and g' shows that $p_c = 4k/3$ and $p_a = 2k$ are global maxima for f and g, respectively. Thus the maximum profit occurs when $p_c = 4k/3$ and $p_a = 2k$. Thus,

$$\frac{p_c}{p_a} = \frac{4k/3}{2k} = \frac{2}{3}.$$

13. The function $f(x, y)$ in Example 3 is given by

$$f(x, y) = \frac{80}{xy} + 10x + 10xy + 20y.$$

This has critical points when $f_x = f_y = 0$.

$$f_x(x, y) = \frac{-80}{x^2 y} + 10 + 10y.$$

$$f_y(x, y) = \frac{-80}{xy^2} + 10x + 20.$$

Substituting $x = 2, y = 1$ gives

$$f_x(2, 1) = \frac{-80}{2^2 \cdot 1} + 10 + 10.1 = 0$$

$$f_y(2, 1) = \frac{-80}{2 \cdot 1^2} + 10.2 + 20 = 0.$$

So our point, $(2, 1)$, is a critical point.

To determine if this critical point is a minimum we use the second derivative test.

$$
\begin{array}{ll}
f_{xx} = \frac{160}{x^3 y}, & f_{xx}(2,1) = 20, \\
f_{yy} = \frac{160}{x y^3}, & f_{yy}(2,1) = 80, \\
f_{xy} = \frac{80}{x^2 y^2} + 10, & f_{xy}(2,1) = 30.
\end{array}
$$

So $D = 20 \cdot 80 - 30^2 = 700 > 0$ and $f_{xx}(2,1) > 0$, therefore the point $(2,1)$ is a local minimum.

14.

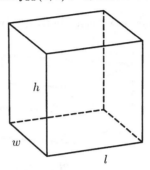

Figure 14.7

The box is shown in Figure 14.7. Cost of four sides $= (2hl + 2wh)(1)\cancel{c}$. Cost of two bottoms $= (2wl)(2)\cancel{c}$. Thus the total cost C (in cents) of the box is

$$
C = 2(hl + wh) + 4wl.
$$

But volume $wlh = 512$, so $l = 512/(wh)$, thus

$$
C = \frac{1024}{w} + 2wh + \frac{2048}{h}.
$$

To minimize C, find the critical points of C by solving

$$
C_h = 2w - \frac{2048}{h^2} = 0,
$$
$$
C_w = 2h - \frac{1024}{w^2} = 0.
$$

We get

$$
2wh^2 = 2048
$$
$$
2hw^2 = 1024.
$$

Since $w, h \neq 0$, we can divide the first equation by the second giving

$$
\frac{2wh^2}{2hw^2} = \frac{2048}{1024},
$$

so

$$
\frac{h}{w} = 2,
$$

thus

$$
h = 2w.
$$

Substituting this in $C_h = 0$, we obtain $h^3 = 2048$, so $h = 12.7$ cm. Thus $w = h/2 = 6.35$ cm, and $l = 512/(wh) = 6.35$ cm. Now we check that these dimensions minimize the cost C. We find that

$$D = C_{hh}C_{ww} - C_{hw}^2 = (\frac{4096}{h^3})(\frac{2048}{w^3}) - 2^2,$$

and at $h = 12.7$, $w = 6.35$, $C_{hh} > 0$ and $D = 16 - 4 > 0$, thus C has a local minimum at $h = 12.7$ and $w = 6.35$. Since C increases without bound as $w, h \to 0$ or ∞, this local minimum must be a global minimum.

Therefore, the dimensions of the box that minimize the cost are $w = 6.35$ cm, $l = 6.35$ cm and $h = 12.7$ cm.

15.

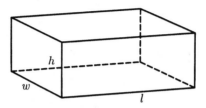

Figure 14.8

Let w, h and l be width, height and length of the suitcase in cm. Then its volume $V = lwh$, and $w + h + l \le 135$. To maximize the volume V, choose $w + h + l = 135$, and thus $l = 135 - w - h$,

$$V = wh(135 - w - h)$$
$$= 135wh - w^2h - wh^2$$

Differentiating gives

$$V_w = 135h - 2wh - h^2,$$
$$V_h = 135w - w^2 - 2wh.$$

Find the critical points by solving $V_w = 0$ and $V_h = 0$:

$$V_w = 0 \quad \text{gives} \quad 135h - h^2 = 2wh,$$
$$V_h = 0 \quad \text{gives} \quad 135w - w^2 = 2wh.$$

As $hw \ne 0$, we cancel h (and w respectively) in the above equations and get

$$135 - h = 2w$$
$$135 - w = 2h$$

Subtracting gives

$$w - h = 2(w - h)$$

hence $w = h$. Therefore, substituting into the equation $v_w = 0$

$$135h - h^2 = 2h^2$$

and therefore

$$3h^2 = 135h.$$

Since $h \neq 0$, we have

$$h = \frac{135}{3} = 45.$$

So $w = h = 45$ cm. Thus, $l = 135 - w - h = 45$ cm. To check that this critical point is a maximum, we find

$$V_{ww} = -2h, \quad V_{hh} = -2w,$$
$$V_{wh} = 135 - 2w - 2h,$$

so

$$D = V_{ww}V_{hh} - V_{wh}^2 = 4hw - (135 - 2w - 2h)^2.$$

At $w = h = 45$, we have $V_{ww} = -2(45) < 0$ and $D = 4(45)^2 - (135 - 90 - 90)^2 = 6075 > 0$, hence V is maximum at $w = h = l = 45$.

Therefore, the suitcase with maximum volume is a cube with dimensions width = height = length = 45 cm.

16. Let $P(K, L)$ be the profit obtained using K units of capital and L units of labor. The cost of production is given by

$$C(K, L) = kK + \ell L,$$

and the revenue function is given by

$$R(K, L) = pQ = pAK^aL^b.$$

Hence, the profit is

$$P = R - C = pAK^aL^b - (kK + \ell L).$$

In order to find local maxima of P, we calculate the partial derivatives and see where they are zero. We have:

$$\frac{\partial P}{\partial K} = apAK^{a-1}L^b - k,$$
$$\frac{\partial P}{\partial L} = bpAK^aL^{b-1} - \ell.$$

The critical points of the function $P(K, L)$ are solutions (K, L) of the simultaneous equations:

$$\frac{k}{a} = pAK^{a-1}L^b,$$
$$\frac{\ell}{b} = pAK^aL^{b-1}.$$

Multiplying the first equation by K and the second by L, we get

$$\frac{kK}{a} = \frac{\ell L}{b},$$

and so

$$K = \frac{\ell a}{kb}L.$$

Substituting for K in the equation $k/a = pAK^{a-1}L^b$, we get:

$$\frac{k}{a} = pA\left(\frac{\ell a}{kb}\right)^{a-1}L^{a-1}L^b.$$

We must therefore have

$$L^{1-a-b} = pA \left(\frac{a}{k}\right)^a \left(\frac{\ell}{b}\right)^{a-1}.$$

Hence, if $a + b \neq 1$,

$$L = \left[pA \left(\frac{a}{k}\right)^a \left(\frac{\ell}{b}\right)^{(a-1)}\right]^{1/(1-a-b)},$$

and

$$K = \frac{\ell a}{kb} L = \frac{\ell a}{kb} \left[pA \left(\frac{a}{k}\right)^a \left(\frac{\ell}{b}\right)^{(a-1)}\right]^{1/(1-a-b)}.$$

To see if this is really a local maximum, we apply the second derivative test. We have:

$$\frac{\partial^2 P}{\partial K^2} = a(a-1)pAK^{a-2}L^b,$$

$$\frac{\partial^2 P}{\partial L^2} = b(b-1)pAK^a L^{b-2},$$

$$\frac{\partial^2 P}{\partial K \partial L} = abpAK^{a-1}L^{b-1}.$$

Hence,

$$\begin{aligned} D &= \frac{\partial^2 P}{\partial K^2} \frac{\partial^2 P}{\partial L^2} - \left(\frac{\partial^2 P}{\partial K \partial L}\right)^2 \\ &= ab(a-1)(b-1)p^2 A^2 K^{2a-2} L^{2b-2} - a^2 b^2 p^2 A^2 K^{2a-2} L^{2b-2} \\ &= ab((a-1)(b-1) - ab)p^2 A^2 K^{2a-2} L^{2b-2} \\ &= ab(1-a-b)p^2 A^2 K^{2a-2} L^{2b-2}. \end{aligned}$$

Now a, b, p, A, K, and L are positive numbers. So, the sign of this last expression is determined by the sign of $1 - a - b$.

(a)   We assumed that $a + b < 1$, so $D > 0$, and as $0 < a < 1$, then $\partial^2 P/\partial K^2 < 0$ and so we have a unique local maximum. To verify that the local maximum is a global maximum, we focus on the cost. Let $C = kK + \ell L$. Since $K \geq 0$ and $L \geq 0$, $K \leq C/k$ and $L \leq C/\ell$. Therefore the profit satisfies:

$$\begin{aligned} P &= pAK^a L^b - (kK + \ell L) \\ &\leq pA \left(\frac{C}{k}\right)^a \left(\frac{C}{\ell}\right)^b - C \\ &= mC^{a+b} - C \end{aligned}$$

where $m = pA(1/k)^a(1/\ell)^b$. Since $a + b < 1$, the profit is negative for large costs C, say $C \geq C_0$ ($C_0 = m^{1-a-b}$ will do). Therefore, in the KL-plane for $K \geq 0$ and $L \geq 0$, the profit is less than or equal to zero everywhere on or above the line $kK + \ell L = C_o$. Thus the global maximum must occur inside the triangle bounded by this line and the K and L axes. Since $P \leq 0$ on the K and L axes as well, the global maximum must be in the interior of the triangle at the unique local maximum we found.

In the case $a + b < 1$, we have decreasing returns to scale. That is, if the amount of capital and labor used is multiplied by a constant $\lambda > 0$, we get less than λ times the production.

(b)   Now suppose $a + b \geq 1$. If we multiply K and L by λ for some $\lambda > 0$, then

$$Q(\lambda K, \lambda L) = A(\lambda K)^a (\lambda L)^b = \lambda^{a+b} Q(K, L).$$

We also see that

$$C(\lambda K, \lambda L) = \lambda C(K, L).$$

So if $a + b = 1$, we have

$$P(\lambda K, \lambda L) = \lambda P(K, L).$$

Thus, if $\lambda = 2$, so we are doubling the inputs K and L, then the profit P is doubled and hence there can be no maximum profit.

If $a + b > 1$, we have increasing returns to scale and there can again be no maximum profit: doubling the inputs will more than double the profit. In this case, the profit increases without bound as K, L go toward infinity.

17.   Let the line be in the form $y = b + mx$. When x equals -1, 0 and 1, then y equals $b - m$, b, and $b + m$, respectively. The sum of the squares of the vertical distances, which is what we want to minimize, is

$$f(m, b) = (2 - (b - m))^2 + (-1 - b)^2 + (1 - (b + m))^2.$$

To find the critical points, we compute the partial derivatives with respect to m and b,

$$\begin{aligned}
f_m &= 2(2 - b + m) + 0 + 2(1 - b - m)(-1) \\
&= 4 - 2b + 2m - 2 + 2b + 2m \\
&= 2 + 4m, \\
f_b &= 2(2 - b + m)(-1) + 2(-1 - b)(-1) + 2(1 - b - m)(-1) \\
&= -4 + 2b - 2m + 2 + 2b - 2 + 2b + 2m \\
&= -4 + 6b.
\end{aligned}$$

Setting both partial derivatives equal to zero, we get a system of equations:

$$2 + 4m = 0,$$
$$-4 + 6b = 0.$$

The solution is $m = -1/2$ and $b = 2/3$. One can check that it is a minimum. Hence, the regression line is $y = \frac{2}{3} - \frac{1}{2}x$.

18.

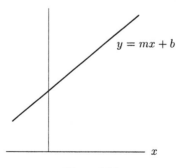

Figure 14.9

(a)   Points which are directly above or below each other share the same x coordinate, therefore, the point on the least squares line which is directly above or below the point in question will have x coordinate x_i and from the formula for the least squares line, it will have y coordinate $b + mx_i$.

(b) The general distance formula in two dimensions is $d = \sqrt{(x_2 - x_1)^2 + (y_2 - y_1)^2}$, so $d^2 = (x_2 - x_1)^2 + (y_2 - y_1)^2$. Since the x coordinates are identical for the two points in question, the first term in the square root is zero. This yields $d^2 = (y_i - (b + mx_i))^2$.

(c) In both cases we use the chain rule and our knowledge of summations to show the relationship.

$$\frac{\partial f}{\partial b} = \frac{\partial}{\partial b}(\sum_{i=1}^{n}(y_i - (b + mx_i))^2) = \sum_{i=1}^{n}\frac{\partial}{\partial b}(y_i - (b + mx_i))^2$$

$$= \sum_{i=1}^{n}2(y_i - (b + mx_i)) \cdot \frac{\partial}{\partial b}(y_i - (b + mx_i))$$

$$= \sum_{i=1}^{n}2(y_i - (b + mx_i)) \cdot (-1)$$

$$= -2\sum_{i=1}^{n}(y_i - (b + mx_i))$$

$$\frac{\partial f}{\partial m} = \frac{\partial}{\partial m}(\sum_{i=1}^{n}(y_i - (b + mx_i))^2) = \sum_{i=1}^{n}\frac{\partial}{\partial m}(y_i - (b + mx_i))^2$$

$$= \sum_{i=1}^{n}2(y_i - (b + mx_i)) \cdot \frac{\partial}{\partial m}(y_i - (b + mx_i))$$

$$= \sum_{i=1}^{n}2(y_i - (b + mx_i)) \cdot (-x_i)$$

$$= -2\sum_{i=1}^{n}(y_i - (b + mx_i)) \cdot x_i$$

(d) We can separate $\frac{\partial f}{\partial b}$ into three sums as shown:

$$\frac{\partial f}{\partial b} = -2\left(\sum_{i=1}^{n}y_i - b\sum_{i=1}^{n}1 - m\sum_{i=1}^{n}x_i\right)$$

Similarly we can separate $\frac{\partial f}{\partial m}$ after multiplying through by x_i:

$$\frac{\partial f}{\partial m} = -2\left(\sum_{i=1}^{n}y_i x_i - b\sum_{i=1}^{n}x_i - m\sum_{i=1}^{n}x_i^2\right)$$

Setting $\frac{\partial f}{\partial b}$ and $\frac{\partial f}{\partial m}$ equal to zero we have:

$$bn + m\sum_{i=1}^{n}x_i = \sum_{i=1}^{n}y_i$$

$$b\sum_{i=1}^{n}x_i + m\sum_{i=1}^{n}x_i^2 = \sum_{i=1}^{n}x_i y_i$$

(e) To solve this pair of linear equations, we multiply the first equation by $\sum_{i=1}^{n} x_i^2$, multiply the second one by $\sum_{i=1}^{n} x_i$, and subtract; we get

$$bn \sum_{i=1}^{n} x_i^2 - b\left(\sum_{i=1}^{n} x_i\right)^2 = \sum_{i=1}^{n} y_i \sum_{i=1}^{n} x_i^2 - \sum_{i=1}^{n} x_i y_i \sum_{i=1}^{n} x_i,$$

So,

$$b = \left(\sum_{i=1}^{n} x_i^2 \sum_{i=1}^{n} y_i - \sum_{i=1}^{n} x_i \sum_{i=1}^{n} x_i y_i\right) \Big/ \left(n \sum_{i=1}^{n} x_i^2 - \left(\sum_{i=1}^{n} x_i\right)^2\right)$$

Similarly,

$$m = \left(n \sum_{i=1}^{n} x_i y_i - \sum_{i=1}^{n} x_i \sum_{i=1}^{n} y_i\right) \Big/ \left(n \sum_{i=1}^{n} x_i^2 - \left(\sum_{i=1}^{n} x_i\right)^2\right)$$

(f) Applying the formula to the given data, we have $b = -\frac{1}{3}$, $m = 1$ which gives $y = -(1/3) + x$, in perfect agreement with the example.

19. (a) Let t be the number of years since 1960 and let $P(t)$ be the population in millions in the year $1960 + t$. We assume that $P = Ce^{at}$, and therefore

$$\ln P = at + \ln C.$$

So, we plot $\ln P$ against t and find the line of best fit. Our data points are $(0, \ln 180)$, $(10, \ln 206)$, and $(20, \ln 226)$. Applying the method of least squares to find the best-fitting line, we find that

$$a = \frac{\ln 226 - \ln 180}{20} \approx 0.0114,$$

$$\ln C = \frac{\ln 206}{3} - \frac{\ln 206}{6} + \frac{5 \ln 180}{6} \approx 5.20$$

Then, $C = e^{5.20} = 181.3$ and so

$$P(t) = 181.3e^{0.0114t}.$$

In 1990, we have $t = 30$ and the predicted population in millions is

$$P(30) = 18.3e^{0.01141(30)} = 255.2.$$

(b) The difference between the actual and the predicted population is about 6 million or $2\frac{1}{2}\%$. Given that only three data points were used to calculate a and c, this discrepancy is not surprising. Thus, the 1990 census, data does not mean that the assumption of exponential growth is unjustified.

(c) In 2010, we have $t = 50$ and $P(50) = 320.6$.

20. (a) Let the line take the form of $y = mx + b$, where x equals the number of years since 1920, which is the original year, and y equals the postage corresponding to the year. When the year is 1920, we have $x = 0$, so the postage equals b. When the year is 1932, the postage equals $b + 12m$, because the difference in years is 12. And, when the year is 1995, the postage equals $b + 75m$. The sum of the squares of the vertical distances, which is what we wish to minimize, is

$$\begin{aligned}f(m, b) = &(0.02 - b)^2 + (0.03 - (b + 12m))^2 + (0.04 - (b + 38m))^2 \\ &+ (0.05 - (b + 43m))^2 + (0.06 - (b + 48m))^2 + (0.08 - (b + 51m))^2 \\ &+ (0.1 - (b + 54m))^2 + (0.13 - (b + 55m))^2 + (0.15 - (b + 58m))^2 \\ &+ (0.2 - (b + 61m))^2 + (0.22 - (b + 65m))^2 + (0.25 - (b + 68m))^2 \\ &+ (0.29 - (b + 71m))^2 + (0.32 - (b + 75m))^2\end{aligned}$$

To find the critical points, calculate the partial derivatives f_m and f_b.

$$f_m = 0 + 2(0.03 - (b + 12m))(-12) + 2(0.04 - (b + 38m))(-38)$$
$$+2(0.05 - (b + 43m))(-43) + 2(0.06 - (b + 48m))(-48) + 2(0.08 - (b + 51m))(-51)$$
$$+2(0.1 - (b + 54m))(-54) + 2(0.13 - (b + 55m))(-55) + 2(0.15 - (b + 58m))(-58)$$
$$+2(0.2 - (b + 61m))(-61) + 2(0.22 - (b + 65m))(-65) + 2(0.25 - (b + 68m))(-68)$$
$$+2(0.29 - (b + 71m))(-71) + 2(0.32 - (b + 75m))(-75)$$
$$= 1398b + 81766m - 240.66$$

$$f_b = 2(0.02 - b)(-1) + 2(0.03 - (b + 12m))(-1) + 2(0.04 - (b + 38m))(-1)$$
$$+2(0.05 - (b + 43m))(-1) + 2(0.06 - (b + 48m))(-1) + 2(0.08 - (b + 51m))(-1)$$
$$+2(0.1 - (b + 54m))(-1) + 2(0.13 - (b + 55m))(-1) + 2(0.15 - (b + 58m))(-1)$$
$$+2(0.2 - (b + 61m))(-1) + 2(0.22 - (b + 65m))(-1) + 2(0.25 - (b + 68m))(-1)$$
$$+2(0.29 - (b + 71m))(-1) + 2(0.32 - (b + 75m)(-1)$$
$$= 28b + 1398m - 3.88$$

Setting both partial derivatives equal to zero, we get a system of two equations:

$$1398b + 81766m = 240.66$$
$$28b + 1398m = 3.88$$

with solutions $m \approx 0.0066$ and $b \approx -0.2135$. Thus the equation of the line is $y = 0.0066x - 0.2135$

To predict the cost of a postage stamp in the year 2010, we substitute $x = 2010 - 1920$ into the equation we just created and obtain:

$$y = 0.0066(2010 - 1920) - 0.2135$$
$$y = 0.3805$$

Therefore, the cost of a postage stamp in the year 2010 would be $0.38.

(b)   Looking at the data in Figure 14.10 you can see that it does not appear linear over the whole graph, but does look linear after about 1972.

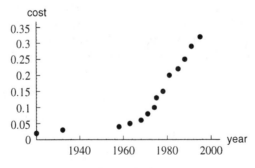

Figure 14.10

(c)

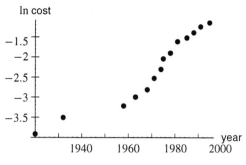

Figure 14.11

Looking at the data in Figure 14.11 you can see that it appears linear after 1960. If it were linear over the entire range, say $\ln y = mx + b$, then $y = e^{mx+b}$ so the price of a stamp is exponential, and is increasing rapidly as time goes on.

To find the line that best fits this data, we use the same method as before. This time

$$f(m, b) = (\ln 0.02 - b)^2 + (\ln 0.03 - (b + 12m))^2 + (\ln 0.04 - (b + 38m))^2$$
$$+ (\ln 0.05 - (b + 43m))^2 + (\ln 0.06 - (b + 48m))^2 + (\ln 0.08 - (b + 51m))^2$$
$$+ (\ln 0.10 - (b + 54m))^2 + (\ln 0.13 - (b + 55m))^2 + (\ln 0.15 - (b + 58m))^2$$
$$+ (\ln 0.20 - (b + 61m))^2 + (\ln 0.22 - (b + 65m))^2 + (\ln 0.25 - (b + 68m))^2$$
$$+ (\ln 0.29 - (b + 71m))^2 + (\ln 0.32 - (b + 75m))^2$$

Setting the partial derivatives equal to zero, we get:

$$f_m = 0 + 2(\ln 0.03 - (b + 12m))(-12) + 2(\ln 0.04 - (b + 38m))(-38)$$
$$+ 2(\ln 0.05 - (b + 43m))(-43) + 2(\ln 0.06 - (b + 48m))(-48)$$
$$+ 2(\ln 0.08 - (b + 51m))(-51) + 2(\ln 0.10 - (b + 54m))(-54)$$
$$+ 2(\ln 0.13 - (b + 55m))(-55) + 2(\ln 0.15 - (b + 58m))(-58)$$
$$+ 2(\ln 0.20 - (b + 61m))(-61) + 2(\ln 0.22 - (b + 65m))(-65)$$
$$+ 2(\ln 0.25 - (b + 68m))(-68) + 2(\ln 0.29 - (b + 71m))(-71)$$
$$+ 2(\ln 0.32 - (b + 75m))(-1)$$
$$= 1398b + 8166m + 2572.72 = 0,$$

$$f_b = 2(\ln 0.02 - b)(-1) + 2(\ln 0.03 - (b + 12m))(-1) + 2(\ln 0.04 - (b + 38m))(-1)$$
$$+ 2(\ln 0.05 - (b + 43m))(-1) + 2(\ln 0.06 - (b + 48m))(-1)$$
$$+ 2(\ln 0.08 - (b + 51m))(-1) + 2(\ln 0.10 - (b + 54m))(-1)$$
$$+ 2(\ln 0.13 - (b + 55m))(-1) + 2(\ln 0.15 - (b + 58m))(-1)$$
$$+ 2(\ln 0.20 - (b + 61m))(-1) + 2(\ln 0.22 - (b + 65m))(-1)$$
$$+ 2(\ln 0.25 - (b + 68m))(-1) + 2(\ln 0.29 - (b + 71m))(-1)$$
$$+ 2(\ln 0.32 - (b + 75m))(-1)$$
$$= 28b + 1398 + 64.20 = 0.$$

These equations have solutions $m \approx 0.0529$ and $b \approx -4.9329$. Thus, the equation of this line is

$$\ln y = 0.0529x - 4.9329,$$

where y is the price of a stamp.

To predict the cost of the stamp in 2010, we substitute $x = 2010 - 1920$ into the equation and get

$$\ln y = 0.0529(2010 - 1920) - 4.9329$$
$$\ln y = -0.1719$$

Therefore, $y = e^{-.1719} \approx 0.8421$, and so the cost would be $0.84.

21. (a) We have the following contour diagram for f:

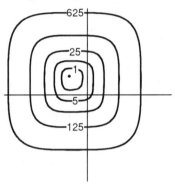

Figure 14.12

(b) We first compute grad f:

$$f_x = 4(x + 1)^3 - \frac{2xy^2}{(x^2y^2 + 1)^2} \quad \text{and} \quad f_y = 4(y - 1)^3 - \frac{2yx^2}{(x^2y^2 + 1)^2}.$$

These equations are difficult to solve simultaneously and this is why we need to use a gradient search method. We choose $(x_0, y_0) = (-1, 1)$ as our starting point and compute

$$\text{grad } f(-1, 1) = 0.5\vec{i} - 0.5\vec{j}.$$

Since we wish to minimize f, we move from (x_0, y_0) in the opposite direction of the gradient to a point

$$(x_1(t), y_1(t)) = (x_0, y_0) - t \text{ grad } f(x_0, y_0) = (-1, 1) - t(0.5, -0.5),$$

such that $f(x_1, y_1) < f(x_0, y_0)$. We should choose t so that the function $f(x_1(t), y_1(t))$ is minimized. Since the function $f(x, y)$ is complicated and only an approximate answer is required, we will try $t = 0.5$ for each iteration of the gradient search method. Therefore,

$$(x_1, y_1) = (-1.25, 1.25),$$
$$\text{grad } f(-1.25, 1.25) \approx 0.27\vec{i} - 0.27\vec{j}.$$

Repeating this step, again with $t = 0.5$, gives

$$(x_2, y_2) = (-1.25, 1.25) - (0.5)(0.27, -0.27) \approx (-1.38, 1.38),$$
$$\text{grad } f(-1.38, 1.38) \approx 0.02\vec{i} - 0.02\vec{j}.$$

Performing one more iteration, we get

$$(x_3, y_3) = (-1.38, 1.38) - (0.5)(0.02, -0.02) \approx (-1.39, 1.39),$$
$$\text{grad } f(-1.39, 1.39) \approx 0.002\vec{i} - 0.002\vec{j},$$

so we are already very close to a critical point. We find that $f(-1.39, 1.39) \approx 0.2575$ and verify that this is a global minimum using the contour diagram from part (a).

22. To execute the gradient search we first evaluate the two first partial derivatives of C:

$$\frac{\partial C}{\partial n} = 0.15 - 3 \left(\frac{4d}{5}\right)^{1.8} n^{-2}$$

$$\frac{\partial C}{\partial d} = -11.688 \left(\frac{4d}{5}\right)^{-5.87} + 1.44 \left(\frac{4d}{5}\right)^{0.8} + 4.32 \left(\frac{4d}{5}\right)^{0.8} n^{-1}$$

If we choose as our starting point for the gradient search a pipe diameter of 1 meter with three pumping stations, after numerous iterations, we find the optimum values to be about 1.61 meter diameter pipe with six pumping stations. This produces a minimum cost of 4.14 million dollars.

23. We have

$$f_x = 2x(y+1)^3 = 0 \quad \text{only when } x = 0 \text{ or } y = -1$$
$$f_y = 3x^2(y+1)^2 + 2y = 0 \quad \text{never when } y = -1 \text{ and only for } y = 0 \text{ when } x = 0$$

We conclude that $f_x = 0$ and $f_y = 0$ only when $x = 0, y = 0$, so f has only one critical point, namely $(0,0)$. The second derivative test at $(0,0)$ gives

$$D = f_{xx}f_{yy} - (f_{xy})^2 = 2(y+1)^3(6x^2(y+1)+2) - (6x(y+1)^2)^2$$
$$= 2(1)(2) - 0 > 0 \quad \text{when } x = 0, y = 0$$

Since $f_{xx} > 0$ at $(0,0)$, this means f has a local minimum at $(0,0)$.

[Alternatively, if we expand $(y+1)^3$, then we can view $f(x,y)$ as $x^2 + y^2 +$ (terms of degree 3 or greater in x and y), which means that f behaves likes $x^2 + y^2$ near $(0,0)$.]

Although $(0,0)$ is a local minimum, it cannot be a global minimum since for fixed x, say $x = 1$, the function $f(x,y)$ is a cubic polynomial in y and cubics take on arbitrarily large positive and negative values.

In the single-variable case, suppose a function f defined on the real line is differentiable and its derivative is continuous. Then if f has only one critical point, say $x = 0$, then if that critical point is a local minimum, it must also be a global minimum. This is because f' cannot change sign without $f' = 0$ so we must have $f' < 0$ for $x < 0$ and $f' > 0$ for $x > 0$. Thus f is decreasing for all $x < 0$ and increasing for all $x > 0$, which makes $x = 0$ the global minimum for f.

Solutions for Section 14.3

1. Our objective function is $f(x,y) = x + y$ and our equation of constraint is $g(x,y) = x^2 + y^2 = 1$. To optimize $f(x,y)$ with Lagrange multipliers, we solve $\nabla f(x,y) = \lambda \nabla g(x,y)$ subject to $g(x,y) = 1$. The gradients of f and g are

$$\nabla f(x,y) = \vec{i} + \vec{j},$$
$$\nabla g(x,y) = 2x\vec{i} + 2y\vec{j}.$$

So the equation $\nabla f = \lambda \nabla g$ becomes

$$\vec{i} + \vec{j} = \lambda(2x\vec{i} + 2y\vec{j})$$

Solving for λ gives

$$\lambda = \frac{1}{2x} = \frac{1}{2y},$$

which tells us that $x = y$. Going back to our equation of constraint, we use the substitution $x = y$ to solve for y

$$g(y,y) = y^2 + y^2 = 1$$
$$2y^2 = 1$$
$$y^2 = \frac{1}{2}$$
$$y = \pm\sqrt{\frac{1}{2}} = \pm\frac{\sqrt{2}}{2}.$$

Since $x = y$, our critical points are $(\frac{\sqrt{2}}{2}, \frac{\sqrt{2}}{2})$ and $(-\frac{\sqrt{2}}{2}, -\frac{\sqrt{2}}{2})$. Evaluating f at these points we find that the maximum value is $f(\frac{\sqrt{2}}{2}, \frac{\sqrt{2}}{2}) = \sqrt{2}$ and the minimum value is $f(-\frac{\sqrt{2}}{2}, -\frac{\sqrt{2}}{2}) = -\sqrt{2}$.

2. Our objective function is $f(x,y) = 3x - 2y$ and our equation of constraint is $g(x,y) = x^2 + 2y^2 = 44$. Their gradients are

$$\nabla f(x,y) = 3\vec{i} - 2\vec{j},$$
$$\nabla g(x,y) = 2x\vec{i} + 4y\vec{j}.$$

So the equation $\nabla f = \lambda\nabla g$ becomes $3\vec{i} - 2\vec{j} = \lambda(2x\vec{i} + 4y\vec{j})$. Solving for λ gives us

$$\lambda = \frac{3}{2x} = \frac{-2}{4y},$$

which we can use to find x in terms of y:

$$\frac{3}{2x} = \frac{-2}{4y}$$
$$-4x = 12y$$
$$x = -3y.$$

Using this relation in our equation of constraint, we can solve for y:

$$x^2 + 2y^2 = 44$$
$$(-3y)^2 + 2y^2 = 44$$
$$9y^2 + 2y^2 = 44$$
$$11y^2 = 44$$
$$y^2 = 4$$
$$y = \pm 2.$$

Thus, the critical points are $(6,2)$ and $(6,-2)$. Evaluating f at these points, we find that the maximum is $f(6,-2) = 18 + 4 = 22$ and the minimum value is $f(-6,2) = -18 - 4 = -22$.

3. Our objective function is $f(x,y) = x^2 + y$ and our equation of constraint is $g(x,y) = x^2 - y^2 = 1$. Their gradients are

$$\nabla f(x,y) = 2x\vec{i} + \vec{j},$$
$$\nabla g(x,y) = 2x\vec{i} - 2y\vec{j}.$$

Thus $\nabla f = \lambda\nabla g$ gives

$$2x = \lambda 2x$$
$$1 = \lambda 2y$$

But x cannot be zero, since the constraint equation, $-y^2 = 1$, would then have no real solution for y. So the equation $\nabla f = \lambda \nabla g$ becomes

$$\lambda = \frac{2x}{2x} = \frac{1}{-2y}$$

$$1 = \frac{1}{-2y}$$

$$-2y = 1$$

$$y = -\frac{1}{2}.$$

Substituting this into our equation of constraint we find

$$g(x, -\frac{1}{2}) = x^2 - \left(-\frac{1}{2}\right)^2 = 1$$

$$x^2 = \frac{5}{4}$$

$$x = \pm\frac{\sqrt{5}}{2}.$$

So the critical points are $(\frac{\sqrt{5}}{2}, -\frac{1}{2})$ and $(-\frac{\sqrt{5}}{2}, -\frac{1}{2})$. Evaluating f at these points we find $f(\frac{\sqrt{5}}{2}, -\frac{1}{2}) = f(-\frac{\sqrt{5}}{2}, -\frac{1}{2}) = \frac{5}{4} - \frac{1}{2} = \frac{3}{4}$. This is the minimum value for $f(x, y)$ constrained to $g(x, y) = 1$. To see that f has no maximum on $g(x, y) = 1$, note that as x and y increase, f increases and that on the part of the graph of $g(x, y) = 1$ in quadrant I $x \to \infty$ and $y \to \infty$.

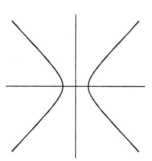

Figure 14.13: Graph of
$x^2 - y^2 = 1$

4. Our objective function is $f(x, y) = xy$ and our equation of constraint is $g(x, y) = 4x^2 + y^2 = 8$. Their gradients are

$$\nabla f(x, y) = y\vec{i} + x\vec{j},$$
$$\nabla g(x, y) = 8x\vec{i} + 2y\vec{j}.$$

So the equation $\nabla f = \lambda \nabla g$ becomes $y\vec{i} + x\vec{j} = \lambda(8x\vec{i} + 2y\vec{j})$. This gives

$$8x\lambda = y \quad \text{and} \quad 2y\lambda = x.$$

Multiplying, we get

$$8x^2\lambda = 2y^2\lambda.$$

If $\lambda = 0$, then $x = y = 0$, which doesn't satisfy the constraint equation. So $\lambda \neq 0$ and we get

$$2y^2 = 8x^2$$
$$y^2 = 4x^2$$
$$y = \pm 2x.$$

To find x, we substitute for y in our equation of constraint.

$$4x^2 + y^2 = 8$$
$$4x^2 + 4x^2 = 8$$
$$x^2 = 1$$
$$x = \pm 1$$

So our critical points are $(1, 2)$, $(1, -2)$, $(-1, 2)$ and $(-1, -2)$. Evaluating $f(x, y)$ at these points we have

$$f(1, 2) = f(-1, -2) = 2$$
$$f(1, -2) = f(1, -2) = -2.$$

So the maximum value of f on $g(x, y) = 8$ is 2, and the minimum value is -2.

5. The objective function is $f(x, y) = x^2 + y^2$ and the equation of constraint is $g(x, y) = x^4 + y^4 = 2$. Their gradients are

$$\nabla f(x, y) = 2x\vec{i} + 2y\vec{j},$$
$$\nabla g(x, y) = 4x^3\vec{i} + 4y^3\vec{j}.$$

So the equation $\nabla f = \lambda \nabla g$ becomes $2x\vec{i} + 2y\vec{j} = \lambda(4x^3\vec{i} + 4y^3\vec{j})$. This tells us that

$$2x = 4\lambda x^3,$$
$$2y = 4\lambda y^3.$$

Now if $x = 0$, the first equation is true for any value of λ. In particular, we can choose λ which satisfies the second equation. Similarly, $y = 0$ is solution.

Assuming both $x \neq 0$ and $y \neq 0$, we can divide to solve for λ and find

$$\lambda = \frac{2x}{4x^3} = \frac{2x}{4y^3}$$
$$\frac{1}{2x^2} = \frac{1}{2y^2}$$
$$y^2 = x^2$$
$$y = \pm x.$$

Going back to our equation of constraint, we find

$$g(0, y) = 0^4 + y^4 = 2 \Rightarrow y = \pm\sqrt[4]{2}$$
$$g(x, 0) = x^4 + 0^4 = 2 \Rightarrow x = \pm\sqrt[4]{2}$$
$$g(x, \pm x) = x^4 + (\pm x)^4 = 2 \Rightarrow x = \pm 1.$$

Thus, the critical points are $(0, \pm\sqrt[4]{2})$, $(\pm\sqrt[4]{2}, 0)$, $(1, \pm 1)$ and $(-1, \pm 1)$. Evaluating f at these points we find

$$f(1, 1) = f(1, -1) = f(-1, 1) = f(-1, -1) = 2,$$
$$f(0, \sqrt[4]{2}) = f(0, -\sqrt[4]{2}) = f(\sqrt[4]{2}, 0) = f(-\sqrt[4]{2}, 0) = \sqrt{2}.$$

So the minimum value of $f(x, y)$ on $g(x, y) = 2$ is $\sqrt{2}$ and the maximum value is 2.

6. Our objective function is $f(x,y) = x^2 - xy + y^2$ and our equation of constraint is $g(x,y) = x^2 - y^2 = 1$. To optimize $f(x,y)$ with Lagrange multipliers, we solve $\nabla f(x,y) = \lambda \nabla g(x,y)$ subject to $g(x,y) = 1$. The gradients of f and g are

$$\nabla f(x,y) = (2x - y)\vec{i} + (-x + 2y)\vec{j},$$
$$\nabla g(x,y) = 2x\vec{i} - 2y\vec{j}.$$

Therefore we get

$$2x - y = 2\lambda x$$
$$-x + 2y = -2\lambda y$$
$$x^2 - y^2 = 1.$$

Let us suppose that $\lambda = 0$. Then $2x = y$ and $2y = x$ give $x = y = 0$. But $(0,0)$ is not a solution of the third equation, so we conclude that $\lambda \neq 0$.

Now let's multiply the first two equations

$$-2\lambda y(2x - y) = 2\lambda x(-x + 2y).$$

As $\lambda \neq 0$, we can cancel it in the equation above and after doing the algebra we get

$$x^2 - 4xy + y^2 = 0$$

which gives $x = (2 + \sqrt{3})y$ or $x = (2 - \sqrt{3})y$.
   If $x = (2 + \sqrt{3})y$, the third equation gives

$$(2 + \sqrt{3})^2 y^2 - y^2 = 1$$

so $y = \pm\frac{1}{\sqrt{6+4\sqrt{3}}} \approx \pm 0.278$ and $x = \pm\frac{2+\sqrt{3}}{\sqrt{6+\sqrt{3}}} \approx \pm 1.038$. These give the solutions $(1.038, 0.278)$, $(-1.038, 0.278)$.
   If $x = (2 - \sqrt{3})y$, from the third equation we get

$$(2 - \sqrt{3})^2 y^2 - y^2 = 1.$$

But $(2 - \sqrt{3})^2 - 1 = 6 - 4\sqrt{3} \approx -0.928 < 0$ so the equation has no solution. Evaluating f gives

$$f(1.038, 0.278) \approx 0.866$$
$$f(-1.038, 0.278) \approx 1.443$$

So the maximum value of f is 1.443 and the minimum is 0.966.

7. The objective function is $f(x,y,z) = x + 3y + 5z$ and the equation of constraint is $g(x,y,z) = x^2 + y^2 + z^2 = 1$. Their gradients are

$$\nabla f(x,y,z) = \vec{i} + 3\vec{j} + 5\vec{k},$$
$$\nabla g(x,y,z) = 2x\vec{i} + 2y\vec{j} + 2z\vec{k}.$$

So the equation $\nabla f = \lambda \nabla g$ becomes $\vec{i} + 3\vec{j} + 5\vec{k} = \lambda(2x\vec{i} + 2y\vec{j} + 2z\vec{k})$. Solving for λ we find

$$\lambda = \frac{1}{2x} = \frac{3}{2y} = \frac{5}{2z}.$$

Which provides us with the equations

$$2y = 6x$$
$$10x = 2z.$$

Solving the first equation for y gives us $y = 3x$. Solving the second equation for z gives us $z = 5x$. Substituting these into the equation of constraint, we can find x:

$$x^2 + (3x)^2 + (5x)^2 = 1$$
$$x^2 + 9x^2 + 25x^2 = 1$$
$$35x^2 = 1$$
$$x^2 = \frac{1}{35}$$
$$x = \pm\sqrt{\frac{1}{35}} = \pm\frac{\sqrt{35}}{35}.$$

Since $y = 3x$ and $z = 5x$, the critical points are at $\pm(\frac{\sqrt{35}}{35}, 3\frac{\sqrt{35}}{35}, \frac{\sqrt{35}}{7})$. Evaluating f at these two points, we find the maximum is $f(\frac{\sqrt{35}}{35}, 3\frac{\sqrt{35}}{35}, \frac{\sqrt{35}}{7}) = \sqrt{35}\frac{35}{35} = \sqrt{35}$, and the minimum value is $f(-\frac{\sqrt{35}}{35}, -3\frac{\sqrt{35}}{35}, -\frac{\sqrt{35}}{7}) = -\sqrt{35}$.

8. Our objective function is $f(x, y, z) = 2x + y + 4z$ and our equation of constraint is $g(x, y, z) = x^2 + y + z^2 = 16$. Their gradients are

$$\nabla f(x, y, z) = 2\vec{i} + 1\vec{j} + 4\vec{k},$$
$$\nabla g(x, y, z) = 2x\vec{i} + 1\vec{j} + 2z\vec{k}.$$

So the equation $\nabla f = \lambda \nabla g$ becomes $2\vec{i} + 1\vec{j} + 4\vec{k} = \lambda(2x\vec{i} + 1\vec{j} + 2z\vec{k})$. Solving for λ we find

$$\lambda = \frac{2}{2x} = \frac{1}{1} = \frac{4}{2z}$$
$$\lambda = \frac{1}{x} = 1 = \frac{2}{z}.$$

Which tells us that $x = 1$ and $z = 2$. Going back to our equation of constraint, we can solve for y.

$$g(1, y, 2) = 16$$
$$1^2 + y + 2^2 = 16$$
$$y = 11.$$

So our one critical point is at $(1, 11, 2)$. The value of f at this point is $f(1, 11, 2) = 2 + 11 + 8 = 21$. This is the maximum value of $f(x, y, z)$ on $g(x, y, z) = 16$. To see this, look at a graph of $x^2 + y + z^2 = 16$. This shape is called a paraboloid. The important thing to note is that as you move farther back on the y-axis, the graph opens up more and more; specifically, $x \to -\infty, y \to -\infty, z \to -\infty$ on the surface. Now $f(x, y, z)$ decreases for decreasing x, y and z, so there is no minimum value for $f(x, y, z)$ on $g(x, y, z) = 16$.

9. Our objective function is $f(x, y, z) = x^2 - y^2 - 2z$ and our equation of constraint is $g(x, y, z) = x^2 + y^2 - z = 0$. To optimize $f(x, y, z)$ with Lagrange multipliers, we solve $\nabla f(x, y, z) = \lambda \nabla g(x, y, z)$ subject to $g(x, y, z) = 0$. The gradients of f and g are

$$\nabla f(x, y, z) = 2x\vec{i} - 2y\vec{j} - 2\vec{k},$$
$$\nabla g(x, y, z) = 2x\vec{i} + 2y\vec{j} - \vec{k}.$$

We get

$$2x = 2\lambda x$$
$$-2y = 2\lambda y$$
$$-2 = -\lambda$$
$$x^2 + y^2 = z.$$

The third equation gives $\lambda = 2$ and from the first $x = 0$, from the second $y = 0$ and from the fourth $z = 0$. So the only solution is $(0, 0, 0)$, and $f(0, 0, 0) = 0$.

To see what kind of extreme point is $(0, 0, 0)$, let (a, b, c) be a point which satisfies the constraint, i.e. $a^2 + b^2 = c$. Then $f(a, b, c) = a^2 - b^2 - 2c = -a^2 - 3b^2 \leq 0$. The conclusion is that 0 is the maximum value of f and that there is no minimum.

10. Our objective function is $f(x, y, z) = x^2 - 2y + 2z^2$ and our equation of constraint is $g(x, y, z) = x^2 + y^2 + z^2 - 1 = 0$. To optimize $f(x, y, z)$ with Lagrange multipliers, we solve $\nabla f(x, y, z) = \lambda \nabla g(x, y, z)$ subject to $g(x, y, z) = 0$. The gradients of f and g are

$$\nabla f(x, y, z) = 2x\vec{i} - 2\vec{j} + 4z\vec{k},$$
$$\nabla g(x, y) = 2x\vec{i} + 2y\vec{j} + 2z\vec{k}.$$

We get,

$$x = \lambda x$$
$$-1 = \lambda y$$
$$2z = \lambda z$$
$$x^2 + y^2 + z^2 = 1.$$

From the first equation we get $x = 0$ or $\lambda = 1$.

If $x = 0$ we have

$$-1 = \lambda y$$
$$2z = \lambda z$$
$$y^2 + z^2 = 1.$$

From the second equation $z = 0$ or $\lambda = 2$. So if $z = 0$, we have $y = \pm 1$ and we get the solutions $(0, 1, 0), (0, -1, 0)$. If $z \neq 0$ then $\lambda = 2$ and $y = -\frac{1}{2}$. So $z^2 = \frac{3}{4}$ which gives the solutions $(0, -\frac{1}{2}, \frac{\sqrt{3}}{2})$, $(0, -\frac{1}{2}, -\frac{\sqrt{3}}{2})$.

If $x \neq 0$, then $\lambda = 1$, so $y = -1$, which implies, from the equation $x^2 + y^2 + z^2 = 1$, that $x = 0$, which contradicts the assumption.

Therefore, evaluating f at these points we get $f(0, 1, 0) = -2$, $f(0, -1, 0) = 2$ and $f(0, -\frac{1}{2}, \frac{\sqrt{3}}{2}) = f(0, -\frac{1}{2}, -\frac{\sqrt{3}}{2}) = 4$. So the maximum value of f is 4 and the minimum is -2.

11. Our objective function is $f(x, y, z) = x + y + z$ and our equations of constraint are $g(x, y, z) = x^2 + y^2 + z^2 - 1 = 0$ and $h(x, y, z) = x - y - 1 = 0$. To optimize $f(x, y, z)$ with Lagrange multipliers, we solve $\nabla f(x, y, z) = \lambda_1 \nabla g(x, y, z) + \lambda_2 \nabla h(x, y, z)$ subject to $g(x, y, z) = 0$ and $h(x, y, z) = 0$. The gradients of f, g and h are

$$\nabla f(x, y, z) = \vec{i} + \vec{j} + \vec{k},$$
$$\nabla g(x, y, z) = 2x\vec{i} + 2y\vec{j} + 2z\vec{k}$$
$$\nabla h(x, y, z) = \vec{i} - \vec{j}$$

Therefore we get

$$1 = 2\lambda_1 x + \lambda_2$$
$$1 = 2\lambda_1 y - \lambda_2$$
$$1 = 2\lambda_1 z$$
$$x^2 + y^2 + z^2 = 1$$
$$x - y = 1$$

Adding the first two equations we get

$$2\lambda_1 (x + y) = 2.$$

From this and the third equation we get

$$4\lambda_1 z = 2\lambda_1 (x + y).$$

As $\lambda_1 \neq 0$, we can cancel it and get $x + y = 2z$. From this and $x - y = 1$ we get

$$x = \frac{2z + 1}{2},$$
$$y = \frac{2z - 1}{2}.$$

Substituting into the equation $x^2 + y^2 + z^2 = 1$ gives $z^2 = \frac{1}{6}$. So we get the solutions $(\frac{1}{2} + \frac{1}{\sqrt{6}}, -\frac{1}{2} + \frac{1}{\sqrt{6}}, \frac{1}{\sqrt{6}})$
and $(\frac{1}{2} - \frac{1}{\sqrt{6}}, -\frac{1}{2} - \frac{1}{\sqrt{6}}, -\frac{1}{\sqrt{6}})$.
Evaluating f at these points gives

$$f(\tfrac{1}{2} + \tfrac{1}{\sqrt{6}}, -\tfrac{1}{2} + \tfrac{1}{\sqrt{6}}, \tfrac{1}{\sqrt{6}}) = \tfrac{3}{\sqrt{6}}$$
$$f(\tfrac{1}{2} - \tfrac{1}{\sqrt{6}}, -\tfrac{1}{2} - \tfrac{1}{\sqrt{6}}, -\tfrac{1}{\sqrt{6}}) = -\tfrac{3}{\sqrt{6}}.$$

Therefore the maximum value of f is $\frac{3}{\sqrt{6}}$ and the minimum is $-\frac{3}{\sqrt{6}}$.

12. The region $x^2 + y^2 \leq 4$ is the shaded disk of radius 2 centered at the origin (including the circle $x^2 + y^2 = 4$) shown in Figure 14.14.

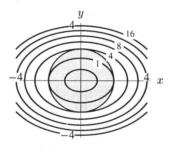

Figure 14.14

We will first find the relative maxima and minima in the interior of the disk. So we need to find the extrema of

$$f(x, y) = x^2 + 2y^2 \quad \text{in the region} \quad x^2 + y^2 < 4.$$

For this we compute the critical points:

$$f_x = 2x = 0$$
$$f_y = 4y = 0$$

So the critical point is $(0,0)$. As $f_{xx}(0,0) = 2$, $f_{yy}(0,0) = 4$ and $f_{xy}(0,0) = 0$ we have

$$D = f_{xx}(0,0) \cdot f_{yy}(0,0) - (f_{xy}(0,0))^2 = 8 > 0 \quad \text{and} \quad f_{xx}(0,0) = 2 > 0.$$

Therefore $(0,0)$ is a minimum point and $f(0,0) = 0$.

Now let's find the relative extrema of f on the boundary of the disk, hence this time we have to solve a constraint problem. We want the extrema of $f(x,y) = x^2 + 2y^2$ subject to $g(x,y) = x^2 + y^2 - 4 = 0$. We use Lagrange multipliers:

$$\text{grad } f = \lambda \text{ grad } g \quad \text{and} \quad x^2 + y^2 = 4,$$

which give

$$2x = 2\lambda x$$
$$4y = 2\lambda y$$
$$x^2 + y^2 = 4.$$

From the first equation we have $x = 0$ or $\lambda = 1$. If $x = 0$, from the last equation $y^2 = 4$ and therefore $(0,2)$ and $(0,-2)$ are solutions.

If $x \neq 0$ then $\lambda = 1$ and from the second equation $y = 0$. Substituting this into the third equation we get $x^2 = 4$ so $(2,0)$ and $(-2,0)$ are the other two solutions.

Therefore, as $f(0,2) = f(0,-2) = 8$ and $f(2,0) = f(-2,0) = 4$, $(0,2)$ and $(0,-2)$ are global maxima and $(0,0)$ is the global minimum on the whole region $x^2 = y^2 \leq 4$. The maximum value of f is 8 and the minimum value of f is 0.

13. The domain $x^2 + 2y^2 \leq 1$ is the shaded interior of the ellipse $x^2 + 2y^2 = 1$ including the boundary, shown in Figure 14.15.

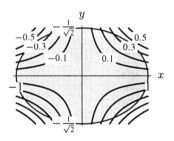

Figure 14.15

First we want to find the relative maxima and minima of f in the interior of the ellipse. So we need to find the extrema of

$$f(x,y) = xy, \quad \text{in the region} \quad x^2 + 2y^2 < 1.$$

For this we compute the critical points:

$$f_x = y = 0 \quad \text{and} \quad f_y = x = 0.$$

So there is one critical point, $(0,0)$. As $f_{xx}(0,0) = 0$, $f_{yy}(0,0) = 0$ and $f_{xy}(0,0) = 1$ we have

$$D = f_{xx}(0,0) \cdot f_{yy}(0,0) - (f_{xy}(0,0))^2 = -1 < 0$$

so $(0,0)$ is a saddle and f doesn't have relative extrema in the interior of the ellipse.

Now let's find the relative extrema of f on the boundary, hence this time we'll have a constraint problem. We want the extrema of $f(x,y) = xy$ subject to $g(x,y) = x^2 + 2y^2 - 1 = 0$. We use Lagrange multipliers:

$$\text{grad } f = \lambda \text{ grad } g \quad \text{and} \quad x^2 + 2y^2 = 1$$

which give

$$y = 2\lambda x$$
$$x = 4\lambda y$$
$$x^2 + 2y^2 = 1$$

From the first two equations we get

$$xy = 8\lambda^2 xy.$$

So $x = 0$ or $y = 0$ or $8\lambda^2 = 1$.

If $x = 0$, from the last equation $2y^2 = 1$ so $y = \pm\frac{\sqrt{2}}{2}$ and we get the solutions $(0, \frac{\sqrt{2}}{2})$ and $(0, -\frac{\sqrt{2}}{2})$.

If $y = 0$, from the last equation we get $x^2 = 1$ and so the solutions are $(1,0)$ and $(-1,0)$.

If $x \neq 0$ and $y \neq 0$ then $8\lambda^2 = 1$, hence $\lambda = \pm\frac{1}{2\sqrt{2}}$. For $\lambda = \frac{1}{2\sqrt{2}}$

$$x = \sqrt{2}y$$

and plugging into the third equation gives $4y^2 = 1$ so we get the solutions $(\frac{\sqrt{2}}{2}, \frac{1}{2})$ and $(-\frac{\sqrt{2}}{2}, -\frac{1}{2})$.

For $\lambda = -\frac{1}{2\sqrt{2}}$ we get

$$x = -\sqrt{2}y$$

and plugging into the third equation gives $4y^2 = 1$, and the solutions $(\frac{\sqrt{2}}{2}, -\frac{1}{2})$ and $(-\frac{\sqrt{2}}{2}, \frac{1}{2})$. So finally we have the solutions: $(1,0)$, $(-1,0)$, $(\frac{\sqrt{2}}{2}, \frac{1}{2})$, $(-\frac{\sqrt{2}}{2}, -\frac{1}{2})$, $(\frac{\sqrt{2}}{2}, -\frac{1}{2})$, $(-\frac{\sqrt{2}}{2}, \frac{1}{2})$.

Evaluating f at these points gives:

$$f(0, \frac{\sqrt{2}}{2}) = f(0, -\frac{\sqrt{2}}{2}) = f(1,0) = f(-1,0) = 0$$
$$f(\frac{\sqrt{2}}{2}, \frac{1}{2}) = f(-\frac{\sqrt{2}}{2}, -\frac{1}{2}) = \frac{\sqrt{2}}{4}$$
$$f(\frac{\sqrt{2}}{2}, -\frac{1}{2}) = f(-\frac{\sqrt{2}}{2}, \frac{1}{2}) = -\frac{\sqrt{2}}{4}.$$

Hence the maximum value of f is $\frac{\sqrt{2}}{4}$ and the minimum value of f is $-\frac{\sqrt{2}}{4}$.

14.   The region $x^2 \geq y$ is the shaded region in Figure 14.16 which includes the parabola $y = x^2$.

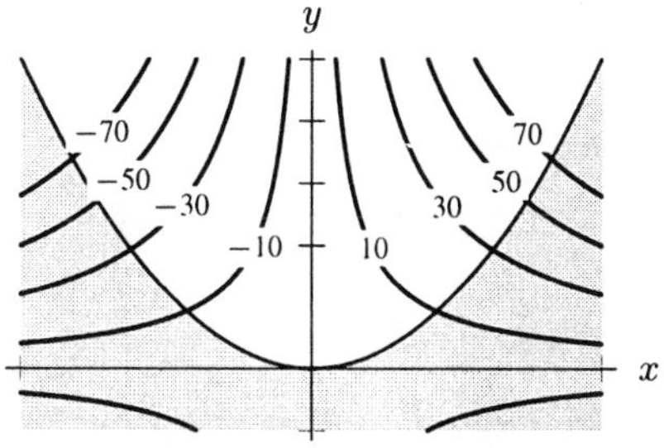

Figure 14.16

We first want to find the relative maxima and minima of f in the interior of our region. So we need to find the extrema of

$$f(x,y) = x^2 - y^2, \quad \text{in the region} \quad x^2 > y.$$

For this we compute the critical points:

$$f_x = 2x = 0$$
$$f_y = -2y = 0.$$

As $(0,0)$ does not belong to the region $x^2 > y$, we have no critical points. Now let's find the relative extrema of f on the boundary of our region, hence this time we have to solve a constraint problem. We want to find the extrema of $f(x,y) = x^2 - y^2$ subject to $g(x,y) = x^2 - y = 0$. We use Lagrange multipliers:

$$\text{grad } f = \lambda \text{ grad } g \quad \text{and} \quad x^2 = y.$$

This gives

$$2x = 2\lambda x$$
$$2y = \lambda$$
$$x^2 = y.$$

From the first equation we get $x = 0$ or $\lambda = 1$.

If $x = 0$, from the third equation we get $y = 0$, so one solution is $(0,0)$. If $x \neq 0$, then $\lambda = 1$ and from the second equation we get $y = \frac{1}{2}$. This gives $x^2 = \frac{1}{2}$ so the solutions $(\frac{1}{\sqrt{2}}, \frac{1}{2})$ and $(-\frac{1}{\sqrt{2}}, \frac{1}{2})$.

So $f(0,0) = 0$ and $f(\frac{1}{\sqrt{2}}, \frac{1}{2}) = f(-\frac{1}{\sqrt{2}}, \frac{1}{2}) = \frac{1}{4}$. From Figure 14.16 showing the level curves of f and the region $x^2 \geq y$, we see that $(0,0)$ is a local minimum of f on $x^2 = y$, but not a global minimum and that $(\frac{1}{\sqrt{2}}, \frac{1}{2})$ and $(-\frac{1}{\sqrt{2}}, \frac{1}{2})$ are global maxima of f on $x^2 = y$ but *not* global maxima of f on the whole region $x^2 \geq y$.

So there are no global extrema of f in the region $x^2 \geq y$.

15. The region $x^2 + y^2 \leq 2$ is the shaded disk of radius $\sqrt{2}$ centered at the origin (including the circle $x^2 + y^2 = 2$) as shown in Figure 14.17.

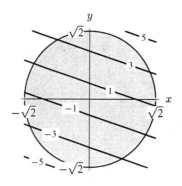

Figure 14.17

We first find the relative maxima and minima of f in the interior of our disk. So we need to find the extrema of

$$f(x,y) = x + 3y, \quad \text{in the region} \quad x^2 + y^2 < 2.$$

As

$$f_x = 1$$
$$f_y = 3$$

f doesn't have critical points. Now let's find the relative extrema of f on the boundary of the disk. We want to find the extrema of $f(x,y) = x + 3y$ subject to the constraint $g(x,y) = x^2 + y^2 - 2 = 0$. We use Lagrange multipliers

$$\text{grad } f = \lambda \text{ grad } g \quad \text{and} \quad x^2 + y^2 = 2,$$

which give

$$1 = 2\lambda x$$
$$3 = 2\lambda y$$
$$x^2 + y^2 = 2.$$

As λ cannot be zero, we solve for x and y in the first two equations and get $x = \frac{1}{2\lambda}$ and $y = \frac{3}{2\lambda}$. Plugging into the third equation gives

$$8\lambda^2 = 10$$

so $\lambda = \pm\frac{\sqrt{5}}{2}$ and we get the solutions $(\frac{1}{\sqrt{5}}, \frac{3}{\sqrt{5}})$ and $(-\frac{1}{\sqrt{5}}, -\frac{3}{\sqrt{5}})$. Evaluating f at these points gives

$$f(\frac{1}{\sqrt{5}}, \frac{3}{\sqrt{5}}) = 2\sqrt{5} \quad \text{and}$$

$$f(-\frac{1}{\sqrt{5}}, -\frac{3}{\sqrt{5}}) = -2\sqrt{5}$$

Therefore $(\frac{1}{\sqrt{5}}, \frac{3}{\sqrt{5}})$ is a global maximum of f and $(-\frac{1}{\sqrt{5}}, -\frac{3}{\sqrt{5}})$ is a global minimum of f on the whole region $x^2 + y^2 \le 2$.

16. The region $x + y \ge 1$ is the shaded half plane (including the line $x + y = 1$) shown in Figure 14.18.

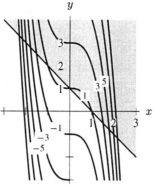

Figure 14.18

Let's look for the critical points of f in the interior of the region. As

$$f_x = 3x^2$$
$$f_y = 1$$

there are no critical points inside the shaded region. Now let's find the extrema of f on the boundary of our region. We want the extrema of $f(x,y) = x^3 + y$ subject to the constraint $g(x,y) = x + y - 1 = 0$. We use Lagrange multipliers

$$\text{grad } f = \lambda \text{ grad } g \quad \text{and} \quad x + y = 1,$$

which give

$$3x^2 = \lambda$$
$$1 = \lambda$$
$$x + y = 1.$$

From the first two equations we get $3x^2 = 1$, so the solutions are

$$(\frac{1}{\sqrt{3}}, 1 - \frac{1}{\sqrt{3}}) \quad \text{and} \quad (-\frac{1}{\sqrt{3}}, 1 + \frac{1}{\sqrt{3}}).$$

Evaluating f at these points we get

$$f(\frac{1}{\sqrt{3}}, 1 - \frac{1}{\sqrt{3}}) = 1 - \frac{2}{3\sqrt{3}}$$
$$f(-\frac{1}{\sqrt{3}}, 1 + \frac{1}{\sqrt{3}}) = 1 + \frac{2}{3\sqrt{3}}.$$

So $(\frac{1}{\sqrt{3}}, 1 - \frac{1}{\sqrt{3}})$ is a local minimum and $(-\frac{1}{\sqrt{3}}, 1 + \frac{1}{\sqrt{3}})$ is a local maximum of f on $x + y = 1$. Are they global extrema as well?

If we take x very big and $y = 1 - x$ then $f(x,y) = x^3 + y = x^3 - x + 1$ which can be made as big as we want (if we choose x big enough). So there will be no global maximum.

Similarly, taking x negative with big absolute value and $y = 1 - x$, $f(x,y) = x^3 + y = x^3 - x + 1$ can be made as small as we want (if we choose x small enough). So there is no global minimum. This can also be seen from Figure 14.18.

17. The region $x^2 + y^2 \le 1$ is the shaded disk of radius 1 centered at the origin (including the circle $x^2 + y^2 = 1$) shown in Figure 14.19.

Let's first compute the critical points of f in the interior of the disk. We have

$$f_x = 3x^2 = 0$$
$$f_y = -2y = 0,$$

whose solution is $x = y = 0$. So the only one critical point is $(0,0)$. As $f_{xx}(0,0) = 0$, $f_{yy}(0,0) = -2$ and $f_{xy}(0,0) = 0$,

$$D = f_{xx}(0,0) \cdot f_{yy}(0,0) - (f_{xy}(0,0))^2 = 0$$

which doesn't tell us anything about the nature of the critical point $(0,0)$.

But, if we choose x,y very small in absolute value and such that $x^3 > y^2$, then $f(x,y) > 0$. If we choose x,y very small in absolute value and such that $x^3 < y^2$, then $f(x,y) < 0$. As $f(0,0) = 0$, we conclude that $(0,0)$ is a saddle point.

We can get the same conclusion looking at the level curves of f around $(0,0)$, as shown in Figure 14.20.

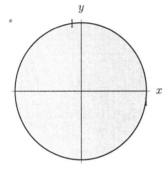

Figure 14.19

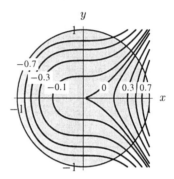

Figure 14.20: Level curves of f

So, f doesn't have extrema in the interior of the disk.

Now, let's find the relative extrema of f on the circle $x^2 + y^2 = 1$. So we want the extrema of $f(x,y) = x^3 - y^2$ subject to the constraint $g(x,y) = x^2 + y^2 - 1 = 0$. Using Lagrange multipliers we get

$$\text{grad } f = \lambda \text{ grad } g \quad \text{and} \quad x^2 + y^2 = 1,$$

which gives

$$3x^2 = 2\lambda x$$
$$-2y = 2\lambda y$$
$$x^2 + y^2 = 1.$$

From the second equation $y = 0$ or $\lambda = -1$.

If $y = 0$, from the third equation we get $x^2 = 1$, which gives the solutions $(1,0)$, $(-1,0)$.

If $y \ne 0$ then $\lambda = -1$ and from the first equation we get $3x^2 = -2x$, hence $x = 0$ or $x = -\frac{2}{3}$. If $x = 0$, from the third equation we get $y^2 = 1$, so the solutions $(0,1),(0,-1)$. If $x = -\frac{2}{3}$, from the third equation we get $y^2 = \frac{5}{9}$, so the solutions $(-\frac{2}{3}, \frac{\sqrt{5}}{3})$, $(-\frac{2}{3}, -\frac{\sqrt{5}}{3})$.

Evaluating f at these points we get

$$f(1,0) = 1, \quad f(-1,0) = f(0,1) = f(0,-1) = -1$$

and

$$f\left(-\frac{2}{3}, -\frac{\sqrt{5}}{3}\right) = f\left(-\frac{2}{3}, \frac{\sqrt{5}}{3}\right) = -\frac{23}{27}.$$

Therefore the maximum value of f is 1 and the minimum value is -1.

18. We wish to minimize the objective function

$$C(x, y, z) = 20x + 10y + 5z$$

subject to the budget constraint

$$Q(x, y, z) = 20x^{1/2}y^{1/4}z^{2/5} = 1,200.$$

Therefore, we solve the equations grad $C = \lambda$ grad Q and $Q = 1,200$:

$$20 = 10\lambda x^{-1/2}y^{1/4}z^{2/5} \quad \text{or} \quad \lambda = 2x^{1/2}y^{-1/4}z^{-2/5},$$
$$10 = 5\lambda x^{1/2}y^{-3/4}z^{2/5}, \quad \text{or} \quad \lambda = 2x^{-1/2}y^{3/4}z^{-2/5},$$
$$5 = 8\lambda x^{1/2}y^{1/4}z^{-3/5}, \quad \text{or} \quad \lambda = 0.625x^{-1/2}y^{-1/4}z^{3/5},$$
$$20x^{1/2}y^{1/4}z^{2/5} = 1,200.$$

The first and second equations imply that

$$x = y,$$

while the second and third equations imply that

$$3.2y = z.$$

Substituting for x and z in the constraint equation gives

$$20y^{1/2}y^{1/4}(3.2y)^{2/5} = 1200$$

$$y \approx 23.47,$$

and so

$$x \approx 23.47 \quad \text{and} \quad z \approx 75.1.$$

19. We want to minimize

$$C = f(q_1, q_2) = 2q_1^2 + q_1 q_2 + q_2^2 + 500$$

subject to the constraint $q_1 + q_2 = 200$ or $g(q_1, q_2) = q_1 + q_2 - 200 = 0$.
Since $\nabla f = (4q_1 + q_2)\vec{i} + (2q_2 + q_1)\vec{j}$ and $\nabla g = \vec{i} + \vec{j}$, $\nabla f = \lambda \nabla g$ gives

$$4q_1 + q_2 = \lambda$$
$$2q_2 + q_1 = \lambda.$$

Solving we get

$$4q_1 + q_2 = 2q_2 + q_1.$$

So

$$3q_1 = q_2.$$

We want

$$q_1 + q_2 = 200$$
$$q_1 + 3q_1 = 4q_1 = 200.$$

Therefore

$$q_1 = 50 \text{ units}, \quad q_2 = 150 \text{ units}.$$

20. Constraint is $G = P_1 x + P_2 y - K = 0$.
    Since $\nabla Q = \lambda \nabla G$, we have

    $$cax^{a-1}y^b = \lambda P_1 \quad \text{and} \quad cbx^a y^{b-1} = \lambda P_2.$$

    Dividing the two equations yields $\dfrac{cax^{a-1}y^b}{cbx^a y^{b-1}} = \dfrac{\lambda P_1}{\lambda P_2}$, or simplifying, $\dfrac{ay}{bx} = \dfrac{P_1}{P_2}$. Hence, $y = \dfrac{bP_1}{aP_2}x$.

    Substitute into the constraint to obtain $P_1 x + P_2 \dfrac{bP_1}{aP_2}x = P_1 \left(\dfrac{a+b}{a}\right)x = K$, giving

    $$x = \frac{aK}{(a+b)P_1} \quad \text{and} \quad y = \frac{bK}{(a+b)P_2}.$$

    We now check that this is indeed the maximization point. Since $x, y \geq 0$, possible maximization points are $(0, \dfrac{K}{P_2})$, $(\dfrac{K}{P_1}, 0)$, and $(\dfrac{aK}{(a+b)P_1}, \dfrac{bK}{(a+b)P_2})$. Since $Q = 0$ for the first two points and Q is positive for the last point, it follows that $(\dfrac{aK}{(a+b)P_1}, \dfrac{bK}{(a+b)P_2})$ gives the maximal value.

21. (a) The curves are shown in Figure 14.21.

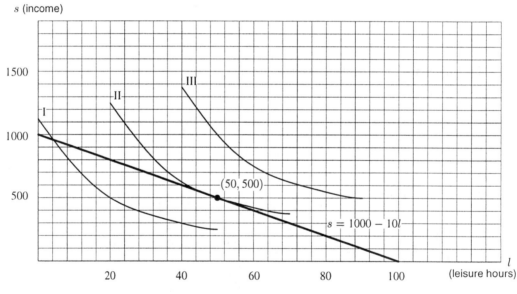

Figure 14.21

    (b) The income equals $10/hour times the number of hours of work:

    $$s = 10(100 - l) = 1000 - 10l.$$

    (c) The graph of this constraint is the straight line in Figure 14.21.
    (d) For any given salary, curve III allows for the most leisure time, curve I the least. Similarly, for any amount of leisure time, any curve III also has the greatest salary, and curve I the least. Thus, any point on curve III is preferable to any point on curve II, which is preferable to any point on curve I. We prefer to be on the outermost curve that our constraint allows. We want to choose the point on $s = 1000 - 10l$ which is on the most preferable curve. Since all the curves are concave up, this occurs at the point where $s = 1000 - 10l$ is *tangent* to curve II. So we choose $l = 50$, $s = 500$, and work 50 hours a week.

22. (a) The gradient vectors, ∇f, point inward around a local maximum. See the two points marked A in Figure 14.22.

(b) Some of the gradient vectors around a saddle are pointing inward toward the point; some are pointing outward away from the point. See the point marked B in Figure 14.22.

(c) The critical points on $g = 1$ are at points where ∇f is perpendicular to the curve $g = 1$. There are four of them, all marked with a dot in Figure 14.22. Imagine the level surfaces of f sketched in everywhere perpendicular to ∇f; the maximum value of f is at the point marked C in Figure 14.22

(d) Again imagine level curves of f. The minimum value of f is at the point marked D.

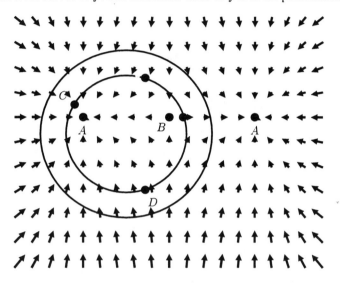

Figure 14.22

(e) At C, the maximum on $g = 1$, the vector ∇g points outwards (because it points towards g = 2), while ∇f points inwards. The Lagrange multiplier, λ, is defined so that $\nabla f = \lambda \nabla g$, so λ must be negative.

23.

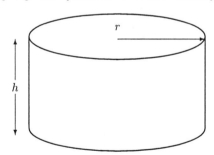

Figure 14.23

Let V be the volume and S be the surface area of the container. Then

$$V = \pi r^2 h \quad \text{and} \quad S = 2\pi r h + 2\pi r^2$$

where h is the height and r is the radius as shown in Figure 14.23. We have $V = 100\,\text{cm}^3$ as our constraint. Since

$$\nabla S = (2\pi h + 4\pi r)\vec{i} + 2\pi r \vec{j} = \pi((2h + 4r)\vec{i} + 2r\vec{j})$$

and $\quad \nabla V = 2\pi r h \vec{i} + \pi r^2 \vec{j} = \pi(2rh\vec{i} + r^2\vec{j}),$

at the optimum

$$\nabla S = \lambda \nabla V, \text{ we have}$$

$$\pi((2h + 4r)\vec{i} + 2r\vec{j}) = \pi\lambda(2rh\vec{i} + r^2\vec{j}),$$

that is $2h + 4r = 2\lambda rh$ and $2r = \lambda r^2$,   hence $\lambda = \dfrac{2}{r}$.

We assume $r \neq 0$ or else we have a very awkward cylinder. Then, plug $\lambda = 2/r$ into the first equation to obtain:

$$2h + 4r = 2\left(\frac{2}{r}\right) rh$$

$$2h + 4r = 4h$$

$$h = 2r.$$

Finally, solve for r and h using the constraint:

$$V = \pi r^2 h = 100$$

$$\pi r^2 (2r) = 100$$

$$r^3 = \frac{50}{\pi}$$

$$r = \sqrt[3]{\frac{50}{\pi}}.$$

Solving for h, we obtain $h = 2r = 2\sqrt[3]{\dfrac{50}{\pi}}$.

24. (a)   We want to minimize C subject to $g = x + y = 39$. Solving $\nabla C = \lambda \nabla g$ gives

$$10x + 2y = \lambda$$

$$2x + 6y = \lambda$$

so $y = 2x$. Solving with $x + y = 39$ gives $x = 13, y = 26, \lambda = 182$. Therefore $C = 4349$.

(b)   Since $\lambda = 182$, increasing production by 1 will cause costs to increase by approximately \$182 (because $\lambda = \dfrac{\|\nabla C\|}{\|\nabla g\|}$ = rate of change of C with g). Similarly, decreasing production by 1 will save approximately \$182.

25.   We want to minimize the function $h(x, y)$ subject to the constraint that

$$g(x, y) = x^2 + y^2 = 1,000^2 = 1,000,000.$$

Using the method of Lagrange multipliers, we obtain the following system of equations:

$$h_x = -\frac{10x + 4y}{10,000} = 2\lambda x,$$

$$h_y = -\frac{4x + 4y}{10,000} = 2\lambda y,$$

$$x^2 + y^2 = 1,000,000.$$

Multiplying the first equation by y and the second by x we get

$$\frac{-y(10x + 4y)}{10,000} = \frac{-x(4x + 4y)}{10,000}.$$

Hence:

$$2y^2 + 3xy - 2x^2 = (2y - x)(y + 2x) = 0,$$

and so the climber either moves along the line $x = 2y$ or $y = -2x$.

We must now choose one of these lines and the direction along that line which will lead to the point of minimum height on the circle. To do this we find the points of intersection of these lines with the circle $x^2 + y^2 = 1,000,000$, compute the corresponding heights, and then select the minimum point.

If $x = 2y$, the third equation gives

$$5y^2 = 1,000^2,$$

so that $y = \pm 1,000/\sqrt{5} \approx \pm 447.21$ and $x = \pm 894.43$. The corresponding height is $h(\pm 894.43, \pm 447.21) = 2400$ m. If $y = -2x$, we find that $x = \pm 447.21$ and $y = \mp 894.43$. The corresponding height is $h(\pm 447.21, \mp 894.43) = 2900$ m. Therefore, she should travel along the line $x = 2y$, in either of the two possible directions.

26. (a) Let c be the cost of producing the product. Then $c = 10W + 20K = 3000$. At optimum production,

$$\nabla q = \lambda \nabla c.$$

$\nabla q = \left(\frac{9}{2} W^{-\frac{1}{4}} K^{\frac{1}{4}} \right) \vec{i} + \left(\frac{3}{2} W^{\frac{3}{4}} K^{-\frac{3}{4}} \right) \vec{j}$, and $\nabla c = 10\vec{i} + 20\vec{j}$. Equating we get

$$\frac{9}{2} W^{-\frac{1}{4}} K^{\frac{1}{4}} = \lambda 10, \quad \text{and} \quad \frac{3}{2} W^{\frac{3}{4}} K^{-\frac{3}{4}} = \lambda 20.$$

Dividing yields $K = \frac{1}{6} W$, so substituting into c gives

$$10W + 20 \left(\frac{1}{6} W \right) = \frac{40}{3} W = 3000.$$

Thus $W = 225$ and $K = 37.5$. Substituting both answers to find λ gives

$$\lambda = \frac{\frac{9}{2}(225)^{-\frac{1}{4}}(37.5)^{\frac{1}{4}}}{10} = 0.2875.$$

We also find the optimum quantity produced, $q = 6(225)^{\frac{3}{4}}(37.5)^{\frac{1}{4}} = 862.57$.

(b) At the optimum values found above, marginal productivity of labor is given by

$$\left. \frac{\partial q}{\partial W} \right|_{(225,37.5)} = \left. \frac{9}{2} W^{-\frac{1}{4}} K^{\frac{1}{4}} \right|_{(225,37.5)} = 2.875,$$

and marginal productivity of capital is given by

$$\left. \frac{\partial q}{\partial K} \right|_{(225,37.5)} = \left. \frac{3}{2} W^{\frac{3}{4}} K^{-\frac{3}{4}} \right|_{(225,37.5)} = 5.750.$$

The ratio of marginal productivity of labor to that of capital is

$$\frac{\frac{\partial q}{\partial W}}{\frac{\partial q}{\partial K}} = \frac{1}{2} = \frac{10}{20} = \frac{\text{cost of a unit of L}}{\text{cost of a unit of K}}.$$

(c) When the budget is increased by one dollar, we substitute the relation $K_1 = \frac{1}{6} W_1$ into $10W_1 + 20K_1 = 3001$ which gives $10W_1 + 20(\frac{1}{6} W_1) = \frac{40}{3} W_1 = 3001$. Solving yields $W_1 = 225.075$ and $K_1 = 37.513$, so $q_1 = 862.86 = q + 0.29$. Thus production has increased by $0.29 \approx \lambda$, the Lagrange Multiplier.

27. (a) The problem is to maximize
$$V = 1000D^{0.6}N^{0.3}$$

subject to the budget constraint in dollars
$$40000D + 10000N \leq 600000$$

or (in thousand dollars)
$$40D + 10N \leq 600$$

(b) Let $B = 40D + 10N = 600$ (thousand dollars) be the budget constraint. At the optimum
$$\nabla V = \lambda \nabla B,$$
$$\text{so} \quad \frac{\partial V}{\partial D} = \lambda \frac{\partial B}{\partial D} = 40\lambda$$
$$\frac{\partial V}{\partial N} = \lambda \frac{\partial B}{\partial N} = 10\lambda.$$
$$\text{Thus} \quad \frac{\frac{\partial V}{\partial D}}{\frac{\partial V}{\partial N}} = 4.$$

Therefore, at the optimum point, the rate of increase in the number of visits to the number of doctors is four times the corresponding rate for nurses. This factor of four is the same as the ratio of the salaries.

(c) Differentiating and setting $\nabla V = \lambda \nabla B$ yields
$$600D^{-0.4}N^{0.3} = 40\lambda$$
$$300D^{0.6}N^{-0.7} = 10\lambda$$

Thus, we get
$$\frac{600D^{-0.4}N^{0.3}}{40} = \lambda = \frac{300D^{0.6}N^{-0.7}}{10}$$

So
$$N = 2D.$$

To solve for D and N, substitute in the budget constraint:
$$600 - 40D - 10N = 0$$
$$600 - 40D - 10 \cdot (2D) = 0$$

So $D = 10$ and $N = 20$.
$$\lambda = \frac{600(10^{-0.4})(20^{0.3})}{40} \approx 14.67$$

Thus the clinic should hire 10 doctors and 20 nurses. With that staff, the clinic can provide
$$V = 1000(10^{0.6})(20^{0.3}) \approx 9{,}779 \text{ visits per year.}$$

(d) From part c), the Lagrange multiplier is $\lambda = 14.67$. At the optimum, the Lagrange multiplier tells us that about 14.67 extra visits can be generated through an increase of $1,000 in the budget. (If we had written out the constraint in dollars instead of thousands of dollars, the Lagrange multiplier would tell us the number of extra visits per dollar.)

(e)   The marginal cost, MC, is the cost of an additional visit. Thus, at the optimum point, we need the reciprocal of the Lagrange multiplier:

$$MC = \frac{1}{\lambda} \approx \frac{1}{14.67} \approx 0.068 \text{ (thousand dollars)}$$

i.e. at the optimum point, an extra visit costs the clinic 0.068 thousand dollars, or $68.

This production function exhibits declining returns to scale (e.g. doubling both inputs less than doubles output, because the two exponents add up to less than one). This means that for large V, increasing V will require increasing D and N by more than when V is small. Thus the cost of an additional visit is greater for large V than for small. In other words, the marginal cost will rise with the number of visits.

28.   The objective function is

$$f(x, y, z) = \sqrt{(x - a)^2 + (y - b)^2 + (z - c)^2},$$

and the constraint is

$$g(x, y, z) = Ax + By + Cz + D = 0.$$

Partial derivatives of f and g are

$$f_x = \frac{\frac{1}{2} \cdot 2 \cdot (x - a)}{f(x, y, z)} = \frac{x - a}{f(x, y, z)},$$

$$f_y = \frac{\frac{1}{2} \cdot 2 \cdot (y - b)}{f(x, y, z)} = \frac{y - b}{f(x, y, z)},$$

$$f_z = \frac{\frac{1}{2} \cdot 2 \cdot (z - c)}{f(x, y, z)} = \frac{z - c}{f(x, y, z)},$$

$$g_x = A, \quad g_y = B, \quad \text{and} \quad g_z = C.$$

Using Lagrange multipliers, we need to solve the equations

$$\text{grad } f = \lambda \text{ grad } g$$

where $\text{grad } f = f_x \vec{i} + f_y \vec{j} + f_z \vec{k}$ and $\text{grad } g = g_x \vec{i} + g_y \vec{j} + g_z \vec{k}$. This gives a system of equations:

$$\frac{x - a}{f(x, y, z)} = \lambda A$$

$$\frac{y - b}{f(x, y, z)} = \lambda B$$

$$\frac{z - c}{f(x, y, z)} = \lambda C$$

$$Ax + By + Cz + D = 0.$$

Now $\frac{x-a}{A} = \frac{y-b}{B} = \frac{z-c}{C} = \lambda f(x, y, z)$ gives

$$x = \frac{A}{B}(y - b) + a,$$

$$z = \frac{C}{B}(y - b) + c,$$

Substitute into the constraint,

$$A \left(\frac{A}{B}(y - b) + a \right) + By + C \left(\frac{C}{B}(y - b) + c \right) + D = 0,$$

$$\left(\frac{A^2}{B} + B + \frac{C^2}{B} \right) y = \frac{A^2}{B}b - Aa + \frac{C^2}{B}b - Cc - D.$$

Hence

$$y = \frac{(A^2 + C^2)b - B(Aa + Cc + D)}{A^2 + B^2 + C^2},$$

$$y - b = \frac{-B(Aa + Bb + Cc + D)}{A^2 + B^2 + C^2}$$

$$x - a = \frac{A}{B}(y - b)$$

$$= \frac{-A(Aa + Bb + Cc + D)}{A^2 + B^2 + C^2}$$

$$z - c = \frac{C}{B}(y - b)$$

$$= \frac{-C(Aa + Bb + Cc + D)}{A^2 + B^2 + C^2}$$

Thus the minimum $f(x, y, z)$ is

$$f(x, y, z) = \sqrt{(x - a)^2 + (y - b)^2 + (z - c)^2}$$

$$= \left[\left(\frac{-A(Aa + Bb + Cc + D)}{A^2 + B^2 + C^2} \right)^2 + \left(\frac{-B(Aa + Bb + Cc + D)}{A^2 + B^2 + C^2} \right)^2 \right.$$

$$\left. + \left(\frac{-C(Aa + Bb + Cc + D)}{A^2 + B^2 + C^2} \right)^2 \right]^{1/2}$$

$$= \frac{|Aa + Bb + Cc + D|}{\sqrt{A^2 + B^2 + C^2}}.$$

The geometric meaning is finding the shortest distance from a point (a, b, c) to the plane $Ax + By + Cz + D = 0$.

29. (a) We draw the level curves (parallel straight lines) of $f(x, y) = ax + by + c$. We can see that the level lines with the maximum and minimum f-values which intersect with the disk are the level lines that are tangent to the boundary of the disk. Therefore, the maximum and minimum occur at the boundary of the disk. See Figure 14.24.

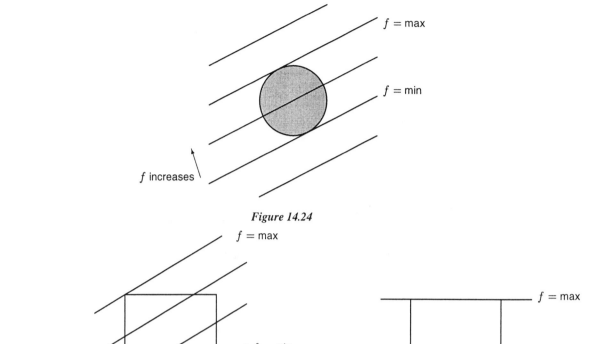

Figure 14.24

Figure 14.25   *Figure 14.26*

(b) Similar to part (a), we see the level lines with the largest and smallest f-values which intersect with the rectangle must pass the corner of the rectangle. So the maximum and minimum occur at the corners of rectangle. See Figure 14.25. When the level curves are parallel to a pair of the sides, then the points on the sides are all maximum or minimum, as shown below in Figure 14.26.

(c) The graph of f is a plane. The part of the graph lying above a disk R is either a flat disk, in which case every point is a maximum, or is a tilted ellipse, in which case you can see that the maximum will be on the edge. Similarly, the part lying above a rectangle is either a rectangle or a tilted parallelogram, in which case the maximum will be at a corner.

30. (a) The solution to Problem 26 gives $\lambda = 0.29$. We recalculate λ with a budget of $4000.
    The condition that $\text{grad } q = \lambda \text{ grad(budget)}$ in Problem 26 gives

$$\frac{9}{2}W^{-1/4}K^{1/4} = \lambda(10) \quad \text{and} \quad \frac{3}{2}W^{3/4}K^{-3/4} = \lambda(20),$$

so $K = \frac{1}{6}W$. Substituting into the budget constraint after replacing the budget of $3000 by $4000 gives

$$10W + 20(\frac{1}{6}W) = \frac{40}{3}W = 4000.$$

Thus, $W = 300$ and $K = 50$ and $q = 1150.098$.
    Multiplying the first equation by W and the second by K and adding gives

$$W(\frac{9}{2}W^{-1/4}K^{1/4}) + K(\frac{3}{2}W^{3/4}K^{-3/4}) = W(10\lambda) + K(20\lambda).$$

So

$$\left(\frac{9}{2} + \frac{3}{2}\right)W^{3/4}K^{1/4} = \lambda(10W + 20K)$$

$$6W^{3/4}K^{1/4} = \lambda(4000)$$

Thus,

$$\lambda = \frac{6W^{3/4}K^{1/4}}{4000} = \frac{1150.098}{4000} = 0.29$$

Thus, the value of λ remains unchanged.

(b) The solution to Problem 27 shows that $\lambda = 14.67$. We solve the problem again with a budget of $700,000.

The condition that grad $V = \lambda$ grad B in Problem 27 gives

$$600D^{-0.4}N^{0.3} = 40\lambda$$
$$300D^{0.6}N^{-0.7} = 10\lambda$$

Thus, $N = 2D$. Substituting in the budget constraint after replacing the budget of 600 by 700 (the budget in measured in thousands of dollars) gives

$$40D + 10(2D) = 700$$

so $D = 11.667$ and $N = 23.337$ and $V = 11234.705$. As in part a), we multiply the first equation by D and the second by N and add:

$$D(600D^{-0.4}N^{0.3}) + N(300D^{0.6}N^{-0.7}) = D(40\lambda) + N(10\lambda),$$

so

$$(600 + 300)D^{0.6}N^{0.3} = \lambda(400 + 10N)$$
$$900D^{0.6}N^{0.3} = \lambda(700)$$

Since $V = 1000D^{0.6}N^{0.3} = 11234.705$, we have

$$\lambda = \frac{900D^{0.6}N^{0.3}}{700} = \frac{9}{7}\left(\frac{V}{1000}\right) = 14.44.$$

Thus, the value of λ has changed with the budget.

(c) We are interested in the marginal increase of production with budget (that is, the value of λ) and whether it is affected by the budget.

Suppose $\$B$ is the budget. In part (a) we found

$$\lambda = \frac{6W^{3/4}K^{1/4}}{B}$$

and in part (b) we found

$$\lambda = \frac{900D^{0.6}N^{0.3}}{B}.$$

In part (a), both W and K are proportional to B. Thus, $W = c_1B$ and $K = c_2B$, so

$$\lambda = \frac{6(c_1B)^{3/4}(c_2B)^{1/4}}{B}$$
$$= \frac{6c_1^{3/4}C_2^{1/4}B^{3/4}B^{1/4}}{B}$$
$$= 6c_1^{3/4}c_2^{1/4}.$$

So we see λ is independent of B.

In part (b), both D and N are proportional to B, so $D = c_3B$ and $N = c_4B$. Thus,

$$\lambda = \frac{900(c_3B)^{0.6}(c_4B)^{0.3}}{B}$$
$$= \frac{900c_3^{0.6}C_4^{0.3}B^{0.6}B^{0.3}}{B}$$
$$= 900c_3^{0.6}c_4^{0.3}\frac{1}{B^{0.1}}.$$

So we see λ is not independent of B.

The crucial difference is that the exponents in Problem 26 add to 1, that is $3/4 + 1/4 = 1$, whereas the exponents in Problem 27 do not add to 1, since $0.6 + 0.3 = 0.9$.

Thus, the condition that must be satisfied by the Cobb-Donglas production function

$$Q = cK^aL^b$$

to ensure that the value of λ is not affected by production is that

$$a + b = 1.$$

This is called constant returns to scale.

Solutions for Chapter 14 Review

1. The partial derivatives are

$$f_x = \cos x + \cos(x + y).$$
$$f_y = \cos y + \cos(x + y).$$

Setting $f_x = 0$ and $f_y = 0$ gives

$$\cos x = \cos y$$

For $0 < x < \pi$ and $0 < y < \pi$, $\cos x = \cos y$ only if $x = y$. Then, setting $f_x = f_y = 0$:

$$\cos x + \cos 2x = 0,$$
$$\cos x + 2\cos^2 x - 1 = 0,$$
$$(2\cos x - 1)(\cos x + 1) = 0.$$

So $\cos x = 1/2$ or $\cos x = -1$, that is $x = \pi/3$ or $x = \pi$. For the given domain $0 < x < \pi, 0 < y < \pi$, we only consider the solution when $x = \pi/3$ then $y = x = \pi/3$. Therefore, the critical point is $\left(\frac{\pi}{3}, \frac{\pi}{3}\right)$.

Since

$$f_{xx}(x,y) = -\sin x - \sin(x+y) \qquad f_{xx}(\tfrac{\pi}{3}, \tfrac{\pi}{3}) = -\sin\tfrac{\pi}{3} - \sin\tfrac{2\pi}{3} = -\sqrt{3}$$
$$f_{xy}(x,y) = -\sin(x+y) \qquad f_{xy}(\tfrac{\pi}{3}, \tfrac{\pi}{3}) = -\sin\tfrac{2\pi}{3} \qquad = -\tfrac{\sqrt{3}}{2}$$
$$f_{yy}(x,y) = -\sin y - \sin(x+y) \qquad f_{yy}(\tfrac{\pi}{3}, \tfrac{\pi}{3}) = -\sin\tfrac{\pi}{3} - \sin\tfrac{2\pi}{3} = -\sqrt{3}$$

the discriminant is

$$D(x,y) = f_{xx}f_{yy} - f_{xy}^2$$
$$= (\sqrt{3})(-\sqrt{3}) - (-\tfrac{\sqrt{3}}{2})^2 = \tfrac{9}{4} > 0.$$

Since $f_{xx}(\tfrac{\pi}{3}, \tfrac{\pi}{3}) = -\sqrt{3} < 0$, $(\tfrac{\pi}{3}, \tfrac{\pi}{3})$ is a local maximum.

2. The partial derivatives are $f_x = 2x - 3y$, $f_y = 3y^2 - 3x$. For critical points, solve $f_x = 0$ and $f_y = 0$ simultaneously. From $2x - 3y = 0$ we get $x = \tfrac{3}{2}y$. Substituting it into $3y^2 - 3x = 0$, we have that

$$3y^2 - 3(\tfrac{3}{2}y) = 3y^2 - \tfrac{9}{2}y = y(3y - \tfrac{9}{2}) = 0.$$

So $y = 0$ or $3y - \tfrac{9}{2} = 0$, that is, $y = 0$ or $y = 3/2$. Therefore the critical points are $(0,0)$ and $(\tfrac{9}{4}, \tfrac{3}{2})$. The contour diagram for f in Figure 14.27 (drawn by a computer), shows that $(0,0)$ is a saddle point and that $(\tfrac{9}{4}, \tfrac{3}{2})$ is a local minimum.

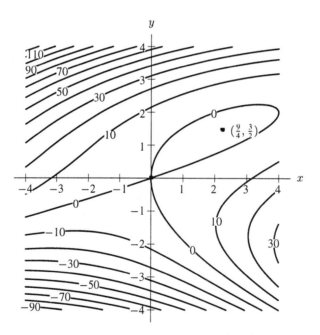

Figure 14.27: Contour map of $f(x,y) = x^2 + y^3 - 3xy$

We can also see that $(0,0)$ is a saddle point and $(\tfrac{9}{4}, \tfrac{3}{2})$ is a local minimum analytically. Since $f_{xx} = 2$, $f_{yy} = 6y$, $f_{xy} = -3$, the discriminant is

$$D(x,y) = f_{xx}f_{yy} - f_{xy}^2 = 12y - (-3)^2 = 12y - 9.$$

$D(0,0) = -9 < 0$, so $(0,0)$ is a saddle point.
$D(\tfrac{9}{4}, \tfrac{3}{2}) = 9 > 0$ and $f_{xx} = 2 > 0$, we know that $(\tfrac{9}{4}, \tfrac{3}{2})$ is a local minimum. The point $(\tfrac{9}{4}, \tfrac{3}{2})$ is not a global minimum since $f(\tfrac{9}{4}, \tfrac{3}{2}) = -1.6875$, whereas $f(0,-2) = -8$.

3. The partial derivatives are

$$f_x = y + \frac{1}{x}, f_y = x + 2y.$$

For critical points, solve $f_x = 0$ and $f_y = 0$ simultaneously. From $f_y = x + 2y = 0$ we get that $x = -2y$. Substituting into $f_x = 0$, we have

$$y + \frac{1}{x} = y - \frac{1}{2y} = \frac{1}{y}(y^2 - \frac{1}{2}) = 0$$

Since $\frac{1}{y} \neq 0$, $y^2 - \frac{1}{2} = 0$, therefore

$$y = \pm\frac{1}{\sqrt{2}} = \pm\frac{\sqrt{2}}{2},$$

and $x = \mp\sqrt{2}$. So the critical points are $\left(-\sqrt{2}, \frac{\sqrt{2}}{2}\right)$ and $\left(\sqrt{2}, -\frac{\sqrt{2}}{2}\right)$. But x must be greater than 0, so $\left(-\sqrt{2}, \frac{\sqrt{2}}{2}\right)$ is not in the domain.

The contour diagram for f in Figure 14.28 (drawn by computer), shows that $\left(\sqrt{2}, -\frac{\sqrt{2}}{2}\right)$ is a saddle point of $f(x, y)$.

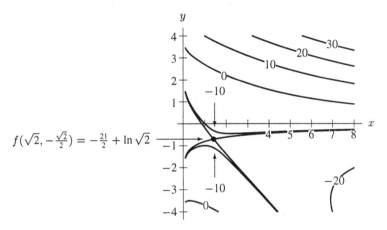

Figure 14.28: Contour map of $f(x, y) = xy + \ln x + y^2 - 10$

We can also see that $\left(\sqrt{2}, -\frac{\sqrt{2}}{2}\right)$ is a saddle point analytically. Since $f_{xx} = -\frac{1}{x^2}, f_{yy} = 2, f_{xy} = 1$, the discriminant is:

$$D(x, y) = f_{xx}f_{yy} - f_{xy}^2$$
$$= -\frac{2}{x^2} - 1.$$

$D\left(\sqrt{2}, -\frac{\sqrt{2}}{2}\right) = -2 < 0$, so $\left(\sqrt{2}, -\frac{\sqrt{2}}{2}\right)$ is a saddle point.

4. Note that the x-axis and the y-axis are not in the domain of f. Since $x \neq 0$ and $y \neq 0$, by setting $f_x = 0$ and $f_y = 0$ we get

$$f_x = 1 - \frac{1}{x^2} = 0 \text{ when } x = \pm 1$$
$$f_y = 1 - \frac{4}{y^2} = 0 \text{ when } y = \pm 2$$

So the critical points are $(1,2), (-1,2), (1,-2), (-1,-2)$. Since $f_{xx} = 2/x^3$ and $f_{yy} = 8/y^3$ and $f_{xy} = 0$, the discriminant is

$$D(x,y) = f_{xx}f_{yy} - f_{xy}^2 = \left(\frac{2}{x^3}\right)\left(\frac{8}{y^3}\right) - 0^2 = \frac{16}{(xy)^3}.$$

Since $D < 0$ at the points $(-1,2)$ and $(1,-2)$, these points are saddle points. Since $D > 0$ at $(1,2)$ and $(-1,-2)$ and $f_{xx}(1,2) > 0$ and $f_{xx}(-1,-2) < 0$, the point $(1,2)$ is a local minimum and the point $(-1,-2)$ is a local maximum. No global maximum or minimum, since $f(x,y)$ increases without bound if x and y increase in the first quadrant; $f(x,y)$ decreases without bound if x and y decrease in the third quadrant.

5. Since $f_{xx} < 0$ and $D = f_{xx}f_{yy} - f_{xy}^2 > 0$, $(1,3)$ is a maximum.

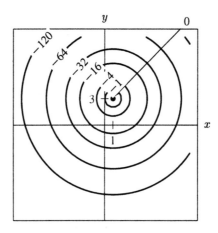

Figure 14.29

6. Let the line be in the form $y = b + mx$. Then, when x equals 0, 1, and 2, y equals b, $b + m$, and $b + 2m$ respectively. The sum of the squares of the vertical distances, which is what we want to minimize, is
$$f(m,b) = (4 - b)^2 + (3 - (b + m))^2 + (1 - (b + 2m))^2$$
To find critical points, set each partial derivative equal to zero.

$$f_m = 0 + 2(3 - (b + m))(-1) + 2(1 - (b + 2m))(-2)$$
$$= 6b + 10m - 10$$
$$f_b = 2(4 - b)(-1) + 2(3 - (b + m))(-1) + 2(1 - (b + 2m))(-1)$$
$$= 6b + 6m - 16$$

Setting both partial derivatives equal to zero and dividing by 2, we get a system of equations:

$$3b + 5m = 5$$
$$3b + 3m = 8$$

with solutions $m = -\frac{3}{2}$ and $b = \frac{25}{6}$. Thus, the line is $y = \frac{25}{6} - \frac{3}{2}x$.

7. We first find the critical points in the disk

$$\nabla z = (8x - y)\vec{i} + (8y - x)\vec{j}$$

Setting $\nabla z = 0$ gives $8x - y = 0$ and $8y - x = 0$. The only solution is $x = y = 0$. So $(0,0)$ is the only critical point in the disk.

Next we find the extremal values on the boundary using Lagrange multipliers. We have objective function $z = 4x^2 - xy + 4y^2$ and constraint $G = x^2 + y^2 - 2 = 0$.

$$\nabla z = (8x - y)\vec{i} + (8y - x)\vec{j}$$
$$\nabla G = 2x\vec{i} + 2y\vec{j}$$

$\nabla z = \lambda \nabla G$ gives

$$8x - y = 2\lambda x$$
$$8y - x = 2\lambda y$$

If $\lambda = 0$ we get

$$8x - y = 0$$
$$8y - x = 0$$

with only solutions $x = y = 0$, which doesn't satisfy the constraint: $x^2 + y^2 - 2 = 0$. Therefore $\lambda \neq 0$ and we get:

$$2\lambda y(8x - y) = 2\lambda x(8y - x)$$

and

$$y(8x - y) = x(8y - x).$$

So $x^2 = y^2$, $x = \pm y$.

Substitute into $G = 0$, we get $2x^2 - 2 = 0$ so $x = \pm 1$. The extremal points on the boundary are therefore $(1, 1), (1, -1), (-1, 1), (-1, -1)$. We check the values of z at these points :

$$z(1, 1) = 7, \quad z(-1, -1) = 7, \quad z(1, -1) = 9, \quad z(-1, 1) = 9, \quad z(0, 0) = 0$$

Thus $(-1, 1)$ and $(1, -1)$ give the maxima over the closed disk and $(0, 0)$ gives the minimum.

8. We will use the Lagrange multipliers with:
    Objective function: $f(x, y) = -3x^2 - 2y^2 + 20xy$
    Constraint: $g(x, y) = x + y - 100$
    We first find

$$\nabla f = (-6x + 20y)\vec{i} + (-4y + 20x)\vec{j}$$
$$\nabla g = \vec{i} + \vec{j}.$$

To optimize f, we must solve the equations

$$\nabla f = \lambda \nabla g$$
$$(-6x + 20y)\vec{i} + (-4y + 20x)\vec{j} = \lambda(\vec{i} + \vec{j}) = \lambda\vec{i} + \lambda\vec{j}$$

We have a vector equation, so we equate the coordinates:

$$-6x + 20y = \lambda$$
$$20x - 4y = \lambda.$$
$$\text{So} \quad -6x + 20y = 20x - 4y$$
$$24y = 26x$$
$$y = \frac{13}{12}x$$

Substituting into the constraint equation $x + y = 100$, we obtain:

$$x + \frac{13}{12}x = 100$$
$$\frac{25}{12}x = 100$$
$$x = 48.$$

Consequently, $y = 52$, and $f(48, 52) = 37,600$. The point $(48, 52)$ leads to the extreme value of $f(x, y)$, given that $x + y = 100$. Note that f has no minimum on the line $x + y = 100$ since $f(x, 100 - x) = -3x^2 - 2(100 - x)^2 + 20x(100 - x)$ which goes to $-\infty$ as x goes to $\pm\infty$. Therefore, the point $(48, 52)$ gives the maximum value for f on the line $x + y = 100$.

9. We first express the revenue R in terms of the prices p_1 and p_2:

$$R(p_1, p_2) = p_1 q_1 + p_2 q_2$$
$$= p_1(517 - 3.5p_1 + 0.8p_2) + p_2(770 - 4.4p_2 + 1.4p_1)$$
$$= 517p_1 - 3.5p_1^2 + 770p_2 - 4.4p_2^2 + 2.2p_1 p_2.$$

At a local maximum we have grad $R = 0$, and so:

$$\frac{\partial R}{\partial p_1} = 517 - 7p_1 + 2.2p_2 = 0,$$
$$\frac{\partial R}{\partial p_2} = 770 - 8.8p_2 + 2.2p_1 = 0.$$

Solving these equations, we find that

$$p_1 = 110 \quad \text{and} \quad p_2 = 115.$$

To see whether or not we have a found a local maximum, we compute the second-order partial derivatives:

$$\frac{\partial^2 R}{\partial p_1^2} = -7, \quad \frac{\partial^2 R}{\partial p_2^2} = -8.8, \quad \frac{\partial^2 R}{\partial p_1 \partial p_2} = 2.2.$$

Therefore,

$$D = \frac{\partial^2 R}{\partial p_1^2}\frac{\partial^2 R}{\partial p_2^2} - \frac{\partial^2 R}{\partial p_1 \partial p_2} = (-7)(-8.8) - (2.2)^2 = 56.76,$$

and so we have found a local maximum point. The graph of $P(p_1, p_2)$ has the shape of an upside down paraboloid. Since P is quadratic in q_1 and q_2, $(110, 115)$ is a global maximum point.

10. (a) Suppose $N = kA^p$. Then the rule of thumb tells us that if A is multiplied by 10, the value of N doubles. Thus

$$2N = k(10A)^p = k10^p A^p.$$

Thus, dividing by $N = kA^p$, we have

$$2 = 10^p$$

so taking logs to base 10 we have

$$p = \log 2 = 0.3010.$$

(where $\log 2$ means $\log_{10} 2$). Thus,

$$N = kA^{0.3010}.$$

(b) Taking natural logs gives

$$\ln N = \ln(kA^p)$$
$$\ln N = \ln k + p\ln A$$
$$\ln N \approx \ln k + 0.301\ln A$$

Thus, $\ln N$ is a linear function of $\ln A$.

(c) Table 14.1 contains the natural logarithms of the data:

TABLE 14.1 $\ln N$ *and* $\ln A$

Island	$\ln A$	$\ln N$
Redonda	1.1	1.6
Saba	3.0	2.2
Montserrat	2.3	2.7
Puerto Rico	9.1	4.3
Jamaica	9.3	4.2
Hispaniola (Haiti and Dominican Rep.)	11.2	4.8
Cuba	11.6	4.8

Using a least squares fit we find the line:

$$\ln N = 1.20 + 0.32\ln A$$

This yields the power function:
$$N = e^{1.20}A^{0.32} = 3.4A^{0.32}$$

Since 0.32 is pretty close to $\log 2 \approx 0.301$, the answer does agree with the biological rule.

11. We want to minimize cost $C = 100L + 200K$ subject to $Q = 900L^{\frac{1}{2}}K^{\frac{2}{3}} = 36000$. Using Lagrange multipliers, we get

$$\nabla Q = \left(450L^{-\frac{1}{2}}K^{\frac{2}{3}}\right)\vec{i} + \left(600L^{\frac{1}{2}}K^{-\frac{1}{3}}\right)\vec{j}.$$

$$\nabla C = 100\vec{i} + 200\vec{j}$$

$\nabla C = \lambda\nabla Q$ gives

$$100 = \lambda 450L^{-\frac{1}{2}}K^{\frac{2}{3}} \quad \text{and} \quad 200 = \lambda 600L^{\frac{1}{2}}K^{-\frac{1}{3}}.$$

Since $\lambda \neq 0$ this gives

$$450L^{-\frac{1}{2}}K^{\frac{2}{3}} = 300L^{\frac{1}{2}}K^{-\frac{1}{3}}.$$

Solving, we get $L = \frac{3}{2}K$. Substituting into $Q = 36{,}000$ gives

$$900\left(\frac{3}{2}K\right)^{\frac{1}{2}}K^{\frac{2}{3}} = 36{,}000.$$

Solving yields $K = \left[40\cdot\left(\frac{2}{3}\right)^{\frac{1}{2}}\right]^{\frac{6}{7}} \approx 19.85$, so $L \approx \frac{3}{2}(19.85) = 29.78$. We can thus calculate cost using $K = 20$ and $L = 30$ which gives $C = \$7{,}000$.

12. Using Lagrange multipliers, let $G = 2000 - 5x - 10y = 0$ be the constraint.
$$\nabla P = \left(1 + \frac{2xy^2}{2 \cdot 10^8}\right)\vec{i} + \left(2 + \frac{2yx^2}{2 \cdot 10^8}\right)\vec{j} = \left(1 + \frac{xy^2}{10^8}\right)\vec{i} + \left(2 + \frac{yx^2}{10^8}\right)\vec{j}.$$
$$\nabla G = -5\vec{i} - 10\vec{j}.$$
Now, $\nabla P = \lambda \nabla G$, so

$$1 + \frac{xy^2}{10^8} = -5\lambda \quad \text{and} \quad 2 + \frac{yx^2}{10^8} = -10\lambda.$$

Thus

$$2 + \frac{2xy^2}{10^8} = 2 + \frac{yx^2}{10^8}.$$

Solving, we get $2y = x$ or $x = 0$ or $y = 0$.

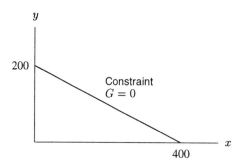

Figure 14.30

From $G = 0$ we have: when $x = 0, y = 200$, when $y = 0, x = 400$, and when $x = 2y, x = 200, y = 100$. So (0,200), (400,0) and (200,100) are the critical points and they include the end points.
Substitute into P: $P(0, 200) = 400, P(400, 0) = 400, P(200, 100) = 402$ so the organization should buy 200 sacks of rice and 100 sacks of beans.

13. (a) To be producing the maximum quantity Q under the cost constraint given, the firm should be using K and L values given by $\nabla Q = \lambda \nabla C, C = 20K + 10L = 150$, so

$$\frac{\partial Q}{\partial K} = 0.6aK^{-0.4}L^{0.4} = 20\lambda$$

$$\frac{\partial Q}{\partial L} = 0.4aK^{0.6}L^{-0.6} = 10\lambda$$

Hence $\dfrac{0.6aK^{-0.4}L^{0.4}}{0.4aK^{0.6}L^{-0.6}} = 1.5\dfrac{L}{K} = \dfrac{20\lambda}{10\lambda} = 2$, so $L = \dfrac{4}{3}K$. Substituting in $20K + 10L = 150$, we obtain $20K + 10\left(\dfrac{4}{3}\right)K = 150$. Then $K = \dfrac{9}{2}$ and $L = 6$, so capital should be reduced by $\dfrac{1}{2}$ unit, and labor should be increased by 1 unit.

(b) $\dfrac{\text{New production}}{\text{Old production}} = \dfrac{a4.5^{0.6}6^{0.4}}{a5^{0.6}5^{0.4}} \approx 1.01$, so tell the board of directors, "Reducing the quantity of capital by 1/2 unit and increasing the quantity of labor by 1 unit will increase production by 1% while holding costs to \$150."

14. Since patient 1 has a visit every x_1 weeks, this patient has $1/x_1$ visits per week. Similarly, patient 2 has $1/x_2$ visits per week. Thus, the constraint is

$$g(x_1, x_2) = \frac{1}{x_1} + \frac{1}{x_2} = m$$

To minimize

$$f(x_1, x_2) = \frac{v_1}{v_1 + v_2} \cdot \frac{x_1}{2} + \frac{v_2}{v_1 + v_2} \cdot \frac{x_2}{2}$$

subject to $g(x_1, x_2)$, we solve the equations

$$\operatorname{grad} f = \lambda \operatorname{grad} g$$
$$g(x_1, x_2) = m.$$

This gives us the equations

$$\frac{\partial f}{\partial x_1} = \frac{v_1}{v_1 + v_2} \cdot \frac{1}{2} = \lambda \left(-\frac{1}{x_1^2} \right) = \lambda \frac{\partial g}{\partial x_1}$$

$$\frac{\partial f}{\partial x_2} = \frac{v_2}{v_1 + v_2} \cdot \frac{1}{2} = \lambda \left(-\frac{1}{x_2^2} \right) = \lambda \frac{\partial g}{\partial x_2}$$

$$\frac{1}{x_1} + \frac{1}{x_2} = m.$$

Dividing the first equation by the second gives

$$\frac{v_1}{v_2} = \frac{x_2^2}{x_1^2}.$$

As v_1, v_2, x_1, x_2, m are strictly positive we have

$$\frac{x_2}{x_1} = \left(\frac{v_1}{v_2} \right)^{\frac{1}{2}}.$$

Substituting for x_2 in the constraint occasion gives

$$\frac{1}{x_1} + \left(\frac{v_2}{v_1} \right)^{\frac{1}{2}} \cdot \frac{1}{x_1} = m$$

solving for x_1 gives

$$\frac{1}{x_1} \left(1 + \left(\frac{v_1}{v_2} \right)^{\frac{1}{2}} \right) = m$$

$$x_1 = \frac{(v_1)^{\frac{1}{2}} + (v_2)^{\frac{1}{2}}}{m \cdot (v_1)^{\frac{1}{2}}},$$

and similarly

$$x_2 = \frac{(v_1)^{\frac{1}{2}} + (v_2)^{\frac{1}{2}}}{m \cdot (v_2)^{\frac{1}{2}}}.$$

15. We want to optimize

$$f(x_1, x_2) = \frac{v_1}{v_1 + v_2} \cdot \frac{x_1}{2} + \frac{v_2}{v_1 + v_2} \cdot \frac{x_2}{2}$$

subject to

$$g(x_1, x_2) = \frac{1}{x_1} + \frac{1}{x_2} = m.$$

At the optimum point, x_1, x_2, and the Lagrange multiplier λ must satisfy the equations

$$\frac{v_1}{v_1 + v_2} \cdot \frac{1}{2} = -\frac{\lambda}{x_1^2}$$

$$\frac{v_2}{v_1 + v_2} \cdot \frac{1}{2} = -\frac{\lambda}{x_2^2}$$

$$\frac{1}{x_1} + \frac{1}{x_2} = m.$$

Solving the first and second equations for $1/x_1$ and $1/x_2$, respectively, gives

$$\frac{1}{x_1^2} = -\frac{1}{2\lambda} \cdot \frac{v_1}{(v_1 + v_2)}$$

$$\frac{1}{x_2^2} = -\frac{1}{2\lambda} \cdot \frac{v_2}{(v_1 + v_2)}$$

substituting into the constraint gives (note that $\lambda < 0$):

$$\left(-\frac{1}{2\lambda} \cdot \frac{v_1}{(v_1 + v_2)} \right)^{\frac{1}{2}} + \left(\frac{-1}{2\lambda} \cdot \frac{v_2}{(v_1 + v_2)} \right)^{\frac{1}{2}} = \left(-\frac{1}{2\lambda} \right)^{\frac{1}{2}} \cdot \frac{(v_1)^{\frac{1}{2}} + (v_2)^{\frac{1}{2}}}{(v_1 + v_2)^{\frac{1}{2}}} = m.$$

So

$$-\frac{1}{2\lambda} \cdot \frac{v_1 + v_2 + 2(v_1 v_2)^{\frac{1}{2}}}{v_1 + v_2} = m^2.$$

and thus

$$\lambda = -\frac{1}{2m^2} \left(1 + \frac{2(v_1 v_2)^{\frac{1}{2}}}{v_1 + v_2} \right).$$

The units of λ are weeks^2 (since the units of m are 1/weeks). The Lagrange multiplier measures df/dm, which represents the rate of change in the expected delay in tumor detection as the available number of visits per week increases. The negative sign represents the fact that as the number of visits per week increases, the delay decreases.

16. Cost of production, C, is given by $C = p_1 W + p_2 K = b$. At the optimal point, $\nabla q = \lambda \nabla C$. Since $\nabla q = \left(c(1-a)W^{-a}K^a \right) \vec{i} + \left(caW^{1-a}K^{a-1} \right) \vec{j}$ and $\nabla C = p_1 \vec{i} + p_2 \vec{j}$, we get

$$c(1-a)W^{-a}K^a = \lambda p_1 \quad \text{and} \quad caW^{1-a}K^{a-1} = \lambda p_2.$$

Now, marginal productivity of labor is given by $\frac{\partial q}{\partial W} = c(1-a)W^{-a}K^a$ and marginal productivity of capital is given by $\frac{\partial q}{\partial K} = caW^{1-a}K^{a-1}$, so their ratio is given by

$$\frac{\frac{\partial q}{\partial W}}{\frac{\partial q}{\partial K}} = \frac{c(1-a)W^{-a}K^a}{caW^{1-a}K^{a-1}} = \frac{\lambda p_1}{\lambda p_2} = \frac{p_1}{p_2}$$

which is the ratio of the cost of one unit of labor to the cost of one unit of capital.

17. The wetted perimeter of the trapezoid is given by the sum of the lengths of the three walls, so

$$p = w + \frac{2d}{\sin \theta}.$$

We want to minimize p subject to the constraint that the area is fixed at 50 m^2. A trapezoid of height h and with parallel sides of lengths b_1 and b_2 has

$$A = \text{Area} = h\frac{(b_1 + b_2)}{2}.$$

In this case, d corresponds to h and b_1 corresponds to w. The b_2 term corresponds to the width of the exposed surface of the canal. We find that $b_2 = w + (2d)/(\tan \theta)$. Substituting into our original equation for the area along with the fact that the area is fixed at 50 m^2, we arrive at the formula:

$$\text{Area} = \frac{d}{2}\left(w + w + \frac{2d}{\tan \theta}\right) = d\left(w + \frac{d}{\tan \theta}\right) = 50$$

We now solve the constraint equation for one of the variables; we will choose w to give

$$w = \frac{50}{d} - \frac{d}{\tan \theta}.$$

Substituting into the expression for p gives

$$p = w + \frac{2d}{\sin \theta} = \frac{50}{d} - \frac{d}{\tan \theta} + \frac{2d}{\sin \theta}.$$

We now take partial derivatives:

$$\frac{\partial p}{\partial d} = -\frac{50}{d^2} - \frac{1}{\tan \theta} + \frac{2}{\sin \theta}$$

$$\frac{\partial p}{\partial \theta} = \frac{d}{\tan^2 \theta} \cdot \frac{1}{\cos^2 \theta} - \frac{2d}{\sin^2 \theta} \cdot \cos \theta$$

From $\partial p/\partial \theta = 0$, we get

$$\frac{d \cdot \cos^2 \theta}{\sin^2 \theta} \cdot \frac{1}{\cos^2 \theta} = \frac{2d}{\sin^2 \theta} \cdot \cos \theta.$$

Since $\sin \theta \neq 0$ and $\cos \theta \neq 0$, canceling gives

$$1 = 2 \cos \theta$$

so

$$\cos \theta = \frac{1}{2}.$$

Since $0 < \theta < \frac{\pi}{2}$,   we get $\theta = \frac{\pi}{3}$.

Substituting into the equation $\partial p/\partial d = 0$ and solving for d gives:

$$\frac{-50}{d^2} - \frac{1}{\sqrt{3}} + \frac{2}{\sqrt{3}/2} = 0$$

which leads to

$$d = \sqrt{\frac{50}{\sqrt{3}}} \approx 5.37 \text{m}.$$

Then

$$w = \frac{50}{d} - \frac{d}{\tan\theta} \approx \frac{50}{5.37} - \frac{5.37}{\sqrt{3}} \approx 6.21 \text{ m}.$$

When $\theta = \pi/3$, $w \approx 6.21$ m and $d \approx 5.37$ m, we have $p \approx 18.61$ m.

Since there is only one critical point, and since p increases without limit as d or θ shrink to zero, the critical point must give the global minimum for p.

18. To make the problem simpler, let's start with $N = 2$. Then,

$$E = \left(\frac{p_2}{p_1}\right)^2 + \left(\frac{p_3}{p_2}\right)^2 - 2.$$

The first step is to differentiate E with respect to its variables, in this case, p_2.

$$\frac{\partial E}{\partial p_2} = \frac{2p_2}{p_1{}^2} + \frac{-2p_3{}^2}{p_2{}^3}.$$

Setting $\frac{\partial E}{\partial p_2}$ equal to zero and solving for p_2, we have

$$p_2{}^4 = p_1{}^2 p_3{}^2$$
$$p_2 = \sqrt{p_1 p_3}.$$

For general N, we can take the $N - 1$ partial derivatives with respect to the variables $p_2, \ldots, p_N$, of E:

$$\frac{\partial E}{\partial p_2} = \frac{2p_2}{p_1{}^2} + \frac{-2p_3{}^2}{p_2{}^3}$$

$$\frac{\partial E}{\partial p_3} = \frac{2p_3}{p_2{}^2} + \frac{-2p_4{}^2}{p_3{}^3}$$

$$\vdots$$

$$\frac{\partial E}{\partial p_i} = \frac{2p_i}{p_{i-1}{}^2} + \frac{-2p_{i+1}{}^2}{p_i{}^3}$$

$$\vdots$$

$$\frac{\partial E}{\partial p_N} = \frac{2p_N}{p_{N-1}{}^2} + \frac{-2p_{N+1}{}^2}{p_N{}^3}.$$

Setting these all equal to zero, similar to $N = 2$ case we can get

$$p_i = \sqrt{p_{i-1} p_{i+1}}, i = 2, 3, \ldots, N$$

and

$$\frac{p_i}{p_{i-1}} = \frac{p_{i+1}}{p_i}, \quad \text{for} \quad i = 2, 3, \ldots, N.$$

Since

$$\frac{p_2}{p_1} \frac{p_3}{p_2} \frac{p_4}{p_3} \cdots \frac{p_{N+1}}{p_N} = \frac{p_{N+1}}{p_1}$$

and since we know that each ratio is equal:

$$\frac{p_{i+1}}{p_i} = \sqrt[N]{\frac{p_{N+1}}{p_1}}$$

We can solve for successive p_i's starting with p_2:

$$\frac{p_2}{p_1} = \sqrt[N]{\frac{p_{N+1}}{p_1}}$$

$$p_2 = p_1 \sqrt[N]{(p_{N+1})p_1^{-1}}$$

$$= \sqrt[N]{(p_1^{N-1})p_{N+1}}$$

$$\frac{p_3}{p_2} = \sqrt[N]{(p_{N+1})p_1^{-1}}$$

$$p_3 = \sqrt[N]{(p_1^{N-1})p_{N+1}} \sqrt[N]{(p_{N+1})p_1^{-1}}$$

$$= \sqrt[N]{(p_1^{N-2})(p_{N+1})^2}$$

$$\frac{p_4}{p_3} = \sqrt[N]{(p_{N+1})p_1^{-1}}$$

$$p_4 = \sqrt[N]{(p_1^{N-2})(p_{N+1})^2} \sqrt[N]{(p_{N+1})p_1^{-1}}$$

$$= \sqrt[N]{p_1^{N-3}(p_{N+1})^3}$$

$$\vdots$$

and finally:
$$p_i = \sqrt[N]{p_1^{N+1-i}(p_{N+1})^{i-1}}$$

19. (a) Let d_i $(i = 1, 2, 3)$ be the distance function from P_i to an arbitrary point Q. Then the contour diagram for d_i (see Figure 14.31) consists of level curves which are circles centered at P_i and the value of d_i on each contour level is the distance from P_i to a point on the contour.

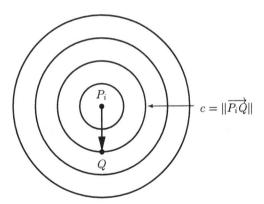

Figure 14.31

The vector $\operatorname{grad} d_i$ at any point Q is in the direction in which d_i is increasing at the greatest rate, that is in the direction of $\overrightarrow{P_i Q}$. Since the value of d_i on the contour on which Q is located is the same as the distance from P_i to Q, it follows that

$$\| \operatorname{grad} d_i \| = \text{Rate of change of } d_i \text{ with respect to distance} = \frac{\text{Change in } d_i}{\text{Distance traveled}} = 1.$$

Thus, $\operatorname{grad} d_i$ is a unit vector pointing directly away from P_i.

(b)   Since $d = d_1 + d_2 + d_3$, we have

$$\text{grad } d = \text{grad } d_1 + \text{grad } d_2 + \text{grad } d_3.$$

From part(a), grad d is a sum of unit vectors grad d_i which point away from P_i (see Figure 14.32).

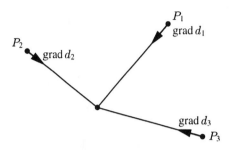

Figure 14.32

Since grad d points in the direction in which d is increasing at the greatest rate, to decrease d the family should move in the direction of $-$ grad d, that is, opposite to the direction of grad d.

(c)   To find the point H where grad $d = 0$, observe that grad $d = $ grad $d_1 + $ grad $d_2 + $ grad $d_3 = 0$ if grad $d_1 + $ grad $d_2 = -$ grad d_3. Thus H must be located at the point that grad $d_1 + $ grad d_2 is a unit vector pointing in the direction opposite to grad d_3 which is the same as the direction from H to P_3.

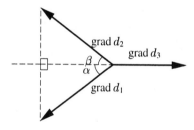

Figure 14.33

Suppose we resolve grad d_1 and grad d_2 into components parallel and perpendicular to grad d_3 (see Figure 14.33). The perpendicular components must cancel out. Since the lengths of both vectors are 1, from Figure 14.33 we have

$$\sin \alpha = \sin \beta,$$

so

$$\alpha = \beta.$$

Since the horizontal components must also balance

$$\cos \alpha + \cos \beta = 1$$

this gives

$$2 \cos \alpha = 1$$
$$\alpha = 60° = \beta.$$

Thus, in equilibrium the three vectors have angles of 120° between them. See Figure 14.34.

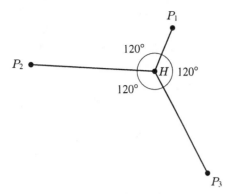

Figure 14.34

Since d is nonnegative, it must have a minimum. Since there is only one point where grad $d = 0$, there is only one critical point, thus, it must be a minimum.

20. (a) The critical points of f are the point(s) at which the partial derivatives, f_x and f_y, are zero. We have

$$f_x = 4x - 3y + 1$$

$$f_y = -3x + 16y - 1.$$

Solving the linear system $f_x = 0$, $f_y = 0$, we find $(x, y) = (-13/55, 1/55)$. To classify this point, we have to find the sign of $f_{xx} \cdot f_{yy} - f_{xy}^2$ there. We calculate

$$f_{xx} \cdot f_{yy} - f_{xy}^2 = 4 \cdot 16 - 9 = 55 > 0.$$

Thus, $(-13/55, 1/55)$ is a local minimum.

(b) We complete the square in the following way:

$$
\begin{aligned}
2x^2 - 3xy + 8y^2 + x - y &= 2x^2 - x(3y - 1) + 8y^2 - y \\
&= 2(x - \frac{1}{4}(3y - 1))^2 - \frac{1}{8}(3y - 1)^2 + 8y^2 - y \\
&= 2(x - \frac{1}{4}(3y - 1))^2 + \frac{1}{8}(55y^2 - 2y - 1) \\
&= 2(x - \frac{1}{4}(3y - 1))^2 + \frac{55}{8}(y - \frac{1}{55})^2 - \frac{7}{55}.
\end{aligned}
$$

Therefore the function has a global minimum located at the point (x, y) where the two "squares" vanish. The coordinates of that point satisfy:

$$x - \frac{1}{4}(3y - 1) = 0 \quad \text{and} \quad y - \frac{1}{55} = 0.$$

The two conditions again give the point $(x, y) = (-13/55, 1/55)$. The contour diagram for f is shown in Figure 14.35. The fact that f can be written in this way as the sum of two squares shows that the point $(x, y) = (-33/55, 1/55)$ is a global minimum.

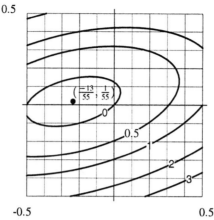

Figure 14.35: The contour diagram for
$f(x,y) = 2x^2 - 3xy + 8y^2 + x - y$

(c) Starting from the point $(x_0, y_0) = (1, 1)$, we move in the direction given by $-\operatorname{grad} f(x_0, y_0)$. We have

$$\operatorname{grad} f(x_0, y_0) = f_x(x_0, y_0)\vec{i} + f_y(x_0, y_0)\vec{i}$$
$$= 2\vec{i} + 12\vec{j},$$

by using the expressions of f_x and f_y derived in part (b). We travel in the direction $-2\vec{i} - 12\vec{j}$ as long as f decreases in this direction. We now compute

$$f((x_0, y_0) - t \operatorname{grad} f(x_0, y_0)) = f(1 - 2t, 1 - 12t)$$
$$= 1088t^2 - 140t + 7.$$

This is minimized when $t = \frac{140}{2 \cdot 1088} = 0.0643$, so we should stop at $(x_1, y_1) = (1, 1) - 0.0643(2, 12) = (0.8713, 0.2279)$. We have to repeat the procedure starting at (x_1, y_1). Using the formulas for f_x and f_y from part (b) we compute

$$f_x(0.8713, 0.2279) = 3.8015$$
$$f_y(0.8713, 0.2279) = 0.0345.$$

As for the values of the function along the line segment starting from (x_1, y_1) in the direction $-\operatorname{grad} f(x_1, y_1)$ we have

$$f((x_1, y_1) - t \operatorname{grad} f(x_1, y_1)) = f(0.8713 - 3.8015t, 0.2279 - 0.0345t)$$
$$= 1.9815 - 14.4525t + 28.5188t^2.$$

the expression is minimized by

$$t = \frac{14.4525}{2 \cdot 28.5188} = 0.2534,$$

the corresponding values of x_1, x_2, and the function being

$$x_1 = -0.0919, \quad x_2 = 0.2192, \quad f(x_2, y_2) = -0.1504.$$

The last value is supposed to represent an approximation of the minimum value of the function f. We know from the expression of f as a sum of squares in part (a) that the actual minimum value of f is $-7/55 = -0.1273$. The two-step approximation is therefore very satisfactory.

21. (a)

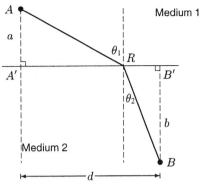

Figure 14.36

See Figure 14.36. The time to travel from A to B is given by

$$T(\theta_1, \theta_2) = \frac{AR}{v_1} + \frac{RB}{v_2} = \frac{a}{v_1 \cos \theta_1} + \frac{b}{v_2 \cos \theta_2}.$$

(b) The distance $d = A'B' = A'R + RB'$. Hence

$$d = a \tan \theta_1 + b \tan \theta_2.$$

(c) We imagine the following extreme case: the light ray first travels through medium 1 to a point R on the boundary far to the left of A', then through medium 2 towards B. The distance traveled this way is very large, hence the travel time is large as well. Similarly, if R is far to the right of B', the travel time will be large. Therefore values of θ_1 near $-\pi/2$ or $\pi/2$ increase the time, T.

(d) The constrained optimization problem is: minimize $T(\theta_1, \theta_2)$ subject to $g(\theta_1, \theta_2) = a \tan \theta_1 + b \tan \theta_2 = d$. According to the method of Lagrange multipliers, the minimum point should be among those satisfying $\operatorname{grad} T = \lambda \operatorname{grad} g$ as well as the constraint. We have

$$\operatorname{grad} T(\theta_1, \theta_2) = \frac{a}{v_1} \frac{\sin \theta_1}{\cos^2 \theta_1} \vec{i} + \frac{b}{v_2} \frac{\sin \theta_2}{\cos^2 \theta_2} \vec{j}$$

and

$$\operatorname{grad} g(\theta_1, \theta_2) = a \frac{1}{\cos^2 \theta_1} \vec{i} + b \frac{1}{\cos^2 \theta_2} \vec{j}.$$

The condition $\operatorname{grad} T = \lambda \operatorname{grad} g$ becomes

$$\frac{a}{v_1} \frac{\sin \theta_1}{\cos^2 \theta_1} = \lambda a \frac{1}{\cos^2 \theta_1} \quad \text{and} \quad \frac{b}{v_2} \frac{\sin \theta_2}{\cos^2 \theta_2} = \lambda b \frac{1}{\cos^2 \theta_2}.$$

Eliminating λ we are left with

$$\frac{\sin \theta_1}{v_1} = \frac{\sin \theta_2}{v_2}$$

or

$$\frac{\sin \theta_1}{\sin \theta_2} = \frac{v_1}{v_2},$$

which is Snell's law. The argument in part (c) shows that the critical point corresponding to θ_1, θ_2 satisfying Snell's law is indeed a minimum.

22.   (a)   If $p = e^{-x}$ where $x \to \infty$ then $p \to 0$ with $p > 0$ and

$$\lim_{\substack{p \to 0 \\ p > 0}} (p \ln p) = \lim_{x \to \infty} (-xe^{-x}) = 0,$$

since the exponential decreases faster than any power of x. Alternatively, use L'Hopital's rule:

$$\lim_{\substack{p \to 0 \\ p > 0}} (p \ln p) = \lim_{\substack{p \to 0 \\ p > 0}} \frac{\ln p}{1/p} = \lim_{\substack{p \to 0 \\ p > 0}} \frac{1/p}{-1/p^2} = 0$$

(b)   We apply the method of Lagrange multipliers to find the critical points of $S(p_1, \cdots, p_{26})$. The constraint function is $g(p_1, \cdots, p_{26}) = p_1 + \cdots + p_{26}$. We have

$$\frac{\partial S}{\partial p_j} = \frac{\partial}{\partial p_j}\left(-\sum_{i=1}^{26} p_i \frac{\ln p_i}{\ln 2}\right) = -\frac{1}{\ln 2}(\ln p_j + 1),$$

therefore

$$\text{grad } S = -\frac{1}{\ln 2}\sum_{j=1}^{26}(\ln p_j + 1)\vec{k_j}$$

where $\vec{k_1}, \cdots, \vec{k_{26}}$ are the unit vectors corresponding 26 independent directions of the p_j-axes. Also,

$$\text{grad } g = \sum_{j=1}^{26}\vec{k_j}$$

so the condition grad $S = \lambda$ grad g becomes

$$-\frac{1}{\ln 2}(\ln p_j + 1) = \lambda, \qquad \text{for} \quad i = 1, \cdots, 26.$$

Thus,

$$\ln p_j = -\lambda \ln 2 - 1$$

and, in particular, all the p_js must be equal. Since the p_js have to satisfy the constraint $g(p_1, \cdots, p_{26}) = 1$ we deduce that the point $(\frac{1}{26}, \frac{1}{26}, \cdots, \frac{1}{26})$ is the only critical point of S. We have $S(\frac{1}{26}, \frac{1}{26}, \cdots, \frac{1}{26}) = -26 \cdot \frac{1}{26}\frac{(-\ln 26)}{\ln 2} = \frac{\ln 26}{\ln 2}$. We will not prove that this is indeed the maximum value of S (this requires a higher-dimensional analogue of the second derivative test). Since in part (c) we show that the minimum value of S is 0, the critical point we have found here is not a global minimum; the maximum of S has to be attained somewhere and it is reasonable to believe that it is attained at the unique critical point. The maximum entropy corresponds to maximum uncertainty in the outcome of the competition.

(c)   We already know that $S \geq 0$. Actually S can be zero: For example, if $p_1 = 1$ and $p_2 = \cdots = p_{26} = 0$, we have $S(1, 0, \cdots, 0) = 0$. Therefore the minimum value of S is 0. We have to determine all the values of p_is for which $S(p_1, \cdots, p_{26}) = 0$. The condition

$$-\sum_{i=1}^{26}\frac{p_i \ln p_i}{2} = 0$$

together with the restrictions $-\ln p_i \geq 0$ shows that, for S to vanish, each individual term in the above sum has to vanish. This means $p_i \ln p_i = 0$ for all $i = 1, \cdots, 26$, i.e. $p_i = 0$ or $p_i = 1$ for $i = 1, \cdots, 26$. Since $\sum_{i=1}^{26} p_i = 1$, only one of the p_is is 1 whereas the other 25 are 0. This corresponds to the case where one of the terms is certain to win, i.e. there is no uncertainty. The result can be interpreted by saying that zero entropy implies zero uncertainty.

23. Let (x, y, z) be a point on the paraboloid. The square of the distance from (x, y, z) to the point $(1, 2, 10)$ is given by

$$f(x, y, z) = (x - 1)^2 + (y - 2)^2 + (z - 10)^2,$$

and so we wish to minimize $f(x, y, z)$ subject to the constraint

$$g(x, y, z) = x^2 + y^2 - z = 0.$$

We look for solutions to the equations grad $f = \lambda$ grad g and $g = 0$:

$$2(x - 1) = 2\lambda x,$$
$$2(y - 2) = 2\lambda y,$$
$$2(z - 10) = -\lambda,$$
$$x^2 + y^2 - z = 0.$$

If $\lambda = 0$, the first three equations would imply that $(x, y, z) = (1, 2, 10)$, which does not satisfy the fourth equation and so $\lambda \neq 0$. The first equation then implies that $x \neq 0$ and the second equation implies that $y \neq 0$, so we can eliminate λ from the first three equations to get:

$$\frac{x - 1}{x} = \frac{y - 2}{y} \quad \text{and} \quad \frac{y - 2}{y} = -2(z - 10).$$

These give

$$y = 2x \quad \text{and} \quad z = \frac{2 - y}{2y} + 10.$$

Substituting for x and z in the equation $z = x^2 + y^2$, we obtain

$$\frac{2 - y}{2y} + 10 = \frac{y^2}{4} + y^2,$$

which simplifies to give

$$5y^3 - 38y - 4 = 0.$$

Let $h(y) = 5y^3 - 38y - 4$. We find that $h(-3) < 0$ and $h(-1) > 0$, and so the cubic $h(y)$ has a root between -3 and -1. Similarly, since $h(-1) > 0$ and $h(0) < 0$, then $h(y)$ has one a between -1 and 0. Finally, as $h(0) < 0$ and $h(3) > 0$, we see that $h(y)$ has a root between 0 and 3. Let's find the root lying between 0 and 3. Using a calculator, we find that this root is approximately given by

$$y_1 \approx 2.808,$$

and so, using $y = 2x$, the corresponding point (x, y, z) on the paraboloid $z = x^2 + y^2$ is given by

$$(x_1, y_1, z_1) \approx (1.404, 2.808, 9.856).$$

The remaining roots of $h(y)$ are given by $y_2 \approx -0.1055$ and $y_3 \approx -2.7026$. The corresponding points on the paraboloid $z = x^2 + y^2$ which satisfy $y = 2x$ are then

$$(x_3, y_3, z_3) \approx (-1.3513, -2.7026, 9.1301),$$
$$(x_2, y_2, z_2) \approx (-0.0528, -0.1055, 0.0139),$$

and so we easily see that the point $(1.404, 2.808, 9.856)$ will be closest to $(1, 2, 10)$. Therefore, the minimum distance is

$$d = \sqrt{(1.404 - 1)^2 + (2.808 - 2)^2 + (9.856 - 10)^2} \approx 0.9148.$$

24. We apply the method of Lagrange multipliers for the case of two constraints. The function to be optimized is $f(x, y, z) = x^2 + y^2 + z^2$ (representing the square of the distance from the point (x, y, z) to the origin) and the constraint functions are $g_1(x, y, z) = x^2 + y^2 - z^2$ and $g_2(x, y, z) = x + y - z$. At the critical points the following condition holds true:

$$\text{grad } f = \lambda_1 \text{ grad } g_1 + \lambda_2 \text{ grad } g_2,$$

where λ_1, λ_2 are two scalars. We have

$$\text{grad } f = 2x\vec{i} + 2y\vec{j} + 2z\vec{k}$$
$$\text{grad } g_1 = 2x\vec{i} + 2y\vec{j} - 2z\vec{k}$$
$$\text{grad } g_2 = \vec{i} + \vec{j} - \vec{k}.$$

so the condition becomes

$$2x = 2x\lambda_1 + \lambda_2$$
$$2y = 2y\lambda_1 + \lambda_2$$
$$2z = -2z\lambda_1 - \lambda_2.$$

We now try to eliminate λ_1 and λ_2. By subtracting the first two equations we get

$$2(x - y) = 2(x - y)\lambda_1, \quad \text{so} \quad 2(x - y)(1 - \lambda_1) = 0.$$

There are two possibilities: $\lambda_1 = 1$ or $x = y$. If $\lambda_1 = 1$ then any of the first two equations implies $\lambda_2 = 0$; the third one, after substituting $\lambda_1 = 1$ and $\lambda_2 = 0$, gives $z = 0$. This can be discarded since the cone $\{x^2 + y^2 = z^2\}$ intersects the plane $\{z = 0\}$ only at the origin, which is not in the plane $\{z = 1 + x + y\}$ and, consequently there are no solutions with $\lambda_1 = 1$ satisfying the given constraints.

We now try $x = y$. Combining this condition and the constraints, we have

$$x = y$$
$$x^2 + y^2 = z^2$$
$$x + y + 1 = z.$$

Substituting $x = y$ in the second and third equations we get $2y^2 = z^2$ and $2y + 1 = z$; substituting z into $2y^2 = z^2$, we end up with

$$2y^2 = (2y + 1)^2, \quad \text{i.e.} \quad 2y^2 + 4y + 1 = 0.$$

The solutions of this equation are $y_1 = \frac{-2-\sqrt{2}}{2}$ and $y_2 = \frac{-2+\sqrt{2}}{2}$, hence $x_1 = y_1 = \frac{-2-\sqrt{2}}{2}$; $x_2 = y_2 = \frac{-2+\sqrt{2}}{2}$; $z_1 = 2y_1 + 1 = -1 - \sqrt{2}$; $z_2 = 2y_2 + 1 = -1 + \sqrt{2}$. Hence the constrained optimization problem has two critical points,

$$A = (\frac{-2 - \sqrt{2}}{2}, \frac{-2 - \sqrt{2}}{2}, -1 - \sqrt{2}) \quad \text{and} \quad B = (\frac{-2 - \sqrt{2}}{2}, \frac{-2 + \sqrt{2}}{2}, -1 + \sqrt{2}).$$

As there is no easy criterion describing the nature of the critical points, we use geometry. We have to intersect

$$z^2 = x^2 + y^2 \quad \text{and} \quad z = 1 + x + y.$$

So we get $(1 + x + y)^2 = x^2 + y^2$, or

$$1 + 2x + 2y + 2xy = 0.$$

So the curve of intersection is given by the equations:

$$1 + 2x + 2y + 2xy = 0.$$

$$z = x + y + 1.$$

In other words, it is a hyperbola in the plane $z = x + y + 1$. A hyperbola being an unbounded curve, there is no point on it farthest from the origin; a point closest to the origin should exist. Since A is a distance of about 3.14 from the origin, and B is about 1.78, the point B has the minimum distance.

CHAPTER FIFTEEN

1. Mark the values of the function on the plane, as shown in Figure 15.1, so that you can guess respectively at the smallest and largest values the function takes on each small rectangle.

$$\text{Lower sum} = \sum f(x_i, y_i)\Delta x \Delta y$$
$$= 4\Delta x \Delta y + 6\Delta x \Delta y + 3\Delta x \Delta y + 4\Delta x \Delta y$$
$$= 17\Delta x \Delta y$$
$$= 17(0.1)(0.2)$$
$$= 0.34.$$

$$\text{Upper sum} = \sum f(x_i, y_i)\Delta x \Delta y$$
$$= 7\Delta x \Delta y + 10\Delta x \Delta y + 6\Delta x \Delta y + 8\Delta x \Delta y$$
$$= 31\Delta x \Delta y$$
$$= 31(0.1)(0.2)$$
$$= 0.62.$$

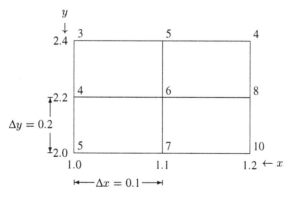

Figure 15.1

2. Partition R into subrectangles with the lines $x = 0$, $x = 0.5$, $x = 1$, $x = 1.5$, and $x = 2$ and the lines $y = 0$, $y = 1$, $y = 2$, $y = 3$, and $y = 4$. Then we have 16 subrectangles, each of which we denote $R_{(a,b)}$, where (a, b) is the location of the lower-left corner of the subrectangle.

    We want to find a lower bound and an upper bound for the volume above each subrectangle. The lower bound for the volume of $R_{(a,b)}$ is

$$0.5(\text{Min of } f \text{ on } R_{(a,b)})$$

because the area of $R_{(a,b)}$ is $0.5 \cdot 1 = 0.5$. The function $f(x, y) = 2 + xy$ increases with both x and y over the whole region R, as shown in Figure 15.2. Thus,

$$\text{Min of } f \text{ on } R_{(a,b)} = f(a, b) = 2 + ab,$$

because the minimum on each subrectangle is at the corner closest to the origin.

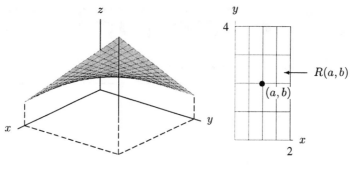

Figure 15.2   **Figure 15.3**

Similarly,

$$\text{Max of } f \text{ on } R_{(a,b)} = f(a+0.5, b+1) = 2 + (a+0.5)(b+1).$$

So we have

$$\text{Lower sum } = \sum_{(a,b)} 0.5(2 + ab) = 0.5 \sum_{(a,b)} (2 + ab)$$

$$= 16 + 0.5 \sum_{(a,b)} ab$$

Since $a = 0, 0.5, 1, 1.5$ and $b = 0, 1, 2, 3$, expanding this sum gives

$$\begin{aligned}
\text{Lower sum } = 16 + 0.5 \; (\; & 0 \cdot 0 + 0 \cdot 1 + 0 \cdot 2 + 0 \cdot 3 \\
& + 0.5 \cdot 0 + 0.5 \cdot 1 + 0.5 \cdot 2 + 0.5 \cdot 3 \\
& + 1 \cdot 0 + 1 \cdot 1 + 1 \cdot 2 + 1 \cdot 3 \\
& + 1.5 \cdot 0 + 1.5 \cdot 1 + 1.5 \cdot 2 + 1.5 \cdot 3)
\end{aligned}$$

$$= 25.$$

Similarly, we can compute the upper sum:

$$\text{Upper sum } = \sum_{(a,b)} 0.5(2 + (a+0.5)(b+1)) = 0.5 \sum_{(a,b)} (2 + (a+0.5)(b+1))$$

$$= 16 + 0.5 \sum_{(a,b)} (a+0.5)(b+1)$$

$$= 41.$$

3.   (a)   If we take the partition of R consisting of just R itself, we get

$$\text{Lower bound for integral } = \min_R f \cdot A_R = 0 \cdot (4-0)(4-0) = 0.$$

Similarly, we get

$$\text{Upper bound for integral } = \max_R f \cdot A_R = 4 \cdot (4-0)(4-0) = 64.$$

(b) The estimates asked for are just the upper and lower sums. We partition R into subrectangles $R_{(a,b)}$ of width 2 and height 2, where (a,b) is the lower-left corner of $R_{(a,b)}$. The subrectangles are then $R_{(0,0)}$, $R_{(2,0)}$, $R_{(0,2)}$, and $R_{(2,2)}$.

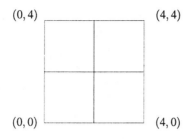

Figure 15.4

Then we find the lower sum

$$\text{Lower sum} = \sum_{(a,b)} A_{R_{(a,b)}} \cdot \min_{R_{(a,b)}} f = \sum_{(a,b)} 4 \cdot (\text{Min of } f \text{ on } R_{(a,b)})$$

$$= 4 \sum_{(a,b)} (\text{Min of } f \text{ on } R_{(a,b)})$$

$$= 4(f(0,0) + f(2,0) + f(0,2) + f(2,2))$$

$$= 4(\sqrt{0 \cdot 0} + \sqrt{2 \cdot 0} + \sqrt{0 \cdot 2} + \sqrt{2 \cdot 2})$$

$$= 8.$$

Similarly, the upper sum is

$$\text{Upper sum} = 4 \sum_{(a,b)} (\text{Max of } f \text{ on } R_{(a,b)})$$

$$= 4(f(2,2) + f(4,2) + f(2,4) + f(4,4))$$

$$= 4(\sqrt{2 \cdot 2} + \sqrt{4 \cdot 2} + \sqrt{2 \cdot 4} + \sqrt{4 \cdot 4})$$

$$= 24 + 16\sqrt{2} \approx 46.63.$$

The upper sum is an overestimate and the lower sum is an underestimate, so we can get a better estimate by averaging them to get $16 + 8\sqrt{2} \approx 27.3$.

4. Let's break up the room into 25 sections, each of which is 1 meter by 1 meter and has area $\Delta A = 1$.

We shall begin our sum as an upper estimate starting with the lower left corner of the room and continue across the bottom and moving upwards using the highest temperature in each case. So the upper Riemann sum becomes

$$\sum_{i=1}^{25} T_i \Delta A = T_1 \Delta A + T_2 \Delta A + T_3 \Delta A + \cdots + T_{25} \Delta A$$

$$= \Delta A (T_1 + T_2 + T_3 + \cdots + T_{25})$$

$$= (1)(30 + 29 + 27 + 26 + 26 +$$

$$28 + 27 + 27 + 26 + 25 +$$

$$27 + 26 + 26 + 25 + 25 +$$

$$25 + 25 + 24 + 24 + 24 +$$

$$24 + 24 + 23 + 24 + 24)$$

$$= (1)(641)$$

$$= 641$$

In the same way, the lower Riemann sum is:

$$
\begin{aligned}
\sum_{i=1}^{25} t_i \Delta A &= t_1 \Delta A + t_2 \Delta A + t_3 \Delta A + \cdots + t_{25} \Delta A \\
&= \Delta A (t_1 + t_2 + t_3 + \cdots + t_{25}) \\
&= (1)(27 + 27 + 27 + 26 + 25 + \\
&\qquad 26 + 26 + 26 + 25 + 25 + \\
&\qquad 25 + 24 + 24 + 24 + 24 + \\
&\qquad 24 + 23 + 23 + 23 + 24 + \\
&\qquad 23 + 21 + 20 + 21 + 22) \\
&= (1)(605) \\
&= 605
\end{aligned}
$$

So, averaging the upper and lower sums we get: 623.

To compute the average temperature, we divide by the area of the room, giving

$$
\text{Average temperature} = \frac{623}{(5)(5)} \approx 24.9°C.
$$

Alternatively we can use the temperature at the central point of each section ΔA. Then the sum becomes

$$
\begin{aligned}
\sum_{i=1}^{25} T_i' \Delta A &= \Delta A \sum_{i=1}^{25} T_i' \\
&= (1)(29 + 28 + 27 + 26 + 25 + \\
&\qquad 27 + 27 + 26 + 26 + 25 + \\
&\qquad 26 + 25 + 25 + 25 + 25 + \\
&\qquad 25 + 24 + 24 + 24 + 24 + \\
&\qquad 24 + 23 + 22 + 22 + 23) \\
&= (1)(627) \\
&= 627.
\end{aligned}
$$

Then we get

$$
\text{Average temperature} = \frac{\sum_{i=1}^{25} T_i' \Delta A}{\text{Area}} = \frac{627}{(5)(5)} \approx 25.1°C.
$$

5. The total area of the square R is $(1.5)(1.5) = 2.25$. See Figure 15.5. On a disk of radius ≈ 0.5 the function has a value of 3 or more, giving a total contribution to the integral of at least $(3) \cdot (\pi \cdot 0.5^2) \approx 2.3$. On less than half of the rest of the square the function has a value between -2 and 0, giving a contribution to the integral of between $(1/2 \cdot 2.25)(-2) = -2.25$ and 0. Since the positive contribution to the integral is therefore greater in magnitude than the negative contribution, $\int_R f \, dA$ is positive.

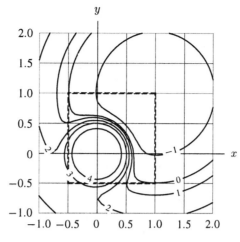

Figure 15.5

6. The question is asking which graph has more volume under it, and from inspection, it appears that it would be the graph for the mosquitos.

7. The exact value of the integral is $40/3$.

8. The value of the integral is around -2.4.

9. Let R be the region $0 \leq x \leq 60, \quad 0 \leq y \leq 8$. Then

$$\text{Volume} = \int_R w(x,y)\, dA$$

Lower estimate:
$10 \cdot 2(1+4+8+10+10+8+0+3+4+6+6+4+0+1+2+3+3+2+0+0+1+1+1+1) = 1580.$
Upper estimate:
$10 \cdot 2(8+13+16+17+17+16+4+8+10+11+11+10+3+4+6+7+7+6+1+2+3+4+4+3) = 3820.$
The average of the two estimates is 2700 cubic feet.

10. (a) The graph of f looks like the graph of g in the xz-plane slid parallel to itself forward and backward in the y direction, since the value of y does not affect the value of f.

    (b) The solid bounded by the graph of f is a cylinder, hence

$$\int_R f\, dA = \text{Volume under surface} = \begin{pmatrix} \text{Length in} \\ y \text{ direction} \end{pmatrix} \begin{pmatrix} \text{Cross-sectional area} \\ \text{in } xz\text{-plane} \end{pmatrix}$$

$$= (d-c) \int_a^b g(x)\, dx$$

11. The total number of tornados, per year, in a certain region, is the integral of the frequency of tornados in that region. In order to approximate it, we first subdivide the states into smaller regions of 100 square miles, as shown in Figure 11. We will find the upper and lower bounds for the frequency of tornados, and then take the average of the two. We do this by finding the highest frequency of tornados in each subdivision, and then add up all the frequencies, and then do the same for the low frequencies.

    (a) For the high frequency in Texas, going left to right, top to bottom, we get: $[(9+8) + (9+8+9+9+8) + (7+7+9+9+8) + (0+3+4+6+8+7+7+6) + (0+1+4+6+7+7) + (2+4+5) + (2+3)]$. This equals 182 tornados. For the low frequency, we get: $[(7+7) + (7+7+7+7+5) + (3+5+6+7+5) + (0+0+0+1+4+6+7+6) + (0+0+0+3+5+5) + (0+1+3) + (0+0)]$.

This equals 114 tornados. The average of the two equals $(182 + 114)/2 = 148$ tornados.

(b) For the high frequency in Florida, going left to right, top to bottom, we get: $[(5+5+5+9)+(9+9)+(7+7)+(5)]$. This equals 61 tornados. For the low frequency, we get: $[(5+5+5+5)+(7+7)+(7+5)+(5)]$. This equals 51 tornados. The average of the two is equal to $(61 + 51)/2 = 56$ tornados.

(c) For the high frequency in Arizona, going left to right, top to bottom, we get: $[(1 + 1 + 0) + (1 + 1 + 0) + (0 + 0 + 0) + (0 + 0)]$. This equals 4 tornados. For the low frequency, we get: $[(0 + 0 + 0) + (0 + 0 + 0) + (0 + 0 + 0) + (0 + 0)]$. This equals 0 tornados. The average of the two is equal to $(4 + 0)/2 = 2$ tornados.

12. First we will subdivide the region into 54 areas, as shown in Figure 12. We will find the upper and lower bounds for the total rainfall, and then take the average of the two. We do this by finding the highest amount of rainfall in that subdivision, multiplying it by the area, and then adding up all these quantities, and then doing the same for the lowest amount of rainfall.

For the high rainfall, going left to right, top to bottom, we get : $[(0+300+175+0+0+25+50+100+50)+(0+300+175+200+40+40+75+100+100)+(0+250+150+175+40+40+75+75+100)+(175+200+125+150+25+25+30+25+30)+(200+175+150+150+40+40+25+25+35)+(200+250+75+75+40+30+25+25+30)](0.00001 \text{ km})(100 \text{ km})(100 \text{ km})$. This equals 501 km^3 for the upper bound. For the low rainfall, we get: $[(0+150+125+0+0+20+20+40+40)+(0+175+100+40+25+25+30+40+40)+(0+100+100+40+20+30+30+30+25)+(175+100+100+30+20+25+20+20+25)+(175+50+50+25+25+25+20+20+30)+(175+75+50+30+30+25+20+25+30)](0.00001 \text{ km})(100 \text{ km})(100 \text{ km})$. This equals 264 km^3 for the lower bound. Taking the average of the two, we get 382.5 km^3 of rain over one year.

13. The region D is symmetric both with respect to x and y axes, while in the interior of R, $x > 0$ and in the interior of B, $y < 0$.

(a) The function being integrated is $f(x, y) = 1$, which is positive everywhere. Thus, its integral over any region is positive.

(b) For the same reason as in part (a), the integral is positive.

(c) The function being integrated is $f(x, y) = 5x$. Since $x > 0$ in R, f is positive in R and thus the integral is positive.

(d) The function being integrated is $f(x, y) = 5x$, which is an odd function in x. Since B is symmetric with respect to x, the contributions to the integral will cancel out, as $f(x, y) = -f(-x, y)$. Thus, the integral is zero.

(e) As in part (d), the function is an odd function in x. Since D is also symmetric with respect to x, the integral of the function over the area D is zero.

(f) The function being integrated, $f(x, y) = y^3 + y^5$, is an odd function in y while D is symmetric with respect to y. Then, by symmetry, the positive and negative contributions of f are equal and thus its integral is zero.

(g) In a region such as R in which $y < 0$, the quantity $y^3 + y^5$ is less than zero. Thus, its integral is negative.

(h) Using the same reasoning as in part (f), we determine that the integral over B is zero.

(i) The function being integrated, $f(x, y) = y - y^3$ is always negative in the region B since in that region $-1 < y < 0$. Thus, the integral is negative.

(j) The function being integrated, $f(x, y) = y - y^3$, is an odd function in y while D is symmetric with respect to y. By symmetry, the integral is zero.

(k) As in part (f), the function being integrated is odd with respect to y in the region D. Thus, its integral is zero.

(l) In the region D, y has range $|y| < 1$. Thus, $-\pi/2 < y < \pi/2$. Thus, $\cos y$ is always positive in the region D and thus its integral is positive.

(m) The function $f(x, y) = e^x$ is positive for any value of x. Thus, its integral is always positive for any region, such as D, with non-zero area.

(n)   Looking at the contributions to the integral of the function $f(x, y) = xe^x$, we can see that any contribution made by the point (x, y), where $x > 0$, is greater than the corresponding contribution made by $(-x, y)$, since $e^x > 1 > e^{-x}$ for $x > 0$. Thus, the integral of f in the region D is positive.

(o)   The function $f(x, y) = xy^2$ is odd with respect to x and thus, using the same reasoning as in part (d), has integral zero in the region D.

(p)   The function $f(x, y)$ is odd with respect to x and thus, using the same reasoning as in part (d), has integral zero in region B, which is symmetric with respect to x.

14.   Take a Riemannian sum approximation to

$$\int_R f \, dA \approx \sum_{i,j} f(x_i, y_i) \Delta A.$$

Then, using the fact that $|a + b| \leq |a| + |b|$ repeatedly, we have:

$$\left| \int_R f \, dA \right| \approx \left| \sum_{i,j} f(x_i, y_i) \Delta A \right| \leq \sum |f(x_i, y_i) \Delta A|.$$

Now $|f(x_i, y_j) \Delta A| = |f(x_i, y_j| \Delta A$ since ΔA is non-negative, so

$$\left| \int_R f \, dA \right| \leq \sum_{i,j} |f(x_i, y_i) \Delta A| = \sum_{i,j} |f(x_i, y_j)| \Delta A.$$

But the last expression on the right is a Riemann sum approximation to the integral $\int_R |f| \, dA$, so we have

$$\left| \int_R f \, dA \right| \approx \left| \sum_{i,j} f(x_i, y_j) \Delta A \right| \leq \sum_{i,j} |f(x_{i,j})| \Delta A \approx \int_R |f| \, dA.$$

Thus,

$$\left| \int_R f \, dA \right| \leq \int_R |f| \, dA.$$

Solutions for Section 15.2

1.

$$\int_R \sqrt{x + y} \, dA = \int_0^2 \int_0^1 \sqrt{x + y} \, dx \, dy$$

$$= \int_0^2 \frac{2}{3} (x + y)^{\frac{3}{2}} \Big|_0^1 \, dy$$

$$= \frac{2}{3} \int_0^2 ((1 + y)^{\frac{3}{2}} - y^{\frac{3}{2}}) \, dy$$

$$= \frac{2}{3} \cdot \frac{2}{5} [(1 + y)^{\frac{5}{2}} - y^{\frac{5}{2}}] \Big|_0^2$$

$$= \frac{4}{15} ((3^{\frac{5}{2}} - 2^{\frac{5}{2}}) - (1 - 0))$$

$$= \frac{4}{15} (9\sqrt{3} - 4\sqrt{2} - 1) = 2.38176$$

2. In the other order, the integral is

$$\int_0^1 \int_0^2 \sqrt{x+y}\, dy\, dx.$$

First we keep x fixed and calculate the inside integral with respect to y:

$$\int_0^2 \sqrt{x+y}\, dy = \frac{2}{3}(x+y)^{3/2}\bigg|_{y=0}^{y=2}$$

$$= \frac{2}{3}\left[(x+2)^{3/2} - x^{3/2}\right].$$

Then the outside integral becomes

$$\int_0^1 \frac{2}{3}\left[(x+2)^{3/2} - x^{3/2}\right]\, dx = \frac{2}{3}\left[\frac{2}{5}(x+2)^{5/2} - \frac{2}{5}x^{5/2}\right]\bigg|_0^1$$

$$= \frac{2}{3}\cdot\frac{2}{5}\left[3^{5/2} - 1 - 2^{5/2}\right]$$

$$= 2.38176$$

Note that the answer is the same as the one we got in Problem 1

3.

$$\int_R (5x^2 + 1)\sin 3y\, dA = \int_{-1}^1 \int_0^{\pi/3} (5x^2 + 1)\sin 3y\, dy\, dx$$

$$= \int_{-1}^1 (5x^2 + 1)\left(-\frac{1}{3}\cos 3y\bigg|_0^{\pi/3}\right)\, dx$$

$$= \frac{2}{3}\int_{-1}^1 (5x^2 + 1)\, dx$$

$$= \frac{2}{3}(\frac{5}{3}x^3 + x)\bigg|_{-1}^1$$

$$= \frac{2}{3}(\frac{10}{3} + 2)$$

$$= \frac{32}{9} \approx 3.556$$

4. It would be easier to integrate first in the x direction from $x = y - 1$ to $x = -y + 1$, because integrating first in the y direction would involve two separate integrals.

$$\int_R (2x + 3y)^2\, dA = \int_0^1 \int_{y-1}^{-y+1} (2x + 3y)^2\, dx\, dy$$

$$= \int_0^1 \int_{y-1}^{-y+1} (4x^2 + 12xy + 9y^2)\, dx\, dy$$

$$= \int_0^1 \left[\frac{4}{3}x^3 + 6x^2 y + 9xy^2\right]_{y-1}^{-y+1}\, dy$$

$$= \int_0^1 [\frac{8}{3}(-y + 1)^3 + 9y^2(-2y + 2)]\, dy$$

$$= \left[-\frac{2}{3}(-y+1)^4 - \frac{9}{2}y^4 + 6y^3 \right]_0^1$$

$$= -\frac{2}{3}(-1) - \frac{9}{2} + 6$$

$$= \frac{13}{6}$$

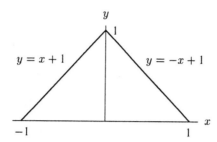

Figure 15.6

5.

$$\int_1^4 \int_1^2 f \, dy \, dx \quad \text{or} \quad \int_1^2 \int_1^4 f \, dx \, dy$$

6. The line connecting $(-1, 1)$ and $(3, -2)$ is

$$3x + 4y = 1$$

or

$$y = \frac{1 - 3x}{4}$$

So the integral becomes

$$\int_{-1}^3 \int_{-2}^{(1-3x)/4} f \, dy \, dx \quad \text{or} \quad \int_{-2}^1 \int_{-1}^{(1-4y)/3} f \, dx \, dy$$

7. The line connecting $(1, 0)$ and $(4, 1)$ is

$$y = \frac{1}{3}(x - 1)$$

So the integral is

$$\int_1^4 \int_{(x-1)/3}^2 f \, dy \, dx$$

8. Two of the sides of the triangle have equations $x = \dfrac{y-1}{2}$ and $x = \dfrac{y-5}{-2}$. So the integral is

$$\int_1^3 \int_{\frac{1}{2}(y-1)}^{-\frac{1}{2}(y-5)} f \, dx \, dy$$

9.

$$\int_1^3 \int_0^4 e^{x+y}\, dx dy = \int_1^3 e^x e^y \Big|_0^4 dx = \int_1^3 e^x(e^4 - 1)\, dx = (e^4 - 1)(e^3 - e)$$

See Figure 15.7.

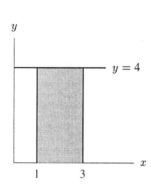

Figure 15.7

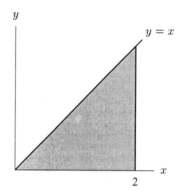

Figure 15.8

10.

$$\int_0^2 \int_0^x e^{x^2}\, dy dx = \int_0^2 e^{x^2} y \Big|_0^x dx = \int_0^2 x e^{x^2}\, dx = \frac{1}{2} e^{x^2} \Big|_0^2 = \frac{1}{2}(e^4 - 1) \approx 26.8$$

See Figure 15.8.

11.

$$\int_1^5 \int_x^{2x} \sin x\, dy\, dx = \int_1^5 \sin x \cdot y \Big|_x^{2x} dx$$

$$= \int_1^5 \sin x \cdot x\, dx$$

$$= (\sin x - x \cos x) \Big|_1^5$$

$$= (\sin 5 - 5 \cos 5) - (\sin 1 - \cos 1) \approx -2.68.$$

See Figure 15.9.

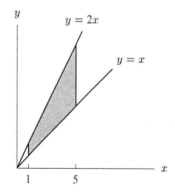

Figure 15.9

12.

$$\int_1^4 \int_{\sqrt{y}}^y x^2 y^3 \, dx dy = \int_1^4 y^3 \frac{x^3}{3} \Big|_{\sqrt{y}}^y \, dy$$

$$= \frac{1}{3} \int_1^4 (y^6 - y^{\frac{9}{2}}) \, dy$$

$$= \frac{1}{3} \left(\frac{y^7}{7} - \frac{y^{11/2}}{11/2} \right) \Big|_1^4$$

$$= \frac{1}{3} \left[\left(\frac{4^7}{7} - \frac{4^{11/2} \times 2}{11} \right) - \left(\frac{1}{7} - \frac{2}{11} \right) \right] \approx 656.082$$

See Figure 15.10.

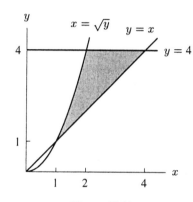

Figure 15.10

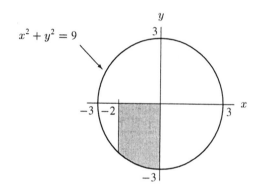

Figure 15.11

13.

$$\int_{-2}^0 \int_{-\sqrt{9-x^2}}^0 2xy \, dy dx = \int_{-2}^0 x \, y^2 \Big|_{-\sqrt{9-x^2}}^0 \, dx$$

$$= -\int_{-2}^0 x(9 - x^2) \, dx$$

$$= \int_{-2}^0 (x^3 - 9x) \, dx$$

$$= \left(\frac{x^4}{4} - \frac{9}{2}x^2 \right) \Big|_{-2}^0$$

$$= -4 + 18 = 14$$

See Figure 15.11.

14.   (a)

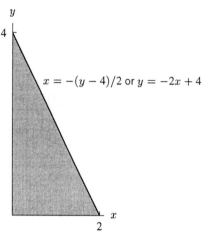

$x = -(y-4)/2$ or $y = -2x + 4$

Figure 15.12

(b) $\int_0^2 \int_0^{-2x+4} g(x,y)\, dy\, dx$.

15.   As given, the region of integration is as shown in Figure 15.13.

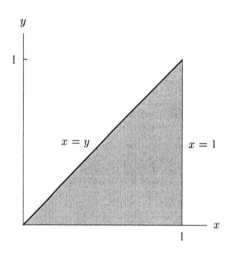

$x = y$ $x = 1$

Figure 15.13

Reversing the limits gives

$$\int_0^1 \int_0^x e^{x^2}\, dy dx = \int_0^1 \left(y e^{x^2} \Big|_0^x \right) dx = \int_0^1 x e^{x^2}\, dx$$
$$= \frac{e^{x^2}}{2} \Big|_0^1 = \frac{e-1}{2}.$$

16. As given, the region of integration is as shown in Figure 15.14.

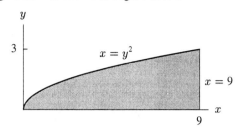

Figure 15.14

Reversing the limits gives

$$\int_0^9 \int_0^{\sqrt{x}} y \sin\left(x^2\right) dy dx = \int_0^9 \left(\left. \frac{y^2 \sin\left(x^2\right)}{2} \right|_0^{\sqrt{x}} \right) dx$$

$$= \frac{1}{2} \int_0^9 x \sin\left(x^2\right) dx$$

$$= \left. \frac{\cos\left(x^2\right)}{4} \right|_0^9$$

$$= \frac{1}{4} - \frac{\cos\left(81\right)}{4} = 0.056.$$

17. As given, the region of integration is as shown in Figure 15.15.

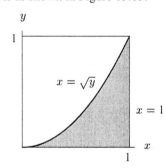

Figure 15.15

Reversing the limits gives

$$\int_0^1 \int_0^{x^2} \sqrt{2 + x^3}\, dy dx = \int_0^1 \left(\left. y\sqrt{2 + x^3} \right|_0^{x^2} \right) dx$$

$$= \int_0^1 x^2 \sqrt{2 + x^3}\, dx$$

$$= \left. \frac{2}{9}(2 + x^3)^{\frac{3}{2}} \right|_0^1$$

$$= \frac{2}{9}(3\sqrt{3} - 2\sqrt{2}).$$

18.

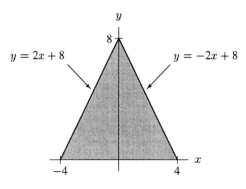

Figure 15.16

Order reversed: $\displaystyle\int_0^8 \int_{(y-8)/2}^{(8-y)/2} f(x,y)\,dx\,dy.$

19.   The intersection of the graph of $f(x,y) = 25 - x^2 - y^2$ and xy-plane is a circle $x^2 + y^2 = 25$. The given solid is shown in Figure 15.17.

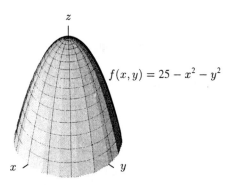

Figure 15.17

Thus the volume of the solid is

$$V = \int_R f(x,y)\,dA$$

$$= \int_{-5}^5 \int_{-\sqrt{25-y^2}}^{\sqrt{25-y^2}} (25 - x^2 - y^2)\,dx\,dy.$$

20.   The intersection of the graph of $f(x,y) = 25 - x^2 - y^2$ and the plane $z = 16$ is a circle, $x^2 + y^2 = 3^2$. The given solid is shown in Figure 15.18.

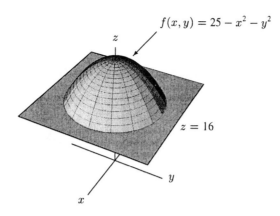

$$f(x, y) = 25 - x^2 - y^2$$

$$z = 16$$

Figure 15.18

Thus, the volume of the solid is

$$V = \int_R (f(x, y) - 16) \, dA$$

$$= \int_{-3}^{3} \int_{-\sqrt{9-y^2}}^{\sqrt{9-y^2}} (9 - x^2 - y^2) \, dx \, dy.$$

21. The solid is shown in Figure 15.19, and the base of the integral is the triangle as shown in Figure 15.20.

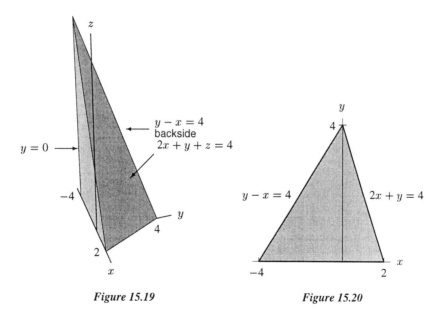

Figure 15.19   **Figure 15.20**

Thus, the volume of the solid is

$$V = \int_R z \, dA$$

$$= \int_R (4 - 2x - y) \, dA$$

$$= \int_0^4 \int_{y-4}^{(4-y)/2} (4 - 2x - y) \, dx \, dy.$$

22.

$$\text{Volume} = \int_0^2 \int_0^2 xy \, dy \, dx = \int_0^2 \frac{1}{2}xy^2 \Big|_0^2 \, dx$$

$$= \int_0^2 2x \, dx$$

$$= x^2 \Big|_0^2$$

$$= 4$$

23. Let R be the triangle with vertices $(1, 0)$, $(2, 2)$ and $(0, 1)$. Note that $(3x + 2y + 1) - (x + y) = 2x + y + 1 > 0$ for $x, y > 0$, so $z = 3x + 2y + 1$ is above $z = x + y$ on R. We want to find

$$\text{Volume} = \int_R ((3x + 2y + 1) - (x + y)) \, dA = \int_R (2x + y + 1) \, dA.$$

We need to express this in terms of double integrals.

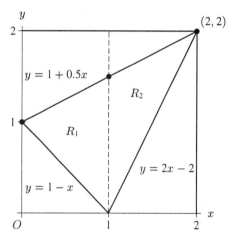

Figure 15.21

To do this, divide R into two regions with the line $x = 1$ to make regions R_1 for $x \le 1$ and R_2 for $x \ge 1$. See Figure 15.21. We want to find

$$\int_R (2x + y + 1) \, dA = \int_{R_1} (2x + y + 1) \, dA + \int_{R_2} (2x + y + 1) \, dA.$$

Note that the line connecting $(0, 1)$ and $(1, 0)$ is $y = 1 - x$, and the line connecting $(0, 1)$ and $(2, 2)$ is $y = 1 + 0.5x$. So

$$\int_{R_1} (2x + y + 1) \, dA = \int_0^1 \int_{1-x}^{1+0.5x} (2x + y + 1) \, dy \, dx.$$

The line between $(1, 0)$ and $(2, 2)$ is $y = 2x - 2$, so

$$\int_{R_2} (2x + y + 1)\, dA = \int_1^2 \int_{2x-2}^{1+0.5x} (2x + y + 1)\, dy\, dx.$$

We can now compute the double integral for R_1:

$$\int_0^1 \int_{1-x}^{1+0.5x} (2x + y + 1)\, dy\, dx = \int_0^1 \left(2xy + \frac{y^2}{2} + y \right) \bigg|_{1-x}^{1+0.5x}\, dx$$

$$= \int_0^1 \left(\frac{21}{8} x^2 + 3x \right)\, dx$$

$$= \left(\frac{7}{8} x^3 + \frac{3}{2} x^2 \right) \bigg|_0^1\, dx$$

$$= \frac{19}{8},$$

and the double integral for R_2:

$$\int_1^2 \int_{2x-2}^{1+0.5x} (2x + y + 1)\, dy\, dx = \int_1^2 (2xy + y^2/2 + y) \bigg|_{2x-2}^{1+0.5x}\, dx$$

$$= \int_0^1 \left(-\frac{39}{8} x^2 + 9x + \frac{3}{2} \right)\, dx$$

$$= \left(-\frac{13}{8} x^3 + \frac{9}{2} x^2 + \frac{3}{2} x \right) \bigg|_1^2$$

$$= \frac{29}{8}.$$

So, Volume $= \dfrac{19}{8} + \dfrac{29}{8} = \dfrac{48}{8} = 6.$

24. We want to calculate the volume of the tetrahedron shown in Figure 15.22.

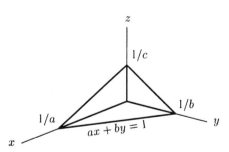

Figure 15.22

We first find the region in the xy-plane where the graph of $ax + by + cz = 1$ is above the xy-plane. When $z = 0$ we have $ax + by = 1$. So the region over which we want to integrate is bounded by $x = 0, y = 0$ and $ax + by = 1$. Integrating with respect to y first, we have

$$\text{Volume} = \int_0^{1/a} \int_0^{(1-ax)/b} z\, dy\, dx = \int_0^{1/a} \int_0^{(1-ax)/b} \frac{1 - by - ax}{c}\, dy\, dx$$

$$= \int_0^{1/a} \left(\frac{y}{c} - \frac{by^2}{2c} - \frac{axy}{c} \right) \Big|_{y=0}^{y=(1-ax)/b} dx$$

$$= \int_0^{1/a} \frac{1}{2bc}(1 - 2ax + a^2x^2)\, dx$$

$$= \frac{1}{6abc}.$$

25.   The region bounded by the x-axis and the graph of $y = x - x^2$ is shown in Figure 15.23. The area of this region is

$$A = \int_0^1 (x - x^2)dx = (\frac{x^2}{2} - \frac{x^3}{3}) \Big|_0^1$$

$$= \frac{1}{2} - \frac{1}{3} = \frac{1}{6}.$$

So the average distance to the x-axis for points in the region is

$$\text{Average distance} = \frac{\int_R y\, dA}{\text{area}(R)}$$

$$\int_R y\, dA = \int_0^1 \left(\int_0^{x-x^2} y\, dy \right) dx$$

$$= \int_0^1 \left(\frac{x^2}{2} - x^3 + \frac{x^4}{2} \right) dx = \frac{1}{6} - \frac{1}{4} + \frac{1}{10} = \frac{1}{60}.$$

Therefore the average distance is $\frac{1/60}{1/6} = 1/10$.

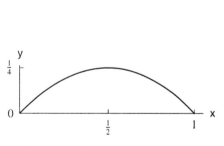

Figure 15.23

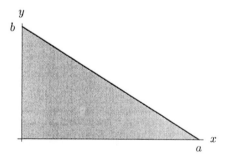

Figure 15.24

26.   Assume the length of the two legs of the right triangle are a and b, respectively. See Figure 15.24. The line through $(a, 0)$ and $(0, b)$ is given by $\frac{y}{b} + \frac{x}{a} = 1$. So the area of this triangle is

$$A = \frac{1}{2}ab.$$

Thus the average distance from the points in the triangle to the y-axis (one of the legs) is

$$\text{Average distance} = \frac{1}{A} \int_0^a \int_0^{-\frac{b}{a}x+b} x \, dy \, dx$$

$$= \frac{2}{ab} \int_0^a \left(-\frac{b}{a}x^2 + bx \right) dx$$

$$= \frac{2}{ab} \left(-\frac{b}{3a}x^3 + \frac{b}{2}x^2 \right) \Big|_0^a$$

$$= \frac{2}{ab} \left(\frac{a^2 b}{6} \right) = \frac{a}{3}.$$

Similarly, the average distance from the points in the triangle to the x-axis (the other leg) is

$$\text{Average distance} = \frac{1}{A} \int_0^b \int_0^{-\frac{a}{b}y+a} y \, dx \, dy$$

$$= \frac{2}{ab} \int_0^b \left(-\frac{a}{b}y^2 + ay \right) dy$$

$$= \frac{2}{ab} \left(\frac{ab^2}{6} \right) = \frac{b}{3}.$$

27. The function $\sin (x^2)$ has no elementary antiderivative, so we try integrating with respect to y first. The region of integration is shown in Figure 15.25. Changing the order of integration, we get

$$\int_0^1 \int_y^1 \sin (x^2) \, dx \, dy = \int_0^1 \int_0^x \sin (x^2) \, dy \, dx$$

$$= \int_0^1 \sin (x^2) \cdot y \Big|_0^x dx$$

$$= \int_0^1 \sin (x^2) \cdot x \, dx$$

$$= -\frac{\cos (x^2)}{2} \Big|_0^1$$

$$= -\frac{\cos 1}{2} + \frac{1}{2} = \frac{1}{2}(1 - \cos 1) = 0.23.$$

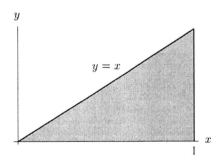

Figure 15.25

28. The region of the integration is shown in Figure 15.26. To make the integration easier, we want to change the order of the integration and get

$$\int_0^1 \int_{e^y}^e \frac{x}{\ln x}\, dx\, dy = \int_1^e \int_0^{\ln x} \frac{x}{\ln x}\, dy\, dx$$

$$= \int_1^e \frac{x}{\ln x} \cdot y \Big|_0^{\ln x} dx$$

$$= \int_1^e x\, dx = \frac{x^2}{2}\Big|_1^e$$

$$= \frac{1}{2}(e^2 - 1).$$

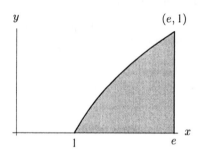

Figure 15.26

29. We want to find the average value of $|x - y|$, over the square $0 \le x \le 1, 0 \le y \le 1$:

$$\text{Average distance between gates} = \int_0^1 \int_0^1 |x - y|\, dy\, dx.$$

Let's fix x, with $0 \le x \le 1$. Then $|x - y| = \begin{cases} y - x & \text{for } y \ge x \\ x - y & \text{for } y \le x \end{cases}$. Therefore

$$\int_0^1 |x - y|\, dy = \int_0^x (x - y)\, dy + \int_x^1 (y - x)\, dy$$

$$= \left(xy - \frac{y^2}{2}\right)\Big|_0^x + \left(\frac{y^2}{2} - xy\right)\Big|_x^1 = x^2 - \frac{x^2}{2} + \frac{1}{2} - x - \frac{x^2}{2} + x^2$$

$$= x^2 - x + \frac{1}{2}.$$

So,

$$\text{Average distance between gates} = \int_0^1 \int_0^1 |x - y|\, dy\, dx$$

$$= \int_0^1 \left(\int_0^1 |x - y|\, dy\right) dx = \int_0^1 \left(x^2 - x + \frac{1}{2}\right) dx$$

$$= \frac{x^3}{3} - \frac{x^2}{2} + \frac{1}{2}x \Big|_0^1 = \frac{1}{3}.$$

30. There are $(n+1)^2$ possible pairs (i,j) of gates, $i=0,\ldots,n$, $j=0,\ldots,n$, so the sum given represents the average distances apart of all such gates. The Riemann sum with $\Delta x = \Delta y = \frac{1}{n}$, if we choose the least x and y-values in each subdivision is

$$\sum_{i=0}^{n-1}\sum_{j=0}^{n-1}\left|\frac{i}{n}-\frac{j}{n}\right|\frac{1}{n^2},$$

which for large n is just about the same as the other sum. For $n=5$ the sum is about 0.389; for $n=10$ the sum is about 0.364.

Solutions for Section 15.3

1.
$$\begin{aligned}
\int_W f\,dV &= \int_0^2\int_{-1}^1\int_2^3 (x^2+5y^2-z)\,dz\,dy\,dx\\
&= \int_0^2\int_{-1}^1 \left(x^2 z+5y^2 z-\frac{1}{2}z^2\right)\Big|_2^3\,dy\,dx\\
&= \int_0^2\int_{-1}^1\left(x^2+5y^2-\frac{5}{2}\right)dy\,dx\\
&= \int_0^2\left(x^2 y+\frac{5}{3}y^3-\frac{5}{2}y\right)\Big|_{-1}^1\,dx\\
&= \int_0^2\left(2x^2+\frac{10}{3}-5\right)dx\\
&= \left(\frac{2}{3}x^3-\frac{5}{3}x\right)\Big|_0^2\\
&= \frac{16}{3}-\frac{10}{3}\\
&= 2
\end{aligned}$$

2.
$$\begin{aligned}
\int_W f\,dV &= \int_0^\pi\int_0^\pi\int_0^\pi \sin x\cos(y+z)\,dz\,dy\,dx\\
&= \int_0^\pi\int_0^\pi \sin x\sin(y+z)\Big|_0^\pi\,dy\,dx\\
&= \int_0^\pi\int_0^\pi \sin x[\sin(y+\pi)-\sin y]\,dy\,dx\\
&= \int_0^\pi\int_0^\pi \sin x(-2\sin y)\,dy\,dx\\
&= -2\int_0^\pi \sin x(-\cos y)\Big|_0^\pi\,dx\\
&= -2\int_0^\pi 2\sin x\,dx\\
&= -4(-\cos x)\Big|_0^\pi\\
&= (-4)(2)\\
&= -8
\end{aligned}$$

3.

$$\int_W f \, dV = \int_0^1 \int_0^1 \int_0^2 (ax + by + cz) \, dz \, dy \, dx$$

$$= \int_0^1 \int_0^1 (2ax + 2by + 2c) \, dy \, dx$$

$$= \int_0^1 (2ax + b + 2c) \, dx$$

$$= a + b + 2c$$

4.

$$\int_W f \, dV = \int_0^a \int_0^b \int_0^c e^{-x-y-z} \, dz \, dy \, dx$$

$$= \int_0^a \int_0^b \int_0^c e^{-x} e^{-y} e^{-z} \, dz \, dy \, dx$$

$$= \int_0^a \int_0^b e^{-x} e^{-y} (-e^{-z}) \Big|_0^c \, dy \, dx$$

$$= \int_0^a \int_0^b e^{-x} e^{-y} (-e^{-c} + 1) \, dy \, dx$$

$$= (1 - e^{-c}) \int_0^a e^{-x} (-e^{-y}) \Big|_0^b \, dx$$

$$= (1 - e^{-b})(1 - e^{-c}) \int_0^a e^{-x} \, dx$$

$$= (1 - e^{-a})(1 - e^{-b})(1 - e^{-c})$$

5.   The region of integration for this integral is shown in Figure 15.27.

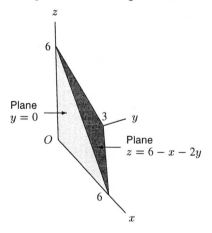

Figure 15.27

This is a three-sided pyramid whose base is the xy-plane and whose three sides are the vertical planes $x = 0$, $y = 0$ and slanted plane $z = 6 - x - 2y$.

6. The region of integration for this integral is shown in Figure 15.28. This is the region bounded by the xz-plane, xy-plane and plane $x = 1$, plane $z = x$ and plane $y = x$.

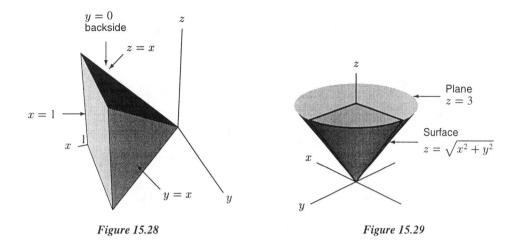

Figure 15.28   **Figure 15.29**

7. To integrate $f(x, y, z)$ in the order $dz\,dy\,dx$, the limits of the middle integral can only contain variable x. Therefore, the limits do not make sense.

8. The region of integration for this integral is a quarter of the conical solid bounded by $z = \sqrt{x^2 + y^2}$, $z = 3$ for $x \geq 0$, $y \geq 0$. See Figure 15.29.

9. The limits do not make sense since the limits for the middle integral involve two variables.

10. The region of integration is a prism shown in Figure 15.30 which is bounded by the two bases $z = 0$ and $z = 3$, and four vertical planes $x = 0$, $y = 0$, $x = 1$ and $y + x = 2$.

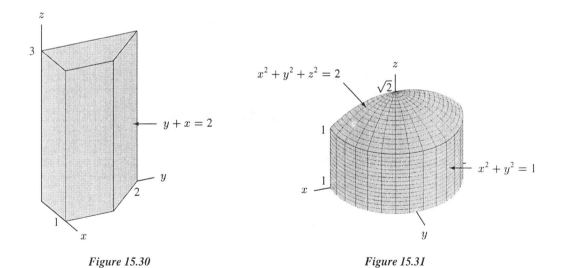

Figure 15.30   **Figure 15.31**

11. The region of integration is the solid below the graph of $z = \sqrt{2 - x^2 - y^2}$ (a sphere) above the disk $x^2 + y^2 \leq 1$ on xy-plane. Figure 15.31.

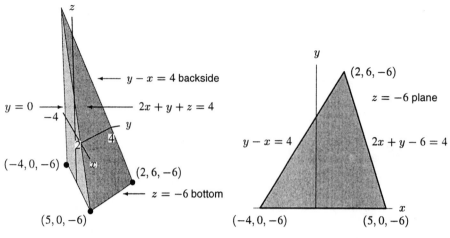

Figure 15.32

Figure 15.33

12.

The pyramid is shown in Figure 15.32. The planes $y = 0$, and $y - x = 4$, and $2x + y + z = 4$ intersect the plane $z = -6$ in the lines $y = 0$, $y - x = 4$, $2x + y = 10$ on the $z = -6$ plane as shown in Figure 15.33. These three lines intersect at the points $(-4, 0, -6)$, $(5, 0, -6)$, and $(2, 6, -6)$. Let R be the triangle in the planes $z = -6$ with the above three points as vertices. Then, the volume of the solid is

$$
\begin{aligned}
V &= \int_0^6 \int_{y-4}^{(10-y)/2} \int_{-6}^{4-2x-y} dz\, dx\, dy \\
&= \int_0^6 \int_{y-4}^{(10-y)/2} (10 - 2x - y)\, dx\, dy = 162 \\
&= \int_0^6 (10x - x^2 - xy) \Big|_{y-4}^{(10-y)/2} dy \\
&= \int_0^6 \left(\frac{9y^2}{2} - 27y + 81 \right) dy \\
&= 162
\end{aligned}
$$

13. The region of integration is shown in Figure 15.34, and the mass of the given solid is given by

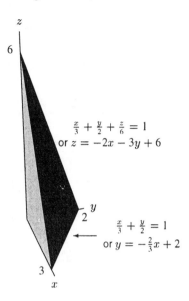

$\frac{x}{3} + \frac{y}{2} + \frac{z}{6} = 1$
or $z = -2x - 3y + 6$

$\frac{x}{3} + \frac{y}{2} = 1$
or $y = -\frac{2}{3}x + 2$

Figure 15.34

$$
\begin{aligned}
\text{mass} &= \int_R \delta\, dV \\
&= \int_0^3 \int_0^{-\frac{2}{3}x+2} \int_0^{-2x-3y+6} (x+y)\, dz\, dy\, dx \\
&= \int_0^3 \int_0^{-\frac{2}{3}x+2} (x+y)z \Big|_0^{-2x-3y+6}\, dy\, dx \\
&= \int_0^3 \int_0^{-\frac{2}{3}x+2} (x+y)(-2x-3y+6)\, dy\, dx \\
&= \int_0^3 \int_0^{-\frac{2}{3}x+2} (-2x^2 - 3y^2 - 5xy + 6x + 6y)\, dy\, dx \\
&= \int_0^3 \left(-2x^2y - y^3 - \frac{5}{2}xy^2 + 6xy + 3y^2 \right) \Big|_0^{-\frac{2}{3}x+2}\, dx \\
&= \int_0^3 \left(\frac{14}{27}x^3 - \frac{8}{3}x^2 + 2x + 4 \right)\, dx \\
&= \left(\frac{7}{54}x^4 - \frac{8}{9}x^3 + x^2 + 4x \right) \Big|_0^3 \\
&= \frac{7}{54} \cdot 3^4 - \frac{8}{9} \cdot 3^3 + 3^2 + 12 = \frac{21}{2} - 3 = \frac{15}{2}.
\end{aligned}
$$

14. From the problem, we know that (x, y, z) is in the cube which is bounded by the three coordinate planes, $x = 0$, $y = 0$, $z = 0$ and the planes $x = 2$, $y = 2$, $z = 2$. We can regard the value $x^2 + y^2 + z^2$ as the density

of the cube. The average value of $x^2 + y^2 + z^2$ is given by

$$\text{average value} = \frac{\int_V (x^2 + y^2 + z^2)\, dV}{\text{volume}(V)}$$

$$= \frac{\int_0^2 \int_0^2 \int_0^2 (x^2 + y^2 + z^2)\, dx\, dy\, dz}{8}$$

$$= \frac{\int_0^2 \int_0^2 \left(\frac{x^3}{3} + (y^2 + z^2)x \right)\Big|_0^2\, dy\, dz}{8}$$

$$= \frac{\int_0^2 \int_0^2 \left(\frac{8}{3} + 2y^2 + 2z^2 \right)\, dy\, dz}{8}$$

$$= \frac{\int_0^2 \left(\frac{8}{3}y + \frac{2}{3}y^3 + 2z^2 y \right)\Big|_0^2\, dz}{8}$$

$$= \frac{\int_0^2 \left(\frac{16}{3} + \frac{16}{3} + 4z^2 \right)\, dz}{8}$$

$$= \frac{\left(\frac{32}{3}z + \frac{4}{3}z^3 \right)\Big|_0^2}{8}$$

$$= \frac{\left(\frac{64}{3} + \frac{32}{3} \right)}{8} = 4.$$

15.   The intersection of two cylinders $x^2 + z^2 = 1$ and $y^2 + z^2 = 1$ is shown in Figure 15.35.

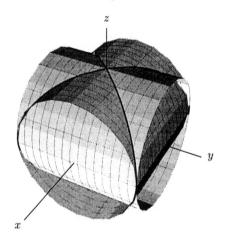

Figure 15.35

This region is bounded by four surfaces:

$$z = -\sqrt{1 - x^2}, \quad z = \sqrt{1 - x^2}, \quad y = -\sqrt{1 - z^2}, \quad \text{and} \quad y = \sqrt{1 - z^2}$$

So the volume of the given solid is

$$V = \int_{-1}^{1} \int_{-\sqrt{1-x^2}}^{\sqrt{1-x^2}} \int_{-\sqrt{1-z^2}}^{\sqrt{1-z^2}} dy\, dz\, dx$$

16.   The mass m is given by

$$m = \int_W 1 \, dV = \int_0^1 \int_0^1 \int_0^{x+y+1} 1 \, dz \, dy \, dx$$

$$= \int_0^1 \int_0^1 (x + y + 1) \, dy \, dx$$

$$= \int_0^1 (xy + y^2/2 + y)\big|_0^1 \, dx$$

$$= \int_0^1 (x + 3/2) \, dx = 2 \text{ gm}.$$

Then the x-coordinate of the center of mass is given by

$$\bar{x} = \frac{1}{2} \int_W x \, dV = \frac{1}{2} \int_0^1 \int_0^1 \int_0^{x+y+1} x \, dz \, dy \, dx$$

$$= \frac{1}{2} \int_0^1 \int_0^1 x(x + y + 1) \, dy \, dx$$

$$= \frac{1}{2} \int_0^1 (x^2 y + xy^2/2 + xy)\big|_0^1 \, dx$$

$$= \frac{1}{2} \int_0^1 (x^2 + 3/2 x) \, dx = 13/24 \text{ cm}.$$

An essentially identical calculation (since the region is symmetric in x and y) gives $\bar{y} = 13/24$ cm. Finally, we compute $\bar{z}$:

$$\bar{z} = \frac{1}{2} \int_W z \, dV = \frac{1}{2} \int_0^1 \int_0^1 \int_0^{x+y+1} z \, dz \, dy \, dx$$

$$= \frac{1}{2} \int_0^1 \int_0^1 (x + y + 1)^2/2 \, dy \, dx$$

$$= \frac{1}{2} \int_0^1 (x + y + 1)^3/6 \big|_0^1 \, dx$$

$$= \frac{1}{12} \int_0^1 ((x + 2)^3 - (x + 1)^3) \, dx = 25/24 \text{ cm}.$$

So $(\bar{x}, \bar{y}, \bar{z}) = (13/24, 13/24, 25/24)$.

17.   The mass m is given by

$$m = \int_W 1 \, dV = \int_0^1 \int_0^{(1-x)/2} \int_0^{(1-x-2y)/3} 1 \, dz \, dy \, dx$$

$$= \int_0^1 \int_0^{(1-x)/2} \frac{1 - x - 2y}{3} \, dy \, dx$$

$$= \frac{1}{3} \int_0^1 (y - xy - y^2)\Big|_0^{(1-x)/2} \, dx$$

$$= \frac{1}{3} \left(\int_0^1 \frac{1-x}{2} - x\frac{1-x}{2} - \left(\frac{1-x}{2}\right)^2 \right) dx = 1/36 \text{ gm}.$$

Then the coordinates of the center of mass are given by

$$\bar{x} = 36 \int_W x \, dV = 36 \int_0^1 \int_0^{(1-x)/2} \int_0^{(1-x-2y)/3} x \, dz \, dy \, dx = 1/4 \text{ cm.}$$

and

$$\bar{y} = 36 \int_W y \, dV = 36 \int_0^1 \int_0^{(1-x)/2} \int_0^{(1-x-2y)/3} y \, dz \, dy \, dx = 1/8 \text{ cm.}$$

and

$$\bar{z} = 36 \int_W z \, dV = 36 \int_0^1 \int_0^{(1-x)/2} \int_0^{(1-x-2y)/3} z \, dz \, dy \, dx = 1/12 \text{ cm.}$$

18. The volume V of the solid is $1 \cdot 2 \cdot 3 = 6$. We need to compute

$$\frac{m}{6} \int_W x^2 + y^2 \, dV = \frac{m}{6} \int_0^1 \int_0^2 \int_0^3 x^2 + y^2 \, dz \, dy \, dx$$

$$= \frac{m}{6} \int_0^1 \int_0^2 3(x^2 + y^2) \, dy \, dx$$

$$= \frac{m}{2} \int_0^1 (x^2 y + y^3/3)\Big|_0^2 \, dx$$

$$= \frac{m}{2} \int_0^1 (2x^2 + 8/3) \, dx = 5m/3$$

19. The volume of the solid is $8abc$, so we need to evaluate

$$\frac{m}{8abc} \int_W (y^2 + z^2) \, dV = \frac{m}{8abc} \int_{-c}^c \int_{-b}^b \int_{-a}^a (y^2 + z^2) \, dx \, dy \, dz$$

$$= \frac{m}{8abc} \int_{-c}^c \int_{-b}^b 2a(y^2 + z^2) \, dy \, dz$$

$$= \frac{m}{4bc} \int_{-c}^c (y^3/3 + yz^2)\Big|_{-b}^b \, dz$$

$$= \frac{m}{2c} \int_{-c}^c (b^2/3 + z^2) \, dz = m(b^2 + c^2)/3$$

20. By the definition, we have that

$$a + b = \frac{m}{V} \int_W (y^2 + z^2) \, dV + \frac{m}{V} \int_W (x^2 + z^2) \, dV$$

$$= \frac{m}{V} \int_W (x^2 + y^2 + 2z^2) \, dV$$

$$= \frac{m}{V} \int_W (x^2 + y^2) \, dV + \frac{m}{V} \int_W (2z^2) \, dV$$

$$= c + \frac{m}{V} \int_W (2z^2) \, dV$$

Since z^2 is always positive, the integral $\int_W (2z^2) \, dV$ will be positive, thus $a + b > c$.

Solutions for Section 15.4

1. We enclose the area represented by the integral in a square of area 1. Here is a table of values of one run of the method for different numbers of points.

    TABLE 15.1

N	10^3	10^4	10^5	10^6	10^7
N_R/N	0.78400	0.78730	0.78544	0.78536	0.78543

    The integral is the area under the graph of $\sqrt{1-x^2}$ for $0 \le x \le 1$, which is a quarter-circle of radius 1, so the integral is $\pi/4$. Since $\pi/4 \approx 0.7853981635$, we see from the table that, for large N, our results are comparable to about the fourth place.

2. We enclose the area represented by the integral in a square of area 1. Here is a table of values of one run of the method for different numbers of points.

    TABLE 15.2

N	10^3	10^4	10^5	10^6	10^7
N_R/N	0.72900	0.74730	0.74749	0.74698	0.74699

    The table seems to have stabilized in about the fourth place, so we expect that

    $$\int_0^1 e^{-x^2} \, dx \approx 0.7469.$$

3. We enclose the volume represented by the integral in a box of volume 1. Here is a table of values of one run of the method for different numbers of points:

    TABLE 15.3

N	10^3	10^4	10^5	10^6	10^7
N_R/N	0.7990	0.7941	0.795620	0.796680	0.7966495

    The table seems to have stabilized at the fourth decimal place, so we expect that $\int_0^1 \int_0^1 e^{-xy} \, dx \, dy \approx$ 0.7966.

4. We enclose the volume represented by the integral in a box of volume 1. Here is a table of values of one run of the method for different numbers of points:

    TABLE 15.4

N	10^3	10^4	10^5	10^6	10^7
N_R/N	0.408	0.4205	0.42426	0.425025	0.4253769

    The table seems to have stabilized at the third decimal place, so we expect that

    $$\int_0^1 \int_0^1 xy^{xy} \, dx \, dy \approx 0.43$$

5. In this problem, we see that the graph of $z = x \sin(y)$ over the region $0 \le x \le 2, 0 \le y \le \pi$ lies between 0 and 2. So we'll have to generate random z and x between 0 and 2, and random y between 0 and π. We enclose the volume represented by the integral in a box of volume $2 \cdot \pi \cdot 2 = 4\pi$. Here is a table of values of one run of the method for different numbers of points:

TABLE 15.5

N	10^3	10^4	10^5	10^6	10^7
N_R/N	0.305	0.31299	0.31785	0.31796	0.31831

Since

$$\frac{N_R}{N} \approx \frac{\int_R f \, dA}{\text{Vol}(C)}$$

our estimate for the integral is

$$\int_R f \, dA \approx \text{Vol}(C) \frac{N_R}{N} = 4\pi \frac{N_R}{N}.$$

Multiplying the last estimate (0.31831) by 4π gives 4.00000143, which is very close to the exact value of 4. The exact value is found by calculating

$$\int_0^\pi \int_0^2 x \sin y \, dx \, dy = \int_0^\pi (\sin y) \frac{x^2}{2} \Big|_0^2 dy = 2 \int_0^\pi \sin y \, dy = -2 \cos y \Big|_0^\pi = 4.$$

6. Here is a table of values of one run of the method for different numbers of points:

TABLE 15.6

N	10^3	10^4	10^5	10^6	10^7
N_R/N	2.76825	2.66573	3.51822	3.41995	3.16816

The estimates found by Monte Carlo method do not seem to converge. The problem is that this integral is really improper (the integrand is undefined at the origin), and in fact the integral does not converge.

7. The area of the region is 1, so we need only compute the average $a = \frac{1}{N} \sum_{i=1}^{N} f(x_i, y_i)$. Here is a table of values of one run of the method for different numbers of points:

TABLE 15.7

N	10^3	10^4	10^5	10^6	10^7
a	0.79047	0.79768	0.79682	0.79661	0.79662

The table seems to have stabilized at the fourth decimal place, so we can say that $\int_0^1 \int_0^1 e^{-xy} \, dx \, dy \approx$ 0.79.

8. The area of the region is 1, so we need only compute the average $a = \frac{1}{N} \sum_{i=1}^{N} f(x_i, y_i)$. Here is a table of values of one run of the method for different numbers of points:

TABLE 15.8

N	10^3	10^4	10^5	10^6	10^7
a	0.412	0.4254	0.42321	0.425198	0.4253769

The table seems to have stabilized at the third decimal place, so we can say that $\int_0^1 \int_0^1 xy^{xy} \, dx \, dy \approx 0.43$.

9. The area of the region is 2π, so we need to compute the average $a = \frac{1}{N} \sum_{i=1}^{N} f(x_i, y_i)$ and multiply by 2π. Here is a table of values of one run of the method for different numbers of points:

TABLE 15.9

N	10^3	10^4	10^5	10^6	10^7
$2\pi a$	3.79003	3.81567	3.760207	3.9900132	4.0000012

The values seem to be getting closer to 4 (which is the exact answer).

10. By using Simpson's rule with $\Delta y = 1$ and $\Delta x = 1$, we get Table 15.10.

TABLE 15.10   *Values of* $f(x, y) = \sqrt{x^2 + y^2}$

		y				
		2	3	4	5	6
	1	2.236	3.162	4.123	5.099	6.083
	2	2.230	3.610	4.472	5.385	6.325
x	3	3.610	4.243	5.000	5.831	6.708
	4	4.472	5.000	5.657	6.403	7.211
	5	5.383	5.831	6.403	7.071	7.810

Thus,

$$\int_1^5 \int_2^6 \sqrt{x^2 + y^2} \, dy \, dx \approx \sum f(x_i, y_i) \Delta y \Delta x = 128.008.$$

Solutions for Section 15.5

1.

$$\int_{\pi/4}^{3\pi/4} \int_0^2 f \, r \, dr \, d\theta$$

2.

$$\int_{\pi/2}^{3\pi/2} \int_1^2 f \, r dr \, d\theta$$

3.

$$\int_0^{2\pi} \int_0^{\sqrt{2}} f \, r dr \, d\theta$$

4.

$$\int_0^{\pi/2} \int_0^{1/2} f \, r dr \, d\theta$$

5.

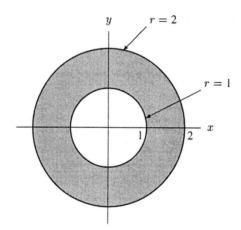

Figure 15.36

6.

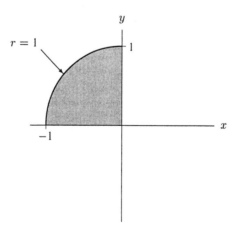

Figure 15.37

7.

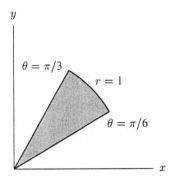

Figure 15.38

8.

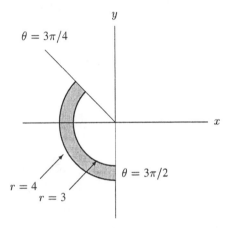

Figure 15.39

9.

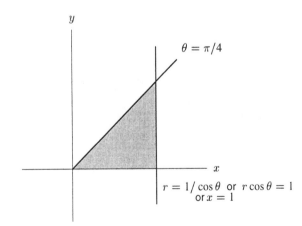

Figure 15.40

10.

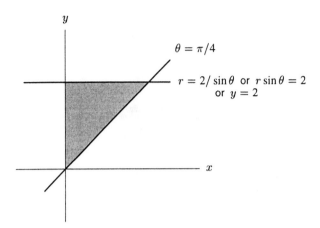

$\theta = \pi/4$

$r = 2/\sin\theta$ or $r\sin\theta = 2$
or $y = 2$

Figure 15.41

11.

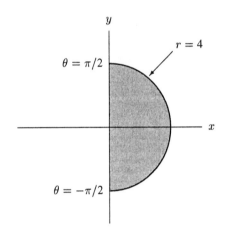

$r = 4$

$\theta = \pi/2$

$\theta = -\pi/2$

Figure 15.42

12. By using polar coordinates, we get

$$\int_R \sin(x^2 + y^2)dA = \int_0^{2\pi} \int_0^2 \sin(r^2)r \, dr \, d\theta$$

$$= \int_0^{2\pi} -\frac{1}{2}\cos(r^2)\Big|_0^2 \, d\theta$$

$$= -\frac{1}{2}\int_0^{2\pi} (\cos 4 - \cos 0) \, d\theta$$

$$= -\frac{1}{2}(\cos 4 - 1) \cdot 2\pi$$

$$= \pi(1 - \cos 4)$$

13. The region is pictured in Figure 15.43.

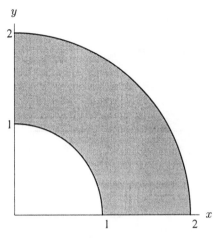

Figure 15.43

By using polar coordinates, we get

$$\int_R (x^2 - y^2)dA = \int_0^{\pi/2} \int_1^2 r^2(\cos^2\theta - \sin^2\theta)r\,dr\,d\theta = \int_0^{\pi/2} (\cos^2\theta - \sin^2\theta)\cdot\frac{1}{4}r^4\Big|_1^2\,d\theta$$

$$= \frac{15}{4}\int_0^{\pi/2} (\cos^2\theta - \sin^2\theta)\,d\theta$$

$$= \frac{15}{4}\int_0^{\pi/2} \cos 2\theta\,d\theta$$

$$= \frac{15}{4}\cdot\frac{1}{2}\sin 2\theta\Big|_0^{\pi/2}$$

$$= 0$$

14. (a)

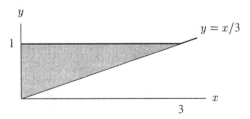

Figure 15.44

(b) $\int_0^1 \int_0^{3y} f(x,y)\,dx\,dy$.

(c) For polar coordinates, on the line $y = x/3$, $\tan\theta = y/x = 1/3$, so $\theta = \tan^{-1}(1/3)$. On the y-axis, $\theta = \pi/2$. The quantity r goes from 0 to the line $y = 1$, or $r\sin\theta = 1$, giving $r = 1/\sin\theta$ and $f(x,y) = f(r\cos\theta, r\sin\theta)$. Thus the integral is

$$\int_{\tan^{-1}(1/3)}^{\pi/2} \int_0^{1/\sin\theta} f(r\cos\theta, r\sin\theta)r\,dr\,d\theta.$$

15.   By the given limits $0 \leq x \leq -1$, and $-\sqrt{1-x^2} \leq y \leq \sqrt{1-x^2}$, the region of integration is in Figure 15.45.

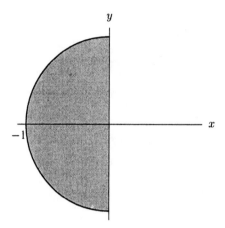

Figure 15.45

In polar coordinates, we have

$$
\int_{\pi/2}^{3\pi/2} \int_0^1 r \cos\theta \, r \, dr \, d\theta = \int_{\pi/2}^{3\pi/2} \cos\theta \left(\frac{1}{3} r^3 \right) \Big|_0^1 \; d\theta
$$

$$
= \frac{1}{3} \int_{\pi/2}^{3\pi/2} \cos\theta \quad d\theta
$$

$$
= \frac{1}{3} \sin\theta \Big|_{\pi/2}^{3\pi/2}
$$

$$
= \frac{1}{3}(-1 - 1) = -\frac{2}{3}
$$

16.   From the given limits, the region of integration is in Figure 15.46.

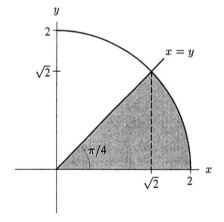

Figure 15.46

So, in polar coordinates, we have,

$$\int_0^{\pi/4} \int_0^2 (r^2 \cos\theta \sin\theta) r \, dr \, d\theta = \int_0^{\pi/4} \cos\theta \sin\theta \left(\frac{1}{4} r^4\right)\Big|_0^2 \, d\theta$$

$$= 4 \int_0^{\pi/4} \frac{\sin(2\theta)}{2} \, d\theta$$

$$= -\cos(2\theta)\Big|_0^{\pi/4}$$

$$= 0 - (-1)$$

$$= 1.$$

17. From the given limits, the region of integration is in Figure 15.47.

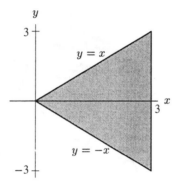

Figure 15.47

In polar coordinates, $-\pi/4 \le \theta \le \pi/4$. Also, $3 = x = r\cos\theta$. Hence, $0 \le r \le 3/\cos\theta$. The integral becomes

$$\int_{-\pi/4}^{\pi/4} \int_0^{3/\cos\theta} \frac{r\cos\theta}{(r\sin\theta)^2} r \, dr \, d\theta$$

$$= \int_{-\pi/4}^{\pi/4} \frac{\cos\theta}{\sin^2\theta} \left(\frac{3}{\cos\theta}\right) d\theta$$

$$= 3 \int_{-\pi/4}^{\pi/4} \frac{d\theta}{\sin^2\theta}$$

$$= -3 \frac{\cos\theta}{\sin\theta}\Big|_{-\pi/4}^{\pi/4}$$

$$= -3 \cdot (-1 - (-1)) = 6.$$

18. The graph of $f(x, y) = 25 - x^2 - y^2$ is an upside down bowl, and the region whose volume we want is contained between the bowl (above) and the xy-plane (below). We must first find the region in the xy-plane where $f(x, y)$ is positive. To do that, we set $f(x, y) \geq 0$ and get $x^2 + y^2 \leq 25$. The disk $x^2 + y^2 \leq 25$ is the region R over which we integrate.

$$\text{Volume} = \int_R (25 - x^2 - y^2)\, dA = \int_0^{2\pi} \int_0^5 (25 - r^2)\, r\, dr\, d\theta$$

$$= \int_0^{2\pi} \left(\frac{25}{2} r^2 - \frac{1}{4} r^4 \right) \Big|_0^5 d\theta$$

$$= \frac{625}{4} \int_0^{2\pi} d\theta$$

$$= \frac{625\pi}{2}$$

19.

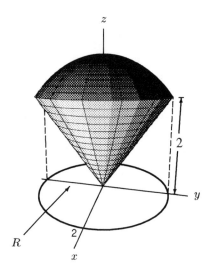

Figure 15.48

First, let's find where the two surfaces intersect.

$$\sqrt{8 - x^2 - y^2} = \sqrt{x^2 + y^2}$$
$$8 - x^2 - y^2 = x^2 + y^2$$
$$x^2 + y^2 = 4$$

So $z = 2$. The volume of the ice cream cone has two parts. The first part (which is the volume of the cone) is the volume of the solid bounded by the plane $z = 2$ and the cone $z = \sqrt{x^2 + y^2}$. Hence, this volume is given by $\int_R (2 - \sqrt{x^2 + y^2})\, dA$, where R is the disk of radius 2 centered at the origin, in the xy-plane. Using polar coordinates, one has:

$$\int_R \left(2 - \sqrt{x^2 + y^2}\right) dA = \int_0^{2\pi} \int_0^2 (2 - r) \cdot r\, dr\, d\theta$$

$$= \int_0^{2\pi} \left[\left(r^2 - \frac{r^3}{3} \right) \Big|_0^2 \right] d\theta$$

$$= \frac{4}{3} \int_0^{2\pi} d\theta$$

$$= 8\pi/3 \; \leftharpoonup \; 8.3^3$$

The second part is the volume of the region above the plane $z = 2$ but inside the sphere $x^2 + y^2 + z^2 = 8$, which is given by $\int_R (\sqrt{8 - x^2 - y^2} - 2)\, dA$ where R is the same disk as before. Now

$$\int_R (\sqrt{8 - x^2 - y^2} - 2)\, dA = \int_0^{2\pi} \int_0^2 (\sqrt{8 - r^2} - 2) r \, dr \, d\theta$$

$$= \int_0^{2\pi} \int_0^2 r\sqrt{8 - r^2}\, dr \, d\theta - \int_0^{2\pi} \int_0^2 2r \, dr \, d\theta$$

$$= \int_0^{2\pi} \left(-\frac{1}{3}(8 - r^2)^{3/2} \Big|_0^2 \right) d\theta - \int_0^{2\pi} r^2 \Big|_0^2 \, d\theta$$

$$= -\frac{1}{3} \int_0^{2\pi} (4^{3/2} - 8^{3/2})\, d\theta - \int_0^{2\pi} 4 \, d\theta$$

$$= -\frac{1}{3} \cdot 2\pi(8 - 16\sqrt{2}) - 8\pi$$

$$= \frac{2\pi}{3}(16\sqrt{2} - 8) - 8\pi$$

$$= \frac{8\pi(4\sqrt{2} - 5)}{3}$$

Thus, the total volume is the sum of the two volumes, which is $32\pi(\sqrt{2} - 1)/3$.

20. The density function is given by

$$\rho(r) = 10 - 2r$$

where r is the distance from the center of the disk. So the mass of the disk in grams is

$$\int_R \rho(r)\, dA = \int_0^{2\pi} \int_0^5 (10 - 2r) r \, dr \, d\theta$$

$$= \int_0^{2\pi} \left[5r^2 - \frac{2}{3}r^3 \right]_0^5 d\theta$$

$$= \int_0^{2\pi} \frac{125}{3}\, d\theta$$

$$= \frac{250\pi}{3} \,(\text{grams})$$

21. (a)

$$\text{Total Population} = \int_{\pi/2}^{3\pi/2} \int_1^4 \delta(r, \theta)\, r \, dr \, d\theta.$$

(b) We know that $\delta(r, \theta)$ decreases as r increases, so that eliminates (iii). We also know that $\delta(r, \theta)$ decreases as the x-coordinate decreases, but $x = r \cos \theta$. With a fixed r, x is proportional to $\cos \theta$. So as the x-coordinate decreases, $\cos \theta$ decreases and (i) $\delta(r, \theta) = (4 - r)(2 + \cos \theta)$ best describes this situation.

(c)

$$\int_{\pi/2}^{3\pi/2}\int_1^4 (4-r)(2+\cos\theta)\,r\,dr\,d\theta = \int_{\pi/2}^{3\pi/2}(2+\cos\theta)\left(2r^2 - \frac{1}{3}r^3\right)\Big|_1^4 d\theta$$

$$= 9\int_{\pi/2}^{3\pi/2}(2+\cos\theta)\,d\theta$$

$$= 9\left[2\theta + \sin\theta\right]_{\pi/2}^{3\pi/2}$$

$$= 18(\pi - 1)$$

$$\approx 39$$

Thus, the population is around $39,000$.

22. A rough graph of the base of the spring is in Figure 15.49, where the coil is roughly of width 0.01 inches. The volume is equal to the product of the base area and the height. To calculate the area we use polar coordinates, taking the following integral:

$$\text{Area} = \int_0^{4\pi}\int_{0.25+0.04\theta}^{0.26+0.04\theta} r\,dr\,d\theta$$

$$= \frac{1}{2}\int_0^{4\pi}(0.26-0.04\theta)^2 - (0.25-0.04\theta)^2 d\theta$$

$$= \frac{1}{2}\int_0^{4\pi} 0.01\cdot(0.51+0.08\theta)d\theta$$

$$= 0.0051\cdot 2\pi + \frac{1}{4}(0.0008\theta^2)\Big|_0^{4\pi}$$

$$= 0.0636$$

Therefore Vol $= 0.0636\cdot 0.2 = 0.0127\text{in}^3$.

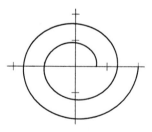

Figure 15.49

Solutions for Section 15.6

1.

$$\int_W f\, dV = \int_{-1}^{1} \int_{\pi/4}^{3\pi/4} \int_0^4 (r^2 + z^2)\, r dr\, d\theta\, dz$$

$$= \int_{-1}^{1} \int_{\pi/4}^{3\pi/4} (64 + 8z^2)\, d\theta\, dz$$

$$= \int_{-1}^{1} \frac{\pi}{2}(64 + 8z^2)\, dz$$

$$= 64\pi + \frac{8}{3}\pi$$

$$= \frac{200}{3}\pi$$

2.

$$\int_W f\, dV = \int_{-1}^{3} \int_0^{2\pi} \int_0^1 (\sin(r^2))\, r dr\, d\theta\, dz$$

$$= \int_{-1}^{3} \int_0^{2\pi} \left(-\frac{1}{2}\cos r^2\right)\Big|_0^1 d\theta\, dz$$

$$= -\frac{1}{2}\int_{-1}^{3} \int_0^{2\pi} (\cos 1 - \cos 0)\, d\theta\, dz$$

$$= -\pi \int_{-1}^{3} (\cos 1 - 1)\, dz$$

$$= -4\pi(\cos 1 - 1)$$

$$= 4\pi(1 - \cos 1) = 5.78$$

3.

$$\int_W f\, dV = \int_0^5 \int_0^{2\pi} \int_{\pi/2}^{\pi} \frac{1}{\rho} \cdot \rho^2 \sin\phi\, d\phi\, d\theta\, d\rho$$

$$= \int_0^5 \int_0^{2\pi} \int_{\pi/2}^{\pi} \rho \sin\phi\, d\phi\, d\theta\, d\rho$$

$$= \int_0^5 \int_0^{2\pi} \rho\, d\theta\, d\rho$$

$$= 2\pi \int_0^5 \rho\, d\rho$$

$$= 25\pi$$

4.

$$
\begin{aligned}
\int_W f \, dV &= \int_0^{2\pi} \int_0^{\pi/4} \int_1^2 (\sin\phi)\rho^2 \sin\phi \, d\rho \, d\phi \, d\theta \\
&= \int_0^{2\pi} \int_0^{\pi/4} \int_1^2 \rho^2 \sin^2\phi \, d\rho \, d\phi \, d\theta \\
&= \frac{7}{3} \int_0^{2\pi} \int_0^{\frac{\pi}{4}} \sin^2\phi \, d\phi \, d\theta \\
&= \frac{7}{3} \int_0^{2\pi} \int_0^{\pi/4} \frac{1 - \cos 2\phi}{2} \, d\phi \, d\theta \\
&= \frac{7}{6} \int_0^{2\pi} \left. \left(\phi - \frac{1}{2}\sin 2\phi \right) \right|_0^{\pi/4} d\theta \\
&= \frac{7}{6} \int_0^{2\pi} \left(\frac{\pi}{4} - \frac{1}{2} \right) d\theta \\
&= \frac{7}{6} \cdot 2\pi \left(\frac{\pi}{4} - \frac{1}{2} \right) \\
&= \frac{7\pi(\pi - 2)}{12}
\end{aligned}
$$

5. Using cylindrical coordinates, we get:

$$
\int_0^1 \int_0^{2\pi} \int_0^4 \delta \cdot r \, dr \, d\theta \, dz
$$

6. Using cylindrical coordinates, we get:

$$
\int_0^4 \int_0^{\pi/2} \int_0^2 \delta \cdot r \, dr \, d\theta \, dz
$$

7. Using spherical coordinates, we get:

$$
\int_0^{2\pi} \int_0^{\pi/6} \int_0^3 \delta \cdot \rho^2 \sin\phi \, d\rho \, d\phi \, d\theta
$$

8. Using spherical coordinates, we get:

$$
\int_0^{\pi} \int_0^{\pi} \int_2^3 \delta \cdot \rho^2 \sin\phi \, d\rho \, d\phi \, d\theta
$$

9. Using Cartesian coordinates, we get:

$$
\int_0^3 \int_0^1 \int_0^5 \delta \, dz \, dy \, dx
$$

10.

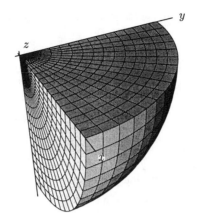

Figure 15.50

R is one eighth of a sphere of radius 1, below the xy-plane and under the first quadrant.

11. The region of integration is half of a ball centered at the origin, radius 1, on the $x \geq 0$ side. Since the integral is symmetric, we can integrate over the quarter unit ball $(x \geq 0, y \geq 0)$ and multiply the result by 2. Use spherical coordinates:

$$\int_0^1 \int_{-\sqrt{1-x^2}}^{\sqrt{1-x^2}} \int_{-\sqrt{1-x^2-z^2}}^{\sqrt{1-x^2-z^2}} \frac{1}{(x^2+y^2+z^2)^{1/2}} \, dy \, dz \, dx$$

$$= 2 \int_0^{\pi/2} \int_0^{\pi} \int_0^1 \frac{1}{\rho} \rho^2 \sin \phi \, d\rho \, d\phi \, d\theta$$

$$= 2 \int_0^{\pi/2} \int_0^{\pi} \left. \frac{\rho^2}{2} \right|_0^1 \sin \phi \, d\phi \, d\theta$$

$$= \int_0^{\pi/2} \left. (-\cos \phi) \right|_0^{\pi} \, d\theta$$

$$= (-(-1) - 0) \cdot \pi$$

$$= \pi.$$

12. The region is a solid cylinder of height 1, radius 1 with base on the xy-plane and axis on the z-axis. Use cylindrical coordinates:

$$\int_0^1 \int_{-1}^1 \int_{-\sqrt{1-x^2}}^{\sqrt{1-x^2}} \frac{1}{(x^2+y^2)^{1/2}} \, dy \, dx \, dz = \int_0^1 \int_0^{2\pi} \int_0^1 \frac{1}{r} r \, dr \, d\theta \, dz$$

$$= \int_0^1 \int_0^{2\pi} \left. r \right|_0^1 \, d\theta \, dz$$

$$= \int_0^1 \int_0^{2\pi} d\theta \, dz$$

$$= 2\pi.$$

13. (a) The angle ϕ takes on values in the range $0 \le \phi \le \pi$. Thus, $\sin \phi$ is nonnegative everywhere in W_1, and so its integral is positive.

   (b) The function ϕ is symmetric across the xy plane, such that for any point (x, y, z) in W_1, with $z \ne 0$, the point $(x, y, -z)$ has a $\cos \phi$ value with the same magnitude but opposite sign of the $\cos \phi$ value for (x, y, z). Furthermore, if $z = 0$, then (x, y, z) has a $\cos \phi$ value of 0. Thus, with $\cos \phi$ positive on the top half of the sphere and negative on the bottom half, the integral will cancel out and be equal to zero.

14. (a) The integral is negative. In W_2, we have $0 < z < 1$. Thus, $z^2 - z$ is negative throughout W_2 and thus its integral is negative.

   (b) On the top half of the sphere, z is nonnegative, but x can be both positive and negative. Thus, since W_2 is symmetric with respect to the yz plane, the contribution of a point (x, y, z) will be canceled out by its reflection $(-x, y, z)$. Thus, the integral is zero.

15. The region whose volume we want is shown in Figure 15.51:

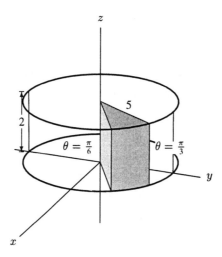

Figure 15.51

Using cylindrical coordinates, the volume is given by the integral:

$$V = \int_0^2 \int_{\pi/6}^{\pi/3} \int_0^5 r \, dr \, d\theta \, dz$$

$$= \int_0^2 \int_{\pi/6}^{\pi/3} \frac{r^2}{2} \Big|_0^5 \, d\theta \, dz$$

$$= \frac{25}{2} \int_0^2 \int_{\pi/6}^{\pi/3} d\theta \, dz$$

$$= \frac{25}{2} \int_0^2 \left(\frac{\pi}{3} - \frac{\pi}{6} \right) dz$$

$$= \frac{25}{2} \cdot \frac{\pi}{6} \cdot 2$$

$$= \frac{25\pi}{6}.$$

16. The density function can be rewritten as $\delta(\rho, \phi, \theta) = \rho$. So the mass is

$$
\int_W \delta(P)\, dV = \int_0^{2\pi} \int_0^{\pi/4} \int_0^3 \rho \cdot \rho^2 \sin\phi\, d\rho\, d\phi\, d\theta
$$

$$
= \int_0^{2\pi} \int_0^{\pi/4} \frac{81}{4} \sin\phi\, d\phi\, d\theta
$$

$$
= \frac{81}{4} \int_0^{2\pi} \left(-\frac{\sqrt{2}}{2} + 1 \right) d\theta
$$

$$
= \frac{81}{4} \cdot 2\pi \cdot \left(-\frac{\sqrt{2}}{2} + 1 \right)
$$

$$
= \frac{81}{4} \pi (-\sqrt{2} + 2)
$$

17. Using spherical coordinates:

$$
M = \int_0^{\pi} \int_0^{2\pi} \int_0^3 (3 - \rho)\rho^2 \sin\phi\, d\rho\, d\theta\, d\phi
$$

$$
= \int_0^{\pi} \int_0^{2\pi} \left[\rho^3 - \frac{\rho^4}{4} \right]_0^3 \sin\phi\, d\theta\, d\phi
$$

$$
= \frac{27}{4} \int_0^{\pi} \int_0^{2\pi} \sin\phi\, d\theta\, d\phi
$$

$$
= \frac{27}{4} \cdot 2\pi \cdot [-\cos\phi]_0^{\pi}
$$

$$
= \frac{27}{4} \cdot 2\pi \cdot [-(-1) + 1]
$$

$$
= 27\pi.
$$

18. Since the hole resembles a cylinder, we will use cylindrical coordinates. Let the center of the sphere be at the origin, and let the center of the hole be the z-axis (see Figure 15.52).

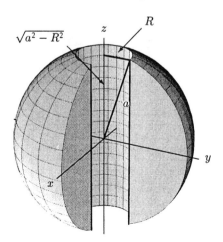

Figure 15.52

Then we will integrate from $z = -\sqrt{a^2 - R^2}$ to $z = \sqrt{a^2 - R^2}$, and each cross-section will be an annulus. So the volume is

$$\int_{-\sqrt{a^2-R^2}}^{\sqrt{a^2-R^2}} \int_0^{2\pi} \int_R^{\sqrt{a^2-z^2}} r\,dr\,d\theta\,dz = \int_{-\sqrt{a^2-R^2}}^{\sqrt{a^2-R^2}} \int_0^{2\pi} \frac{1}{2}(a^2 - z^2 - R^2)\,d\theta\,dz$$

$$= \pi \int_{-\sqrt{a^2-R^2}}^{\sqrt{a^2-R^2}} (a^2 - z^2 - R^2)\,dz$$

$$= \pi \left[(a^2 - R^2)(2\sqrt{a^2 - R^2}) - \frac{1}{3}(2(a^2 - R^2)^{\frac{3}{2}}) \right]$$

$$= \frac{4\pi}{3}(a^2 - R^2)^{\frac{3}{2}}$$

19. The distance from a point (x, y, z) to the origin is given by $\sqrt{x^2 + y^2 + z^2}$. Thus we want to evaluate

$$\frac{\int_R \sqrt{x^2 + y^2 + z^2}\,dV}{\text{Vol}(R)}$$

where R is the region bounded by the hemisphere $z = \sqrt{8 - x^2 - y^2}$ and the cone $z = \sqrt{x^2 + y^2}$. See Figure 15.53. We will use spherical coordinates.

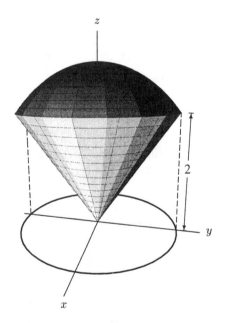

Figure 15.53

In spherical coordinates, the quantity ρ goes from 0 to $\sqrt{8}$, and θ goes from 0 to 2π, and ϕ goes from 0 to $\pi/4$ (because the angle of the cone is $\pi/4$). Thus we have

$$\int_R \sqrt{x^2 + y^2 + z^2}\,dV = \int_0^{2\pi} \int_0^{\pi/4} \int_0^{\sqrt{8}} \rho(\rho^2 \sin\phi)\,d\rho\,d\phi\,d\theta$$

$$= \int_0^{2\pi} \int_0^{\pi/4} \sin\phi \cdot \frac{\rho^4}{4}\Big|_0^{\sqrt{8}} \, d\phi d\theta$$

$$= \int_0^{2\pi} \int_0^{\pi/4} 16 \sin\phi \, d\phi d\theta$$

$$= \int_0^{2\pi} 16(-\cos\phi)\Big|_0^{\pi/4} \, d\theta$$

$$= \int_0^{2\pi} 16\left(1 - \frac{\sqrt{2}}{2}\right) d\theta$$

$$= 32\left(1 - \frac{\sqrt{2}}{2}\right)\pi$$

From Problem 19 of Section 15.5 we know that $\text{Vol}(R) = 32\pi(\sqrt{2} - 1)/3$, therefore

$$\text{Average distance} = \frac{\int_R \sqrt{x^2 + y^2 + z^2} \, dV}{\text{Vol}(R)}$$

$$= \frac{32\left(1 - \frac{\sqrt{2}}{2}\right)\pi}{[32(\sqrt{2} - 1)\pi/3]}$$

$$= \frac{3}{\sqrt{2}}.$$

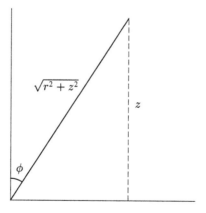

Figure 15.54

20. Assume the base of the cylinder sits on the xy-plane with center at the origin. Because the cylinder is symmetric about the z-axis, the force in the horizontal x or y direction is 0. Thus we need only compute the vertical z component of the force. We are going to use cylindrical coordinates; since the force is $G \cdot \text{mass}/(\text{distance})^2$, a piece of the cylinder of volume dV located at (r, θ, z) exerts on the unit mass a force with magnitude $G(\delta \, dV)/(r^2 + z^2)$. The vertical component of this force is (See Figure 15.54)

$$(G(\delta \, dV)/(r^2 + z^2)) \cdot \cos\phi = (G(\delta \, dV)/(r^2 + z^2)) \cdot \frac{z}{\sqrt{r^2 + z^2}} = \frac{G\delta z \, dV}{(r^2 + z^2)^{3/2}}$$

Adding up all the contributions of all the dV's, we get a vertical force of

$$\int_0^H \int_0^{2\pi} \int_0^R \frac{G\delta zr}{(r^2 + z^2)^{3/2}}\, dr d\theta dz$$

$$= \int_0^H \int_0^{2\pi} (G\delta z) \left(-\frac{1}{\sqrt{r^2 + z^2}}\right)\bigg|_0^R d\theta dz$$

$$= \int_0^H \int_0^{2\pi} (G\delta z) \cdot \left(-\frac{1}{\sqrt{R^2 + z^2}} + \frac{1}{z}\right) d\theta dz$$

$$= \int_0^H 2\pi G\delta \left(1 - \frac{z}{\sqrt{R^2 + z^2}}\right) dz$$

$$= 2\pi G\delta(z - \sqrt{R^2 + z^2})\bigg|_0^H$$

$$= 2\pi G\delta(H - \sqrt{R^2 + H^2} + R) = 2\pi G\delta(H + R - \sqrt{R^2 + H^2})$$

21. The total volume of the cone is $\frac{1}{3}\pi r^2 h = \frac{1}{3}\pi \cdot 1^2 \cdot 1 = \frac{1}{3}\pi$, so the total mass is $\frac{1}{3}\pi$ (since the density is always 1). The center of mass z-coordinate is given by

$$\bar{z} = \frac{3}{\pi} \int_C z\, dV$$

Using cylindrical coordinates to evaluate this integral gives

$$\bar{z} = \frac{3}{\pi} \int_0^{2\pi} \int_0^1 \int_0^z zr\, dr\, dz\, d\theta$$

$$= \frac{3}{\pi} \int_0^{2\pi} \int_0^1 \frac{z^3}{2}\, dz\, d\theta$$

$$= \frac{3}{\pi} \int_0^{2\pi} \frac{1}{8}\, d\theta = \frac{3}{4}$$

22. (a) The mass m of the cone is given by $\int_C \rho\, dV$. In cylindrical coordinates this is

$$m = \int_0^{2\pi} \int_0^1 \int_0^z z^2 r\, dr\, dz\, d\theta$$

$$= \int_0^{2\pi} \int_0^1 \frac{z^4}{2}\, dz\, d\theta$$

$$= \int_0^{2\pi} \frac{1}{10}\, d\theta = \frac{\pi}{5}$$

(b) The center of mass z-coordinate is given by

$$\bar{z} = \frac{5}{\pi} \int_C z \cdot z^2\, dV$$

Using cylindrical coordinates to evaluate this integral gives

$$\bar{z} = \frac{5}{\pi} \int_0^{2\pi} \int_0^1 \int_0^z z^3 r\, dr\, dz\, d\theta$$

$$= \frac{5}{\pi} \int_0^{2\pi} \int_0^1 \frac{z^5}{2} \, dz \, d\theta$$

$$= \frac{5}{\pi} \int_0^{2\pi} \frac{1}{12} \, d\theta = \frac{5}{6}$$

Comparing this answer with the center of mass in Problem 21, where the density was constant, it makes sense that the center of mass would be higher in this problem, since more mass is concentrated near the top of the cone.

23. We first need to find the mass of the solid, using cylindrical coordinates:

$$m = \int_0^{2\pi} \int_0^1 \int_0^{\sqrt{z/a}} r \, dr \, dz \, d\theta$$

$$= \int_0^{2\pi} \int_0^1 \frac{z}{2a} \, dz \, d\theta$$

$$= \int_0^{2\pi} \frac{1}{4a} \, d\theta = \frac{\pi}{2a}$$

It makes sense that the mass would vary inversely with a, since increasing a makes the paraboloid skinnier. Now for the z-coordinate of the center of mass, again using cylindrical coordinates:

$$\bar{z} = \frac{2a}{\pi} \int_0^{2\pi} \int_0^1 \int_0^{\sqrt{z/a}} zr \, dr \, dz \, d\theta$$

$$= \frac{2a}{\pi} \int_0^{2\pi} \int_0^1 \frac{z^2}{2a} \, dz \, d\theta$$

$$= \frac{2a}{\pi} \int_0^{2\pi} \frac{1}{6a} \, d\theta = \frac{2}{3}$$

24. The volume of the hemisphere is $\frac{2}{3}\pi a^3$ so its mass is $\frac{2}{3}\pi a^3 b$. To find the location of the center of mass, we place the base of the hemisphere on the xy-plane with the origin at its center, so we can describe it in spherical coordinates by $0 \le \rho \le a$, $0 \le \phi \le \frac{\pi}{2}$ and $0 \le \theta \le 2\pi$. Then the x-coordinate of the center of mass is, integrating using spherical coordinates:

$$\bar{x} = \frac{3}{2\pi a^3 b} \int_0^a \int_0^{\frac{\pi}{2}} \int_0^{2\pi} \rho \sin(\phi)\cos(\theta) \cdot \rho^2 \sin(\phi) \, d\theta \, d\phi \, d\rho = 0$$

since the first integral $\int_0^{2\pi} \cos(\theta) \, d\theta$ is zero. A similar computation shows that $\bar{y} = 0$. Now for the z-coordinate:

$$\bar{z} = \frac{3}{2\pi a^3 b} \int_0^a \int_0^{\frac{\pi}{2}} \int_0^{2\pi} \rho\cos(\phi) \cdot \rho^2 \sin(\phi) \, d\theta \, d\phi \, d\rho$$

$$= \frac{3}{2\pi a^3 b} \cdot 2\pi \int_0^a \int_0^{\frac{\pi}{2}} \rho^3 \cos(\phi)\sin(\phi) \, d\phi \, d\rho$$

$$= \frac{3}{a^3 b} \int_0^a \rho^3 \left.\frac{\sin^2(\phi)}{2}\right|_0^{\frac{\pi}{2}} d\rho$$

$$= \frac{3}{2a^3 b} \int_0^a \rho^3 \, d\rho = \frac{3a}{8b}$$

So the x and y-coordinates are located at the center of the base, while the z-coordinate is located $\frac{3a}{8b}$ above the center of the base.

25.   The sum of the three moments of inertia I for the ball B will be

$$3I = \frac{3}{4\pi a^3} \int_B (y^2 + z^2)\, dV + \frac{3}{4\pi a^3} \int_B (x^2 + z^2)\, dV + \frac{3}{4\pi a^3} \int_B (x^2 + y^2)\, dV$$

$$= \frac{3}{4\pi a^3} \int_B (2x^2 + 2y^2 + 2z^2)\, dV,$$

which, in spherical coordinates is

$$\frac{3}{2\pi a^3} \int_B (x^2 + y^2 + z^2)\, dV = \frac{3}{2\pi a^3} \int_0^a \int_0^\pi \int_0^{2\pi} \rho^2 \cdot \rho^2 \sin(\phi)\, d\theta\, d\phi\, d\rho$$

$$= \frac{3}{a^3} \int_0^a \int_0^\pi \rho^4 \sin(\phi)\, d\phi\, d\rho$$

$$= \frac{6}{a^3} \int_0^a \rho^4\, d\rho = \frac{6}{5}a^2.$$

Thus $3I = \frac{6}{5}a^2$, so $I = \frac{2}{5}a^2$.

26.   First we need to find the volume of the cone. In spherical coordinates we find:

$$V = \int_0^a \int_0^{\frac{\pi}{3}} \int_0^{2\pi} \rho^2 \sin(\phi)\, d\theta\, d\phi\, d\rho = \frac{\pi a^3}{3}$$

Now, to find the moment of inertia about the z-axis we need to compute the integral $\frac{3}{\pi a^3} \int_W x^2 + y^2\, dV$. We can do this in spherical coordinates as

$$\frac{3}{\pi a^3} \int_W x^2 + y^2\, dV = \frac{3}{\pi a^3} \int_0^a \int_0^{\frac{\pi}{3}} \int_0^{2\pi} (\rho^2 \sin^2(\phi) \cos^2(\theta) + \rho^2 \sin^2(\phi) \sin^2(\theta)) \cdot \rho^2 \sin(\phi)\, d\theta\, d\phi\, d\rho$$

$$= \frac{3}{\pi a^3} \int_0^a \int_0^{\frac{\pi}{3}} \int_0^{2\pi} \rho^4 \sin^3(\phi)\, d\theta\, d\phi\, d\rho$$

$$= \frac{6}{a^3} \int_0^a \int_0^{\frac{\pi}{3}} \rho^4 \sin^3(\phi)\, d\phi\, d\rho$$

$$= \frac{6}{a^3} \frac{5}{24} \int_0^a \rho^4\, d\rho = \frac{a^2}{4}.$$

Solutions for Section 15.7

1.   (a)

$$\int_0^1 \int_{1/3}^1 \frac{2}{3}(x + 2y)\, dx\, dy = \int_0^1 \frac{2}{3}(\frac{1}{2}x^2 + 2xy)\Big|_{1/3}^1\, dy$$

$$= \int_0^1 \frac{2}{3} \left[(\frac{1}{2} + 2y) - (\frac{1}{18} + \frac{2}{3}y) \right] dy$$

$$= \frac{2}{3} \int_0^1 (\frac{4}{9} + \frac{4}{3}y) \, dy$$

$$= \frac{2}{3} \left(\frac{4}{9}y + \frac{2}{3}y^2 \right) \Big|_0^1$$

$$= \frac{2}{3} \left(\frac{10}{9} \right) = \frac{20}{27}.$$

(b)   It is easier to calculate the probability that $x < (1/3) + y$ does not happen, that is, the probability that $x \geq (1/3) + y$, and subtract it from 1. The probability that $x \geq (1/3) + y$ is

$$\int_{1/3}^1 \int_0^{x-(1/3)} \frac{2}{3}(x + 2y) \, dy \, dx = \int_{1/3}^1 \frac{2}{3}(xy + y^2) \Big|_0^{x-(1/3)} \, dx$$

$$= \frac{2}{3} \int_{1/3}^1 \left(x(x - \frac{1}{3}) + (x - \frac{1}{3})^2 \right) dx$$

$$= \frac{2}{3} \int_{1/3}^1 \left(2x^2 - x + \frac{1}{9} \right) dx$$

$$= \frac{2}{3} (\frac{2}{3}x^3 - \frac{1}{2}x^2 + \frac{1}{9}x) \Big|_{1/3}^1$$

$$= \frac{2}{3} \left[(\frac{2}{3} - \frac{1}{2} + \frac{1}{9}) - (\frac{2}{81} - \frac{1}{18} + \frac{1}{27}) \right]$$

$$= 44/243.$$

Thus, the probability that $x < (1/3) + y$ is $1 - (44/243) = 199/243$.

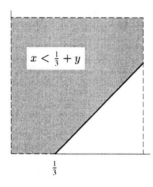

Figure 15.55

2.   (a)   The entries in this table are non-negative and their sum is 1.
   (b)   Add up the values along the row $x = 2$: $0.2 + 0.1 + 0 = 0.3$.
   (c)   Add up the columns with $y = 1$ and $y = 2$: $0.3 + 0.2 + 0.1 + 0 + 0.2 + 0.1 + 0 + 0 = 0.9$.
   (d)   Add up the values in the grid corresponding $x = 1, 2, 3$ and $y = 1, 2$: $0.3 + 0.2 + 0.2 + 0.1 + 0.1 + 0 + 0 + 0 = 0.9$.

3. (a) We know that $\int_{-\infty}^{\infty}\int_{-\infty}^{\infty} f(x,y)dydx = 1$ for a joint density function. So,

$$1 = \int_{-\infty}^{\infty}\int_{-\infty}^{\infty} f(x,y)dydx = \int_0^1 \int_x^1 kxy\,dydx$$
$$= \frac{1}{8}k$$

hence $k = 8$.

(b) The region where $0 \le \sqrt{x} < y$ is sketched in Figure 15.56

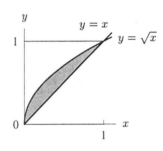

Figure 15.56

So the probability that (x,y) satisfies $0 \le \sqrt{x} \le y$ is given by:

$$\int_0^1 \int_x^{\sqrt{x}} 8xy\, dy\, dx = \int_0^1 4x(y^2)\Big|_x^{\sqrt{x}} dx$$
$$= \int_0^1 4x(x - x^2)dx$$
$$= 4\left(\frac{1}{3}x^3 - \frac{1}{4}x^4\right)\Big|_0^1$$
$$= 4\left(\frac{1}{3} - \frac{1}{4}\right)$$
$$= \frac{1}{3}$$

This tells us that in choosing points from the region defined by $0 \le x \le y \le 1$, that 1/3 of the time we would pick a point from the region defined by $0 \le \sqrt{x} \le y \le 1$. These regions are shown in Figure 15.56.

4. (a) For a density function,

$$1 = \int_{-\infty}^{\infty}\int_{-\infty}^{\infty} f(x,y)\, dy\, dx = \int_0^2 \int_0^1 kx^2\, dy\, dx$$
$$= \int_0^2 kx^2\, dx$$
$$= \frac{kx^3}{3}\Big|_0^2 = \frac{8k}{3}.$$

So $k = 3/8$.

(b)
$$\int_0^1 \int_0^{2-y} \frac{3}{8}x^2 \, dx \, dy = \int_0^1 \frac{1}{8}(2-y)^3 \, dy = \frac{-1}{32}(2-y)^4 \Big|_0^1 = \frac{15}{32}$$

(c)
$$\int_0^{1/2} \int_0^1 \frac{3}{8}x^2 \, dx \, dy = \int_0^{1/2} \frac{1}{8}x^3 \Big|_0^1 \, dy = \int_0^{1/2} \frac{1}{8} \, dy = \frac{1}{16}.$$

5. Since
$$\sum_x \sum_y f(x,y) \, \Delta x \, \Delta y \approx \int_R f(x,y) \, dx \, dy$$

and since x never exceeds 1, and we can assume that no one lives to be over 100, so y does not exceed 100, we have
$$\begin{array}{c} \text{Fraction of} \\ \text{policies} \end{array} = \int_R f(x,y) \, dx \, dy = \int_{65}^{100} \int_{0.8}^1 f(x,y) \, dx \, dy,$$

where R is the rectangle: $0.8 \le x \le 1$, $65 \le y \le 100$.

6. (a) The area of S is $(2)(4) = 8$. Because the density function $p(x,y)$ is constant on S and the total volume under a density function above the xy-plane is 1, $p(x,y) = 1/8$ for (x,y) in S, and $p(x,y) = 0$ for (x,y) outside S.

    (b) The probability that (x,y) is in T is
$$\int_T f(x,y) \, dy \, dx = \frac{1}{8} \int_T dy \, dx = \frac{\text{area}(T)}{8} = \frac{\alpha}{8}.$$

7.
$$f(x,y) = \frac{10}{\sqrt{2\pi}} e^{-50(x-5)^2} \frac{6}{\sqrt{2\pi}} e^{-18(y-15)^2}$$
$$= \frac{30}{\pi} e^{-50(x-5)^2 - 18(y-15)^2}$$

8. (a) Since the exponential function is always positive and λ is positive, $p(t) \ge 0$ for all t, and
$$\int_0^\infty p(t) dt = \lim_{b \to \infty} -e^{-\lambda t} \Big|_0^b = \lim_{b \to \infty} -e^{-bt} + 1 = 1.$$

    (b) The density function for the probability that the first substance decays at time t and the second decays at time s is
$$p(t,s) = \lambda e^{-\lambda t} \mu e^{-\mu s} = \lambda \mu e^{-\lambda t - \mu s},$$
    for $s \ge 0$ and $t \ge 0$, and is zero otherwise.

    (c) We want the probability that the decay time t of the first substance is less than or equal to the decay time s of the second, so we want to integrate the density function over the region $0 \le t \le s$. Thus, we compute
$$\int_0^\infty \int_t^\infty \lambda \mu e^{-\lambda t} e^{-\mu s} \, ds \, dt = \int_0^\infty \lambda e^{-\lambda t} (-e^{-\mu s}) \Big|_t^\infty \, dt$$
$$= \int_0^\infty \lambda e^{-\lambda t} e^{-\mu t} \, dt$$

$$= \int_0^\infty \lambda e^{(-\lambda + \mu)t}\, dt$$

$$= \frac{-\lambda}{\lambda + \mu} e^{-(\lambda + \mu)t}\bigg|_0^\infty = \frac{\lambda}{\lambda + \mu}.$$

So for example, if $\lambda = 1$ and $\mu = 4$, then the probability that the first substance decays first is $1/5$.

9. (a)

$$\int_{\theta=0}^{\frac{\pi}{6}} \int_{r=\frac{1}{\cos\theta}}^{4} p(r,\theta)r\, dr\, d\theta$$

(b)

$$\int_{\theta=\frac{\pi}{6}}^{\frac{\pi}{6}+\frac{\pi}{12}} \int_{r=\frac{1}{\cos\theta}}^{4} p(r,\theta)r\, dr\, d\theta + \int_{\theta=\frac{\pi}{6}+\frac{\pi}{12}}^{\frac{2\pi}{6}} \int_{r=\frac{1}{\sin\theta}}^{4} p(r,\theta)r\, dr\, d\theta$$

Solutions for Section 15.8

1. Given $T = \{(s,t) \mid 0 \le s \le 3, 0 \le t \le 2\}$ and

$$\begin{cases} x = 2s - 3t \\ y = s - 2t \end{cases}$$

The shaded area in Figure 15.57 is the corresponding region R in the xy-plane.

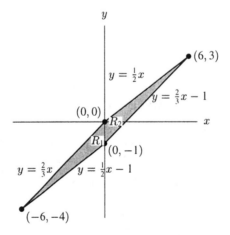

Figure 15.57

Since

$$\frac{\partial(x,y)}{\partial(s,t)} = \begin{vmatrix} \frac{\partial x}{\partial s} & \frac{\partial x}{\partial t} \\ \frac{\partial y}{\partial s} & \frac{\partial y}{\partial t} \end{vmatrix} = \begin{vmatrix} 2 & -3 \\ 1 & -2 \end{vmatrix} = -1,$$

$$\left| \frac{\partial(x,y)}{\partial(s,t)} \right| = 1.$$

Thus we get

$$\int_T \left|\frac{\partial(x,y)}{\partial(s,t)}\right| ds\,dt = \int_0^3 ds \int_0^2 dt = 6.$$

Since

$$\int_R dx\,dy = \int_{R_1} dx\,dy + \int_{R_2} dx\,dy = \int_{-6}^0 dx \int_{\frac12 x-1}^{\frac23 x} dy + \int_0^6 dx \int_{\frac23 x-1}^{\frac12 x} dy$$

$$= \int_{-6}^0 (\frac16 x + 1)\,dx + \int_0^6 (\frac{-1}{6}x + 1)\,dx = 3 + 3 = 6,$$

thus

$$\int_R dx\,dy = \int_T \left|\frac{\partial(x,y)}{\partial(s,t)}\right| ds\,dt.$$

2.

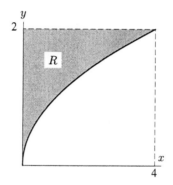

Figure 15.58

Given $T = \{(s,t)\,|\,0 \le s \le 2,\ s \le t \le 2\}$ and

$$\begin{cases} x = s^2 \\ y = t, \end{cases}$$

$$R = \{(x,y)\,|\,0 \le x \le 4,\ \sqrt{x} \le y \le 2\}.$$

$$\frac{\partial(x,y)}{\partial(s,t)} = \begin{vmatrix} \frac{\partial x}{\partial s} & \frac{\partial x}{\partial t} \\ \frac{\partial y}{\partial s} & \frac{\partial y}{\partial t} \end{vmatrix} = \begin{vmatrix} 2s & 0 \\ 0 & 1 \end{vmatrix} = 2s,$$

$$\left|\frac{\partial(x,y)}{\partial(s,t)}\right| = 2s \quad \text{since} \quad 0 \le s \le 2.$$

$$\int_T \left|\frac{\partial(x,y)}{\partial(s,t)}\right| ds\,dt = 2\int_0^2 s\,ds \int_s^2 dt = 2\int_0^2 s(2-s)\,ds = 2\left[s^2 - \frac{s^3}{3}\right]_0^2 = \frac83.$$

So,

$$\int_R dx\,dy = \int_0^4 dx \int_{\sqrt{x}}^2 dy = \int_0^4 (2 - \sqrt{x})\,dx$$

$$= \left[2x - \frac23 x^{3/2}\right]_0^4 = 8 - \frac{16}{3} = \frac83.$$

Thus

$$\int_R dx\, dy = \int_T \left| \frac{\partial(x,y)}{\partial(s,t)} \right| ds\, dt.$$

3. Given

$$\begin{cases} x = \rho \sin\phi \cos\theta \\ y = \rho \sin\phi \sin\theta \\ z = \rho \cos\phi, \end{cases}$$

$$\frac{\partial(x,y,z)}{\partial(\rho,\phi,\theta)} = \begin{vmatrix} \frac{\partial x}{\partial \rho} & \frac{\partial x}{\partial \phi} & \frac{\partial x}{\partial \theta} \\ \frac{\partial y}{\partial \rho} & \frac{\partial y}{\partial \phi} & \frac{\partial y}{\partial \theta} \\ \frac{\partial z}{\partial \rho} & \frac{\partial z}{\partial \phi} & \frac{\partial z}{\partial \theta} \end{vmatrix} = \begin{vmatrix} \sin\phi \cos\theta & \rho \cos\phi \cos\theta & -\rho \sin\phi \sin\theta \\ \sin\phi \sin\theta & \rho \cos\phi \sin\theta & \rho \sin\phi \cos\theta \\ \cos\phi & -\rho \sin\phi & 0 \end{vmatrix}$$

$$= \cos\phi \begin{vmatrix} \rho \cos\phi \cos\theta & -\rho \sin\phi \sin\theta \\ \rho \cos\phi \sin\theta & \rho \sin\phi \cos\theta \end{vmatrix} + \rho \sin\phi \begin{vmatrix} \sin\phi \cos\theta & -\rho \sin\phi \sin\theta \\ \sin\phi \sin\theta & \rho \sin\phi \cos\theta \end{vmatrix}$$

$$= \cos\phi(\rho^2 \cos^2\theta \cos\phi \sin\phi + \rho^2 \sin^2\theta \cos\phi \sin\phi)$$

$$\quad + \rho \sin\phi(\rho \sin^2\phi \cos^2\theta + \rho \sin^2\phi \sin^2\theta)$$

$$= \rho^2 \cos^2\phi \sin\phi + \rho^2 \sin^3\phi$$

$$= \rho^2 \sin\phi.$$

4. Given

$$\begin{cases} x = 3s - 4t \\ y = 5s + 2t, \end{cases}$$

we have

$$\begin{cases} s = \frac{1}{13}(x + 2y) \\ t = \frac{1}{26}(3y - 5x). \end{cases}$$

Since

$$\frac{\partial(x,y)}{\partial(s,t)} = \begin{vmatrix} \frac{\partial x}{\partial s} & \frac{\partial x}{\partial t} \\ \frac{\partial y}{\partial s} & \frac{\partial y}{\partial t} \end{vmatrix} = \begin{vmatrix} 3 & -4 \\ 5 & 2 \end{vmatrix} = 26,$$

$$\frac{\partial(s,t)}{\partial(x,y)} = \begin{vmatrix} \frac{\partial s}{\partial x} & \frac{\partial s}{\partial y} \\ \frac{\partial t}{\partial x} & \frac{\partial t}{\partial y} \end{vmatrix} = \begin{vmatrix} \frac{1}{13} & \frac{2}{13} \\ -\frac{5}{26} & \frac{3}{26} \end{vmatrix} = \left(\frac{3}{26}\right)\left(\frac{1}{13}\right) + \left(\frac{5}{26}\right)\left(\frac{2}{13}\right) = \frac{1}{26}.$$

So

$$\frac{\partial(x,y)}{\partial(s,t)} \cdot \frac{\partial(s,t)}{\partial(x,y)} = 26 \cdot \frac{1}{26} = 1.$$

5. Given

$$\begin{cases} x = 2s + t \\ y = s - t, \end{cases}$$

we have

$$\frac{\partial(x,y)}{\partial(s,t)} = \begin{vmatrix} \frac{\partial x}{\partial s} & \frac{\partial x}{\partial t} \\ \frac{\partial y}{\partial s} & \frac{\partial y}{\partial t} \end{vmatrix} = \begin{vmatrix} 2 & 1 \\ 1 & -1 \end{vmatrix} = -3,$$

hence

$$\left|\frac{\partial(x, y)}{\partial(s, t)}\right| = 3.$$

We get

$$\int_R (x + y) \, dA = \int_T 3s \left|\frac{\partial(x, y)}{\partial(s, t)}\right| ds dt = \int_T (3s)(3) \, ds \, dt = 9 \int_T s \, ds \, dt,$$

where T is the region in the st-plane corresponding to R.

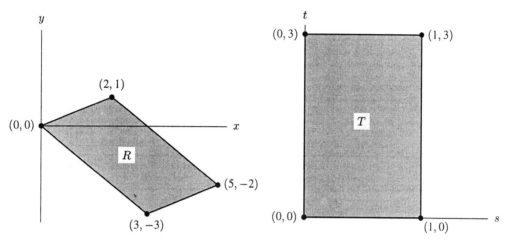

Figure 15.59   **Figure 15.60**

Now, we need to find T.

As

$$\begin{cases} x = 2s + t \\ y = s - t \end{cases} \quad \text{or} \quad \begin{cases} s = \frac{1}{3}(x + y) \\ t = \frac{1}{3}(x - 2y), \end{cases}$$

so from the above transformation and Figure 15.59, T is the shaded area in Figure 15.60. Therefore

$$\int_R (x + y) \, dA = 9 \int_0^1 s \, ds \int_0^3 dt = (27)(\frac{1}{2}) = 13.5.$$

6. Given

$$\begin{cases} x = \frac{s}{2} \\ y = \frac{t}{3}, \end{cases}$$

we have

$$\frac{\partial(x, y)}{\partial(s, t)} = \begin{vmatrix} \frac{\partial x}{\partial s} & \frac{\partial x}{\partial t} \\ \frac{\partial y}{\partial s} & \frac{\partial y}{\partial t} \end{vmatrix} = \begin{vmatrix} \frac{1}{2} & 0 \\ 0 & \frac{1}{3} \end{vmatrix} = \frac{1}{6}$$

hence

$$\left|\frac{\partial(x, y)}{\partial(s, t)}\right| = \frac{1}{6}.$$

So

$$\int_R (x^2 + y^2) \, dA = \int_T \left(\frac{s^2}{4} + \frac{t^2}{9}\right) \left(\frac{1}{6}\right) ds \, dt,$$

where R is the region bounded by the curve $4x^2 + 9y^2 = 36$ and T is the corresponding region in the st-plane.

Since the curve $4x^2 + 9y^2 = 36$ corresponds to the curve $s^2 + t^2 = 36$ under the change of coordinates, T is the region bounded by the curve $s^2 + t^2 = 36$. So

$$\int_R (x^2 + y^2)\, dA = \frac{1}{24} \int_T s^2\, ds\, dt + \frac{1}{54} \int_T t^2\, ds\, dt.$$

Now polar coordinates (r, θ) are used to evaluate the above integral. As

$$\begin{cases} s = r \cos \theta \\ t = r \sin \theta \end{cases} \quad \text{and} \quad \begin{cases} s^2 = r^2 \cos^2 \theta \\ t^2 = r^2 \sin^2 \theta, \end{cases}$$

$$\left| \frac{\partial(s,t)}{\partial(r,\theta)} \right| = \begin{vmatrix} \cos \theta & -r \sin \theta \\ \sin \theta & r \cos \theta \end{vmatrix} = r.$$

So

$$\frac{1}{24} \int_T s^2\, ds\, dt = \frac{1}{24} \int_0^{2\pi} \cos^2 \theta\, d\theta \int_0^6 r^3\, dr = \frac{27}{2} \int_0^{2\pi} \cos^2 \theta\, d\theta$$

$$= \left(\frac{27}{2} \right) \left(\frac{1}{2} \right) \int_0^{2\pi} (1 + \cos 2\theta)\, d\theta = \left(\frac{27}{2} \right) \left(\frac{1}{2} \right) (2\pi) = 13.5\pi$$

$$\frac{1}{54} \int_T t^2\, ds\, dt = \frac{1}{54} \int_0^{2\pi} \sin^2 \theta\, d\theta \int_0^6 r^3\, dr = 6 \int_0^{2\pi} \sin^2 \theta\, d\theta$$

$$= 3 \int_0^{2\pi} (1 - \cos 2\theta)\, d\theta = (3)(2\pi) = 6\pi.$$

$$\int_R (x^2 + y^2)\, dA = 13.5\pi + 6\pi = 19.5\pi.$$

7. Given

$$\begin{cases} s = xy \\ t = xy^2, \end{cases}$$

we have

$$\frac{\partial(s,t)}{\partial(x,y)} = \begin{vmatrix} \frac{\partial s}{\partial x} & \frac{\partial s}{\partial y} \\ \frac{\partial t}{\partial x} & \frac{\partial t}{\partial y} \end{vmatrix} = \begin{vmatrix} y & x \\ y^2 & 2xy \end{vmatrix} = xy^2 = t.$$

Since

$$\frac{\partial(s,t)}{\partial(x,y)} \cdot \frac{\partial(x,y)}{\partial(s,t)} = 1,$$

$$\frac{\partial(x,y)}{\partial(s,t)} = t \quad \text{so} \quad \left| \frac{\partial(x,y)}{\partial(s,t)} \right| = \frac{1}{t}$$

So

$$\int_R xy^2\, dA = \int_T t \left| \frac{\partial(x,y)}{\partial(s,t)} \right|\, ds\, dt = \int_T t \left(\frac{1}{t} \right)\, ds\, dt = \int_T ds\, dt,$$

where T is the region bounded by $s = 1$, $s = 4$, $t = 1$, $t = 4$. Then

$$\int_R xy^2\, dA = \int_1^4 ds \int_1^4 dt = 9.$$

8. Let

$$\begin{cases} s = x - y \\ t = x + y, \end{cases} \quad \text{that is} \quad \begin{cases} x = \frac{1}{2}(s + t) \\ y = \frac{1}{2}(t - s), \end{cases}$$

we get

$$\frac{\partial(x,y)}{\partial(s,t)} = \begin{vmatrix} \frac{1}{2} & \frac{1}{2} \\ -\frac{1}{2} & \frac{1}{2} \end{vmatrix} = \frac{1}{2}.$$

Hence

$$I = \int_R \cos\left(\frac{x-y}{x+y}\right) dx\, dy = \int_T \cos\left(\frac{s}{t}\right)\left|\frac{\partial(x,y)}{\partial(s,t)}\right| ds\, dt = \frac{1}{2}\int_T \cos\left(\frac{s}{t}\right) ds\, dt,$$

where R is the triangle bounded by $x + y = 1, x = 0, y = 0$ and T is its image which is the triangle bounded by $t = 1, s = -t, s = t$.
Then

$$I = \frac{1}{2}\int_0^1 \int_{-t}^t \cos\left(\frac{s}{t}\right) ds\, dt = \frac{1}{2}\int_0^1 t[\sin(1) - \sin(-1)]\, dt$$

$$= \frac{1}{2}\int_0^1 t \cdot 2\sin 1\, dt = \sin 1 \int_0^1 t\, dt = \frac{\sin 1}{2} = 0.42.$$

Solutions for Chapter 15 Review

1. First we will subdivide the area into 70 squares, as shown. We will find the upper and lower bounds for the total rainfall, and then take the average of the two. We do this by finding the highest amount of rainfall in that subdivision, multiplying it by the area, and then adding up all the contributions from each subdivisions, and then doing the same for the lowest amount of rainfall.

   For an upper bound on the rainfall, going left to right, top to bottom, we get: $[(80+40+40+40+0+0+0)+(80+80+80+40+40+0+0)+(80+80+80+40+40+40+80)+(80+80+80+80+40+40+80)+(12+40+40+40+80+80+40)+(20+12+20+80+80+80+40)+(20+20+20+40+80+40+0)+(40+40+40+20+0+40+40)+(0+0+40+20+40+40+0)+(0+0+40+40+40+0+0)] = 2804$ in. This gives a volume of $(2804 \text{ in})(0.00001578 \text{ miles/in})(500 \text{ miles})(500 \text{ miles})$. This equals about 11061 cubic miles for the upper bound. For the low rainfall we get: $[(40+20+20+40+0+0+0)+(80+40+40+40+40+0+0)+(12+80+80+40+40+40+40)+(4+4+80+80+40+40+40)+(4+4+4+40+40+40+40)+(12+4+4+40+80+40+40)+(20+4+4+40+40+40+0)+(40+40+12+20+0+40+40)+(0+0+12+12+40+40+0)+(0+0+12+12+40+0+0)](0.00001578 \text{ miles})(500 \text{ miles})(500 \text{ miles})$. This equals about 7417 cubic miles for the lower bound. Taking the average of these two, we get $(11061 + 7417)/2 = 9239$ cubic miles of rain over a year.

2. We use the average low temperature over the grid squares which cover the city to approximate the average low temperature over Washington, D.C. There are 39 such grid squares, as shown in Figure 15.8. We will find the upper and lower sums for the total temperature, then take the average of the two, and divide it by the area of those 39 grid squares. For the upper sum, going from left to right, top to bottom, we get: $[(28) + (28.25 + 28.75 + 28.75) + (28.25 + 29 + 30 + 30 + 29) + (29 + 30.25 + 30.25 + 29.25 + 29 + 29 + 29) + (30.25 + 31 + 31 + 30 + 29.25 + 29.25 + 29.25 + 29) + (31 + 31 + 30.75 + 30.5 + 30 + 29) + (31 + 31 + 30.5 + 30) + (30.75 + 30.75 + 30.25) + (30.5 + 30.5)](1)(1) = 1162.25$. For the lower sum, we get:$[(27.25) + (27.5 + 27.75 + 28) + (27.5 + 27.75 + 28.25 + 28 + 28) + (27.75 + 28.25 + 29 + 28 + 28.25 + 28 + 28) + (28.25 + 29 + 29.25 + 28 + 28 + 28 + 28.25 + 28.5) + (30 + 29.5 + 29.5 + 29.25 + 29.25 + 28.75) + (30.75 + 30.5 + 30 + 29.5) + (30.5 + 30.25 + 29.75) + (30.25 + 30)](1)(1) = 1122.25$. So the average of the two sums is: $(1162.25 + 1122.25)/2 = 1142.25$. Dividing it by the area of the 39 grid squares, we get the average low temperature over the city of Washington, D.C., which is $1142.25/[(39)(1)(1)] = 29.3°F$.

3.

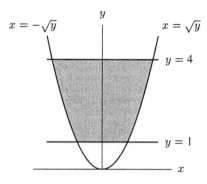

Figure 15.61

4.

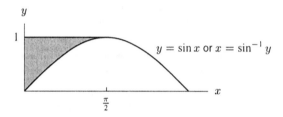

Figure 15.62

5.

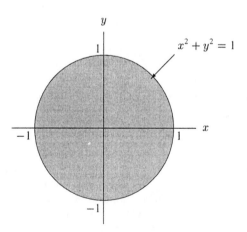

Figure 15.63

6.

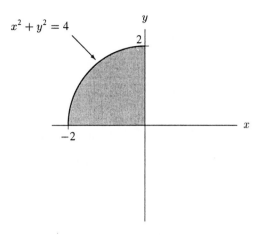

$x^2 + y^2 = 4$

Figure 15.64

7.

$$\int_0^1 \int_0^z \int_0^2 (y + z)^7 \, dx \, dy \, dz = \int_0^1 \int_0^z 2(y + z)^7 \, dy \, dz$$

$$= \int_0^1 \frac{2(y + z)^8}{8} \bigg|_0^z \, dz$$

$$= \int_0^1 \frac{(2z)^8 - z^8}{4} \, dz$$

$$= \frac{255}{4} \int_0^1 z^8 \, dz$$

$$= \frac{255}{4} \left[\frac{z^9}{9} \right]_0^1$$

$$= \frac{255}{4} \cdot \frac{1}{9}$$

$$= \frac{85}{12}.$$

8.

$$\int_0^1 \int_3^4 (\sin(2 - y)) \cos(3x - 7) \, dx \, dy = \int_0^1 (\sin(2 - y)) \left[\frac{\sin(3x - 7)}{3} \right] \bigg|_3^4 \, dy$$

$$= \frac{1}{3} (\sin 5 - \sin 2) \int_0^1 \sin(2 - y) \, dy$$

$$= \frac{1}{3} (\sin 5 - \sin 2) \left[\cos(2 - y) \right]_0^1$$

$$= \frac{1}{3} (\sin 5 - \sin 2)(\cos 1 - \cos 2).$$

9.

$$\int_0^{10} \int_0^{0.1} x e^{xy} \, dy \, dx = \int_0^{10} e^{xy} \big|_0^{0.1} \, dx$$

$$= \int_0^{10} (e^{0.1x} - e^0) \, dx$$

$$= \left(\left(\frac{e^{0.1x}}{0.1} \right) - x \right) \Big|_0^{10}$$

$$= (10e^1 - 10 - 10e^0)$$

$$= 10e - 20 = 10(e - 2).$$

10.

$$\int_0^1 \int_0^y (\sin^3 x)(\cos x)(\cos y) \, dx \, dy = \int_0^1 (\cos y) \left[\frac{\sin^4 x}{4} \right] \Big|_0^y \, dy$$

$$= \frac{1}{4} \int_0^1 (\sin^4 y)(\cos y) \, dy$$

$$= \frac{\sin^5 y}{20} \Big|_0^1$$

$$= \frac{\sin^5 1}{20}.$$

11.   First use integration by parts, with y as the variable, $u = x^2 y$, $u' = x^2$, $v = \frac{\sin(xy)}{x}$, $v' = \cos(xy)$. Then,

$$\int_3^4 \int_0^1 x^2 y \cos(xy) \, dy \, dx = \int_3^4 \left([xy \sin(xy)]_0^1 - \int_0^1 x \sin(xy) \, dy \right) dx$$

$$= \int_3^4 \left(x \sin x + [\cos(xy)]_0^1 \right) dx$$

$$= \int_3^4 (x \sin x + \cos x - 1) \, dx.$$

Now use integration by parts again, with $u = x$, $u' = 1$, $v = -\cos x$, $v' = \sin x$. Then,

$$\int_3^4 (x \sin x + \cos x - 1) \, dx = [-x \cos x]_3^4 + \int_3^4 \cos x \, dx + \int_3^4 (\cos x - 1) \, dx$$

$$= (-x \cos x + 2 \sin x - x)|_3^4$$

$$= -4 \cos 4 + 2 \sin 4 + 3 \cos 3 - 2 \sin 3 - 1.$$

Thus,

$$\int_3^4 \int_0^1 x^2 y \cos(xy) \, dy \, dx = -4 \cos 4 + 2 \sin 4 + 3 \cos 3 - 2 \sin 3 - 1.$$

12.   The region is shown in Figure 15.65.

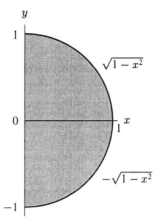

Figure 15.65

The integral has the same values in the upper and lower quarter circles, so we integrate over just the upper circle and multiply by 2. We convert the integral to polar coordinates.

$$\int_0^1 \int_{-\sqrt{1-x^2}}^{\sqrt{1-x^2}} e^{-(x^2+y^2)}\, dy\, dx = 2\int_0^{\pi/2} \int_0^1 e^{-r^2} r\, dr\, d\theta = \int_0^{\pi/2} (-e^{-r^2})\Big|_0^1 \, d\theta$$

$$= \int_0^{\pi/2} 1 - e^{-1} \, d\theta$$

$$= \frac{\pi}{2}(1 - e^{-1}).$$

13.

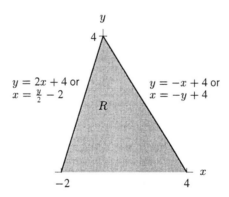

Figure 15.66

Integrating with respect to x first we get

$$\int_0^4 \int_{\frac{y}{2}-2}^{-y+4} f(x,y)\, dx\, dy$$

Integrating with respect to y first we get

$$\int_{-2}^0 \int_0^{2x+4} f(x,y)\, dy\, dx + \int_0^4 \int_0^{-x+4} f(x,y)\, dy\, dx.$$

14. Compute in polar coordinates:

$$
\begin{aligned}
\int_R \sqrt{x^2 + y^2}\, dA &= \int_0^\pi \int_1^2 r \cdot r\, dr\, d\theta \\
&= \int_0^\pi \left[\frac{r^3}{3}\right]_1^2 d\theta \\
&= \int_0^\pi \left(\frac{8}{3} - \frac{1}{3}\right) d\theta \\
&= \frac{7\pi}{3}.
\end{aligned}
$$

15. From Figure 15.67, we have

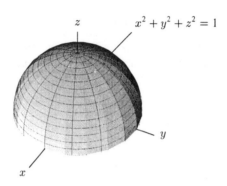

Figure 15.67

(a) $\displaystyle \int_R f\, dV = \int_{-1}^1 \int_{-\sqrt{1-x^2}}^{\sqrt{1-x^2}} \int_0^{\sqrt{1-x^2-y^2}} f(x,y,z)\, dz\,dy\,dx$

(b) $\displaystyle \int_R f\, dV = \int_{-1}^1 \int_{-\sqrt{1-y^2}}^{\sqrt{1-y^2}} \int_0^{\sqrt{1-x^2-y^2}} f(x,y,z)\, dz\,dx\,dy$

(c) $\displaystyle \int_R f\, dV = \int_{-1}^1 \int_0^{\sqrt{1-y^2}} \int_{-\sqrt{1-y^2-z^2}}^{\sqrt{1-y^2-z^2}} f(x,y,z)\, dx\,dz\,dy$

(d) $\displaystyle \int_R f\, dV = \int_{-1}^1 \int_0^{\sqrt{1-x^2}} \int_{-\sqrt{1-x^2-z^2}}^{\sqrt{1-x^2-z^2}} f(x,y,z)\, dy\,dz\,dx$

(e) $\displaystyle \int_R f\, dV = \int_0^1 \int_{-\sqrt{1-z^2}}^{\sqrt{1-z^2}} \int_{-\sqrt{1-x^2-z^2}}^{\sqrt{1-x^2-z^2}} f(x,y,z)\, dy\,dx\,dz$

(f) $\displaystyle \int_R f\, dV = \int_0^1 \int_{-\sqrt{1-z^2}}^{\sqrt{1-z^2}} \int_{-\sqrt{1-y^2-z^2}}^{\sqrt{1-y^2-z^2}} f(x,y,z)\, dx\,dy\,dz$

16. The integral is over the region $0 \leq x^2 + y^2 \leq 3$, $1 \leq z \leq 4 - x^2 - y^2$. Using cylindrical coordinates, we get

$$
\int_0^{2\pi} \int_0^{\sqrt{3}} \int_1^{4-r^2} \frac{1}{z^2} r\, dz\, dr\, d\theta = \int_0^{2\pi} \int_0^{\sqrt{3}} \left(-\frac{r}{z}\right)\Bigg|_1^{4-r^2} dr\, d\theta
$$

$$= \int_0^{2\pi} \int_0^{\sqrt{3}} (-\frac{r}{4-r^2} + \frac{r}{1})dr\, d\theta$$

$$= \int_0^{2\pi} \left[\frac{1}{2}\ln(4-r^2) + \frac{1}{2}r^2 \right]_0^{\sqrt{3}} d\theta$$

$$= \int_0^{2\pi} (\frac{1}{2}\ln 1 + \frac{3}{2} - \frac{1}{2}\ln 4 - 0)d\theta$$

$$= 2\pi(\frac{3}{2} - \ln 2)$$

$$= \pi(3 - 2\ln 2)$$

17. The region is a hemisphere $0 \le x^2 + y^2 + z^2 \le 3^2$, $z \ge 0$, so spherical coordinates are appropriate. Recall the conversion formula $x = \rho\sin\phi\cos\theta$. Then the integral in spherical coordinates becomes

$$\int_0^{2\pi} \int_0^{\pi/2} \int_0^3 (\rho\sin\phi\cos\theta)^2 \rho^2 \sin\phi\, d\rho\, d\phi\, d\theta$$

$$= \int_0^{2\pi} \int_0^{\pi/2} \int_0^3 \rho^4 \sin^3\phi\cos^2\theta\, d\rho\, d\phi\, d\theta$$

$$= \int_0^{2\pi} \int_0^{\pi/2} \frac{243}{5} \sin^3\phi\cos^2\theta\, d\phi\, d\theta$$

$$= \frac{243}{5} \int_0^{2\pi} \int_0^{\pi/2} \cos^2\theta \cdot \sin\phi(1 - \cos^2\phi)\, d\phi\, d\theta$$

$$= \frac{243}{5} \int_0^{2\pi} \cos^2\theta \left[-\cos\phi + \frac{1}{3}\cos^3\phi \right]_0^{\frac{\pi}{2}} d\theta$$

$$= \frac{243}{5} \int_0^{2\pi} \cos^2\theta[-(-1) + \frac{1}{3}(-1)]\, d\theta$$

$$= \frac{243}{5} \cdot \frac{2}{3} \int_0^{2\pi} \frac{1 + \cos 2\theta}{2} d\theta$$

$$= \frac{81}{5}(\theta + \frac{1}{2}\sin 2\theta)\Big|_0^{2\pi}$$

$$= \frac{81}{5}(2\pi + 0)$$

$$= \frac{162\pi}{5}$$

18. The integral is over the region $x, y \ge 0$, $x^2 + y^2 \le 1$, $0 \le z \le \sqrt{x^2 + y^2}$. Using cylindrical coordinates, we get

$$\int_0^{\pi/2} \int_0^1 \int_0^r (z + r)\, r\, dz\, dr\, d\theta = \int_0^{\pi/2} \int_0^1 \int_0^r (rz + r^2)\, dz\, dr\, d\theta$$

$$= \int_0^{\pi/2} \int_0^1 (\frac{1}{2}r^3 + r^3)\, dr\, d\theta$$

$$= \int_0^{\pi/2} \frac{3}{8} r^4 \Big|_0^1 \, d\theta$$

$$= \frac{3}{8} \cdot \frac{\pi}{2}$$

$$= \frac{3\pi}{16}$$

19. W is a cylindrical shell, so cylindrical coordinates should be used. See Figure 15.68.

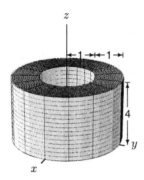

Figure 15.68

$$\int_W \frac{z}{(x^2+y^2)^{3/2}} \, dV = \int_0^4 \int_0^{2\pi} \int_1^2 \frac{z}{r^3} \, r \, dr \, d\theta \, dz$$

$$= \int_0^4 \int_0^{2\pi} \int_1^2 \frac{z}{r^2} \, dr \, d\theta \, dz$$

$$= \int_0^4 \int_0^{2\pi} \left(-\frac{z}{r}\right)\Big|_1^2 \, d\theta \, dz$$

$$= \int_0^4 \int_0^{2\pi} \frac{z}{2} \, d\theta \, dz$$

$$= \int_0^4 \frac{z}{2} \cdot 2\pi \, dz$$

$$= \frac{1}{2}\pi \cdot z^2 \Big|_0^4$$

$$= 8\pi$$

20. We use spherical coordinates because we are integrating over a sphere and the density has spherical symmetry. $D = 2\rho$.

$$M = \int_0^{2\pi} \int_0^\pi \int_0^3 (2\rho)\rho^2 \sin\phi \, d\rho \, d\phi \, d\theta.$$

21. Let the lower left part of the forest be at $(0, 0)$. Then the other corners have coordinates as shown. The population density function is then given by

$$\rho(x, y) = 10 - 2y$$

The equations of the two diagonal lines are $x = -2y/5$ and $x = 6 + 2y/5$. So the total rabbit population in the forest is

$$
\begin{aligned}
\int_0^5 \int_{-\frac{2}{5}y}^{6+\frac{2}{5}y} (10 - 2y) \, dx \, dy &= \int_0^5 (10 - 2y)(6 + \frac{4}{5}y) \, dy \\
&= \int_0^5 (60 - 4y - \frac{8}{5}y^2) \, dy \\
&= (60y - 2y^2 - \frac{8}{15}y^3)\Big|_0^5 \\
&= 300 - 50 - \frac{8}{15} \cdot 125 \\
&= \frac{2750}{15} = \frac{550}{3} \\
&\approx 183
\end{aligned}
$$

22. Suppose the brick is set up as shown in Figure 15.69.

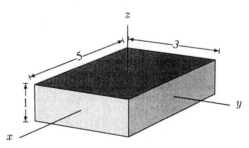

Figure 15.69

The brick has $m/v = $ density $= 1$. The moment of inertia about the z-axis is

$$
\begin{aligned}
I_z &= \int_{-5/2}^{5/2} \int_{-3/2}^{3/2} \int_{-1/2}^{1/2} 1(x^2 + y^2) \, dz \, dy \, dx \\
&= \int_{-5/2}^{5/2} \int_{-3/2}^{3/2} (x^2 + y^2) \, dy \, dx \\
&= \int_{-5/2}^{5/2} (3x^2 + \frac{9}{4}) \, dx \\
&= \frac{125}{4} + \frac{45}{4} = \frac{85}{2}
\end{aligned}
$$

The moment of inertia about the y-axis is

$$
I_y = \int_{-5/2}^{5/2} \int_{-3/2}^{3/2} \int_{-1/2}^{1/2} 1(x^2 + z^2) \, dz \, dy \, dx
$$

$$= \int_{-5/2}^{5/2} \int_{-3/2}^{3/2} (x^2 + \frac{1}{12}) \, dy \, dx$$

$$= \int_{-5/2}^{5/2} (3x^2 + \frac{1}{4}) \, dx$$

$$= \frac{125}{4} + \frac{5}{4} = \frac{65}{2}$$

The moment of inertia about the x-axis is

$$I_x = \int_{-5/2}^{5/2} \int_{-3/2}^{3/2} \int_{-1/2}^{1/2} 1(y^2 + z^2) \, dz \, dy \, dx$$

$$= \int_{-5/2}^{5/2} \int_{-3/2}^{3/2} (y^2 + \frac{1}{12}) \, dy \, dx$$

$$= \int_{-5/2}^{5/2} (\frac{9}{4} + \frac{1}{4}) \, dx$$

$$= 5 \cdot \frac{10}{4} = \frac{25}{2}$$

23. Let the ball be centered at the origin. Since a ball looks the same from all directions, we can choose the axis of rotation; in this case, let it be the z-axis. It is best to use spherical coordinates, so then

$$x^2 + y^2 = (\rho \sin \phi \cos \theta)^2 + (\rho \sin \phi \sin \theta)^2$$
$$= \rho^2 \sin^2 \phi$$

Then $m/v = \text{Density} = 1$, so the moment of inertia is

$$I_z = \int_0^R \int_0^{2\pi} \int_0^{\pi} 1(\rho^2 \sin^2 \phi) \rho^2 \sin \phi \, d\phi \, d\theta \, d\rho$$

$$= \int_0^R \int_0^{2\pi} \int_0^{\pi} \rho^4 (\sin \phi)(1 - \cos^2 \phi) \, d\phi \, d\theta \, d\rho$$

$$= \int_0^R \int_0^{2\pi} \rho^4 (-\cos \phi + \frac{1}{3} \cos^3 \phi) \Big|_0^{\pi} \, d\theta \, d\rho$$

$$= \int_0^R \int_0^{2\pi} \frac{4}{3} \rho^4 \, d\theta \, d\rho$$

$$= \int_0^R \frac{8\pi}{3} \rho^4 \, d\rho$$

$$= \frac{8}{15} \pi R^5$$

24. (a) We are integrating over the whole plane, so converting to polar coordinates gives

$$\int_{-\infty}^{\infty} \int_{-\infty}^{\infty} e^{-(x^2+y^2)} \, dx \, dy = \int_0^{2\pi} \int_0^{\infty} e^{-r^2} r \, dr \, d\theta = \int_0^{2\pi} -\frac{1}{2} e^{-r^2} \Big|_0^{\infty} \, d\theta = \int_0^{2\pi} \frac{1}{2} \, d\theta = \pi.$$

(b) Rewriting the integrand as a product gives

$$\int_{-\infty}^{\infty}\int_{-\infty}^{\infty}e^{-(x^2+y^2)}\,dx\,dy = \int_{-\infty}^{\infty}\int_{-\infty}^{\infty}e^{-x^2}e^{-y^2}\,dx\,dy.$$

Now e^{-y^2} is a constant as far as the integral with respect to x is concerned, so

$$\int_{-\infty}^{\infty}\int_{-\infty}^{\infty}e^{-x^2}e^{-y^2}\,dx\,dy = \int_{-\infty}^{\infty}e^{-y^2}\left(\int_{-\infty}^{\infty}e^{-x^2}\,dx\right)dy.$$

We assume that the integral with respect to x converges, and so is a constant as far as the integral with respect to y is concerned. Thus, we have

$$\int_{-\infty}^{\infty}e^{-y^2}\left(\int_{-\infty}^{\infty}e^{-x^2}\,dx\right)dy = \left(\int_{-\infty}^{\infty}e^{-x^2}\,dx\right)\left(\int_{-\infty}^{\infty}e^{-y^2}\,dy\right).$$

But $\int_{-\infty}^{\infty}e^{-x^2}\,dx$ and $\int_{-\infty}^{\infty}e^{-y^2}\,dy$ are the same number, so we can write

$$\int_{-\infty}^{\infty}\int_{-\infty}^{\infty}e^{-(x^2+y^2)}\,dx\,dy = \left(\int_{-\infty}^{\infty}e^{-x^2}\,dx\right)\left(\int_{-\infty}^{\infty}e^{-y^2}\,dy\right) = \left(\int_{-\infty}^{\infty}e^{-x^2}\,dx\right)^2.$$

(c) Using the results of parts (a) and (b), we have

$$\left(\int_{-\infty}^{\infty}e^{-x^2}\,dx\right)^2 = \int_{-\infty}^{\infty}\int_{-\infty}^{\infty}e^{-(x^2+y^2)}\,dx\,dy = \pi.$$

Taking square roots and observing that the integral we are looking for is positive, we have

$$\int_{-\infty}^{\infty}e^{-x^2}\,dx = \sqrt{\pi}.$$

25. Set up the cylinder with the base centered at the origin on the xy plane, facing up. (See Figure 15.70.) Newton's Law of Gravitation states that the force exerted between two particles is

$$F = G\frac{m_1 m_2}{\rho^2}$$

where G is the gravitational constant, m_1 and m_2 are the masses, and ρ is the distance between the particles. We take a small volume element, so $m_1 = m$, and $m_2 = \delta dV$. In cylindrical coordinates, if m is at $(0,0,0)$ and δdV is at (r, θ, z), (see Figure 15.70), then the distance from m to δdV is given by $\rho = \sqrt{r^2 + z^2}$ for $r_1 \le r \le r_2$ and $0 \le z \le h$.

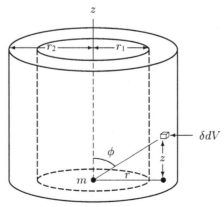

Figure 15.70

Due to the symmetry of the cylinder the sum of all the horizontal forces is zero; the net force on m is vertical.

The force acting on the particle as a result of the small piece dV makes an angle ϕ with the vertical and therefore has vertical component

$$\begin{matrix}\text{Vertical force on}\\ \text{particle from small}\\ \text{piece of cylinder}\end{matrix} = \frac{Gm\delta dV}{(\sqrt{r^2+z^2})^2}\cdot\cos\phi = \frac{Gm\delta dV}{r^2+z^2}\cdot\frac{z}{\sqrt{r^2+z^2}} = \frac{Gm\delta z}{(r^2+z^2)^{\frac{3}{2}}}\,dV.$$

Thus, since $dV = rdzdrd\theta$,

$$\begin{aligned}\text{Total force} &= \int_0^{2\pi}\int_{r_1}^{r_2}\int_0^h \frac{Gm\delta z}{(r^2+z^2)^{3/2}}r\,dzdrd\theta\\ &= 2\pi Gm\delta\int_{r_1}^{r_2}\int_0^h \frac{zr}{(r^2+z^2)^{3/2}}drdz\\ &= 2\pi Gm\delta\int_{r_1}^{r_2}\left(1-\frac{r}{(r^2+h^2)^{\frac{1}{2}}}\right)dr\\ &= 2\pi Gm\delta(r-(r^2+h^2)^{\frac{1}{2}})\Big|_{r_1}^{r_2}\\ &= 2\pi Gm\delta(r_2-r_1-\sqrt{r_2^2+h^2}+\sqrt{r_1^2+h^2}).\end{aligned}$$

26. The outer circle is a semicircle of radius 4. This is shown in Figure 15.71, with center at D. Thus, $CE = 2$ and $DC = 2$, while $AD = 4$. Notice that angle ADO is a right angle.

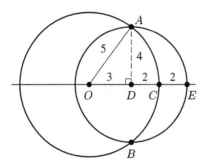

Figure 15.71

Suppose the large circle has center O and radius r. Then $OA = r$ and $OD = OC - DC = r - 2$. Applying Pythagoras' Theorem to triangle OAD gives

$$\begin{aligned}r^2 &= 4^2+(r-2)^2\\ r^2 &= 16+r^2-4r+4\\ r &= 5.\end{aligned}$$

If we put the origin at O, the equation of the large circle is $x^2+y^2 = 25$. In the same coordinates, the equation of the small circle, which has center at $D = (3,0)$, is $(x-3)^2+y^2 = 16$. The right hand side of the two circles are given by

$$x = \sqrt{25-y^2} \quad\text{and}\quad x = 3+\sqrt{16-y^2}.$$

Since the y-coordinate of A is 4 and the y-coordinate of B is -4, we have

$$\text{Area} = \int_{-4}^{4} \int_{\sqrt{25-y^2}}^{3+\sqrt{16-y^2}} 1\, dx\, dy$$

$$= \int_{-4}^{4} \left(3 + \sqrt{16 - y^2} - \sqrt{25 - y^2}\right) dy$$

$$= 13.95.$$

27. The (x, y) coordinates of the vertices of the frame are

$$(0,0), (6\sqrt{2}, 6\sqrt{2}), (12 + 6\sqrt{2}, 6\sqrt{2}), (12, 0).$$

In the new coordinates $u = x - y, v = y$ the frame will have edges:

$$(0,0), (0, 6\sqrt{2}), (12, 6\sqrt{2}), (12, 0).$$

Thus, in the uv-plane the region is a rectangle with length 12 and height $6\sqrt{2}$. So the area of the frame in the (u, v) coordinates is $12 \cdot 6\sqrt{2} = 72\sqrt{2}$. To find its area in the (x, y) coordinates we compute the determinant

$$\begin{vmatrix} \frac{\partial x}{\partial u} & \frac{\partial y}{\partial u} \\ \frac{\partial x}{\partial v} & \frac{\partial y}{\partial v} \end{vmatrix} = \begin{vmatrix} 1 & 0 \\ 1 & 1 \end{vmatrix} = 1.$$

So the areas in the uv-plane and the xy-plane are the same. Therefore the area of the frame with one cutout is

$$12 \cdot 6\sqrt{2} - 8 \cdot 4\sqrt{2} = 40\sqrt{2}.$$

The area of the frame with four cutouts is:

$$12 \cdot 6\sqrt{2} - 4 \cdot 3 \cdot \frac{3}{2}\sqrt{2} = 54\sqrt{2}.$$

28. Let's denote the (x, y) coordinates of the points in the lagoon by L. Since x and y are measured in kilometers and d is measured in meters, and 1 km = 1000 m, the volume of a small piece of the lagoon is given by

$$\Delta V \approx d(x, y)(1000\Delta x)(1000\Delta y)\text{m}^3.$$

Thus, the total volume of the lagoon is given by

$$V = 1000^2 \int_L d(x, y)\, dx\, dy.$$

Changing coordinates using $u = x/2$ and $v = y - f(x)$ converts the depth function to:

$$d(x(u, v), y(u, v)) = 40 - 160v^2 - 160u^2 = 160(\frac{1}{4} - u^2 - v^2) \text{ meters.}$$

Thus, the points in the lagoon have (u, v) coordinates in the disk, D, given by $u^2 + v^2 \leq 1/4$.
The Jacobian of the transformation is:

$$\begin{vmatrix} \frac{\partial x}{\partial u} & \frac{\partial y}{\partial u} \\ \frac{\partial x}{\partial v} & \frac{\partial y}{\partial v} \end{vmatrix} = \begin{vmatrix} 2 & 2f'(2u) \\ 0 & 1 \end{vmatrix} = 2.$$

Thus, the integral in u, v coordinates is

$$V = 1000^2 \int_L d(x, y) \, dx dy = 10^6 \int_D 160(\frac{1}{4} - u^2 - v^2)2 \, du dv = 320 \cdot 10^6 \int_D (\frac{1}{4} - u^2 - v^2) \, du dv.$$

Converting to polar coordinates, we have

$$V = 320 \cdot 10^6 \int_0^{2\pi} \int_0^{1/2} (\frac{1}{4} - r^2)r \, dr d\theta = 320 \cdot 10^6 2\pi (\frac{1}{4}\frac{r^2}{2} - \frac{r^4}{4})\Big|_0^{1/2} = 10^7 \pi \text{ m}^3.$$

CHAPTER SIXTEEN

Solutions for Section 16.1

1. Between times $t = 0$ and $t = 1$, x goes at a constant rate from 0 to 1 and y goes at a constant rate from 1 to 0. So the particle moves in a straight line from $(0, 1)$ to $(1, 0)$. Similarly, between times $t = 1$ and $t = 2$, it goes in a straight line to $(0, -1)$, then to $(-1, 0)$, then back to $(0, 1)$. So it traces out the diamond shown in Figure 16.1.

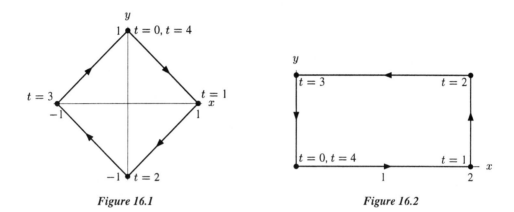

| Figure 16.1 | Figure 16.2 |

2. This is like Example 2, except that the x-coordinate goes all the way to 2 and back. So the particle traces out the rectangle shown in Figure 16.2.

3. Between times $t = 0$ and $t = 1$, x goes from -1 to 1, while y stays fixed at 1. So the particle goes in a straight line from $(-1, 1)$ to $(1, 1)$. Then both the x- and y-coordinates decrease at a constant rate from 1 to -1. So the particle goes in a straight line from $(1, 1)$ to $(-1, -1)$. Then it moves across to $(1, -1)$, then back diagonally to $(-1, 1)$. See Figure 16.3.

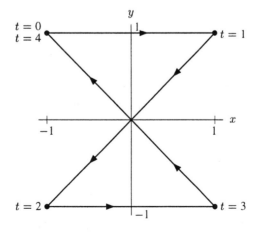

Figure 16.3

4. As the x-coordinate goes at a constant rate from 2 to 0, the y-coordinate goes from 0 to 1, then down to -1, then back to 0. So the particle zigs and zags from $(2,0)$ to $(1.5, 1)$ to $(1, 0)$ to $(.5, -1)$ to $(0, 0)$. Then it zigs and zags back again, forming the shape in Figure 16.4.

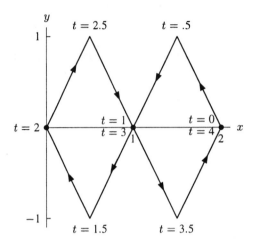

Figure 16.4

5. The particle moves clockwise: For $0 \leq t \leq \frac{\pi}{2}$, we have $x = \cos t$ decreasing and $y = -\sin t$ decreasing. Similarly, for the time intervals $\frac{\pi}{2} \leq t \leq \pi, \pi \leq t \leq \frac{3\pi}{2}$, and $\frac{3\pi}{2} \leq t \leq 2\pi$, we see that the particle moves clockwise.

6. For $0 \leq t \leq \frac{\pi}{2}$, we have $x = \sin t$ increasing and $y = \cos t$ decreasing, so the motion is clockwise for $0 \leq t \leq \frac{\pi}{2}$. Similarly, we see that the motion is clockwise for the time intervals $\frac{\pi}{2} \leq t \leq \pi, \pi \leq t \leq \frac{3\pi}{2}$, and $\frac{3\pi}{2} \leq t \leq 2\pi$.

7. Let $f(t) = t^2$. The particle is moving clockwise when $f(t)$ is decreasing, that is, when $f'(t) = 2t < 0$, so when $t < 0$. The particle is moving counterclockwise when $f'(t) = 2t > 0$, so when $t > 0$.

8. Let $f(t) = t^3 - t$. The particle is moving clockwise when $f(t)$ is decreasing, that is, when $f'(t) = 3t^2 - 1 < 0$, and counterclockwise when $f'(t) = 3t^2 - 1 > 0$. That is, it moves clockwise when $-\sqrt{\frac{1}{3}} < t < \sqrt{\frac{1}{3}}$, between $(\cos((-\sqrt{\frac{1}{3}})^3 + \sqrt{\frac{1}{3}}), \sin((-\sqrt{\frac{1}{3}})^3 + \sqrt{\frac{1}{3}}))$ and $(\cos((\sqrt{\frac{1}{3}})^3 - \sqrt{\frac{1}{3}}), \sin((\sqrt{\frac{1}{3}})^3 - \sqrt{\frac{1}{3}})$, and counterclockwise when $t < -\sqrt{\frac{1}{3}}$ or $t > \sqrt{\frac{1}{3}}$.

9. Let $f(t) = \ln t$. Then $f'(t) = \frac{1}{t}$. The particle is moving counterclockwise when $f'(t) > 0$, that is, when $t > 0$. Any other time, when $t \leq 0$, the position is not defined.

10. Let $f(t) = \cos t$. Then $f'(t) = -\sin t$. The particle is moving clockwise when $f'(t) < 0$, or $-\sin t < 0$, that is, when

$$2k\pi < t < (2k+1)\pi,$$

where k is an integer. The particle is otherwise moving counterclockwise, that is, when

$$(2k-1)\pi < t < 2k\pi,$$

where k is an integer. Actually, the particle does not fully trace out a circle. The range of $f(t)$ is $[-1, 1]$ so the particle oscillates between the points $(\cos(-1), \sin(-1))$ and $(\cos 1, \sin 1)$.

11. In all three cases, $y = x^2$, so that the motion takes place on the parabola $y = x^2$.

    In case (a), the x-coordinate always increases at a constant rate of one unit distance per unit time, so the equations describe a particle moving to the right on the parabola at constant horizontal speed.

    In case (b), the x-coordinate is never negative, so the particle is confined to the right half of the parabola. As t moves from $-\infty$ to $+\infty$, $x = t^2$ goes from ∞ to 0 to ∞. Thus the particle first comes down the right half of the parabola, reaching the origin $(0,0)$ at time $t = 0$, where it reverses direction and goes back up the right half of the parabola.

    In case (c), as in case (a), the particle traces out the entire parabola $y = x^2$ from left to right. The difference is that the horizontal speed is not constant. This is because a unit change in t causes larger and larger changes in $x = t^3$ as t approaches $-\infty$ or ∞. The horizontal motion of the particle is faster when it is farther from the origin.

12. One possible answer is $x = 3\cos t, y = -3\sin t, 0 \leq t \leq 2\pi$.

13. One possible answer is $x = -2, y = t$.

14. One possible answer is $x = 2 + 5\cos t, y = 1 + 5\sin t, 0 \leq t \leq 2\pi$.

15. The parameterization $x = 2\cos t$, $y = 2\sin t$, $0 \leq t \leq 2\pi$, is a circle of radius 2 traced out counterclockwise starting at the point $(2,0)$. To start at $(-2,0)$, put a negative in front of the first coordinate

    $$x = -2\cos t \quad y = 2\sin t, \qquad 0 \leq t \leq 2\pi.$$

    Now we must check whether this parameterization traces out the circle clockwise or counterclockwise. Since when t increases from 0, $\sin t$ is positive, the point (x, y) moves from $(-2, 0)$ into the second quadrant. Thus, the circle is traced out clockwise and so this is one possible parameterization.

16. The slope of the line is

    $$m = \frac{3 - (-1)}{1 - 2} = -4.$$

    The equation of the line with slope -4 through the point $(2, -1)$ is $y - (-1) = (-4)(x - 2)$, so one possible parameterization is $x = t$ and $y = -4t + 8 - 1 = -4t + 7$.

17. The ellipse $x^2/25 + y^2/49 = 1$ can be parameterized by $x = 5\cos t, y = 7\sin t, 0 \leq t \leq 2\pi$.

18. The parameterization $x = -3\cos t, y = 7\sin t, 0 \leq t \leq 2\pi$, starts at the right point but sweeps out the ellipse in the wrong direction (the y-coordinate becomes positive as t increases). Thus, a possible parameterization is $x = -3\cos(-t) = -3\cos t, y = 7\sin(-t) = -7\sin t, 0 \leq t \leq 2\pi$.

19. (a) If $t \geq 0$, we have $x \geq 2, y \geq 4$, so we get the part of the line to the right of and above the point $(2, 4)$.
    (b) When $t = 0, (x, y) = (2, 4)$. When $t = -1, (x, y) = (-1, -3)$. Restricting t to the interval $-1 \leq t \leq 0$ gives the part of the line between these two points.
    (c) If $x < 0$, giving $2 + 3t < 0$ or $t < -2/3$. Thus $t < -2/3$ gives the points on the line to the left of the y-axis.

20. (I) has a positive slope and so must be l_1 or l_2. Since its y-intercept is negative, these equations must describe l_2. (II) has a negative slope and positive x-intercept, so these equations must describe l_3.

21. (a) C_1 has center at the origin and radius 5, so $a = b = 0, k = 5$ or -5.
    (b) C_2 has center at $(0, 5)$ and radius 5, so $a = 0, b = 5, k = 5$ or -5.
    (c) C_3 has center at $(10, -10)$, so $a = 10, b = -10$. The radius of C_3 is $\sqrt{10^2 + (-10)^2} = \sqrt{200}$, so $k = \sqrt{200}$ or $k = -\sqrt{200}$.

22. It is a straight line through the point $(3, 5, 7)$ parallel to the vector $\vec{i} - \vec{j} + 2\vec{k}$. A linear parameterization of the same line is $x = 3 + t, y = 5 - t, z = 7 + 2t$.

23. The xz-plane is $y = 0$, so one possible answer is

$$x = 2\cos t, \quad y = 0, \quad z = 2\sin t.$$

24. The circle lies in the plane $z = 2$, so one possible answer is

$$x = 3\cos t, \quad y = 3\sin t, \quad z = 2.$$

25. The vector connecting the two points is $3\vec{i} - \vec{j} + \vec{k}$. So a possible parameterization is

$$x = 2 + 3t, \quad y = 3 - t, \quad z = -1 + t.$$

26. One possible parameterization is

$$x = 1 + 3t, \quad y = 2 - 3t, \quad z = 3 + t.$$

27. One possible parameterization is

$$x = 1, \quad y = 0, \quad z = t.$$

28. Add the two equations to get $3x = 8$, or $x = \frac{8}{3}$. Then we have

$$-y + z = \frac{1}{3}.$$

So a possible parameterization is

$$x = \frac{8}{3}, \quad y = t, \quad z = \frac{1}{3} + t.$$

29. Let $f(x, y, z) = x^2 + y^2 - z$. Then the surface $z = x^2 + y^2$ is a level surface of f at the value 0. The gradient of f is perpendicular to the level surface.

$$\text{grad } f = 2x\vec{i} + 2y\vec{j} - \vec{k} = 2\vec{i} + 4\vec{j} - \vec{k}.$$

So a possible parameterization is

$$x = 1 + 2t, \quad y = 2 + 4t, \quad z = 5 - t.$$

30. If the two lines intersect, there must be times t_1, t_2 such that each of the following three equations are satisfied:

$$x = 2 + 3t_1 = 1 + 3t_2,$$
$$y = 3 - t_1 = 2 - 3t_2,$$
$$z = -1 + t_1 = 3 + t_2.$$

(Note that the lines need not to go through the intersection point at the same time, so t_1 and t_2 may be different. Adding the last two equations gives $-2 = 5 - 2t_2$ and so $t_2 = \frac{3}{2}$, while the last equation gives $-1 + t_1 = 3 + \frac{3}{2}$, so $t_1 = \frac{11}{2}$. But these don't satisfy the first equation, therefore the two lines don't intersect.)

31. The question is equivalent to asking if the line through $(-3, -4, 2)$ and $(4, 5, 0)$ enters the sphere $x^2+y^2+z^2 = 1$. A parameterization for this line is given by

$$x = -3 + 7t, \quad y = -4 + 9t, \quad z = 2 - 2t.$$

We want to see whether the line intersects the sphere $x^2 + y^2 + z^2 = 1$. Substituting we have

$$(-3 + 7t)^2 + (-4 + 9t)^2 + (2 - 2t)^2 = 1$$

$$29 - 122t + 134t^2 = 0$$

Since $(122)^2 - 4(29)134 < 0$, this equation has no real solutions. Thus, the line does not enter the sphere and the point is visible.

32. These equations parameterize a line. Since $(3 + t) + (2t) + 3(1 - t) = 6$, we have $x + y + 3z = 6$. Similarly, $x - y - z = (3 + t) - 2t - (1 - t) = 2$. That is, the curve lies entirely in the plane $x + y + 3z = 6$ and in the plane $x - y - z = 2$. Since the normals to the two planes, $\vec{n_1} = \vec{i} + \vec{j} + 3\vec{k}$ and $\vec{n_2} = \vec{i} - \vec{j} - \vec{k}$ are not parallel, the line is the intersection of two nonparallel planes, which is a straight line in 3-dimensional space.

33. (a) Both paths are straight lines, the first passes through the point $(-1, 4, -1)$ in the direction of the vector $\vec{i} - \vec{j} + 2\vec{k}$ and the second passes through $(-7, -6, 1)$ in the direction of the vector $2\vec{i} + 2\vec{j} + \vec{k}$. The two paths are not parallel.

(b) Is there a time t when the two particles are at the same place at the same time? If so, then their coordinates will be the same, so equating coordinates we get

$$-1 + t = -7 + 2t$$
$$4 - t = -6 + 2t$$
$$-1 + 2t = -1 + t.$$

Since the first equation is solved by $t = 6$, the second by $t = 10/3$, and the third by $t = 0$, no value of t solves all three equations. The two particles never arrive at the same place at the same time, and so they do not collide.

(c) Are there any times t_1 and t_2 such that the position of the first particle at time t_1 is the same as the position of the second particle at time t_2? If so then

$$-1 + t_1 = -7 + 2t_2$$
$$4 - t_1 = -6 + 2t_2$$
$$-1 + 2t_1 = -1 + t_2.$$

We solve the first two equations and get $t_1 = 2$ and $t_2 = 4$. This is a solution for the third equation as well, so the three equations are satisfied by $t_1 = 2$ and $t_2 = 4$. At time $t = 2$ the first particle is at the point $(1, 2, 3)$, and at time $t = 4$ the second is at the same point. The paths cross at the point $(1, 2, 3)$, and the first particle gets there first.

34. The three shadows appear as a circle, a cosine wave and a sine wave, respectively.

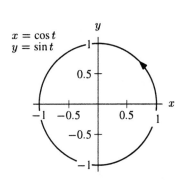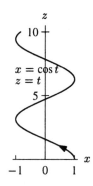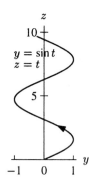

Figure 16.5

35. For $0 \leq t \leq 2\pi$

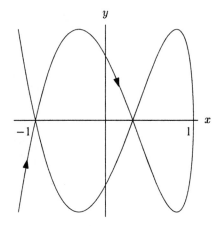

Figure 16.6

36. For $0 \leq t \leq 2\pi$

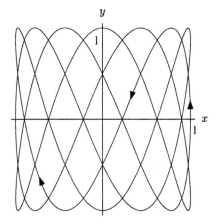

Figure 16.7

37. For $0 \leq t \leq 2\pi$

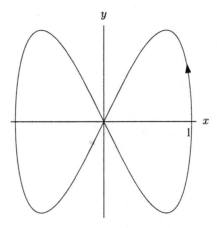

Figure 16.8

38. This curve never closes on itself. Figure 16.9 shows how it starts out.
     The plot for $0 \leq t \leq 8\pi$ is in Figure 16.9.

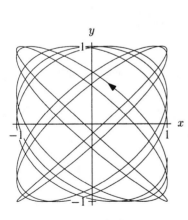

Figure 16.9

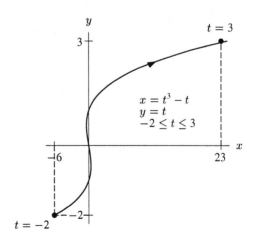

Figure 16.10

39. The particle starts moving from left to right, then reverses its direction for a short time, then continues motion left to right. See Figure 16.10.

40.

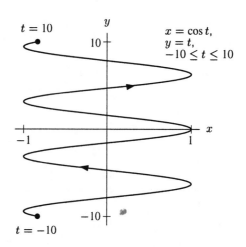

Figure 16.11

The particle moves back and forth between -1 and 1.

41.

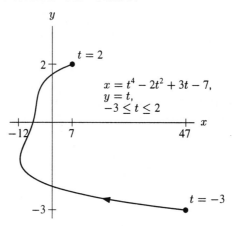

Figure 16.12

The particle starts moving to the left, reverses direction three times, then ends up moving to the right.

42.

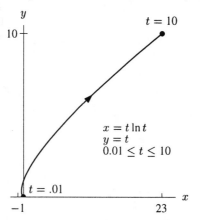

Figure 16.13

After a short move to the left, the particle moves steadily to the right.

Solutions for Section 16.2

1. (a) The first equation gives

$$x = 2 + t, \quad y = 4 + 3t.$$

Eliminating t between this pair of equations gives

$$y - 4 = 3(x - 2),$$
$$y = 3x - 2.$$

The second equation gives

$$x = 1 - 2t, \quad y = 1 - 6t.$$

Eliminating t between this pair of equations gives

$$y - 1 = 3(x - 1),$$
$$y = 3x - 2.$$

Since both parametric equations give rise to the same equation in x and y, they both parameterize the same line.

   (b) Slope $= 3$, y-intercept $= -2$.

2. (a) We get the part of the line with $x < 0$ and $y < 0$ and $z < 10$.
   (b) We get the part of the line between the points $(0, 0, 10)$ and $(1, 2, 13)$.

3. (a) A vector on the line will lie in both planes and will therefore be orthogonal to both normal vectors. To produce a vector orthogonal to two given vectors, you can take their cross product.
   (b) The vector $(\vec{i} + 2\vec{j} - 3\vec{k}) \times (3\vec{i} - \vec{j} + \vec{k}) = -\vec{i} - 10\vec{j} - 7\vec{k}$ is parallel to the line.
   (c) We need a point on the line and a vector parallel to the line. We found a vector in part (b). To find a point, we set $z = 0$ and solve for x and y in the equations for the planes. We have $x + 2y = 7$ and $3x - y = 0$ from which $x = 1$ and $y = 3$. Hence, the point $(1, 3, 0)$ is on the line. Finally, a parametric equation for the line is $\vec{r} = (1 - t)\vec{i} + (3 - 10t)\vec{j} - 7t\vec{k}$. Other answers are possible.

4. Table 16.1 shows values near $t = 1$ with t changing by increments of 0.01.

TABLE 16.1
Values for the position vector
$\vec{r}(t) = \cos t\,\vec{i} + \sin t\,\vec{j}$ *near* $t = 1$

t	$\vec{r}$
0.98	$0.5570\vec{i} + 0.8305\vec{j}$
0.99	$0.5487\vec{i} + 0.8360\vec{j}$
1.00	$0.5403\vec{i} + 0.8415\vec{j}$
1.01	$0.5319\vec{i} + 0.8468\vec{j}$
1.02	$0.5234\vec{i} + 0.8521\vec{j}$

As we go down the table, the x-values are decreasing by about 0.0084 and the y-values are increasing by about 0.0054. Thus, a change in time of $\Delta t = 0.01$ produces the change in position vector $\Delta x\vec{i} + \Delta y\vec{j} \approx -0.0084\vec{i} + 0.0054\vec{j}$. So, the velocity vector is approximately

$$\vec{v} \approx \frac{1}{\Delta t}\left(\Delta x\vec{i} + \Delta y\vec{j}\right) \approx -0.84\vec{i} + 0.54\vec{j}.$$

Note that this velocity vector is perpendicular to the radius vector from the origin $(0,0)$ to the position $(0.54, 0.84)$ at time $t = 1$. Finally, the x and y-values in the table are almost indistinguishable from those of linear motion given by $x = 0.5403 - 0.84(t - 1)$, $y = 0.8415 + 0.54(t - 1)$, which are the parametric equations for the tangent line through the point $(0.5403, 0.8415)$.

5. (a) The curve is a spiral as shown in Figure 16.14.
    (b) We have:

$$\vec{v}(2) \approx \frac{2.001 \cos 2.001 - 2 \cos 2}{0.001}\vec{i} + \frac{2.001 \sin 2.001 - 2 \sin 2}{0.001}\vec{j}$$
$$= -2.24\vec{i} + 0.08\vec{j},$$
$$\vec{v}(4) \approx \frac{4.001 \cos 4.001 - 4 \cos 4}{0.001}\vec{i} + \frac{4.001 \sin 4.001 - 4 \sin 4}{0.001}\vec{j}$$
$$= 2.38\vec{i} - 3.37\vec{j},$$
$$\vec{v}(6) \approx \frac{6.001 \cos 6.001 - 6 \cos 6}{0.001} + \frac{6.001 \sin 6.001 - 6 \sin 6}{0.001}\vec{j}$$
$$= 2.63\vec{i} + 5.48\vec{j}.$$

(c) Evaluating the exact formula $\vec{v}(t) = (\cos t - t \sin t)\vec{i} + (\sin t + t \cos t)\vec{j}$ gives :

$$\vec{v}(2) = -2.235\vec{i} + 0.077\vec{j},$$
$$\vec{v}(4) = 2.374\vec{i} - 3.371\vec{j},$$
$$\vec{v}(6) = 2.637\vec{i} + 5.482\vec{j}.$$

See Figure 16.15.

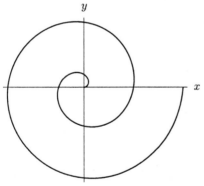

Figure 16.14: The spiral
$x = t \cos t, y = t \sin t$ for $0 \le t \le 4\pi$

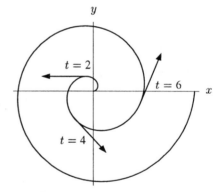

Figure 16.15: The spiral
$x = t \cos t, y = t \sin t$ and three velocity
vectors

6. To find $\vec{v}(t)$ we first find $dx/dt = 2t$ and $dy/dt = 3t^2$. Therefore, the velocity vector is $\vec{v} = 2t\vec{i} + 3t^2\vec{j}$. The speed of the particle is given by the magnitude of the vector,

$$\|\vec{v}\| = \sqrt{(\frac{dx}{dt})^2 + (\frac{dy}{dt})^2} = \sqrt{((2t)^2 + (3t^2)^2)} = |t| \cdot \sqrt{(4 + 9t^2)}.$$

The particle comes to a complete stop when its speed is 0, that is, if $t\sqrt{4 + 9t^2} = 0$, and so when $t = 0$.

7. To find $\vec{v}(t)$ we first find $dx/dt = -2t\sin(t^2)$ and $dy/dt = 2t\cos(t^2)$. Therefore, the velocity is $\vec{v} = -2t\sin(t^2)\vec{i} + 2t\cos(t^2)\vec{j}$. The speed of the particle is given by

$$\begin{aligned}\|\vec{v}\| &= \sqrt{(-2t\sin(t^2))^2 + (2t\cos(t^2))^2} \\ &= \sqrt{4t^2(\sin(t^2))^2 + 4t^2(\cos(t^2))^2} \\ &= 2|t|\sqrt{\sin^2(t^2) + \cos^2(t^2)} \\ &= 2|t|.\end{aligned}$$

The particle comes to a complete stop when speed is 0, that is, if $2|t| = 0$, and so when $t = 0$.

8. The velocity vector $\vec{v}$ is given by:

$$\vec{v} = \frac{d}{dt}(\cos 2t)\vec{i} + \frac{d}{dt}(\sin t)\vec{j} = -2\sin 2t\vec{i} + \cos t\vec{j}.$$

The speed is given by:

$$\|\vec{v}\| = \sqrt{4\sin^2(2t) + \cos^2 t}.$$

Thus, $\|\vec{v}\| = 0$ when $\sin(2t) = \cos t = 0$, and so the particle stops when $t = \pm\pi/2, \pm 3\pi/2, \ldots$ or $t = (2n+1)\frac{\pi}{2}$, for any integer n.

9. The velocity vector $\vec{v}$ is given by:

$$\begin{aligned}\vec{v} &= \frac{d}{dt}(t^2 - 2t)\vec{i} + \frac{d}{dt}(t^3 - 3t)\vec{j} + \frac{d}{dt}(3t^4 - 4t^3)\vec{k} \\ &= (2t - 2)\vec{i} + (3t^2 - 3)\vec{j} + (12t^3 - 12t^2)\vec{k}.\end{aligned}$$

The speed is given by:

$$\vec{v} = \sqrt{(2t - 2)^2 + (3t^2 - 3)^2 + (12t^3 - 12t^2)^2}.$$

The particle stops when $2t - 2 = 0$ and $3t^2 - 3 = 0$ and $12t^3 - 12t^2 = 0$. Since these are all satisfied only by $t = 1$, this is the only time that the particle stops.

10. At $t = 2$, the position and velocity vectors are

$$\begin{aligned}\vec{r}(2) &= 2^2\vec{i} + 2^3\vec{j} = 4\vec{i} + 8\vec{j}, \\ \vec{v}(2) &= 2 \cdot 2\vec{i} + 3 \cdot 2^2\vec{j} = 4\vec{i} + 12\vec{j}.\end{aligned}$$

So we want the line going through the point $(4, 8)$ at the time $t = 2$, in the direction $4\vec{i} + 12\vec{j}$:

$$x = 4 + 4(t - 2), \quad y = 8 + 12(t - 2).$$

11. The velocity vector $\vec{v}$ is given by:

$$\vec{v} = \frac{d}{dt}(3\cos t)\vec{i} + \frac{d}{dt}(4\sin t)\vec{j} = -3\sin t\vec{i} + 4\cos t\vec{j}.$$

The acceleration vector $\vec{a}$ is given by:

$$\vec{a} = \frac{d\vec{v}}{dt} = \frac{d}{dt}(-3\sin t)\vec{i} + \frac{d}{dt}(4\cos t)\vec{j} = -3\cos t\vec{i} - 4\sin t\vec{j}.$$

12. The velocity vector $\vec{v}$ is given by:

$$\vec{v} = \frac{d(t)}{dt}\vec{i} + \left(\frac{d}{dt}(t^3 - t)\right)\vec{j} = \vec{i} + (3t^2 - 1)\vec{j}.$$

The acceleration vector $\vec{a}$ is given by:

$$\vec{a} = \frac{d\vec{v}}{dt} = \frac{d(1)}{dt}\vec{i} + \left(\frac{d}{dt}(3t^2 - 1)\right)\vec{j} = 6t\vec{j}.$$

13. The velocity vector $\vec{v}$ is given by:

$$\vec{v} = \frac{d}{dt}(2 + 3t)\vec{i} + \frac{d}{dt}(4 + t)\vec{j} + \frac{d}{dt}(1 - t)\vec{k} = 3\vec{i} + \vec{j} - \vec{k}.$$

The acceleration vector $\vec{a}$ is given by:

$$\vec{a} = \frac{d\vec{v}}{dt} = \frac{d(3)}{dt}\vec{i} + \frac{d(1)}{dt}\vec{j} - \frac{d(1)}{dt}\vec{k} = \vec{0}$$

14. Since $\vec{r}(t) = 3\cos(t^2)\vec{i} + 3\sin(t^2)\vec{j} + t^2\vec{k}$, we have

$$\vec{v}(t) = -6t\sin(t^2)\vec{i} + 6t\cos(t^2)\vec{j} + 2t\vec{k},$$
$$\vec{a}(t) = (-6\sin(t^2) - 12t^2\cos(t^2))\vec{i} + (6\cos(t^2) - 12t^2\sin(t^2))\vec{j} + 2\vec{k}.$$

15. We have

$$D = \int_1^2 \sqrt{(x'(t))^2 + (y'(t))^2 + (z'(t))^2}\,dt = \int_1^2 \sqrt{5^2 + 4^2 + (-1)^2}\,dt = \sqrt{42}.$$

This is the length of a straight line from the point $(8, 5, 2)$ to $(13, 9, 1)$.

16. We have

$$D = \int_0^1 \sqrt{(-e^t\sin(e^t))^2 + (e^t\cos(e^t))^2}\,dt$$
$$= \int_0^1 \sqrt{e^{2t}}\,dt = \int_0^1 e^t\,dt$$
$$= e - 1.$$

This is the length of the arc of a unit circle from the point $(\cos 1, \sin 1)$ to $(\cos e, \sin e)$—in other words between the angles $\theta = 1$ and $\theta = e$. The length of this arc is $(e - 1)$.

17. We have

$$D = \int_0^{2\pi} \sqrt{(-3\sin 3t)^2 + (5\cos 5t)^2}\,dt.$$

We cannot find this integral symbolically, but numerical methods show $D \approx 24.6$.

18. The parameterization is

$$\vec{r}(t) = 5\vec{i} + 4\vec{j} - 2\vec{k} + (t-4)(2\vec{i} - 3\vec{j} + \vec{k})$$

or equivalently

$$x = 5 + 2(t-4), \quad y = 4 - 3(t-4), \quad z = -2 + (t-4).$$

19. At time t the particle is $s = t - 7$ seconds from P, so the displacement vector from the point P to the particle is $\vec{d} = s\vec{v}$. To find the position vector of the particle at time t, we add this to the position vector $\vec{r}_0 = 5\vec{i} + 4\vec{j} + 3\vec{k}$ for the point P. Thus the vector equation for the motion is:

$$\begin{aligned} \vec{r} &= \vec{r}_0 + s\vec{v} \\ &= (5\vec{i} + 4\vec{j} + 3\vec{k}) + (t-7)(3\vec{i} + \vec{j} + 2\vec{k}), \end{aligned}$$

or equivalently,

$$x = 5 + 3(t-7), \quad y = 4 + 1(t-7), \quad z = 3 + 2(t-7).$$

Notice that these equations are linear. They describe motion on a straight line through the point $(5, 4, 3)$ that is parallel to the velocity vector $\vec{v} = 3\vec{i} + \vec{j} + 2\vec{k}$.

20. The displacement vector from $(1, 1, 1)$ to $(2, -1, 3)$ is $\vec{d} = (2\vec{i} - \vec{j} + 3\vec{k}) - (\vec{i} + \vec{j} + \vec{k}) = \vec{i} - 2\vec{j} + 2\vec{k}$ meters. The velocity vector has the same direction as $\vec{d}$ and is given by

$$\vec{v} = \frac{\vec{d}}{5} = 0.2\vec{i} - 0.4\vec{j} + 0.4\vec{k} \text{ meters/sec.}$$

Since $\vec{v}$ is constant, the acceleration $\vec{a} = \vec{0}$.

21. Plotting the positions on the xy plane and noting their times gives the following:

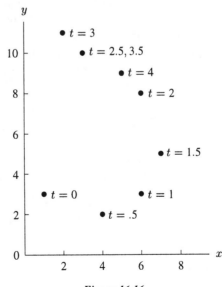

Figure 16.16

(a) We approximate dx/dt by $\Delta x/\Delta t$ calculated between $t = 1.5$ and $t = 2.5$:

$$\frac{dx}{dt} \approx \frac{\Delta x}{\Delta t} = \frac{3-7}{2.5-1.5} = \frac{-4}{1} = -4.$$

Similarly,

$$\frac{dy}{dt} \approx \frac{\Delta y}{\Delta t} = \frac{10 - 5}{2.5 - 1.5} = \frac{5}{1} = 5.$$

So,

$$\vec{v}(2) \approx -4\vec{i} + 5\vec{j} \quad \text{and} \quad \text{Speed} = \|\vec{v}\| = \sqrt{41}.$$

(b) The particle is moving vertically at about time $t = 1.5$.

(c) The particle stops at about time $t = 3$ and reverses course.

22. (a) In order for the particle to stop, its velocity $\vec{v} = (dx/dt)\vec{i} + (dy/dt)\vec{j}$ must be zero, so we solve for t such that $dx/dt = 0$ and $dy/dt = 0$, that is

$$\frac{dx}{dt} = 3t^2 - 3 = 3(t - 1)(t + 1) = 0,$$

$$\frac{dy}{dt} = 2t - 2 = 2(t - 1) = 0.$$

The value $t = 1$ is the only solution. Therefore, the particle stops when $t = 1$ at the point $(t^3 - 3t, t^2 - 2t)|_{t=1} = (-2, -1)$.

(b) In order for the particle to be traveling straight up or down, the x-component of the velocity vector must be 0. Thus, we solve $dx/dt = 3t^2 - 3 = 0$ and obtain $t = \pm 1$. However, at $t = 1$ the particle has no vertical motion, as we saw in part (a). Thus, the particle is moving straight up or down only when $t = -1$. Since the velocity at time $t = -1$ is

$$\vec{v}(-1) = \left.\frac{dx}{dt}\right|_{t=-1} \vec{i} + \left.\frac{dy}{dt}\right|_{t=-1} \vec{j} = -4\vec{j},$$

the motion is straight down. The position at that time is $(t^3 - 3t, t^2 - 2t)|_{t=-1} = (2, 3)$.

(c) For horizontal motion we need $dy/dt = 0$. That happens when $dy/dt = 2t - 2 = 0$, and so $t = 1$. But from part (a) we also have $dx/dt = 0$ also at $t = 1$, so the particle is not moving at all when $t = 1$. Thus, there is no time when the motion is horizontal.

23. (a) No. The height of the particle is given by $2t$; the vertical velocity is the derivative $d(2t)/dt = 2$. Because this is a positive constant, the vertical component of the velocity vector is upward at a constant speed of 2.

(b) When $2t = 10$, so $t = 5$.

(c) The velocity vector is given by

$$\vec{v}(t) = \frac{d\vec{r}}{dt} = \frac{dx}{dt}\vec{i} + \frac{dy}{dt}\vec{j} + \frac{dz}{dt}\vec{k}$$
$$= -(\sin t)\vec{i} + (\cos t)\vec{j} + 2\vec{k}.$$

From (b), the particle is at 10 units above the ground when $t = 5$, so at $t = 5$,

$$\vec{v}(5) = 0.959\vec{i} + 0.284\vec{j} + 2\vec{k}.$$

Therefore, $\vec{v}(5) = -\sin(5)\vec{i} + \cos(5)\vec{j} + 2\vec{k}$.

(d) At this point, $t = 5$, the particle is located at

$$\vec{r}(5) = (\cos(5), \sin(5), 10) = (0.284, -0.959, 10).$$

The tangent vector to the spiral at this point is given by the velocity vector found in part (c), that is, $\vec{v}(5) = 0.959\vec{i} + 0.284\vec{j} + 2\vec{k}$. So, the equation of the tangent line is

$$\vec{r}(t) = 0.284\vec{i} - 0.959\vec{j} + 10\vec{k} + (t - 5)(0.959\vec{i} + 0.284\vec{j} + 2\vec{k}).$$

24. The ship's velocity at time t is given by

$$\vec{v}(t) = \vec{r}\,'(t) = -6e^{3t}\vec{i} - 5\sin t\vec{j} - 6\cos 2t\vec{k}$$

So $\vec{r}(0) = -2\vec{i} + 5\vec{j}$ and $\vec{v}(0) = -6\vec{i} - 6\vec{k}$. At $t = 0$, when Mr. Skywalker turns off the power, he is at $(-2, 5, 0)$ going in the direction $-6\vec{i} - 6\vec{k}$. Xardon is at $(1.5, 5, 3.5)$, so the vector in the direction of Xardon is $(1.5\vec{i} + 5\vec{j} + 3.5\vec{k}) - (-2\vec{i} + 5\vec{j}) = 3.5\vec{i} + 3.5\vec{k}$. Note that $3.5\vec{i} + 3.5\vec{k} = -\frac{7}{12}(-6\vec{i} - 6\vec{k})$. Mr. Skywalker is aiming in the right direction, but he will not get to Xardon because he is leaving, not approaching it. (No positive multiple of $-6\vec{i} - 6\vec{k}$ can be added to $-2\vec{i} + 5\vec{j}$ to get $3.5\vec{i} + 3.5\vec{k}$.)

25. Since the acceleration due to gravity is -9.8 m/sec^2, we have $\vec{r}\,''(t) = -9.8\vec{k}$. Integrating gives

$$\vec{r}\,'(t) = C_1\vec{i} + C_2\vec{j} + (-9.8t + C_3)\vec{k},$$
$$\vec{r}(t) = (C_1 t + C_4)\vec{i} + (C_2 t + C_5)\vec{j} + (-4.9t^2 + C_3 t + C_6)\vec{k}.$$

The initial condition, $\vec{r}(0) = \vec{0}$, implies that $C_4 = C_5 = C_6 = 0$, thus

$$\vec{r}(t) = C_1 t\vec{i} + C_2 t\vec{j} + (-4.9t^2 + C_3 t)\vec{k}.$$

To find the position vector, we need to find the values of C_1, C_2, and C_3. This we do using the coordinates of the highest point. When the rocket reaches its peak, the vertical component of the velocity is zero, so $-9.8t + C_3 = 0$. Thus, at the highest point, $t = C_3/9.8$. At that time

$$\vec{r}(t) = 1000\vec{i} + 3000\vec{j} + 10000\vec{k},$$

so, for the same value of t:

$$C_1 t = 1000,$$
$$C_2 t = 3000,$$
$$-4.9t^2 + C_3 t = 10,000,$$

Substituting $t = C_3/9.8$ into the third equation gives

$$-4.9\left(\frac{C_3}{9.8}\right)^2 + \frac{C_3^2}{9.8} = 10,000$$
$$C_3^2 = 2(9.8)10,000$$
$$C_3 = 442.7$$

Then $C_1 = \frac{1000}{C_3/9.8} = 22.1$ and $C_2 = \frac{3000}{C_3/9.8} = 66.4$. Thus,

$$\vec{r}(t) = 22.1t\vec{i} + 66.4t\vec{j} + (442.7t - 4.9t^2)\vec{k}.$$

26. (a) Since $x = R\cos(\omega t)$ and $y = R\sin(\omega t)$, and $x^2 + y^2 = R^2\cos^2(\omega t) + R^2\sin^2(\omega t) = R^2$, we have motion around a circle of radius R centered at the origin. The particle moves counterclockwise, completing one revolution in time $2\pi/\omega$. Thus, the period $= 2\pi/\omega$.
     (b) The velocity vector is

$$\vec{v} = \frac{d\vec{r}}{dt} = -\omega R\sin(\omega t)\vec{i} + \omega R\cos(\omega t)\vec{j}.$$

We expect the velocity, $\vec{v}$, to be tangent to the circle. To verify that this, we compute

$$\vec{v} \cdot \vec{r} = (-\omega R \sin(\omega t)\vec{i} + \omega R \cos(\omega t)\vec{j}) \cdot (R\cos(\omega t)\vec{i} + R\sin(\omega t)\vec{j})$$
$$= -\omega R^2 \sin(\omega t)\cos(\omega t) + \omega R^2 \cos(\omega t)\sin(\omega t) = 0.$$

This shows that the velocity vector is perpendicular to the radius from the center of the circle to the particle, which moves counterclockwise.

The speed is $\|\vec{v}\| = \omega R$, which is constant. Notice that this makes sense, because in time $2\pi/\omega$, the particle travels a distance of $2\pi R$, giving a speed of $2\pi R/(2\pi/\omega) = \omega R$.

(c) The acceleration vector is

$$\vec{a} = \frac{d\vec{v}}{dt} = -\omega^2 R\cos(\omega t)\vec{i} - \omega^2 R\sin(\omega t)\vec{j} = -\omega^2 \vec{r}.$$

The acceleration vector points in the direction opposite to the position vector $\vec{r}$, and thus points towards the center of the circle. It has constant magnitude $\|\vec{a}\| = \omega^2 R = \|v\|^2/R$.

27. (a) With the center at $(0, 0, 8)$ and a point of the circle at $(0, 5, 8)$, we know that the radius is 5. When $t = 0$, we have $x = 0$ and $y = 5$. Since the stone is rotating horizontally, $z = 0$ for all t. The period is 2π. Thus, the parameterization is:

$$x(t) = 5\sin t$$
$$y(t) = 5\cos t$$
$$z(t) = 8$$

This parameterization has the correct period (if t is in seconds) and satisfies the initial conditions.

(b) From our parameterization with t in seconds, we can see that the stone reaches $(5, 0, 8)$ at time $\pi/2$. Thus at $t = \pi/2$,

$$\vec{v} = x_t(\pi/2)\vec{i} + y_t(\pi/2)\vec{j}$$
$$= 5\cos(\pi/2)\vec{i} - 5\sin(\pi/2)\vec{j}$$
$$= -5\vec{j}.$$

The acceleration of an object is the second derivative of its position. Thus, at $t = \pi/2$,

$$\vec{a} = x_{tt}(\pi/2)\vec{i} + y_{tt}(\pi/2)\vec{j}$$
$$= -5\sin(\pi/2)\vec{i} - 5\cos(\pi/2)\vec{j}$$
$$= -5\vec{i}$$

(c) At the moment in which the stone has left the circle, the only acceleration that acts on the stone is that of gravity. From that, assuming a gravity vector field oriented in the $-z$ direction, we get the differential equations

$$z_{tt}(t) = -g$$
$$x_{tt}(t) = y_{tt}(t) = 0.$$

If we now measure t from the instant the string breaks, then the initial conditions are the velocity and position of the stone at $t = 0$. Since the velocity at the moment of release is $\vec{v} = -5\vec{j}$, we have

$$x_t(0) = 0, \quad y_t(0) = -5, \quad z_t(0) = 0.$$

The initial position at $t = 0$ is:

$$x(0) = 5, \quad y(0) = 0, \quad z(0) = 8.$$

28. (a) The parametric equation describing Emily's motion is

$$x = 10\cos\left(\frac{2\pi}{20}t\right) = 10\cos\left(\frac{\pi}{10}t\right), \quad y = 10\sin\left(\frac{2\pi}{20}t\right) = 10\sin\left(\frac{\pi}{10}t\right) \quad z = \text{constant}.$$

Her velocity vector is

$$\vec{v} = \frac{dx}{dt}\vec{i} + \frac{dy}{dt}\vec{j} + \frac{dz}{dt}\vec{k} = -\pi\sin\left(\frac{\pi}{10}t\right)\vec{i} + \pi\cos\left(\frac{\pi}{10}t\right)\vec{j}.$$

Her speed is given by:

$$\|\vec{v}\| = \sqrt{\left(-\pi\sin\left(\frac{\pi}{10}t\right)\right)^2 + \left(\pi\cos\left(\frac{\pi}{10}t\right)\right)^2 + 0^2}$$

$$= \pi\sqrt{\sin^2\left(\frac{\pi}{10}t\right) + \cos^2\left(\frac{\pi}{10}t\right)}$$

$$= \pi\sqrt{1} = \pi \text{ m/sec},$$

which is independent of time (as we expected). This is certainly the long way to solve this problem though, since we could have simply divided the circumference of the circle (20π) by the time taken for a single rotation (20 seconds) to arrive at the same answer.

(b) When Emily drops the ball, it initially has Emily's velocity vector, but it immediately begins accelerating in the z-direction due to the force of gravity. The motion of the ball will then be tangential to the merry-go-round, curving down to the ground. In order to find the tangential component of the ball's motion, we must know Emily's velocity at the moment she dropped the ball. Then we can integrate the velocity and obtain the position of the ball. Assuming Emily drops the ball at time $t = 0$, her position and velocity vector are

$$\vec{r}(0) = 10\vec{i} + 3\vec{k} \text{ and } \vec{v}(0) = \pi\vec{j}.$$

Thus, the ball has velocity only in the y-direction when it is dropped. In the z-direction, we have

$$\text{Acceleration} = \frac{d^2z}{dt^2} = -9.8 \text{ m/sec}^2.$$

Since the initial velocity 0 and initial height 3, we have

$$z = 3 - 4.9t^2.$$

The ball touches the ground when $z = 0$, that is, when $t = 0.78$ sec. In that time, the ball also travels $\pi(0.78) = 2.45$ meters in the y-direction. So, the final position is $(10, 2.45, 0)$. The distance between this point and $P = (10, 0, 0)$ is 2.45 meters.

(c) The distance of the ball from Emily when it hits the ground is found by finding Emily's position at $t = 0.78$ sec and using the distance formula. Emily's position when the ball hits the ground is $(10\cos(0.078\pi), 10\sin(0.078\pi), 3) = (9.70, 2.43, 3)$. The distance between this point and the point where the ball struck the ground is:

$$d \approx \sqrt{(10 - 9.70)^2 + (2.45 - 2.43)^2 + (0 - 3)^2} = 3.01 \text{ meters}.$$

Note that the merry-go-round doesn't rotate very much in the 0.78 sec needed for the ball to reach the ground, so our answer makes sense.

29. (a) Consider a time Δt in which a fixed angle at the center is swept out. To find the largest and smallest speeds look for points at which the length of coastline cut off by this beam is greatest, which appears to be C, and smallest, which appears to be E.

(b) Now we draw beams of light that sweep out the same area, and again look to see which cuts off the longest length on the shoreline, E, and the smallest, C.

(c) Yes, when the light beam is tangential to the shoreline (when the light house is close to A or C).

(d) Consider the right-hand side of the rectangular lake shown in Figure 16.17. We know that $\frac{d}{dt}$(Area of triangle) = constant, where the Area of the triangle $= xk/2$ and the constant k is the distance from L to the right-hand side.

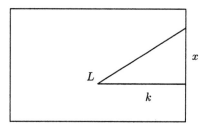

Figure 16.17

Thus,

$$\frac{d}{dt}\left(\frac{xk}{2}\right) = \frac{k}{2}\frac{dx}{dt} = \text{constant},$$

so $dx/dt = $ constant. Therefore the speed is constant along the right-hand side and similarly for the other three sides; it may be different on each side because k varies from side to side. The speed is not defined at the corners.

30. (a) To find the equations of the moon's motion relative to the star, you must first calculate the equation of the planet's motion relative to the star, and then the moon's motion relative to the planet, and then add the two together.

The distance from the planet to the star is R, and the time to make one revolution is one unit, so the parametric equations for the planet relative to the star are $x = R\cos t, y = R\sin t$.

The distance from the moon to the planet is 1, and the time to make one revolution is twelve units, therefore, the parametric equations for the moon relative to the planet are $x = \cos 12t, y = \sin 12t$.

Adding these together, we get:

$$x = R\cos t + \cos 12t,$$
$$y = R\sin t + \sin 12t.$$

(b) For the moon to stop completely at time t, the velocity of the moon must be equal to zero. Therefore,

$$\frac{dx}{dt} = -R\sin t - 12\sin 12t = 0,$$
$$\frac{dy}{dt} = R\cos t + 12\cos 12t = 0.$$

There are many possible values to choose for R and t that make both of these equations equal to zero. We choose $t = \pi$, and $R = 12$.

(c)   The graph with $R = 12$ is shown in Figure 16.18.

Figure 16.18

31.   Substituting $x = t$, and $y = t^2$ into the equation gives us $F = 1/(t^2 + t^4 + 1)$. To find the rate of change of temperature at time t, we differentiate F with respect to t:

$$\text{Rate of change of temperature} = \frac{dF}{dt} = -(t^2 + t^4 + 1)^{-2}(2t + 4t^3).$$

32.   Setting $z = F(x(t), y(t))$, so z is the temperature of the ladybug at time t, we use the chain rule to compute the rate of change of the temperature of the bug at the time t:

$$\frac{dz}{dt} = \frac{\partial z}{\partial x}\frac{dx}{dt} + \frac{\partial z}{\partial y}\frac{dy}{dt}$$
$$= \frac{\partial F}{\partial x}\frac{dx}{dt} + \frac{\partial F}{\partial y}\frac{dy}{dt}$$
$$= \left(\frac{\partial F}{\partial x}\vec{i} + \frac{\partial F}{\partial y}\vec{j}\right) \cdot \left(\frac{dx}{dt}\vec{i} + \frac{dy}{dt}\vec{j}\right)$$
$$= \text{grad } F(x, y) \cdot \vec{r}\,'(t),$$

as required.

33.   (a)   If $\Delta t = t_{i+1} - t_i$ is small enough so that C_i is approximately a straight line, then we can make the linear approximations

$$x(t_{i+1}) \approx x(t_i) + x'(t_i)\Delta t,$$
$$y(t_{i+1}) \approx y(t_i) + y'(t_i)\Delta t,$$
$$z(t_{i+1}) \approx z(t_i) + z'(t_i)\Delta t,$$

and so

$$\text{Length of } C_i \approx \sqrt{(x(t_{i+1}) - x(t_i))^2 + (y(t_{i+1}) - y(t_i))^2 + (z(t_{i+1}) - z(t_i))^2}$$
$$\approx \sqrt{x'(t_i)^2(\Delta t)^2 + y'(t_i)^2(\Delta t)^2 + z'(t_i)^2(\Delta t)^2}$$
$$= \sqrt{x'(t_i)^2 + y'(t_i)^2 + z'(t_i)^2}\,\Delta t.$$

(b) From point (a) we obtain the approximation

$$\text{Length of } C = \sum \text{length of } C_i$$
$$\approx \sum \sqrt{x'(t_i)^2 + y'(t_i)^2 + z'(t_i)^2} \Delta t.$$

The approximation gets better and better as Δt approaches zero, and in the limit the sum becomes a definite integral:

$$\text{Length of } C = \lim_{\Delta t \to 0} \sum \sqrt{x'(t_i)^2 + y'(t_i)^2 + z'(t_i)^2} \, \Delta t$$
$$= \int_a^b \sqrt{x'(t)^2 + y'(t)^2 + z'(t)^2} \, dt.$$

Solutions for Section 16.3

1. A horizontal disk of radius 5 in the plane $z = 7$.

2. A circle of radius 5 in the plane $z = 7$.

3. A cylinder of radius 5 centered around the z-axis and stretching around from $z = 0$ to $z = 7$.

4. A helix (curve) of radius 5 which makes one turn about the z-axis, starting at the point $(5, 0, 0)$ and ending at the point $(5, 0, 10\pi)$.

5. Since $z = r = \sqrt{x^2 + y^2}$, we have a cone around the z-axis. Since $0 \le r \le 5$, we have $0 \le z \le 5$, so the cone has height and maximum radius of 5.

6. Since $x^2 + y^2 = 4z^2$, we have $z = \frac{1}{2}\sqrt{x^2 + y^2}$. Thus we have a cone of height 7 and maximum radius 14, centered around the z-axis.

7. Since $(\frac{x}{3})^2 + (\frac{y}{2})^2 = 1$, so $\frac{x^2}{9} + \frac{y^2}{4} = 1$, which is the equation for an ellipse, we have a cylinder with an elliptical cross-section. The ellipse is centered around the z-axis from $z = 0$ to $z = 7$.

8. This is a parabolic cylinder $y = x^2$, between $x = -5$ and $x = 5$, with its axis along the z-axis, stretching from $z = 0$ to $z = 7$.

9. The cross sections of the cylinder perpendicular to the z-axis are circles which are vertical translates of the circle $x^2 + y^2 = a^2$, which is given parametrically by $x = a \cos \theta$, $y = a \sin \theta$. The vector $a \cos \theta \vec{i} + a \sin \theta \vec{j}$ traces out the circle, at any height. We get to a point on the surface by adding that vector to the vector $z\vec{k}$. Hence, the parameters are θ, with $0 \le \theta \le 2\pi$, and z, with $0 \le z \le h$. The parametric equations for the cylinder are

$$x\vec{i} + y\vec{j} + z\vec{k} = a \cos \theta \vec{i} + a \sin \theta \vec{j} + z\vec{k},$$

which can be written as

$$x = a \cos \theta, \quad y = a \sin \theta, \quad z = z.$$

10. Since you walk 5 blocks east and 1 block west, you walk 5 blocks in the direction of $\vec{v}_1$, and 1 block in the opposite direction. Thus,

$$s = 5 - 1 = 4,$$

Similarly,

$$t = 4 - 2 = 2.$$

Hence

$$
\begin{aligned}
x\vec{i} + y\vec{j} + z\vec{k} &= (x_0\vec{i} + y_0\vec{j} + z_0\vec{k}) + 4\vec{v_1} + 2\vec{v_2} \\
&= (x_0\vec{i} + y_0\vec{j} + z_0\vec{k}) + 4(2\vec{i} - 3\vec{j} + 2\vec{k}) + 2(\vec{i} + 4\vec{j} + 5\vec{k}) \\
&= (x_0 + 10)\vec{i} + (y_0 - 4)\vec{j} + (z_0 + 18)\vec{k}.
\end{aligned}
$$

Thus the coordinates are:

$$
x = x_0 + 10, \quad y = y_0 - 4, \quad z = z_0 + 18.
$$

11. The plane through $(1, 3, 4)$ and orthogonal to $\vec{n} = 2\vec{i} + \vec{j} - \vec{k}$ is given by $2(x - 1) + (y - 3) - (z - 4) = 0$, that is,

$$
2x + y - z - 1 = 0.
$$

Thus, thinking of the plane as $z = 2x + y - 1$, one possible parameterization is

$$
x = u, \quad y = v, \quad z = 2u + v - 1.
$$

12. (a) We want to find s and t so that

$$
\begin{aligned}
2 + s &= 4 \\
3 + s + t &= 8 \\
4t &= 12
\end{aligned}
$$

Since $s = 2$ and $t = 3$ satisfy these equations, the point $(4, 8, 12)$ lies on this plane.

(b) Are there values of s and t corresponding to the point $(1, 2, 3)$? If so, then

$$
\begin{aligned}
1 &= 2 + s \\
2 &= 3 + s + t \\
3 &= 4t
\end{aligned}
$$

From the first equation we must have $s = -1$ and from the third we must have $t = 3/4$. But these values of s and t do not satisfy the second equation. Therefore, no value of s and t corresponds to the point $(1, 2, 3)$, and so $(1, 2, 3)$ is not on the plane.

13. If the planes are parallel, then their normal vectors will also be parallel. The equation of the first plane can be written

$$
\vec{r} = 2\vec{i} + 4\vec{j} + \vec{k} + s(\vec{i} + \vec{j} + 2\vec{k}) + t(\vec{i} - \vec{j}).
$$

A normal vector to the first plane is $\vec{n}_1 = (\vec{i} + \vec{j} + 2\vec{k}) \times (\vec{i} - \vec{j}) = 2\vec{i} + 2\vec{j} - 2\vec{k}$. The second plane can be written

$$
\vec{r} = 2\vec{i} + s(\vec{i} + \vec{k}) + t(2\vec{i} + \vec{j} - \vec{k}).
$$

A normal vector to the second plane is $\vec{n}_2 = (\vec{i} + \vec{k}) \times (2\vec{i} + \vec{j} - \vec{k}) = -\vec{i} + 3\vec{j} + \vec{k}$. Since $\vec{n}_1$ and $\vec{n}_2$ are not parallel, neither are the two planes.

14. (a) Nearer to the equator.
(b) Farther from the north pole.
(c) Farther from Greenwich.

15. A horizontal circle in the northern hemisphere at a latitude of 45° north of the equator.

16. A vertical half-circle, going from the north to south poles.

17. We use spherical coordinates ϕ and θ as the two parameters. Since the radius is 5, we can take

$$x = 5\sin\phi\cos\theta, \quad y = 5\sin\phi\sin\theta, \quad z = 5\cos\phi.$$

18. We use spherical coordinates ϕ and θ as the two parameters. The parameterization of the sphere center at the origin and radius 5 is:

$$x = 5\sin\phi\cos\theta, \quad y = 5\sin\phi\sin\theta, \quad z = 5\cos\phi.$$

We have to shift the center of the sphere from the origin to the point $(2, -1, 3)$. This gives

$$x = 2 + 5\sin\phi\cos\theta, \quad y = -1 + 5\sin\phi\sin\theta, \quad z = 3 + 5\cos\phi.$$

19. The sphere $(x - a)^2 + (y - b)^2 + (z - c)^2 = d^2$ has center at the point (a, b, c) and radius d. We use spherical coordinates θ and ϕ as the two parameters. The parameterization of the sphere with center at the origin and radius d is

$$x = d\sin\phi\cos\theta, \quad y = d\sin\phi\sin\theta, \quad z = d\cos\phi.$$

Since the given sphere has center at the point (a, b, c) we add the displacement vector $a\vec{i} + b\vec{j} + c\vec{k}$ to the radial vector corresponding to a parameterization of the sphere with center at the origin and radius d to give

$$x = a + d\sin\phi\cos\theta, \quad 0 \le \phi \le \pi,$$
$$y = b + d\sin\phi\sin\theta, \quad 0 \le \theta \le 2\pi,$$
$$z = c + d\cos\phi.$$

To check that this is a parameterization for the given, we substitute for x, y, z:

$$(x - a)^2 + (y - b)^2 + (z - c)^2$$
$$= d^2\sin^2\phi\cos^2\theta + d^2\sin^2\phi\sin^2\theta + d^2\cos^2\phi$$
$$= d^2\sin^2\phi + d^2\cos^2\phi = d^2.$$

20. The parameterization for a sphere of radius a using spherical coordinates is

$$x = a\sin\phi\cos\theta, \quad y = a\sin\phi\sin\theta, \quad z = a\cos\phi.$$

Think of the ellipsoid as a sphere whose radius is different along each axis and you get the parameterization:

$$\begin{cases} x = a\sin\phi\cos\theta, & 0 \le \phi \le \pi, \\ y = b\sin\phi\sin\theta, & 0 \le \theta \le 2\pi, \\ z = c\cos\phi. \end{cases}$$

To check this parameterization, substitute into the equation for the ellipsoid:

$$\frac{x^2}{a^2} + \frac{y^2}{b^2} + \frac{z^2}{c^2} = \frac{a^2\sin^2\phi\cos^2\theta}{a^2} + \frac{b^2\sin^2\phi\sin^2\theta}{b^2} + \frac{c^2\cos^2\phi}{c^2}$$
$$= \sin^2\phi(\cos^2\theta + \sin^2\theta) + \cos^2\phi = 1.$$

21. Let $(\theta, \pi/2)$ be the original coordinates. If $\theta < \pi$, then the new coordinates will be $(\theta + \pi, \pi/4)$. If $\theta \geq \pi$, then the new coordinates will be $(\theta - \pi, \pi/4)$.

22. (a) The top half of the sphere $(z \geq 0)$.
    (b) The half of the sphere with $y \leq 0$.
    (c) A vertical segment lying between two longitudinal lines ($\theta = \frac{\pi}{4}$ and $\theta = \frac{\pi}{3}$) and stretching between the poles.
    (d) Half the horizontal ring around the sphere between two latitude lines ($\phi = \frac{\pi}{4}$ and $\phi = \frac{\pi}{3}$) in the northern hemisphere.

23. If we set $z = u$, $x^2 + y^2 = u^2$ is the equation of a circle with radius $|u|$. Hence a parameterization of the cone is:

$$x = u \cos v,$$
$$y = u \sin v, \qquad 0 \leq v \leq 2\pi,$$
$$z = u.$$

24. Since the parameterization in Example 6 on page 300 was $r = (1 - \frac{z}{h})a$ and since the cone is given by $z = r$, we have $z = (1 - \frac{r}{a})h$. The parameterization we want is

$$x = r \cos \theta, \qquad 0 \leq r \leq a,$$
$$y = r \sin \theta, \qquad 0 \leq \theta \leq 2\pi,$$
$$z = \left(1 - \frac{r}{a}\right)h.$$

25. (a) The cone of height h, maximum radius a, vertex at the origin and opening upward is shown in Figure 16.19.

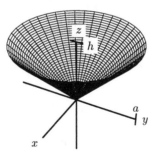

Figure 16.19

By similar triangles, we have

$$\frac{r}{z} = \frac{a}{h},$$

so

$$z = \frac{hr}{a}.$$

Therefore, one parameterization is

$$x = r \cos \theta, \qquad 0 \leq r \leq a,$$
$$y = r \sin \theta, \qquad 0 \leq \theta < 2\pi,$$
$$z = \frac{hr}{a}.$$

(b) Since $r = az/h$, we can write the parameterization in part (a) as

$$x = \frac{az}{h}\cos\theta, \qquad 0 \leq z \leq h,$$
$$y = \frac{az}{h}\sin\theta, \qquad 0 \leq \theta < 2\pi,$$
$$z = z.$$

26. In cylindrical coordinates, the paraboloid has equation $z = r^2$. Thus, a point on the surface is given by

$$x\vec{i} + y\vec{j} + z\vec{k} = r\cos\theta\vec{i} + r\sin\theta\vec{j} + z\vec{k}$$
$$= \sqrt{z}\cos\theta\vec{i} + \sqrt{z}\sin\theta\vec{j} + z\vec{k}.$$

Alternatively, we could write

$$x\vec{i} + y\vec{j} + z\vec{k} = r\cos\theta\vec{i} + r\sin\theta\vec{j} + r^2\vec{k}.$$

27. The vase obtained by rotating the curve $z = 10\sqrt{x-1}$, $1 \leq x \leq 2$, around the z-axis is shown in Figure 16.20.

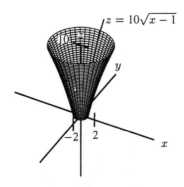

$z = 10\sqrt{x-1}$

y

-2 2

x

Figure 16.20

At height z, the cross-section is a horizontal circle of radius a. Thus, a point on this horizontal circle is given by

$$\vec{r} = a\cos\theta\vec{i} + a\sin\theta\vec{j} + z\vec{k}.$$

However, the radius a varies, so we need to express it in terms of the other parameters θ and z. If you look at the xz-plane, the radius of this circle is given by x, so solving for x in $z = 10\sqrt{x-1}$ gives

$$a = x = \left(\frac{z}{10}\right)^2 + 1.$$

Thus, a parameterization is

$$\vec{r} = \left(\left(\frac{z}{10}\right)^2 + 1\right)\cos\theta\vec{i} + \left(\left(\frac{z}{10}\right)^2 + 1\right)\sin\theta\vec{j} + z\vec{k}$$

so

$$x = \left(\left(\frac{z}{10}\right)^2 + 1\right)\cos\theta, \quad y = \left(\left(\frac{z}{10}\right)^2 + 1\right)\sin\theta, \quad z = z,$$

where $0 \leq \theta \leq 2\pi$, $0 \leq z \leq 10$.

28. (a) Add second and third equations to get $y + z = 1 + 2s$. Thus, $y + z = 1 + x$ or $-x + y + z = 1$, which is the equation of a plane. Now, $s = x/2$, and $t = (y - z + 1)/2$, so the conditions $0 \le s \le 1, 0 \le t \le 1$ are equivalent to $0 \le x \le 2, 0 \le y - z + 1 \le 2$ or $0 \le x \le 2, -1 \le y - z \le 1$.
    (b) The surface is shown in Figure 16.21.

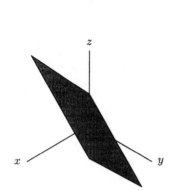

Figure 16.21: The surface
$x = 2s, y = s + t, z = 1 + s - t$,
for $0 \le s \le 1, 0 \le t \le 1$

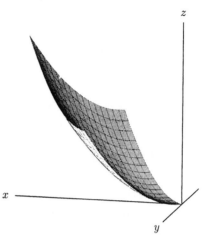

Figure 16.22: The surface $x = s + t$,
$y = s - t, z = s^2 + t^2$ for $0 \le s \le 1$,
$0 \le t \le 1$

29. (a) From the first two equations we get:

$$s = \frac{x + y}{2}, \qquad t = \frac{x - y}{2}.$$

Hence the equation of our surface is:

$$z = \left(\frac{x + y}{2}\right)^2 + \left(\frac{x - y}{2}\right)^2 = \frac{x^2}{2} + \frac{y^2}{2},$$

which is the equation of a paraboloid.

The conditions: $0 \le s \le 1, 0 \le t \le 1$ are equivalent to: $0 \le x + y \le 2, 0 \le x - y \le 2$. So our surface is defined by:

$$z = \frac{x^2}{2} + \frac{y^2}{2}, \qquad 0 \le x + y \le 2 \quad 0 \le x - y \le 2$$

    (b) The surface is shown in Figure 16.22.

30. (a) As $x^2 + y^2 = 9$ and $s \in [0, \pi]$ is equivalent to $x \ge 0$, and $t \in [0, 1]$ is equivalent to $z \in [1, 2]$. So, $x^2 + y^2 = 9$ is the equation of a cylinder, and our surface is defined by:

$$x^2 + y^2 = 9, \qquad x \ge 0, \quad 1 \le z \le 2.$$

(b)   The surface $x = 3\sin s$, $y = 3\cos s$, $z = t + 1$ for $0 \le s \le \pi$, $0 \le t \le 1$ is shown in Figure 16.23.

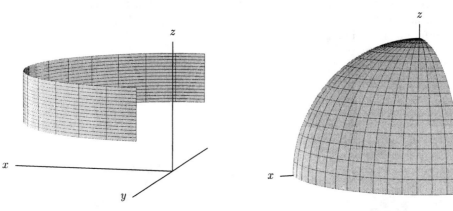

Figure 16.23   **Figure 16.24**

31.   (a) $z^2 = 1 - s^2 - t^2 = 1 - x^2 - y^2$. So $x^2 + y^2 + z^2 = 1$ which is the equation of a sphere. The conditions $s^2 + t^2 \le 1$, $s, t \ge 0$ are equivalent to $x^2 + y^2 \le 1$ and $x, y \ge 0$. But if $x^2 + y^2 + z^2 = 1$, then $x^2 + y^2 \le 1$ is satisfied automatically, so our surface is defined by:

$$x^2 + y^2 + z^2 = 1, \qquad x, y, z \ge 0.$$

(b)   The surface $x = s$, $y = t$, $z = \sqrt{1 - s^2 - t^2}$ for $s^2 + t^2 \le 1$, $s, t \ge 0$ is shown in Figure 16.24.

32.   (a)   The surface is the cylinder $x^2 + y^2 = 1$ of radius 1 centered on the z-axis.

(b)   The parameter curves with constant s and varying t are helices that wind clockwise around the cylinder as they advance up the cylinder with increasing t. See Figure 16.25.

The parameter curves with constant t and varying s are helices that wind counterclockwise up the cylinder. See Figure 16.26.

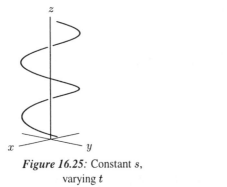

Figure 16.25: Constant s,
varying t

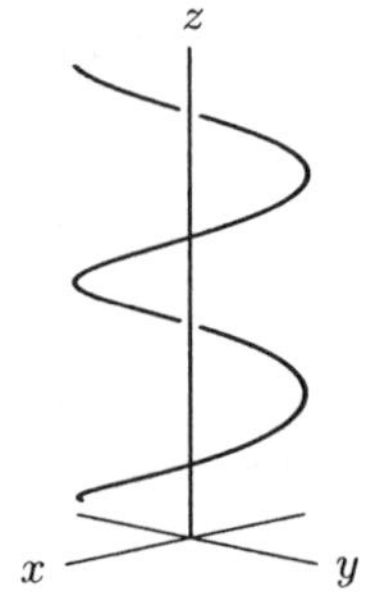

Figure 16.26: Constant t,
varying s

33.   The plane in which the circle lies is parameterized by

$$\vec{r}(p, q) = x_0\vec{i} + y_0\vec{j} + z_0\vec{k} + p\vec{u} + q\vec{v}.$$

Because $\vec{u}$ and $\vec{v}$ are perpendicular unit vectors, the parameters p and q establish a rectangular coordinate system on this plane exactly analogous to the usual xy-coordinate system, with $(p, q) = (0, 0)$ corresponding to the point (x_0, y_0, z_0). Thus the circle we want to describe, which is the circle of radius a centered at $(p, q) = (0, 0)$, can be parameterized by

$$p = a \cos t, \qquad q = a \sin t.$$

Substituting into the equation of the plane gives the desired parameterization of the circle in 3-space,

$$\vec{r}(t) = x_0\vec{i} + y_0\vec{j} + z_0\vec{k} + a\cos t\vec{u} + a\sin t\vec{v},$$

where $0 \leq t \leq 2\pi$.

34. (a) $\vec{R} = b\cos\theta\vec{i} + b\sin\theta\vec{j}$
    (b) One vector is always $\vec{k}$. The other vector we need is in the same direction as $\vec{R}$ in part (a) but has length 1. Therefore we take the other vector to be $\vec{m} = \cos\theta\vec{i} + \sin\theta\vec{j}$. See Figure 16.3. Thus, relative to the center of the small circle, we have

$$\vec{r} = a\cos\phi\vec{m} + a\sin\phi\vec{k} = a\cos\phi(\cos\theta\vec{i} + \sin\theta\vec{j}) + a\sin\phi\vec{k}$$

    (c) The parameterization of the torus is given by

$$\begin{aligned} x\vec{i} + y\vec{j} + z\vec{k} &= \vec{R} + \vec{r} \\ &= (b\cos\theta + a\cos\phi\cos\theta)\vec{i} + (b\sin\theta + a\cos\phi\sin\theta)\vec{j} + a\sin\phi\vec{k} \end{aligned}$$

35.

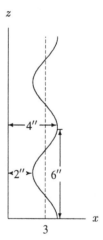

Figure 16.27   Figure 16.28

Set up the coordinates as in Figure 16.27. The surface is the revolution surface obtained by revolving the curve shown in Figure 16.28 about the z axis. From the measurements given, we obtain the equation of the curve in Figure 16.28:

$$x = \cos\left(\frac{\pi}{3}z\right) + 3, \qquad 0 \leq z \leq 48$$

(a) Rotating this around the z-axis, and taking $z = t$ as the parameter, we get the parametric equations

$$x = \left(\cos\left(\frac{\pi}{3}t\right) + 3\right)\cos\theta$$

$$y = \left(\cos\left(\frac{\pi}{3}t\right) + 3\right)\sin\theta$$

$$z = t \qquad 0 \leq \theta \leq 2\pi, \; 0 \leq t \leq 48$$

(b)   We know that the points in the curve consists of cross-sections of circles parallel to the xy plane. Thus, to find the volume, we integrate over these cross-sections. As a function of z, the area of the cross-sections are

$$\text{Area of cross-section} = \int_0^{2\pi} \int_0^{\cos(\frac{\pi}{3}z)+3} r \, dr \, d\theta.$$

Then the volume is

$$\text{Volume} = \int_0^{48} \int_0^{2\pi} \int_0^{\cos(\frac{\pi}{3}z)+3} r \, dr \, d\theta \, dz$$

$$= \pi \int_0^{48} \left(\cos\frac{\pi}{3}z + 3\right)^2 dz$$

$$= \pi \int_0^{48} \left(\cos^2\frac{\pi}{3}z + 6\cos\frac{\pi}{3}z + 9\right) dz$$

$$= 456\pi \text{in.}^3.$$

Solutions for Section 16.4

1.   The circle $(x-2)^2 + (y-2)^2 = 1$.

2.   The line segment $y + x = 4$, for $1 \leq x \leq 3$.

3.   The parabola $y = (x-2)^2$, for $1 \leq x \leq 3$.

4.   Implicit: $xy = 1$, $x > 0$
      Explicit: $y = \frac{1}{x}$, $x > 0$
      Parametric: $x = t$, $y = \frac{1}{t}$, $t > 0$

5.   Implicit: $x^2 - 2x + y^2 = 0$, $y < 0$. Explicit: $y = -\sqrt{-x^2 + 2x}$, $0 \leq x \leq 2$. Parametric: The curve is the lower half of a circle centered at $(1,0)$ with radius 1, so $x = 1 + \cos t$, $y = \sin t$, for $\pi \leq t \leq 2\pi$.

6.   Parametric: $x = e^t$, $y = e^{2t}$ for all t. Explicit: $y = x^2$, for $x > 0$. Implicit: $x^2 - y = 0$, for $x > 0$.

7.   Let $f(x,y) = xe^y + 2ye^x$. We will replace $f(x,y)$ by its linear approximation near $(0,0)$. Since $f_x = e^y + 2ye^x$ and $f_y = xe^y + 2e^x$, we have

$$f_x(0,0) = 1 \quad \text{and} \quad f_y(0,0) = 2.$$

The linear approximation of $f(x,y) = xe^y + 2ye^x$ at $(0,0)$ is $L(x,y) = x + 2y$. Hence the equation of the tangent line is $x + 2y = 0$.

8.   (a)   The linear approximation of the function $f(x,y) = x\cos y + e^x + y$ at $(0,0)$ is $f(x,y) \approx 1 + 2x + y$. Hence the linear approximation of the equation $f(x,y) = 1$ at $(0,0)$ is $1 + 2x + y = 1$. This linear equation can be solved for y, giving $y = -2x$.

      (b)   The equation $1 + 2x + y = 1$, or equivalently $y = -2x$, is the equation of the line tangent to the implicitly given curve $x\cos y + e^x + y = 1$ at the point $(0,0)$.

9.   To find $f_1(-0.02, 0.98)$, we must substitute $x = -0.02$ and $y = 0.98$ into the equation

$$z^3 - 7yz + 6e^x = 0,$$

solve for z, and pick the solution near $z = 2$. Substituting gives

$$z^3 - 6.86z + 6e^{-0.02} = 0.$$

Solving for z gives

$$z = 1.96741, \quad z = 1.00551, \quad z = -2.97292.$$

We pick $z = 1.96741$, because f_1 gives values of z near $z = 2$. Thus,

$$f(-0.02, 0.98) = 1.96741.$$

This gives us the first entry in Table 16.2.

TABLE 16.2   *Values of $f_1(x, y)$*

y/x	-0.02	-0.02	0.00	0.01	0.02
0.98	1.96741	1.95477	1.94158	1.92778	1.9133
0.99	1.99580	1.98385	1.97142	1.95847	1.94492
1.00	2.02312	2.01177	2	1.98776	1.97501
1.01	2.04949	2.03868	2.02747	2.01586	2.00379
1.02	2.07502	2.06468	2.05398	2.04292	2.03145

10.   From Example 6, we have

$$l(x, y) = -0.8 - 1.2x + 2.8y.$$

Values of this function are in Table 16.3.

TABLE 16.3   *Linear approximation values of $f(x, y)$*

			y		
	0.98	0.99	1.00	1.01	1.02
-0.02	1.968	1.996	2.024	2.052	2.080
-0.01	1.956	1.984	2.012	2.040	2.068
x 0.00	1.944	1.972	2.000	2.028	2.056
0.01	1.932	1.960	1.988	2.016	2.044
0.02	1.920	1.948	1.976	2.004	2.03

Compare the values of f_1 in Table 16.3.

11. (a)   The value of $f_2(0.01, 0.98)$ is the solution near 1 of the equation $z^3 - 7(0.98)z + 6e^{0.01} = 0$. Using a computer or calculator, we find that $z = 1.054217$.

   (b)   We approximate the equation $z^3 - 7yz + 6e^x = 0$, which we can not solve explicitly for z, by a much simpler linear equation which we can solve for z. Evaluating the partial derivatives m_x, m_y, and m_z of $m(x, y, z) = z^3 - 7yz + 6e^x$ at the point $(0, 1, 1)$ gives

$$m(x, y, z) \approx 0 + 6x - 7(y - 1) - 4(z - 1)$$

   for (x, y, z) near $(0, 1, 1)$. Solving $0 + 6x - 7(y - 1) - 4(z - 1) = 0$ for z gives

$$z = 1 + (3/2)x - (7/4)(y - 1).$$

   So, the desired linear approximation is

$$f_2(x, y) \approx 1 + (3/2)x - (7/4)(y - 1),$$

   for (x, y) near $(0, 1)$. This gives $f(0.01, 0.98) \approx 1 + (3/2)(0.01) - (7/4)(0.98 - 1) = 1.05$ which can be compared with the true value found in part (a).

   (c)   We can read the partial derivatives directly from the local linearization found in part (b). Thus, $\partial f_2/\partial x$ at $(0, 1) = 3/2$ and $\partial f_2/\partial x$ at $(0, 1)$ is $-7/4$.

12.   (a)   The value of $f_3(0.01, 0.98)$ is the solution near -3 of the equation $z^3 - 7(0.98)z + 6e^{0.01} = 0$. Using a computer or calculator, we find $z = -2.981996$.

(b)   We approximate the equation $z^3 - 7yz + 6e^x = 0$, which we can not solve explicitly for z, by a much simpler linear equation which we can solve for z. Evaluating partial derivatives m_x, m_y, and m_z of $m(x, y, z) = z^3 - 7yz + 6e^x$ at $(0, 1, -3)$ gives

$$m(x, y, z) \approx 0 + 6x + 21(y - 1) + 20(z + 3),$$

for (x, y, z) near $(0, 1, -3)$. Solving $0 + 6x + 21(y - 1) + 20(z + 3) = 0$ for z gives

$$z = -3 - (6/20)x - (21/20)(y - 1).$$

The desired linear approximation is

$$f_3(x, y) \approx -3 - (3/10)x - (21/20)(y - 1).$$

for (x, y) near $(0, 1)$. This gives $f_3(0.01, 0.98) \approx -3 - (6/20)(0.01) - (21/20)(0.98 - 1) = -2.982$ which can be compared with the true value found in part (a).

(c)   We can read the partial derivatives directly from the local linearization found in part (b). Thus, $\partial f_3/\partial x$ at $(0, 1) = -3/10$ and $\partial f_3/\partial y$ at $(0, 1) = -21/20$.

13.   (a)   The plane $2(x - 3) + 4(y - 5) + 5(z - 7) = 0$ will be tangent to the graph of the surface $f(x, y, z) = 0$ at the point $(3, 5, 7)$. Near the point $(3, 5, 7)$ the graph will be very well approximated by this plane. We do not have any information about the graph far from the point $(3, 5, 7)$.

(b)   The solutions of the equation $f(x, y, z) = 0$ near $(3, 5, 7)$ will be well approximated by the solutions of the linear equation $2(x - 3) + 4(y - 5) + 5(z - 7) = 0$. Solving this linear equation for z, we get $z = 7 - (2/5)(x - 3) - (4/5)(y - 5)$. We conclude that for every (x, y) near $(3, 5)$ there is one and only one solution of the equation $f(x, y, z) = 0$ that is near 7. That solution is well approximated by the formula $z \approx 7 - (2/5)(x - 3) - (4/5)(y - 5)$.

14.   The local linearization of the function $f(x, y, z) = z^2 + x^2 - y$ at the point $(1, 1, 0)$ is the function $L(x, y, z) = 0 + 2(x - 1) - (y - 1) + 0z = 2(x - 1) - (y - 1)$. Hence the equation of the tangent plane at this point is $2(x - 1) - (y - 1) = 0$. Notice that this is a vertical plane, because z is absent from its equation.

15.   (a)   If the prices are p_1 and p_2 and the budget is b, the quantities consumed are constrained by

$$p_1 x_1 + p_2 x_2 \leq b.$$

We want to maximize

$$f(x_1, x_2) = a \ln x_1 + (1 - a) \ln x_2$$

subject to the constraint

$$p_1 x_1 + p_2 x_2 = b.$$

Using Lagrange multipliers, we solve

$$f_{x_1} = \frac{a}{x_1} = \lambda p_1$$

$$f_{x_2} = \frac{1 - a}{x_2} = \lambda p_2,$$

giving $x_1 = a/(\lambda p_1)$ and $x_2 = (1 - a)/(\lambda p_2)$. Substituting into the constraint, we get

$$\frac{a}{\lambda} + \frac{(1 - a)}{\lambda} = b$$

so

$$\lambda = \frac{1}{b}.$$

Thus

$$x_1 = \frac{ab}{p_1} \qquad x_2 = \frac{(1-a)b}{p_2}$$

so the maximum satisfaction is given by

$$
\begin{aligned}
S = f(x_1, x_2) &= a \ln \left(\frac{ab}{p_1} \right) + (1-a) \ln \left(\frac{(1-a)b}{p_2} \right) \\
&= a \ln a + a \ln b - a \ln p_1 + (1-a) \ln(1-a) + (1-a) \ln b + (1-a) \ln p_2 \\
&= a \ln a + (1-a) \ln(1-a) + \ln b - a \ln p_1 - (1-a) \ln p_2 \\
&= \ln(a^a (1-a)^{(1-a)}) + \ln b - a \ln p_1 - (1-a) \ln p_2.
\end{aligned}
$$

(b) We want to calculate the value of b needed to achieve $u(x_1, x_2) = c$. Thus, we solve for b in the equation

$$c = a \ln a + (1-a) \ln(1-a) + \ln b - a \ln p_1 - (1-a) \ln p_2$$

Since

$$\ln b = c - a \ln a - (1-a) ln(1-a) + a \ln p_1 + (1-a) \ln p_2,$$

we have

$$b = \frac{e^c \cdot e^{a \ln p_1} \cdot e^{(1-a) \ln p_2}}{e^{a \ln a} \cdot e^{(1-a) \ln(1-a)}} = \frac{e^c p_1^a p_2^{1-a}}{a^a (1-a)^{(1-a)}}.$$

(c) We consider the implicit function $F(b, p_1, p_2, c) = a \ln a + (1-a) \ln(1-a) + \ln b - a \ln p_1 - (1-a) \ln p_2 - c = 0$. This function is used to determine c in terms of the other variables. Since for $b > 0$

$$\frac{\partial F}{\partial b} = \frac{1}{b} \neq 0$$

the implicit function theorem applies, tell us that we can locally solve for b. In fact, the solution to part (b) shows that we can solve for b globally.

16. (a) The fact that utility is an increasing function of m tells us that

$$\frac{\partial f}{\partial m} > 0.$$

The fact that $f_m > 0$ means that utility, or satisfaction, increases with budget. In other words, the more money you have, the greater your satisfaction.

(b) Consider

$$F(p_1, p_2, m, u) = f(p_1, p_2, m) - u = 0$$

since

$$\frac{\partial F}{\partial m} = \frac{\partial f}{\partial m} > 0$$

we can apply the implicit function theorem and solve for m, getting

$$m = g(p_1, p_2, u).$$

Then $F(p_1, p_2, g(p_1, p_2, u)) = 0$.

(c) The function g gives the budget – the amount of money which must be spent to give a particular level of satisfaction. Thus g is the expenditure necessary to achieve a certain utility.

Solutions for Section 16.5

1. The north-south distance between Alexandria and Syene corresponds to 1/50 of a full circle and measures 5,000 stadia. Thus if C is the circumference of the earth, we have

$$\frac{5000}{C} = \frac{1}{50}$$

giving us

$$C = 250,000 \text{ stadia.}$$

Using the fact that 1 stadium = 185 meters, we get

$$C = 250,000 \text{ stadia} \cdot 185 \text{ meters/stadia} = 46,250,000 \text{ meters} \approx 46,000 \text{ km.}$$

(In fact, the circumference is about 40,000 km.)

2. Since the acceleration at a point on the edge of a rotating body (centripetal acceleration) is $a = v^2/r$ (where v is the constant speed and r is the radius of the motion), we need to calculate the speed of a point on the equator. The point moves through a distance of $2\pi \cdot (4000) = 8000\pi$ miles in 24 hours, so the speed is given by

$$v = \frac{8000\pi}{24} \text{ mph.}$$

Thus, using $r = 4000$ miles, we have

$$a = \left(\frac{8000\pi}{24}\right)^2 \cdot \frac{1}{4000} = \frac{16000\pi^2}{(24)^2} \text{ mi/hr}^2.$$

Using the facts that 1 mile = 5280 feet and 1 hour = 3600 sec, we convert to ft/sec²:

$$a = \frac{16000\pi^2}{(24)^2} \cdot \frac{5280}{(3600)^2} = 0.112 \text{ ft/sec}^2.$$

   This is about 1/300 the acceleration due to gravity. To get this equivalent acceleration on a merry-go-round of radius 25 feet, the speed of the point on the merry-go-round satisfies $0.112 = v^2/25$, so $v = 1.67$ ft/s. The period of such a merry-go-round is $(2\pi \cdot 25)/1.67$ or about 94 seconds.

3. Since the sum of the distances to the two foci is constant, the closest point on the ellipse to one focus must be the farthest point to the other focus, and vice versa. Call the foci A and B and let P and Q be the closest and farthest points on the ellipse to A (hence farthest and closest points to B, respectively). (See Figure 16.29.) By symmetry, $\overline{PA} = \overline{QB}$ and $\overline{PB} = \overline{QA}$. By definition of mean distance,

$$d = \frac{1}{2}(\overline{PA} + \overline{QA}) = \frac{1}{2}(\overline{PA} + \overline{PB})$$

so $\overline{PA} + \overline{PB} = 2d$, i.e. the sums of the distances from P to the two foci is $2d$. Since the sum of the distances from any point on the ellipse to the two foci is constant, the value of that constant must be $2d$.

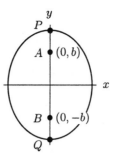

Figure 16.29

The distance from an arbitrary point (x, y) on the ellipse to the focus at $(0, b)$ is $\sqrt{x^2 + (y - b)^2}$. The distance from (x, y) to the focus at $(0, -b)$ is $\sqrt{x^2 + (y + b)^2}$. The sum of these distances is equal to $2d$, so we have that

$$\sqrt{x^2 + (y - b)^2} + \sqrt{x^2 + (y + b)^2} = 2d$$
$$\sqrt{x^2 + (y + b)^2} = 2d - \sqrt{x^2 + (y - b)^2}$$

Squaring both sides, we have

$$x^2 + y^2 + 2by + b^2 = 4d^2 - 4d\sqrt{x^2 + (y - b)^2} + x^2 + y^2 - 2by + b^2$$
$$4by - 4d^2 = -4d\sqrt{x^2 + (y - b)^2}$$
$$by - d^2 = -d\sqrt{x^2 + (y - b)^2}$$

Squaring again,

$$d^4 - 2d^2by + b^2y^2 = d^2x^2 + d^2y^2 - 2d^2by + d^2b^2$$
$$y^2(d^2 - b^2) + x^2d^2 = d^2(d^2 - b^2)$$

Letting $c^2 = d^2 - b^2$, we have that

$$y^2c^2 + x^2d^2 = d^2c^2$$

and finally,

$$\frac{x^2}{c^2} + \frac{y^2}{d^2} = 1.$$

4. If the particle comes no closer to the origin than the point $(a, 0)$, at this point the trajectory must be tangential to the circle of radius a, centered at the origin, $(0, 0)$. Thus the velocity must be perpendicular to the x-axis at the point $(a, 0)$.

Saying

$$\frac{d^2x}{dt^2} = -kx, \quad \frac{d^2y}{dt^2} = -ky, \ k > 0$$

restates the condition that the acceleration vector always points in the direction of the origin (and is proportional to the distance from the origin), so these differential equations specify the motion of the particle correctly. (Recognize that the acceleration in the direction of x is simply some negative constant times x, and that the acceleration in the direction of y is simply that same constant times y).

The general solution to the differential equation $d^2x/dt^2 = -kx, \ k > 0$, is

$$x = A\cos(\sqrt{k}t) + B\sin(\sqrt{k}t).$$

The initial conditions for $t = 0$ are $x = a$, and $dx/dt = 0$, as the velocity is perpendicular to the x-axis. So for $t = 0$, we have

$$a = A$$

$$0 = \frac{dx}{dt}\bigg|_{t=0} = \left[-A\sqrt{k}\sin\left(\sqrt{k}t\right) + B\sqrt{k}\cos\left(\sqrt{k}t\right)\right]_{t=0} = B\sqrt{k}$$

Hence $B = 0$. So

$$x = a\cos\left(\sqrt{k}t\right).$$

Similarly, we take $d^2y/dt^2 = -ky, \ k > 0$ with initial conditions: for $t = 0, y = 0$, and $dy/dt = c$. We have

$$y = A_1\cos(\sqrt{k}t) + B_1\sin(\sqrt{k}t).$$

Hence, for $t = 0$, we have $0 = A_1$ and

$$c = \frac{dy}{dt}\bigg|_{t=0} = \left[-A_1\sqrt{k}\sin\left(\sqrt{k}t\right) + B_1\sqrt{k}\cos\left(\sqrt{k}t\right)\right]_{t=0} = B_1\sqrt{k}$$

Hence $B_1 = \dfrac{c}{\sqrt{k}}$, so $y = \dfrac{c}{\sqrt{k}}\sin(\sqrt{k}t)$. Since $b = \dfrac{c}{\sqrt{k}}$, we get

$$y = b\sin\left(\sqrt{k}t\right).$$

5. Differentiating the equations given, we see that

$$\frac{d^2x}{dt^2} = \frac{du}{dt} = \frac{-kx}{x^2 + y^2}.$$

Similarly

$$\frac{d^2y}{dt^2} = \frac{dv}{dt} = \frac{-ky}{x^2 + y^2}.$$

The acceleration vector $\vec{a}$ is given by

$$\vec{a} = \frac{d^2x}{dt^2}\vec{i} + \frac{d^2y}{dt^2}\vec{j} = \frac{-kx}{x^2 + y^2}\vec{i} - \frac{ky}{x^2 + y^2}\vec{j} = \frac{-k}{x^2 + y^2}(x\vec{i} + y\vec{j}).$$

Thus,

$$\vec{a} = -\frac{k\vec{r}}{r^2}.$$

So $\vec{a}$ points toward the origin and $\|\vec{a}\| = k/r$.

Orbits are not always closed for solutions to differential equations representing a general centripetal acceleration.

6. We are given that

$$x = \frac{c}{2}(e^{kt} - e^{-kt}) \quad \text{and} \quad y = \frac{d}{2}(e^{kt} + e^{-kt}).$$

Then

$$x^2 = \frac{c^2}{4}(e^{2kt} - 2 + e^{-2kt}) \quad \text{and} \quad y^2 = \frac{d^2}{4}(e^{2kt} + 2 + e^{-2kt}).$$

Thus

$$-\frac{x^2}{c^2} + \frac{y^2}{d^2} = \frac{1}{4}[-(e^{2kt} - 2 + e^{-2kt}) + (e^{2kt} + 2 + e^{-2kt})] = 1.$$

Therefore the parameterization of x, y satisfies $-\dfrac{x^2}{c^2} + \dfrac{y^2}{d^2} = 1$.

Now we differentiate, obtaining

$$\frac{dx}{dt} = \frac{ck}{2}(e^{kt} + e^{-kt})$$

$$\frac{d^2x}{dt^2} = \frac{ck^2}{2}(e^{kt} - e^{-kt}) = k^2x.$$

Similarly, we have

$$\frac{dy}{dt} = \frac{d}{2}k(e^{kt} - e^{-kt})$$

$$\frac{d^2y}{dt^2} = \frac{d}{2}k^2(e^{kt} + e^{-kt}) = k^2y.$$

Thus we have that

$$\frac{d^2x}{dt^2} = k^2x \quad \text{and} \quad \frac{d^2y}{dt^2} = k^2y.$$

Solutions for Chapter 16 Review

1. $x = t, y = 5.$
2. The parameterization $x\vec{i} + y\vec{j} = 2\cos t\vec{i} + 2\sin t\vec{j}$ has the right radius but starts at the point $(2, 0)$. To start at $(0, 2)$, we need $x\vec{i} + y\vec{j} = 2\cos(t + \frac{\pi}{2})\vec{i} + 2\sin(t + \frac{\pi}{2})\vec{j} = 2\sin t\vec{i} + 2\cos t\vec{j}$.
3. The parameterization $x\vec{i} + y\vec{j} = (4 + 4\cos t)\vec{i} + (4 + 4\sin t)\vec{j}$ gives the correct circle, but starts at $(8, 4)$. To start on the x-axis we need

$$x\vec{i} + y\vec{j} = (4 + 4\cos(t - \frac{\pi}{2}))\vec{i} + (4 + 4\sin(t - \frac{\pi}{2}))\vec{j} = (4 + 4\sin t)\vec{i} + (4 - 4\cos t)\vec{j}.$$

4. The parametric equation of a circle is

$$x = \cos t, y = \sin t.$$

When $t = 0$, $x = 1, y = 0$, and when $t = \frac{\pi}{2}$, $x = 0, y = 1$. This shows a counterclockwise movement, so our original equation is correct.

5. The vector $(\vec{i} + 2\vec{j} + 5\vec{k}) - (2\vec{i} - \vec{j} + 4\vec{k}) = -\vec{i} + 3\vec{j} + \vec{k}$ is parallel to the line, so a possible parameterization is

$$x = 2 - t, \quad y = -1 + 3t, \quad z = 4 + t.$$

6. A line perpendicular to the xz-plane will have x = constant, z = constant, y = anything: This is given by $x = 1, y = t, z = 2$.

7. Since the vector $\vec{n} = \mathrm{grad}(2x - 3y + 5z) = 2\vec{i} - 3\vec{j} + 5\vec{k}$ is perpendicular to the plane, this vector is parallel to the line. Thus the equation of the line is

$$x = 1 + 2t, \quad y = 1 - 3t, \quad z = 1 + 5t.$$

8. Since the radius is 2, the circle must be of the form $x = 2\cos t, y = 2\sin t, z = 1$. But this parameterization traces out the circle clockwise when viewed from below. Therefore, the parameterization we want is $x = 2\cos t, y = -2\sin t, z = 1$.

9. Since the circle has radius 3, the equation must be of the form $x = 3\cos t, y = 5, z = 3\sin t$. But since the circle is being viewed from farther out on the y-axis, the circle we have now would be seen going clockwise. To correct this, we add a negative to the third component, giving us the equation $x = 3\cos t, y = 5, z = -3\sin t$.

10. We take two unit vectors, $\vec{u}$ and $\vec{v}$, parallel to the plane, and perpendicular to each other, say $\vec{u} = \vec{j}$ and $\vec{v} = \frac{\vec{i}}{\sqrt{2}} - \frac{\vec{k}}{\sqrt{2}}$. Then,

$$\begin{aligned}
\vec{r} &= \vec{j} + 2\cos\theta\vec{u} + 2\sin\theta\vec{v} \\
&= \vec{j} + 2\cos\theta\vec{j} + 2\sin\theta\left(\frac{\vec{i}}{\sqrt{2}} - \frac{\vec{k}}{\sqrt{2}}\right) \\
&= \sqrt{2}\sin\theta\vec{i} + (1 + 2\cos\theta)\vec{j} - \sqrt{2}\sin\theta\vec{k}, \quad \text{for } 0 \le \theta \le 2\pi,
\end{aligned}$$

or

$$\begin{aligned}
x &= \sqrt{2}\sin\theta \\
y &= (1 + 2\cos\theta) \\
z &= -\sqrt{2}\sin\theta \quad 0 \le \theta \le 2\pi
\end{aligned}$$

11. (a) (I)   has radius 1 and traces out a complete circle, so $I = C_4$.
    (II)   has radius 2 and traces out the top half of a circle, so $II = C_1$
    (III)   has radius 1 and traces out a quarter circle, so $III = C_2$.
    (IV)   has radius 2 and traces out the bottom half of a circle, so $IV = C_6$.
    (b) C_3 has radius $1/2$ and traces out a half circle below the x-axis, so

$$\vec{r} = 0.5\cos t\vec{i} - 0.5\sin t\vec{j}.$$

   C_5 has radius 2 and traces out a quarter circle below the x-axis starting at the point $(-2, 0)$. Thus we have

$$\vec{r} = -2\cos(t/2)\vec{i} - 2\sin(t/2)\vec{j}.$$

12. All of the points lie on the unit circle. (You can check this since $x^2 + y^2 = 1$.) The problem is that there is no value of t that gives the point $x = 0$, $y = 1$. This is because

$$y = \frac{t^2 - 1}{t^2 + 1} = 1$$

has no real solution. Only when t approaches positive or negative infinity does the point get close to $(0, 1)$. Technically, it is not a circle.

13. (a) $f_x = \dfrac{[2x(x^2 + y^2) - 2x(x^2 - y^2)]}{(x^2 + y^2)^2} = \dfrac{4xy^2}{(x^2 + y^2)^2}.$

$f_y = \dfrac{[-2y(x^2 + y^2) - 2y(x^2 - y^2)]}{(x^2 + y^2)^2} = \dfrac{-4yx^2}{(x^2 + y^2)^2}.$

$\nabla f(1, 1) = \vec{i} - \vec{j}$, i.e., south-east.

(b) We need a vector $\vec{u}$ such that $\nabla f(1, 1) \cdot \vec{u} = 0$, i.e., such that $(\vec{i} - \vec{j}) \cdot \vec{u} = 0$. The vector $\vec{u} = \vec{i} + \vec{j}$ clearly works; so does $\vec{u} = -\vec{i} - \vec{j}$. Dividing by the length to get a unit vector, we have $\vec{u} = \frac{1}{\sqrt{2}}\vec{i} + \frac{1}{\sqrt{2}}\vec{j}$ or $\vec{u} = -\frac{1}{\sqrt{2}}\vec{i} - \frac{1}{\sqrt{2}}\vec{j}$.

(c) f is a function of x and y, which in turn are functions of t. Thus, the chain rule can be used to show how f changed with t.

$$\frac{df}{dt} = \frac{\partial f}{\partial x} \cdot \frac{dx}{dt} + \frac{\partial f}{\partial y} \cdot \frac{dy}{dt} = \frac{4xy^2}{(x^2 + y^2)^2} \cdot 2e^{2t} - \frac{4x^2y}{(x^2 + y^2)^2} \cdot (6t^2 + 6).$$

At $t = 0$, $x = 1$, $y = 1$; so, $\dfrac{df}{dt} = \dfrac{4}{4} \cdot 2 - \dfrac{4}{4} \cdot 6 = -4.$

14. We can find this equation in two ways. First we could find two points on the line of intersection and then proceed as in Example 2 on page 285. To find two points just substitute two different values for z and solve for x and y for each value of z. Alternatively, assuming the line isn't horizontal (which it turns out not to be), we could take z to be the parameter t, so $z = t$. To find x and y as functions of t we solve the two equations for x and y in terms of t. We have

$$t = 4 + 2x + 5y$$
$$t = 3 + x + 3y.$$

Eliminating x we get

$$-t = -2 - y \quad \text{and} \quad y = -2 + t.$$

Substituting $-2 + t$ for y in the second equation and solving for x, we get

$$x = 3 - 2t.$$

Our equations are therefore

$$x = 3 - 2t, \; y = -2 + t, \; z = t.$$

or

$$\vec{r} = x\vec{i} + y\vec{j} + z\vec{k} = 3\vec{i} - 2\vec{j} + t(-2\vec{i} + \vec{j} + \vec{k}).$$

15. The displacement from the point $(1, 2, 3)$ to the point $(3, 5, 7)$ is $3\vec{i} + 5\vec{j} + 7\vec{k} - (\vec{i} + 2\vec{j} + 3\vec{k}) = 2\vec{i} + 3\vec{j} + 4\vec{k}$. So the equation of the line is

$$x\vec{i} + y\vec{j} + z\vec{k} = 1\vec{i} + 2\vec{j} + 3\vec{k} + t(2\vec{i} + 3\vec{j} + 4\vec{k})$$

or
$$x\vec{i} + y\vec{j} + z\vec{k} = (1 + 2t)\vec{i} + (2 + 3t)\vec{j} + (3 + 4t)\vec{k}.$$

The square of the distance from a point (x, y, z) on the line to the origin, denoted by $D(t)$ is

$$
\begin{aligned}
D(t) &= (x - 0)^2 + (y - 0)^2 + (z - 0)^2 \\
&= (1 + 2t)^2 + (2 + 3t)^2 + (3 + 4t)^2 \\
&= 1 + 4t + 4t^2 + 4 + 12t + 9t^2 + 9 + 24t + 16t^2 \\
&= 14 + 40t + 29t^2 \\
&= 29\left(t^2 + \frac{40}{29}t + \frac{14}{29}\right) \\
&= 29\left(\left(t + \frac{20}{29}\right)^2 - \left(\frac{20}{29}\right)^2 + \frac{14}{29}\right).
\end{aligned}
$$

Clearly, $D(t)$ is minimum when $t = -20/29$, and

$$
\begin{aligned}
D(-20/29) &= 29\left(-\left(\frac{20}{29}\right)^2 + \frac{14}{29}\right) \\
&= \frac{6}{29}.
\end{aligned}
$$

So the shortest distance is $\sqrt{\frac{6}{29}} = \frac{\sqrt{174}}{29}$.

16. No. The first is parallel to the vector $2\vec{i} - \vec{j} + 3\vec{k}$ and the second is parallel to $\vec{i} + 2\vec{j} + 2\vec{k}$.

17. The plot looks like Figure 16.30.

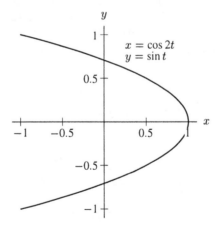

Figure 16.30

which does appear to be part of a parabola. To prove that it is, we note that we have

$$x = \cos 2t$$

$$y = \sin t$$

and must somehow find a relationship between x and y. Recall the trig identity

$$\cos 2t = 1 - 2\sin^2 t.$$

Thus we have $x = 1 - 2y^2$, which is a parabola lying along the x-axis, for $-1 \le y \le 1$.

18. (a) Since P moves in a circle we have

$$x = 10\cos t$$
$$y = 10\sin t.$$

This completes a revolution in time 2π.

(b) First, consider the planet as stationary at (x_0, y_0). Then the equations for M are

$$x = x_0 + 3\cos 8t$$
$$y = y_0 + 3\sin 8t.$$

The factor of 8 is inserted because for every $2\pi/8$ units of time, $8t$ covers 2π, which is one orbit. But since P moves, we must replace (x_0, y_0) by the position of P. So we have

$$x = 10\cos t + 3\cos 8t$$
$$y = 10\sin t + 3\sin 8t.$$

(c)

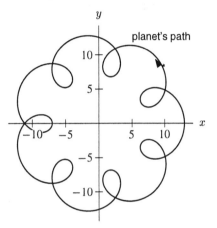

Figure 16.31

(d) What happens if the moon doesn't do an integer number of revolutions in time 2π?

19. (a) To find where the particle is at time equal to 0, we simply substitute 0 in for all t in the equation. Therefore, the particle is at the point with position vector

$$\vec{r}(0) = [2 + 5(0)]\,\vec{i} + (3 + 0)\vec{j} + 2(0)\vec{k}$$
$$= 2\vec{i} + 3\vec{j} + 0\vec{k}.$$

Thus, the particle is at the point $(2, 3, 0)$.

(b) To find the time at which the particle is at the point $(12, 5, 4)$, we solve for t for each component, and the t should be the same, if the curve goes through this point. For the x-component, we get

$$2 + 5t = 12$$
$$t = 2.$$

For the y-component, we get

$$3 + t = 5$$
$$t = 2.$$

And for the z-component, we get

$$2t = 4$$
$$t = 2.$$

Therefore, at $t = 2$ the particle reaches $(12, 5, 4)$.

(c)   The particle never reaches $(12, 4, 4)$, because the equation

$$\vec{r} = (2 + 5t)\vec{i} + (3 + t)\vec{j} + 2t\vec{k} = 12\vec{i} + 4\vec{j} + 4\vec{k}$$

has no solution. Thus, the point does not lie on the line.

20.   (a)   Separate the ant's path into three parts: from $(0, 0)$ to $(1, 0)$ along the x-axis; from $(1, 0)$ to $(0, 1)$ via the circle; and from $(0, 1)$ to $(0, 0)$ along the y-axis. (See Figure 16.32.) The lengths of the paths are 1, $\frac{2\pi}{4} = \frac{\pi}{2}$, and 1 respectively. Thus, the time it takes for the ant to travel the three paths are (using the formula $t = \frac{d}{v}$) $\frac{1}{2}$, $\frac{1}{3}$, and $\frac{1}{2}$ seconds.

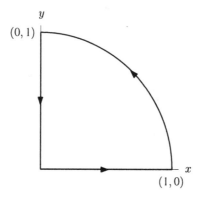

Figure 16.32

From $t = 0$ to $t = \frac{1}{2}$, the ant is heading toward $(1, 0)$ so its coordinate is $(2t, 0)$. From $t = \frac{1}{2}$ to $t = \frac{1}{2} + \frac{1}{3} = \frac{5}{6}$, the ant is veering to the left and heading toward $(0, 1)$. At $t = \frac{1}{2}$, it is at $(1, 0)$ and at $t = \frac{5}{6}$, it is at $(0, 1)$. Thus its position is $(\cos[\frac{3\pi}{2}(t - \frac{1}{2})], \sin[\frac{3\pi}{2}(t - \frac{1}{2})])$. Finally, from $t = \frac{5}{6}$ to $t = \frac{5}{6} + \frac{1}{2} = \frac{4}{3}$, the ant is headed home. Its coordinates are $(0, -2(t - \frac{4}{3}))$.

In summary, the function expressing the ant's coordinates is

$$(x(t), y(t)) = \begin{cases} (2t, 0) & \text{when } 0 \leq t \leq \frac{1}{2} \\ (\cos(\frac{3\pi}{2}(t - \frac{1}{2})), \sin(\frac{3\pi}{2}(t - \frac{1}{2}))) & \text{when } \frac{1}{2} < t \leq \frac{5}{6} \\ (0, -2(t - \frac{4}{3})) & \text{when } \frac{5}{6} \leq t \leq \frac{4}{3}. \end{cases}$$

(b)   To do the reverse path, observe that we can reverse the ant's path by interchanging the x and y coordinates (flipping it with respect to the line $y = x$), so the function is

$$(x(t), y(t)) = \begin{cases} (0, 2t) & \text{when } 0 \leq t \leq \frac{1}{2} \\ (\sin(\frac{3\pi}{2}(t - \frac{1}{2})), \cos(\frac{3\pi}{2}(t - \frac{1}{2}))) & \text{when } \frac{1}{2} < t \leq \frac{5}{6} \\ (-2(t - \frac{4}{3}), 0) & \text{when } \frac{5}{6} < t \leq \frac{4}{3}. \end{cases}$$

21. Set up the coordinate system as shown in Figure 16.33.

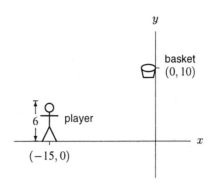

Figure 16.33

(a) We separate the initial velocity vector into its x and y components.

$$V_x = V \cos A$$

$$V_y = V \sin A.$$

Since there is no force acting in the x direction, the x-coordinate of the basketball is just

$$x = (V \cos A)t - 15.$$

For the y-coordinate, we know that
$$y''(t) = -32,$$

so
$$y'(t) = -32t + C_1$$

and
$$y(t) = -16t^2 + C_1 t + C_2.$$

We also know that $y'(0) = V \sin A$ and $y(0) = 6$. Substituting these values in, we get $C_1 = V \sin A$, $C_2 = 6$ and thus
$$y = -16t^2 + (V \sin A)t + 6.$$

(b) Many pairs of V and A work. For example, $V = 26, A = 60°, V = 32, A = 30°$.

(c) Now that we have the equations, we need to find a relationship between V and A that ensures that the basketball goes through the hoop (i.e., the curve passes through $(0, 10)$). So we set

$$x = (V \cos A)t - 15 = 0$$

$$y = -16t^2 + (V \sin A)t + 6 = 10.$$

From the first equation, we get $t = \frac{15}{V \cos A}$. Then we substitute that into the second equation:

$$-16\left(\frac{15}{V \cos A}\right)^2 + (V \sin A)(\frac{15}{V \cos A}) = 4$$

$$-\frac{3600}{V^2 \cos^2 A} + 15 \tan A = 4$$

$$V^2 = \frac{3600}{\cos^2 A(15\tan A - 4)}.$$

Keeping in mind that $\tan^2\theta + 1 = \frac{1}{\cos^2\theta}$, one has:

$$V^2 = \frac{3600(1 + \tan^2 A)}{15\tan A - 4}.$$

We can minimize V by minimizing V^2 (since $V > 0$).

$$\frac{d(V^2)}{dA} = \frac{2\tan A(15\tan A - 4) - 15(\tan^2 A + 1)}{(15\tan A - 4)^2} \cdot \frac{3600}{\cos^2 A} = 0$$

$$\frac{3600}{\cos^2 A}\left[\frac{15\tan^2 A - 8\tan A - 15}{(15\tan A - 4)^2}\right] = 0$$

$$15\tan^2 A - 8\tan A - 15 = 0$$

$$\tan A = \frac{8 + \sqrt{964}}{30}$$

$$\approx 1.30$$

$$A \approx 52°.$$

22. (a) Suppose axes are fixed with the origin on the ground below the point at which the baton is thrown. There is no horizontal acceleration, so if x measures horizontal displacement of the center in meters

$$\frac{d^2x}{dt^2} = 0.$$

Since the initial velocity is 8 m/sec, integrating gives:

$$\frac{dx}{dt} = 8,$$

and since $x = 0$ when $t = 0$,

$$x = 8t.$$

The vertical acceleration is due to gravity. So, if y is vertical displacement of the center in meters:

$$\frac{d^2y}{dt^2} = -g.$$

So

$$\frac{dy}{dt} = -gt + 10,$$

and

$$y = -\frac{gt^2}{2} + 10t + 1.5.$$

Thus, the parametric equations for the center of the baton are

$$x = 8t, \quad y = -\frac{gt^2}{2} + 10t + 1.5.$$

(b)   We put a new origin at the center of the baton. Suppose (h, k) are the coordinates of the end of the baton relative to the center. Since the radius of the circular motion is 0.2 m and the angular velocity is $2(2\pi) = 4\pi$ radians/sec and since $x = 0.2$ and $y = 0$ when $t = 0$, we have

$$h = 0.2\cos(4\pi t) \quad k = 0.2\sin(4\pi t).$$

(c)   To find the coordinates of the end of the baton, we add the results from parts (a) and (b), so if x and y represent the position of the end of the baton relative to the ground, we have

$$x = 8t + 0.2\cos(4\pi t) \quad y = -\frac{gt^2}{2} + 10t + 1.5 + 0.2\sin(4\pi t).$$

(d)   To sketch this, use $g = 9.8$ meters/sec ^{2}.

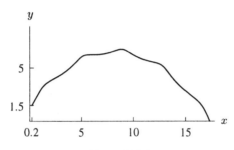

Figure 16.34

23.   (a)   The center of the wheel moves horizontally, so its y-coordinate will never change; it will equal 1 at all times. In one second, the wheel rotates 1 radian, which corresponds to 1 meter on the rim of a wheel of radius 1 meter, and so the rolling wheel advances at a rate of 1 meter/sec. Thus the x-coordinate of the center, which equals 0 at $t = 0$, will equal t at time t. At time t the center will be at the point $(x, y) = (t, 1)$.

(b)   By time t the spot on the rim will have rotated t radians clockwise, putting it at angle $-t$ as in Figure 16.35. The coordinates of the spot with respect to the center of the wheel are $(\cos(-t), \sin(-t))$. Adding these to the coordinates $(t, 1)$ of the center gives the location of the spot as $(x, y) = (t + \cos t, 1 - \sin t)$. See Figure 16.36.

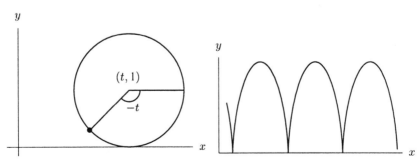

Figure 16.35   *Figure 16.36*

24. (a) The equation of the line is $y = t(x + 2)$. Substitution into the equation of the ellipse gives

$$2x^2 + 3t^2(x + 2)^2 = 8$$

which simplifies to the quadratic equation in x

$$(2 + 3t^2)x^2 + 12t^2x + 12t^2 - 8 = 0$$

There are two solutions $x = x_1$ and $x = x_2$ to this equation, namely the x-coordinates of the two points $P = (-2, 0)$ and $Q = (x_2, y_2)$ that lie on the intersection of the line and the ellipse. The first solution is $x = x_1 = -2$, and the simplest way to find the second is to remark that $x_1 x_2 = (12t^2 - 8)/(2 + 3t^2)$. Therefore

$$x_2 = \frac{4 - 6t^2}{2 + 3t^2}$$

and so

$$y_2 = t(x_2 + 2) = \frac{8t}{2 + 3t^2}$$

Thus the desired point is

$$Q = (x_2, y_2) = \left(\frac{4 - 6t^2}{2 + 3t^2}, \frac{8t}{2 + 3t^2} \right)$$

(b) From (a),

$$x = \frac{4 - 6t^2}{2 + 3t^2}, \quad y = \frac{8t}{2 + 3t^2}$$

will parameterize the ellipse. One reason that this parameterization is interesting is that it allows you to find easily points on the ellipse whose x- and y-coordinates are both rational numbers, simply by choosing rational values for t. For example, $t = 1$ corresponds to the point $(-2/5, 8/5)$ and $t = 1/2$ corresponds to the point $(10/11, 16/11)$.

25. To parameterize a plane we must find one point in the plane and two nonparallel vectors in the plane. Since there is an infinity of possibilities, there are many possible parameterizations.

Points (x, y, z) in the plane can be found by picking values for two coordinates and using the equation of the plane to solve for the third. For example, with $x = 2$ and $y = 3$ we have $3(2) + 4(3) + 5z = 10$, and so $z = -1.6$. Thus the point $P = (2, 3, -1.6)$ is in the plane.

Vectors in the plane can be found as displacement vectors between two points in the plane. To get two displacement vectors, we need three points in the plane. If $x = y = 0$, then $z = 2$, giving the point $Q = (0, 0, 2)$ in the plane. And if $x = z = 0$, the $y = 2.5$, giving the point $R = (0, 2.5, 0)$ in the plane. The vectors $\vec{v}_1 = \overrightarrow{PQ} = -2\vec{i} - 3\vec{j} + 3.6\vec{k}$ and $\vec{v}_2 = \overrightarrow{PR} = -2\vec{i} - 0,5\vec{j} + 1.6\vec{k}$ lie in the plane. Hence one parameterization of the plane is

$$\vec{r} = 2\vec{i} + 3\vec{j} - 1.6\vec{k} + s\vec{v}_1 + t\vec{v}_2$$
$$= (2 - 2s - 2t)\vec{i} + (3 - 3s - 0.5t)\vec{j} + (-1.6 + 3.6s + 1.6t)\vec{k}$$

or, equivalently

$$x = 2 - 2s - 2t$$
$$y = 3 - 3s - 0.5t$$
$$z = -1.6 + 3.6s + 1.6t.$$

Notice how the parametric equations satisfy the equation of the plane:

$$3(2 - 2s - 2t) + 4(3 - 3s - 0.5t) + 5(-1.6 + 3.6s + 1.6t) = 10.$$

26. (a) Reading the coefficients of s and t in the parametric equations shows that the two vectors $-5\vec{i} + \vec{j} + \vec{k}$ and $2\vec{i} + 3\vec{j} - \vec{k}$ are in the plane. Their cross product $(-5\vec{i} + \vec{j} + \vec{k}) \times (2\vec{i} + 3\vec{j} - \vec{k}) = 4\vec{i} - 3\vec{j} - 17\vec{k}$ is a normal vector.

(b) Since the plane contains the point $(3, 1, 0)$ (corresponding to $s = t = 0$) and is normal to the vector $-4\vec{i} - 3\vec{j} - 17\vec{k}$, an equation for the plane is

$$-4(x - 3) - 3(y - 1) - 17(z - 0) = 0$$

or equivalently

$$4x + 3y + 17z = 15.$$

27. The plane through the point with position vector $\vec{r}_0$ and containing two non-parallel vectors $\vec{v}_1$ and $\vec{v}_2$ has a parameterization given by

$$\vec{r}(s, t) = \vec{r}_0 + s\vec{v}_1 + t\vec{v}_2,$$

so

$$\vec{r}(0, 0) = \vec{r}_0, \quad \vec{r}(1, 0) = \vec{r}_0 + \vec{v}_1, \quad \text{and} \quad \vec{r}(0, 1) = \vec{r}_0 + \vec{v}_2.$$

See Figure 16.37.

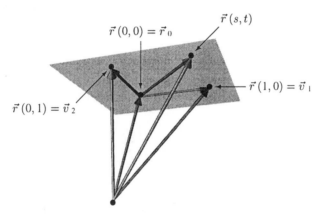

Figure 16.37

My position vector is $-5\vec{i} + 5\vec{k}$, Jane's position vector is $-5\vec{j} - 5\vec{k}$, and Jo's position vector is $10\vec{i} + 5\vec{j}$. For the first parameterization, we want

$$\vec{r}_0 = -5\vec{i} + 5\vec{k} \quad \text{and} \quad \vec{r}(1, 0) = \vec{r}_0 + \vec{v}_1 = 10\vec{i} + 5\vec{j} \quad \text{and} \quad \vec{r}(0, 1) = \vec{r}_0 + \vec{v}_2 = -5\vec{j} - 5\vec{k}.$$

Thus, we choose

$$\vec{r}_0 = -5\vec{i} + 5\vec{k} \quad \text{and} \quad \vec{v}_1 = 15\vec{i} + 5\vec{j} - 5\vec{k} \quad \text{and} \quad \vec{v}_2 = 5\vec{i} - 5\vec{j} - 10\vec{k}.$$

For the second parameterization, we switch $\vec{v}_1$ and $\vec{v}_2$ and so we choose

$$\vec{r}_0 = -5\vec{i} + 5\vec{k} \quad \text{and} \quad \vec{v}_1 = 5\vec{i} - 5\vec{j} - 10\vec{k} \quad \text{and} \quad \vec{v}_2 = 15\vec{i} + 5\vec{j} - 5\vec{k}.$$

For the third parameterization we want

$$\vec{r}_0 = \vec{r}_0 + \vec{v}_2 = -5\vec{j} - 5\vec{k} \quad \text{and} \quad \vec{r}(1, 0) = \vec{r}_0 + \vec{v}_1 = 10\vec{i} + 5\vec{j} \quad \text{and} \quad \vec{r}(0, 1) = \vec{r}_0 + \vec{v}_2 = -5\vec{i} + 5\vec{k}$$

so we choose

$$\vec{r}_0 = -5\vec{j} - 5\vec{k} \quad \text{and} \quad \vec{v}_1 = 10\vec{i} + 10\vec{j} + 5\vec{k} \quad \text{and} \quad \vec{v}_2 = -5\vec{i} + 5\vec{j} + 10\vec{k}.$$

28.   The sphere $x^2 + y^2 + z^2 = 1$ is shown in Figure 16.38.

   (a)   The origin corresponds to the south pole.
   (b)   The circle $x^2 + y^2 = 1$ corresponds to the equator.
   (c)   We get all the points of the sphere by this parameterization except the north pole itself.
   (d) $x^2 + y^2 > 1$ corresponds to the upper hemisphere.
   (e) $x^2 + y^2 < 1$ corresponds to the lower hemisphere.

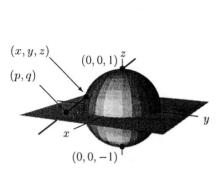

Figure 16.38

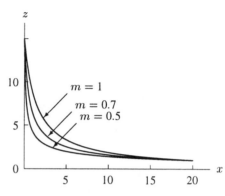

Figure 16.39

29.   We are given the conditions $f(0) = 15$ and $f(20) = 1$ and asked to solve for a and b. The equations are

$$b = 15a^m \quad b = (20 + a)^m$$

from which

$$15^{1/m}a = 20 + a$$

and so

$$a = \frac{20}{(15^{1/m} - 1)}.$$

The values for a and b corresponding to $m = 0.5, 0.7$, and 1 are given in Table 16.4.

TABLE 16.4

m	a	b
0.5	0.0893	4.48
0.7	0.427	8.26
1.0	1.43	21.4

Figure 16.39 shows the graphs of f when $m = 0.5, 0.7$, and 1. Notice that the surface obtained by rotating the graph of $z = f(x)$ in the xz-plane, for $x > 0$, about the x-axis will be more "flared" when $m = 0.5$ that it is for $m = 1$.

30.   We use the distance x from the open end of the horn for one parameter, and an angle θ on the circular cross section for the other, getting

$$x = x,$$
$$y = (4.48/(p + 0.0893)^{0.5}) \cos \theta,$$
$$z = (4.48/(p + 0.0893)^{0.5}) \sin \theta,$$

where $0 \le x \le 20$ and $0 \le \theta \le 2\pi$.

31. When a increases, the absolute value of the x-coordinate will increase while the y- and z-coordinates remain the same. So the surface will spread out and open wider.

When b increases, the absolute value of the y-coordinate will increase while the x- and z-coordinates remain invariant. So the surface will spread out.

When c increases, the absolute value of the z-coordinate will increase while the x- and y-coordinates remain invariant. So the surface will be compressed to a thinner shape.

If c is negative, for example if $c = -1$, the surface will be flipped upside down.

32. (a) The surface obtained by rotating the curve $x^2 z = 1$, for $x > 0$, in the xz-plane about the x-axis is shown in Figure 16.40.

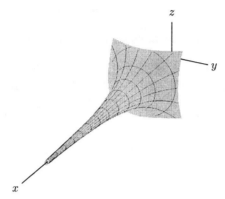

Figure 16.40

For fixed x, the cross sections of the surface parallel to the yz-plane are circles of radius $1/x^2$. so, one parameterization of the surface is given by

$$x = x, \quad y = (1/x^2)\cos\theta, \quad z = (1/x^2)\sin\theta,$$

for $x > 0, 0 \le \theta \le 2\pi$.

(b) The surface obtained by ratating the curve $x^2 z = 1$, for $z > 0$, in the xz-plane about the z-axis is shown in Figure 16.41.

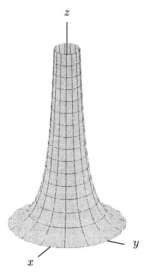

Figure 16.41

For fixed z, the cross sections of the surface parallel to the xy-plane are circles of radius $1/\sqrt{z}$. Thus, a parameterization of the surface is given by

$$x = (1/\sqrt{z})\cos\theta, \quad y = (1/\sqrt{z})\sin\theta, \quad z = z,$$

for $z > 0, 0 \le \theta \le 2\pi$.

33. Suppose we choose a parameterization in which the particle is at the point (a, b, c) when $t = 0$ and at the point (A, B, C) when $t = 1$. Then the x-coordinate changes by $(A - a)$ in unit time, the y-coordinate changes by $(B - b)$, and the z-coordinate by $(C - c)$. Thus, parametric equations for the line of sight are

$$x = a + t(A - a), \quad y = b + t(B - b), \quad z = c + t(C - c).$$

We now substitute these expressions into the equation $Ax + By + Cz = D$ and solve for t:

$$A(a + t(A - a)) + B(b + t(B - b)) + C(c + t(C - c)) = D$$
$$Aa + Bb + Cc + t(A^2 + B^2 + C^2 - Aa - Bb - Cc) = D.$$

Writing $E = Aa + Bb + Cc$ and $F = A^2 + B^2 + C^2$, we get

$$t = \frac{D - E}{F - E}.$$

Then we have:

$$x = a + \frac{D - E}{F - E}(A - a)$$
$$y = b + \frac{D - E}{F - E}(B - b)$$
$$z = c + \frac{D - E}{F - E}(C - c)$$

The problem of converting the xyz-coordinates of the point of intersection into two-dimensional screen coordinates is considered in Problem 34.

34. (a) The equation of the line between the viewer (A, B, C) and the origin is

$$x = At, \quad y = Bt, \quad z = Ct.$$

Substituting these expressions into the equation for the plane gives us the solution in t for the intersection:

$$A(At) + B(Bt) + C(Ct) = D,$$
$$(A^2 + B^2 + C^2)t = D,$$
$$t = \frac{D}{A^2 + B^2 + C^2}.$$

Substituting this back into the equations for x, y and z of the line gives us the solution in terms of those coordinates $Q = (x, y, z)$ where

$$x = \frac{AD}{A^2 + B^2 + C^2}, \quad y = \frac{BD}{A^2 + B^2 + C^2}, \quad z = \frac{CD}{A^2 + B^2 + C^2}.$$

(b) Since $\vec{u}$ is parallel to the xy-plane, its z component is 0. It also must lie within the viewing screen, thus it must be perpendicular to $A\vec{i} + B\vec{j} + C\vec{k}$, so:

$$(u_x\vec{i} + u_y\vec{j}) \cdot (A\vec{i} + B\vec{j} + C\vec{k}) = 0.$$

One solution to this equation is $\vec{u} = -B\vec{i} + A\vec{j}$. This vector points in the right direction, but needs to be normalized by dividing by its length:

$$\vec{u} = \frac{-B}{\sqrt{A^2 + B^2}}\vec{i} + \frac{A}{\sqrt{A^2 + B^2}}\vec{j}.$$

(c) For $\vec{v}$ we seek a vector perpendicular to both $A\vec{i} + B\vec{j} + C\vec{k}$ and $\vec{u}$ and so we compute their cross product. We may use the unnormalized value for $\vec{u}$ since we will have to normalize the resulting vector anyway, so

vector perpendicular to $A\vec{i} + B\vec{j} + C\vec{k}$ and $\vec{u}$
$$= (-B\vec{i} + A\vec{j}) \times (A\vec{i} + B\vec{j} + C\vec{k}) = AC\vec{i} + BC\vec{j} - (A^2 + B^2)\vec{k}.$$

Normalizing this vector gives

$$\vec{v} = (\frac{-AC}{\sqrt{(AC)^2 + (BC)^2 + (A^2 + B^2)^2}}\vec{i} + \frac{-BC}{\sqrt{(AC)^2 + (BC)^2 + (A^2 + B^2)^2}}\vec{j}$$
$$+ \frac{A^2 + B^2}{\sqrt{(AC)^2 + (BC)^2 + (A^2 + B^2)^2}}\vec{k})$$

(d) From Problem 33, we have

$$x = \frac{A(D - Bb - Cc) + a(B^2 + C^2 - D)}{A^2 + B^2 + C^2 - Aa - Bb - Cc},$$
$$y = \frac{B(D - Aa - Cc) + b(A^2 + C^2 - D)}{A^2 + B^2 + C^2 - Aa - Bb - Cc},$$
$$z = \frac{C(D - Aa - Bb) + c(A^2 + B^2 - D)}{A^2 + B^2 + C^2 - Aa - Bb - Cc},$$

where we have substituted the expressions for E and F in the solution to Problem 33 into the expressions for x, y, and z.

(e) Now we calculate the u-coordinate as

$$r = \vec{u} \cdot (\vec{P} - \vec{Q})$$
$$= (\vec{u} \cdot \vec{P}) - (\vec{u} \cdot \vec{Q})$$
$$= \vec{u} \cdot \vec{P} - (\vec{u} \cdot (A\vec{i} + B\vec{j} + C\vec{k}))\frac{D}{A^2 + B^2 + C^2}$$
$$= \vec{u} \cdot \vec{P} - 0$$
$$= \frac{1}{\sqrt{A^2 + B^2}(A^2 + B^2 + C^2 - Aa - Bb - Cc)}(-AB(D - Bb - Cc)$$
$$-Ba(B^2 + C^2 - D) + AB(D - Aa - Cc) + Ab(A^2 + C^2 - D))$$
$$= \frac{A^3b + A^2Ba + AB^2b + AC^2b - ADb - B^3a + BC^2a - BDa}{\sqrt{A^2 + B^2}(A^2 + B^2 + C^2 - Aa - Bb - Cc)}.$$

Similarly, we solve for the v component:

$$
\begin{aligned}
s &= \vec{v} \cdot (\vec{P} - \vec{Q}\,) \\
&= \vec{v} \cdot \vec{P} - \vec{v} \cdot \vec{Q} \\
&= \vec{v} \cdot \vec{P} - 0 \\
&= -(A^2 C(D - Bb - Cc) + ACa(B^2 + C^2 - D) + B^2 C(D - Aa - Cc) \\
&\quad + BCb(A^2 + C^2 - D) - (A^2 + B^2)(C(D - Aa - Bb) \\
&\quad + c(A^2 + B^2 - D)))(\sqrt{(AC)^2 + (BC)^2 + (A^2 + B^2)^2}(A^2 + B^2 + C^2 - Aa - Bb - Cc)^{-1}
\end{aligned}
$$

These are the equations for finding the (u, v) coordinate on the screen lying in the plane given by $Ax + By + Cz = D$ of any point (a, b, c) as perceived by a viewer at (A, B, C).

CHAPTER SEVENTEEN

Solutions for Section 17.1

1. Notice that for a repulsive force, the vectors point outward, away from the particle at the origin, for an attractive force, the vectors point toward the particle. So we can match up the vector field with the description as follows:

   (a) IV
   (b) III
   (c) I
   (d) II

2.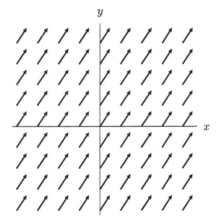

   Figure 17.1: $\vec{F}(x,y) = 2\vec{i} + 3\vec{j}$

3.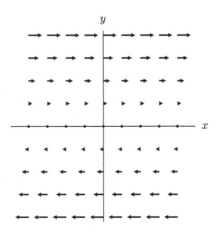

   Figure 17.2: $\vec{F}(x,y) = y\vec{i}$

4.

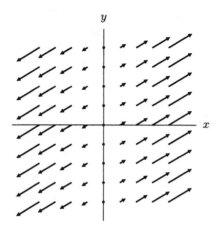

Figure 17.3: $\vec{F}(x,y) = 2x\vec{i} + x\vec{j}$

5.

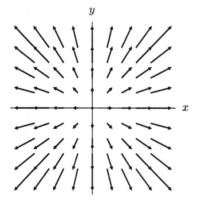

Figure 17.4: $\vec{F}(\vec{r}) = 2\vec{r}$

6.

Figure 17.5: $\vec{F}(\vec{r}) = \dfrac{\vec{r}}{\|\vec{r}\|}$

7.

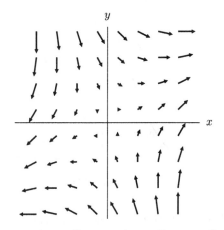

Figure 17.6: $\vec{F}(x,y) = (x+y)\vec{i} + (x-y)\vec{j}$

8. $\vec{V} = x\vec{i}$

9. $\vec{V} = -y\vec{i}$

10. $\vec{V} = x\vec{i} + y\vec{j} = \vec{r}$

11. $\vec{V} = -x\vec{i} - y\vec{j} = -\vec{r}$

12. $\vec{V} = \dfrac{\vec{r}}{\|\vec{r}\|}$: vectors are of unit length and point outwards.

13. $\vec{V} = -y\vec{i} + x\vec{j}$

14. (a) The gradient is perpendicular to the level curves:

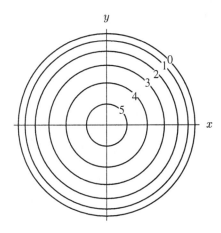

Figure 17.7: Level curves $z = f(x, y)$

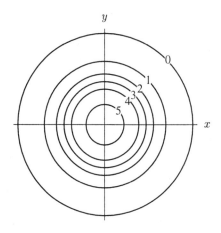

Figure 17.8: Level curves $z = g(x, y)$

A function always increases in the direction of its gradient; this is why the values on the level curves of f and g increase as we approach the origin.

(b) f climbs faster at outside, slower at center; g climbs slower at outside, faster at center:

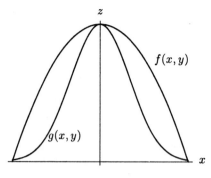

Figure 17.9

This can be understood if we notice that the magnitude of the gradient of f decreases as one approaches the origin whereas the magnitude of the gradient of g increases (at least for a while - what happens very close to the origin depends on the behavior of grad g in the region. One possibility for g is shown in Figure 17.9; the graph of g could also have a sharp peak at 0 or even blow up.)

15.

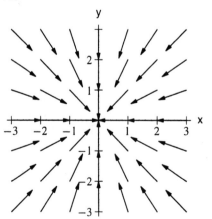

One possible solution is $\vec{F}(x, y) = x\vec{i}$.

16.

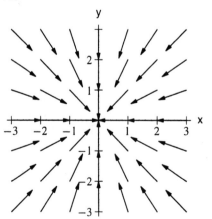

If we let $\vec{F}(x, y) = \frac{-x\vec{i} - y\vec{j}}{\sqrt{x^2 + y^2}}$, then all vectors will be of unit length and will point towards the origin.

17.

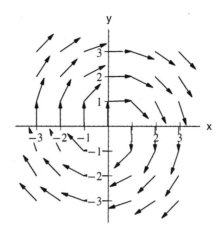

The position vector at each point is $\vec{r} = x\vec{i} + y\vec{j}$. We want to find $\vec{F}(x, y) = A\vec{i} + B\vec{j}$ such that $\vec{F} \cdot \vec{r} = Ax + By = 0$. One possible answer is let $A = y$ and $B = -x$. So $\vec{F}(x, y) = y\vec{i} - x\vec{j}$. Since the vectors are of unit length, we get $\vec{F}(x, y) = \dfrac{y\vec{i} - x\vec{j}}{\sqrt{x^2 + y^2}}$.

18. (a) Since the velocity of the water is the sum of the velocities of the individual fields, then the total field should be

$$\vec{v} = \vec{v}_{\text{stream}} + \vec{v}_{\text{fountain}}.$$

It is reasonable to represent $\vec{v}_{\text{stream}}$ by the vector field $\vec{v}_{\text{stream}} = A\vec{i}$, since $A\vec{i}$ is a constant vector field flowing in the i-direction (provided $A > 0$). It is reasonable to represent $\vec{v}_{\text{fountain}}$ by

$$\vec{v}_{\text{fountain}} = K\vec{r}_r/r^2 = K(x^2 + y^2)^{-1}(x\vec{i} + y\vec{j}),$$

since this is a vector field flowing radially outward (provided $K > 0$), with decreasing velocity as r gets larger. We would expect the velocity to decrease as the water from the fountain spreads out. Adding the two vector fields together, we get

$$\vec{v} = A\vec{i} + K(x^2 + y^2)^{-1}(x\vec{i} + y\vec{j}), \quad A > 0, K > 0.$$

(b) The constants A and K signify the strength of the individual components of the field. A is the strength of the flow of the stream alone (in fact it is the speed of the stream), and K is the strength of the fountain acting alone.

(c)

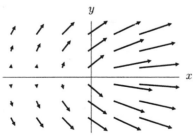

Figure 17.10: $A = 1, K = 1$

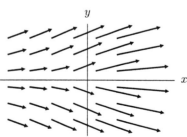

Figure 17.11: $A = 2, K = 1$

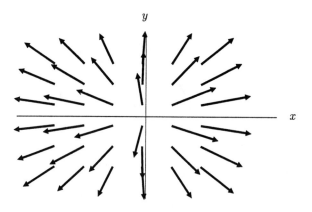

Figure 17.12: $A = 0.2, K = 2$

Solutions for Section 17.2

1. Since $x'(t) = 3$ and $y'(t) = 0$, we have $x = 3t + x_0$ and $y = y_0$. Thus, the solution curves are $y = $ constant.

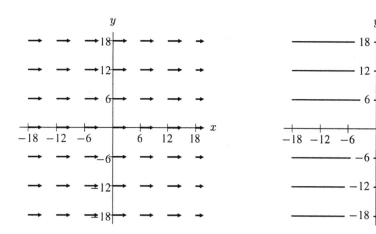

Figure 17.13: The field $\vec{v} = 3\vec{i}$ *Figure 17.14:* The flow $y =$ constant

2. Since $x'(t) = 0$ and $y'(t) = 2$, we have $x = x_0$ and $y = 2t + y_0$. Thus, the solution curves are $x = $ constant.

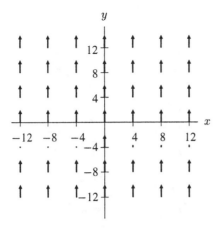

Figure 17.15: The field $\vec{v} = 2\vec{j}$

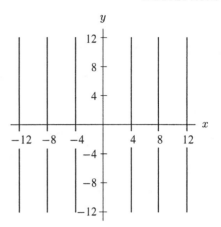

Figure 17.16: The flow $x = $ constant

3. Since $x'(t) = 3$ and $y'(t) = -2$, we have $x = 3t + x_0$ and $y = -2t + y_0$. Thus the flow lines are straight lines parallel to the vector $3\vec{i} - 2\vec{j}$. Alternatively, we have $\frac{dy}{dx} = -\frac{2}{3}$. Thus, $y = -\frac{2}{3}x + c$, where c is a constant.

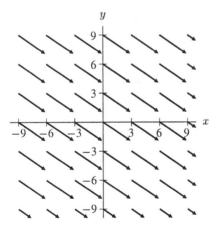

Figure 17.17: The field $\vec{v} = 3\vec{i} - 2\vec{j}$

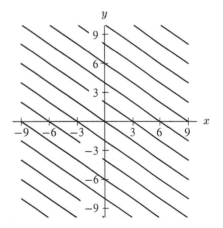

Figure 17.18: The flow $y = -\frac{2}{3}x + c$

4.

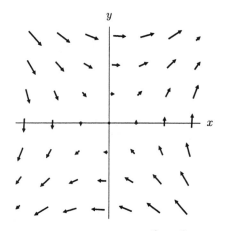

Figure 17.19: $\vec{v}(t) = y\vec{i} + x\vec{j}$

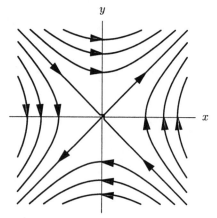

Figure 17.20: The flow
$x(t) = a(e^t + e^{-t}),\ y(t) = a(e^t - e^{-t})$

As

$$\vec{v}(t) = \frac{dx}{dt}\vec{i} + \frac{dy}{dt}\vec{j},$$

the system of differential equations is

$$\begin{cases} \frac{dx}{dt} = y \\ \frac{dy}{dt} = x. \end{cases}$$

Since

$$\frac{dx(t)}{dt} = \frac{d}{dt}[a(e^t + e^{-t})] = a(e^t - e^{-t}) = y(t)$$

and

$$\frac{dy(t)}{dt} = \frac{d}{dt}[a(e^t - e^{-t})] = a(e^t + e^{-t}) = x(t),$$

the given flow satisfies the system. By eliminating the parameter t in $x(t)$ and $y(t)$, the solution curves obtained are $x^2 - y^2 = 4a^2$.

5.

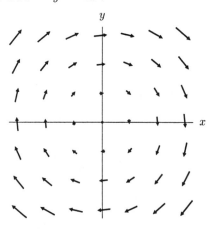

Figure 17.21: $\vec{v}(t) = y\vec{i} - x\vec{j}$

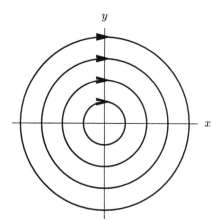

Figure 17.22: The flow $x = a\sin t$,
$y = a\cos t$

As

$$\vec{v}(t) = \frac{dx}{dt}\vec{i} + \frac{dy}{dt}\vec{j},$$

the system of differential equations is

$$\begin{cases} \frac{dx}{dt} = y \\ \frac{dy}{dt} = -x. \end{cases}$$

Since

$$\frac{dx(t)}{dt} = \frac{d}{dt}[a \sin t] = a \cos t = y(t)$$

and

$$\frac{dy(t)}{dt} = \frac{d}{dt}[a \cos t] = -a \sin t = -x(t),$$

the given flow satisfies the system. By eliminating the parameter t in $x(t)$ and $y(t)$, the solution curves obtained are $x^2 + y^2 = a^2$.

6.

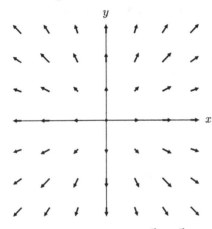

Figure 17.23: $\vec{v}(t) = x\vec{i} + y\vec{j}$

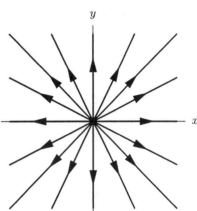

Figure 17.24: The flow $x = ae^t, y = be^t$.

As

$$\vec{v}(t) = \frac{dx}{dt}\vec{i} + \frac{dy}{dt}\vec{j},$$

the system of differential equations is

$$\begin{cases} \frac{dx}{dt} = x \\ \frac{dy}{dt} = y. \end{cases}$$

Since

$$\frac{dx(t)}{dt} = \frac{d}{dt}[ae^t] = ae^t = x(t)$$

and

$$\frac{dy(t)}{dt} = \frac{d}{dt}[be^t] = be^t = y(t),$$

the given flow satisfies the system. By eliminating the parameter t in $x(t)$ and $y(t)$, the solution curves obtained are $y = \frac{b}{a}x$.

7.

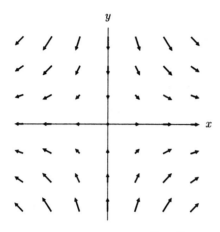

Figure 17.25: $\vec{v}(t) = x\vec{i} - y\vec{j}$

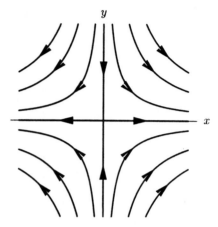

Figure 17.26: The flow $x = ae^t, y = be^{-t}$

As

$$\vec{v}(t) = \frac{dx}{dt}\vec{i} + \frac{dy}{dt}\vec{j},$$

the system of differential equations is

$$\begin{cases} \frac{dx}{dt} = x \\ \frac{dy}{dt} = -y. \end{cases}$$

Since

$$\frac{dx(t)}{dt} = \frac{d}{dt}[ae^t] = ae^t = x(t)$$

and

$$\frac{dy(t)}{dt} = \frac{d}{dt}[be^{-t}] = -be^{-t} = y(t),$$

the given flow satisfies the system. By eliminating the parameter t in $x(t)$ and $y(t)$, the solution curves obtained are $xy = ab$.

8. The vector field is given by $\vec{v} = y^2\vec{i} + 2x^2\vec{j}$, that is, the flow line $(x(t), (y(t))$ satisfies

$$x'(t) = y^2$$
$$y'(t) = 2x^2$$

We'll use Euler's method with $\Delta t = 0.1$ to find the parameterized curve $(x(t), y(t))$ through $(1, 2)$. So

$$x_{n+1} = x_n + 0.1y_n^2$$
$$y_{n+1} = y_n + (0.1)2x_n^2$$

Initially, that is when $t = 0$, we have $(x_0, y_0) = (1, 2)$. Then

$$x_1 = x_0 + 0.1y_0^2 = 1 + 0.1 \cdot 2^2 = 1.4$$
$$y_1 = y_0 + 0.1 \cdot 2x_0^2 = 2 + 0.1 \cdot 2 \cdot 1^2 = 2.2$$

Thus, we see that after one step, $x_1 = 1.4$ and $y_1 = 2.2$. Further values are given in the Table 17.1.

TABLE 17.1

x	1.4	1.884	2.556	3.646	5.770
y	2.2	2.592	3.302	4.609	7.268

9. The directions of the flow lines are as shown.

   (a) III
   (b) I
   (c) II
   (d) V
   (e) VI
   (f) IV

(I)
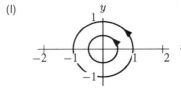

(II)
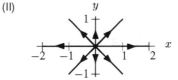

(III)
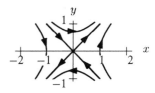

(IV)
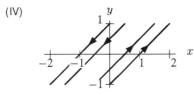

(V)
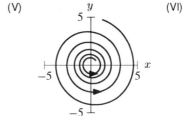

(VI)
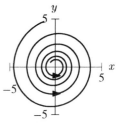

10. (a) Each vector in the vector field $\vec{v}$ is horizontal, tangent to a circle whose center is on the z-axis, and pointing counterclockwise when viewed from above. Thus, $\vec{v}$ is parallel to $-y\vec{i} + x\vec{j}$. The point (x, y, z) is moving on a circle of radius $r = \sqrt{x^2 + y^2}$ and has

$$\text{Speed} = \frac{2\pi r}{24} = \frac{\pi r}{12}.$$

Since the vector at the point (x, y, z) has magnitude $\pi r/12$ and is parallel to the unit vector $(-y\vec{i} + x\vec{j})/\sqrt{x^2 + y^2}$, we have

$$\vec{v} = \frac{\pi r}{12}\left(\frac{-y\vec{i} + x\vec{j}}{\sqrt{x^2 + y^2}}\right) = \frac{\pi}{12}(-y\vec{i} + x\vec{j}) \text{ meters/hr.}$$

   (b) A point moves in a horizontal circle, centered on the z-axis, and oriented counter-clockwise when viewed from above. These circles are the flow lines.

Solutions for Chapter 17 Review

1. (a) A vector field associates a vector to every point in a region of the space. In other words, a vector field is a vector-valued function of position given by $\vec{v} = \vec{f}(\vec{r}) = \vec{f}(x, y, z)$

   (b) (i) Yes, $\vec{r} + \vec{a} = (x + a_1)\vec{i} + (y + a_2)\vec{j} + (z + a_3)\vec{k}$ is a vector-valued function of position.
   (ii) No, $\vec{r} \cdot \vec{a}$ is a scalar.
   (iii) Yes.
   (iv) $x^2 + y^2 + z^2$ is a scalar.

2.   The vector field points in a clockwise direction around the origin. Since

$$\left\| \left(\frac{y}{\sqrt{x^2+y^2}} \right) \vec{i} - \left(\frac{x}{\sqrt{x^2+y^2}} \right) \vec{j} \right\| = \frac{\sqrt{x^2+y^2}}{\sqrt{x^2+y^2}} = 1$$

the length of the vectors is constant everywhere.

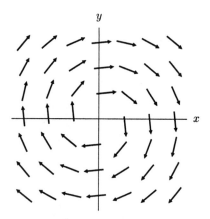

Figure 17.27

3.   At each point, all these vector fields point in the same direction (rotating clockwise around the origin). Since
$\|\vec{F}\| = \frac{1}{x^2+y^2}\|y\vec{i} - x\vec{j}\| = \frac{\sqrt{y^2+x^2}}{x^2+y^2} = \frac{1}{\sqrt{x^2+y^2}}$, the vectors in the field shrink as you go away from the origin.

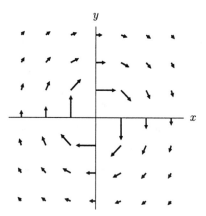

Figure 17.28

4.   The vector field points in a clockwise direction around the origin. Since $\|y\vec{i} - x\vec{j}\| = \sqrt{y^2+x^2}$, the vectors get longer as you go away from the origin.

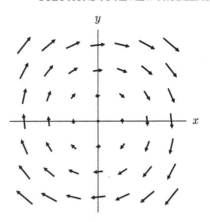

Figure 17.29

5. (a) Since $\vec{F} = \frac{\vec{r}}{\|\vec{r}\|^3}$, the magnitude of $\vec{F}$ is given by

$$\|\vec{F}\| = \frac{\|\vec{r}\|}{\|\vec{r}\|^3} = \frac{1}{\|\vec{r}\|^2}.$$

Now $\vec{r} = x\vec{i} + y\vec{j} + z\vec{k}$, so the magnitude of $\vec{r}$ is given by

$$\|\vec{r}\| = \sqrt{x^2 + y^2 + z^2}.$$

Thus,

$$\|\vec{F}\| = \frac{1}{\|\vec{r}\|^2} = \frac{1}{x^2 + y^2 + z^2}.$$

(b) $\vec{F} \cdot \vec{r} = \frac{\vec{r}}{\|\vec{r}\|^3} \cdot \vec{r} = \frac{\|\vec{r}\|^2}{\|\vec{r}\|^3} = \frac{1}{\|\vec{r}\|} = \frac{1}{\sqrt{x^2+y^2+z^2}}.$

(c) A unit vector parallel to $\vec{F}$ and pointing in the same direction is given by $\vec{U} = \frac{\vec{F}}{\|\vec{F}\|}$.

$\vec{F} = \frac{\vec{r}}{\|\vec{r}\|^3}$, and $\|\vec{F}\| = \frac{1}{\|\vec{r}\|^2}$. Putting these into the expression for $\vec{U}$ we have

$$\vec{U} = \frac{\vec{F}}{\|\vec{F}\|} = \frac{\frac{\vec{r}}{\|\vec{r}\|^3}}{\frac{1}{\|\vec{r}\|^2}} = \frac{\vec{r}}{\|\vec{r}\|}$$

$$= \frac{x}{\sqrt{x^2 + y^2 + z^2}}\vec{i} + \frac{y}{\sqrt{x^2 + y^2 + z^2}}\vec{j} + \frac{z}{\sqrt{x^2 + y^2 + z^2}}\vec{k}.$$

(d) A unit vector parallel to $\vec{F}$ and pointing in the opposite direction is given by:

$$\vec{V} = -\frac{\vec{F}}{\|\vec{F}\|} = -\frac{\vec{r}}{\|\vec{r}\|}$$

$$= \frac{-x}{\sqrt{x^2 + y^2 + z^2}}\vec{i} + \frac{-y}{\sqrt{x^2 + y^2 + z^2}}\vec{j} + \frac{-z}{\sqrt{x^2 + y^2 + z^2}}\vec{k}.$$

(e) If $\vec{r} = \cos t\vec{i} + \sin t\vec{j} + \vec{k}$, then $\|\vec{r}\| = \sqrt{\cos^2 t + \sin^2 t + 1} = \sqrt{2}$.
So, $\vec{F} = \frac{\vec{r}}{\|\vec{r}\|^3} = \frac{\cos t}{\sqrt{8}}\vec{i} + \frac{\sin t}{\sqrt{8}}\vec{j} + \frac{1}{\sqrt{8}}\vec{k} = \frac{\cos t}{2\sqrt{2}}\vec{i} + \frac{\sin t}{2\sqrt{2}}\vec{j} + \frac{1}{2\sqrt{2}}\vec{k}.$

(f)   We know that $\vec{F} \cdot \vec{r} = \frac{1}{\|\vec{r}\|}$, so if $\vec{r} = \cos t \vec{i} + \sin t \vec{j} + \vec{k}$, $\vec{F} \cdot \vec{r} = \frac{1}{\sqrt{2}}$.

6.   This corresponds to area A in Figure 17.30.

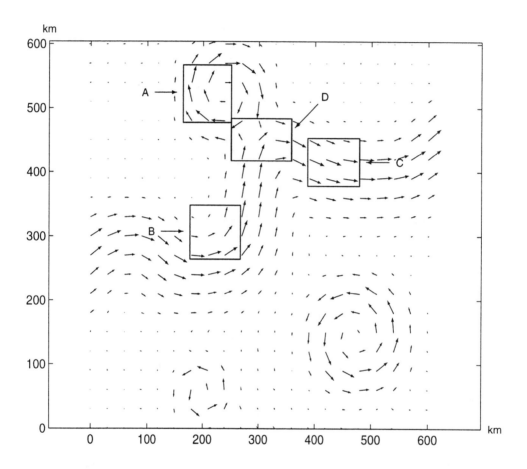

Figure 17.30

7.   This corresponds to area B in Figure 17.30 in Problem 6.
8.   This corresponds to area C in Figure 17.30 in Problem 6.
9.   This corresponds to area D in Figure 17.30 in Problem 6.

10. (a)   The current, and path that the iceberg would travel, would look like:

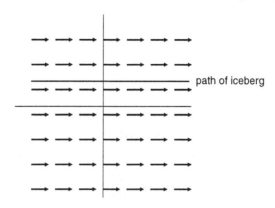

Figure 17.31

To determine the location of the iceberg at time $t = 7$, we must first determine the velocity in the x and y direction. In this current, $V_x = 1$, and $V_y = 0$. To obtain the position, we must integrate the velocity in terms of t. For this current we get

$$\frac{dx}{dt} = 1$$

Hence

$$x(7) = x(0) + \int_0^7 1 \cdot dt$$
$$= 1 + 7$$
$$= 8.$$

Since $V_y = dy/dt = 0$, y is a constant. Thus at $t = 7$, x has moved from $x = 1$ to $x = 8$ and y has stayed at $y = 3$. Therefore the location at $t = 7$ is $(8, 3)$.

(b)   The current and path that the iceberg would travel, would look like:

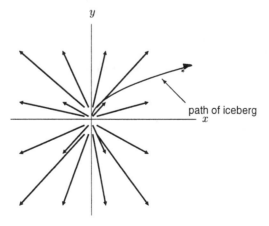

Figure 17.32

Assuming that the iceberg follows the current exactly, we find the position of the iceberg at time $t = 7$ by recognizing that the velocity must be equal to the given vector field.

$$\frac{dx}{dt} = 2x$$
$$\frac{dy}{dt} = y$$

These are separable equations that are solved for x and y as follows:

$$\frac{dx}{dt} = 2x$$
$$\int \frac{dx}{2x} = \int 1 \, dt$$
$$\frac{\ln x}{2} = t + C$$
$$x = k_x e^{2t}$$

and for y

$$\frac{dy}{dt} = y$$
$$\int \frac{dy}{y} = \int 1 \, dt$$
$$\ln y = t + C$$
$$y = k_y e^t$$

We can solve for k_x and k_y, the arbitrary constants, because we know the position of the iceberg at $t = 0$.

$$1 = x(0) = k_x$$
$$3 = y(0) = k_y$$

so

$$x = e^{2t}, \; y = 3e^t.$$

We now substitute $t = 7$:

$$x = e^{2 \cdot 7} = e^{14} \text{ and } y = 3e^7$$

(c)   The current, and path that the iceberg would travel, would look like:

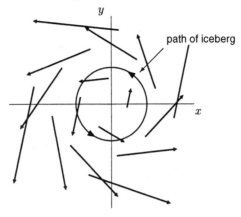

Figure 17.33

Since
$$\vec{v} = -y\vec{i} + x\vec{j},$$
the system of differential equations satisfied by $x(t)$ and $y(t)$ is
$$\frac{dx}{dt} = -y, \frac{dy}{dt} = x.$$
We differentiate one of the equations and substitute into the other, giving a second order equation
$$\frac{dx}{dt} = -y$$
$$\frac{d^2x}{dt^2} = -\frac{dy}{dt}$$
$$\frac{d^2x}{dt^2} = -x$$
$$\frac{d^2x}{dt^2} - x = 0$$

This differential equation has a solution of this form :
$$x = A\cos t + B\sin t$$
By taking the derivative and using the fact that $y = -dx/dt$, we get:
$$y = A\sin t - B\cos t$$
We know the initial position of the iceberg, so we can find the constants A and B with the simultaneous equations:
$$1 = x(0) = A\cos 0 + B\sin 0$$
$$3 = y(0) = A\sin 0 - B\cos 0$$
Thus, $A = 1$ and $B = -3$. Now we evaluate the two expressions for $t = 7$:
$$x = \cos 7 - 3\sin 7 = -1.217$$
$$y = \sin 7 + 3\cos 7 = 2.919$$
and find the position of the iceberg, $(-1.217, 2.919)$.

11. (a)

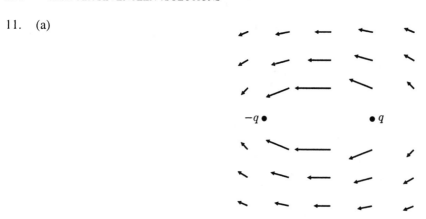

Figure 17.34

(b)

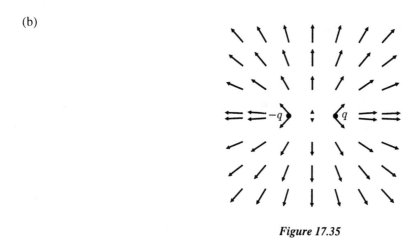

Figure 17.35

12. (a)

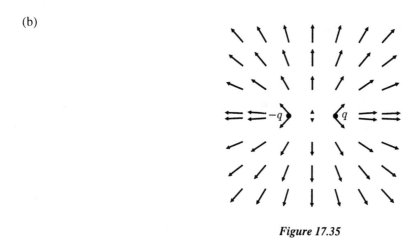

$p = 10$ $p = -10$ $p = 20$

Figure 17.36: Vector Field $\vec{D}$

(b) (i)

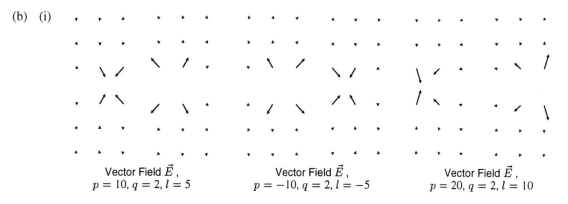

Vector Field $\vec{E}$,
$p = 10, q = 2, l = 5$

Vector Field $\vec{E}$,
$p = -10, q = 2, l = -5$

Vector Field $\vec{E}$,
$p = 20, q = 2, l = 10$

Figure 17.37

(ii) The vector field $\vec{D}$ approximates $\vec{E}$ well far away from the charges q and $-q$. As one gets closer to these, the approximation becomes less accurate.

(iii) The discrepancy is only apparent since, although each term in the expression of $\vec{E}$ decays like $1/|\vec{r}|^2$, their difference decays like $1/|\vec{r}|^3$. This can be seen as follows:

$$\vec{E}(\vec{r}) = q\left(\frac{\vec{r} - \vec{r}_2}{|\vec{r} - \vec{r}_2|^3} - \frac{\vec{r} - \vec{r}_1}{|\vec{r} - \vec{r}_1|^3}\right)$$

$$= q\frac{|\vec{r} - \vec{r}_1|^3(\vec{r} - \vec{r}_2) - |\vec{r} - \vec{r}_2|^3(\vec{r} - \vec{r}_1)}{|\vec{r} - \vec{r}_1|^3|\vec{r} - \vec{r}_2|^3}$$

$$= q\frac{(|\vec{r} - \vec{r}_1|^3 - |\vec{r} - \vec{r}_2|^3)\vec{r} - |\vec{r} - \vec{r}_1|^3\vec{r}_2 + |\vec{r} - \vec{r}_2|^3\vec{r}_1}{|\vec{r} - \vec{r}_1|^3|\vec{r} - \vec{r}_2|^3}$$

The absolute value of the expression $|\vec{r} - \vec{r}_1|^3 - |\vec{r} - \vec{r}_2|^3$ is at most *quadratic* in $|\vec{r}|$ since

$$|\vec{r} - \vec{r}_1|^3 - |\vec{r} - \vec{r}_2|^3 = \left||\vec{r} - \vec{r}_1| - |\vec{r} - \vec{r}_2|\right| \cdot \left||\vec{r} - \vec{r}_1|^2 + |\vec{r} - \vec{r}_1| \cdot |\vec{r} - \vec{r}_2| + |\vec{r} - \vec{r}_2|^2\right|$$

$$\leq |\vec{r}_2 - \vec{r}_1|(|\vec{r} - \vec{r}_1|^2 + |\vec{r} - \vec{r}_1| \cdot |\vec{r} - \vec{r}_2| + |\vec{r} - \vec{r}_2|^2).$$

This shows that the numerator of the above expression of $\vec{E}(\vec{r})$ is in fact of order $|\vec{r}|^3$, whereas the numerator is of order $|\vec{r}|^6$; i.e. $\vec{E}(\vec{r})$ decays like $1/|\vec{r}|^3$.

CHAPTER EIGHTEEN

Solutions for Section 18.1

1. Positive, because the vectors are longer on the portion of the path that goes in the same direction as the vector field.

2. Negative because the vector field points in the opposite direction to the path.

3. Positive, because the vector field points in the same direction as the path.

4. Zero, because, by symmetry, the positive integral along the left half of the path cancels the negative integral along the right half.

5. Since it appears that C_1 is everywhere perpendicular to the vector field, all of the dot products in the line integral are zero, hence $\int_{C_1} \vec{F} \cdot d\vec{r} \approx 0$. Along the path C_2 the dot products of $\vec{F}$ with $\Delta \vec{r_i}$ are all positive, so their sum is positive and we have $\int_{C_1} \vec{F} \cdot d\vec{r} < \int_{C_2} \vec{F} \cdot d\vec{r}$. For C_3 the vectors $\Delta \vec{r_i}$ are in the opposite direction to the vectors of $\vec{F}$, so the dot products $\vec{F} \cdot \Delta \vec{r_i}$ are all negative; so, $\int_{C_3} \vec{F} \cdot d\vec{r} < 0$. Thus, we have

    $$\int_{C_3} \vec{F} \cdot d\vec{r} < \int_{C_1} \vec{F} \cdot d\vec{r} < \int_{C_2} \vec{F} \cdot d\vec{r}$$

6. The vector field is $F(\vec{r}) = \vec{r}$. See Figure 18.1. The vector field is perpendicular to the circular arcs at every point, so

    $$\int_{C_1} \vec{F} \cdot d\vec{r} = \int_{C_3} \vec{F} \cdot d\vec{r} = 0.$$

    Also, since it is radially symmetric,

    $$\int_{C_2} \vec{F} \cdot d\vec{r} = -\int_{C_4} \vec{F} \cdot d\vec{r}.$$

    So,

    $$\int_C = \int_{C_1} + \int_{C_2} + \int_{C_3} + \int_{C_4} = 0.$$

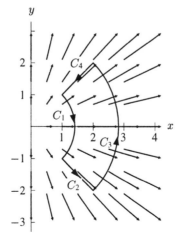

Figure 18.1

7. This vector field is illustrated in Figure 18.2. It is perpendicular to C_2 and C_4 at every point, since $\vec{F}(x, y) \cdot \vec{r}(x, y) = 0$ and C_2 and C_4 are radial line segments, then

$$\int_{C_2} \vec{F} \cdot d\vec{r} = \int_{C_4} \vec{F} \cdot d\vec{r} = 0.$$

Since C_3 is longer than C_1, and the vector field is larger in magnitude along C_3, the line integral along C_3 has greater absolute value than that along C_1. The line integral along C_3 is positive and the line integral along C_1 is negative, so

$$\int_C \vec{F} \cdot d\vec{r} = \int_{C_3} \vec{F} \cdot d\vec{r} + \int_{C_1} \vec{F} \cdot d\vec{r} > 0.$$

See Figure 18.2.

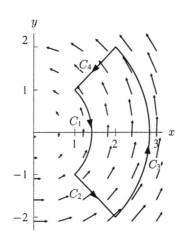

Figure 18.2

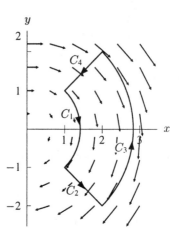

Figure 18.3

8. This vector field is illustrated in Figure 18.3. It is perpendicular to C_2 and C_4 at every point, since $\vec{F}(x, y) \cdot \vec{r}(x, y) = 0$ and C_2 and C_4 are radial line segments, then

$$\int_{C_2} \vec{F} \cdot d\vec{r} = \int_{C_4} \vec{F} \cdot d\vec{r} = 0.$$

Since C_3 is longer than C_1, and the vector field is larger in magnitude along C_3, the line integral along C_3 has greater absolute value than that along C_1. The line integral along C_1 is positive and the line integral along C_3 is negative, so

$$\int_C \vec{F} \cdot d\vec{r} = \int_{C_3} \vec{F} \cdot d\vec{r} + \int_{C_1} \vec{F} \cdot d\vec{r} < 0.$$

See Figure 18.3.

9. Since it does not depend on y, this vector field is constant along vertical lines, $x =$ constant. Now let us consider two points P and Q on C_1 which lie on the same vertical line. Because C_1 is symmetric with respect to the x-axis, the tangent vectors at P and Q will be symmetric with respect to the vertical axis so their sum is a vertical vector. But $\vec{F}$ has only horizontal component and thus $\vec{F} \cdot (\Delta\vec{r}(P) + \Delta\vec{r}(Q)) = 0$. As $\vec{F}$ is constant along vertical lines (so $\vec{F}(P) = \vec{F}(Q)$), we obtain

$$\vec{F}(P) \cdot \Delta\vec{r}(P) + \vec{F}(Q) \cdot \Delta\vec{r}(Q) = 0.$$

Summing these products and making $\|\Delta\vec{r}\| \to 0$ gives us

$$\int_{C_1} \vec{F} \cdot d\vec{r} = 0.$$

The same thing happens on C_3, so $\int_{C_3} \vec{F} \cdot d\vec{r} = 0$.

Now let P be on C_2 and Q on C_4 lying on the same vertical line. The respective tangent vectors are symmetric with respect to the vertical axis hence they add up to a vertical vector and a similar argument as before gives

$$\vec{F}(P) \cdot \Delta\vec{r}(P) + \vec{F}(Q) \cdot \Delta\vec{r}(Q) = 0$$

and

$$\int_{C_2} \vec{F} \cdot d\vec{r} + \int_{C_4} \vec{F} \cdot d\vec{r} = 0$$

and so

$$\int_{C} \vec{F} \cdot d\vec{r} = 0.$$

See Figure 18.4.

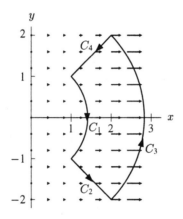

Figure 18.4

10. First of all, $\vec{F}(x,y)$ is perpendicular to the position vector $\vec{r}(x,y) = x\vec{i} + y\vec{j}$ because

$$\vec{F}(x,y) \cdot \vec{r}(x,y) = \frac{-xy}{x^2 + y^2} + \frac{xy}{x^2 + y^2} = 0.$$

Also the magnitude of $\vec{F}$ is inversely proportional to the distance from the origin because

$$\|\vec{F}(x,y)\| = \frac{\sqrt{x^2 + y^2}}{x^2 + y^2} = \frac{1}{\|\vec{r}(x,y)\|}.$$

So $\vec{F}$ is perpendicular to C_2 and C_4 and therefore

$$\int_{C_2} \vec{F} \cdot d\vec{r} = \int_{C_4} \vec{F} \cdot d\vec{r} = 0.$$

Suppose R is the radius of C_3. On C_3, the vector field $\vec{F}$ has the same direction as the tangent vector which is approximated by $\Delta\vec{r}$, so we have

$$\vec{F} \cdot \Delta\vec{r} = \|\vec{F}\| \cdot \|\Delta\vec{r}\| = \frac{1}{R}\|\Delta\vec{r}\|.$$

When all these products are summed and the limit is taken as $\|\Delta\vec{r}\| \to 0$, we get

$$\int_{C_3} \vec{F} \cdot d\vec{r} = \frac{1}{R} \int_{C_3} \|d\vec{r}\|$$
$$= \frac{1}{R}(\text{length of } C_3) = \text{measure of the arc } C_3 \text{ in radians.}$$

Similarly, suppose r is the radius of C_1. On C_1, the vector field $\vec{F}$ is in the opposite direction to the tangent vector which is approximated by $\Delta\vec{r}$. Hence we have

$$\int_{C_1} \vec{F} \cdot d\vec{r} = -\frac{1}{r} \int_{C_1} \|d\vec{r}\|$$
$$= -(\frac{1}{r}(\text{length of } C_1)) = -(\text{measure of } C_1 \text{ in radians}).$$

Since C_1 and C_3 have the same measure in radians, we have

$$\int_C \vec{F} \cdot d\vec{r} = \int_{C_1} \vec{F} \cdot d\vec{r} + \int_{C_2} \vec{F} \cdot d\vec{r} + \int_{C_3} \vec{F} \cdot d\vec{r} + \int_{C_4} \vec{F} \cdot d\vec{r}$$
$$= -\frac{\pi}{2} + 0 + (+\frac{\pi}{2}) + 0 = 0.$$

See Figure 18.5.

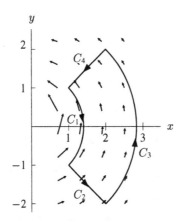

Figure 18.5

11. Since $\vec{F}$ is perpendicular to the curve at every point along it,

$$\int_C \vec{F} \cdot d\vec{r} = 0.$$

12. At every point along the curve, $\vec{F} = 2\vec{j}$ and is parallel to the curve. Thus,

$$\int_C \vec{F} \cdot d\vec{r} = 2 \cdot \text{Length of curve} = 2 \cdot 5 = 10.$$

13. At every point, the vector field is parallel to segments $\Delta \vec{r} = \Delta x \vec{i}$ of the curve. Thus,

$$\int_C \vec{F} \cdot d\vec{r} = \int_2^6 x\vec{i} \cdot dx\vec{i} = \int_2^6 x\,dx = \frac{x^2}{2}\bigg|_2^6 = 16.$$

14. The $\vec{j}$-component of $\vec{F}$ does not contribute to the line integral. Since $\Delta \vec{r} = \Delta x \vec{i}$, we have

$$\int_C \vec{F} \cdot d\vec{r} = \int_2^6 (x\vec{i} + y\vec{j}) \cdot dx\vec{i} = \int_2^6 x\,dx = \frac{x^2}{2}\bigg|_2^6 = 16.$$

15. At every point on the path, $\vec{F}$ is parallel to $\Delta \vec{r}$. Suppose r is the distance from the point (x, y) to the origin, so $\|\vec{r}\| = r$. Then $\vec{F} \cdot \Delta \vec{r} = \|\vec{F}\|\|\Delta \vec{r}\| = r\Delta r$. At the start of the path, $r = \sqrt{2^2 + 2^2} = 2\sqrt{2}$ and at the end $r = 6\sqrt{2}$. Thus,

$$\int_C \vec{F} \cdot d\vec{r} = \int_{2\sqrt{2}}^{6\sqrt{2}} r\,dr = \frac{r^2}{2}\bigg|_{2\sqrt{2}}^{6\sqrt{2}} = 32.$$

16. Since $\vec{F}$ is a constant vector field and the curve is a line, $\int_C \vec{F} \cdot d\vec{r} = \vec{F} \cdot \Delta \vec{r}$, where $\Delta \vec{r} = 7\vec{j}$. Therefore,

$$\int_C \vec{F} \cdot d\vec{r} = (3\vec{i} + 4\vec{j}) \cdot 7\vec{j} = 28$$

17. See Figure 18.6. The example chosen is the vector field $\vec{F}(x, y) = y\vec{j}$ and the path C is the line from $(0, -1)$ to $(0, 1)$. Since the vectors are symmetric about the x-axis, the dot products $\vec{F} \cdot \Delta \vec{r}$ cancel out along C to give 0 for the line integral. Many other answers are possible.

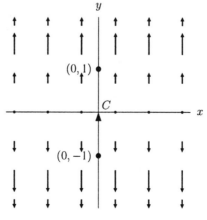

Figure 18.6

18. (a)

TABLE 18.1

(x,y)	$\vec{F}(x,y)$
(0,-1)	$-\vec{i}$
(1,-1)	$-\vec{i}+\vec{j}$
(2,-1)	$-\vec{i}+4\vec{j}$
(3,-1)	$-\vec{i}+9\vec{j}$
(4,-1)	$-\vec{i}+16\vec{j}$
(4,0)	$16\vec{j}$
(4,1)	$\vec{i}+16\vec{j}$
(4,2)	$2\vec{i}+16\vec{j}$
(4,3)	$3\vec{i}+16\vec{j}$

(b)

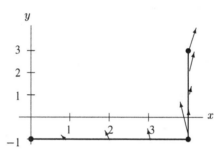

Figure 18.7

(c)   From the point $(0,-1)$ to the point $(4,-1)$, the x-component of the force field is always -1, i.e., it is pushing the object backwards with a constant force of 1. Thus, the work done on that part of the path is $-1\cdot 4 = -4$, because only the horizontal component of the force field contributes to work.

   From the point $(4,-1)$ to the point $(4,3)$, the y-component of the force field is always 16, so it is pushing the object forward with force of 16. Thus, the work done on that part of the path is $16\cdot 4 = 64$, because only the vertical component of the force field contributes to work.

   So the total work done is $-4+64 = 60$.

19.   The force has no horizontal component. Therefore the (positive) work done in the first half of C_1 will be exactly canceled by the (negative) work done in the second half, so the total work over the path C_1 is zero. The same holds true for C_2, again by virtue of the vertical symmetry of the path and the fact that $\vec{F}$ is constant and because the horizontal part of C_2 contributes zero work. For C_3, the total work will be greater than zero, since the diagonal part of C_3 is in the same general direction as $\vec{F}$ and the horizontal part of C_3 contributes zero work.

20.   (a)   You will find that

$$\int_C \vec{F}\cdot d\vec{r} = 0$$

   for all closed curves C.

   (b)   For any curve C, which starts at the origin and ends at the point $(1/2, 1/2)$, you will find that

$$\int_C \vec{F}\cdot d\vec{r} = 1/4.$$

   In other words, the value of any such integral is the same no matter what path is taken.

21. (a)   Different closed curves C, will give different values for the line integral

$$\int_C \vec{F} \cdot d\vec{r}.$$

(b)   The value of the line integral takes on different values, depending on the path taken.

22. (a)   You will find that

$$\int_C \vec{F} \cdot d\vec{r} = 0$$

for all closed curves C (a rectangle, an ellipse or a polygon).

(b)   For any curve, C, which starts at the origin and ends at $(1/2, 1/2)$, you will find that

$$\int_C \vec{F} \cdot d\vec{r} = 5/8.$$

23. (a)   Different closed curves, C, will give different values for the line integral

$$\int_C \vec{F} \cdot d\vec{r}.$$

(b)   The value of the line integral takes on different values, depending upon the path taken.

24. Suppose $\int_C \vec{F} \cdot d\vec{r} = 0$ for every closed curve C. Pick any two fixed points P_1, P_2 and curves C_1, C_2 each going from P_1 to P_2. See Figure 18.8. Define $-C_2$ to be the same curve as C_2 except in the opposite direction. Therefore, the curve formed by traversing C_1, followed by C_2 in the opposite direction, written as $C_1 - C_2$, is a closed curve, so by our assumption, $\int_{C_1-C_2} \vec{F} \cdot d\vec{r} = 0$. However, we can write

$$\int_{C_1-C_2} \vec{F} \cdot d\vec{r} = \int_{C_1} \vec{F} \cdot d\vec{r} - \int_{C_2} \vec{F} \cdot d\vec{r}$$

since C_2 and $-C_2$ are the same except for direction. Therefore,

$$\int_{C_1} \vec{F} \cdot d\vec{r} - \int_{C_2} \vec{F} \cdot d\vec{r} = 0,$$

so

$$\int_{C_1} \vec{F} \cdot d\vec{r} = \int_{C_2} \vec{F} \cdot d\vec{r}.$$

Since C_1 and C_2 are any two curves with the endpoints P_1, P_2, this gives the desired result – namely, that fixing endpoints and direction uniquely determines the value of $\int_C \vec{F} \cdot d\vec{r}$. In other words, the value of the integral $\int_C \vec{F} \cdot d\vec{r}$ does not depend on the path taken. We say the line integral is *path-independent*.

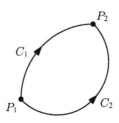

Figure 18.8

25. Pick any closed curve C. Choose two distinct points P_1, P_2 on C. Let C_1, C_2 be the two curves from P_1 to P_2 along C. See Figure 18.9. Let $-C_2$ be the same as C_2, except in the opposite direction. Thus, $C_1 - C_2 = C$. Therefore,

$$\int_C \vec{F} \cdot d\vec{r} = \int_{C_1 - C_2} \vec{F} \cdot d\vec{r} = \int_{C_1} \vec{F} \cdot d\vec{r} - \int_{C_2} \vec{F} \cdot d\vec{r}$$

since C_2 and $-C_2$ differ only in direction. But C_1 and C_2 have the same endpoints (P_1 and P_2) and same direction (P_1 to P_2), so by assumption we have $\int_{C_1} \vec{F} \cdot d\vec{r} = \int_{C_2} \vec{F} \cdot d\vec{r}$. Therefore,

$$\int_C \vec{F} \cdot d\vec{r} = \int_{C_1} \vec{F} \cdot d\vec{r} - \int_{C_2} \vec{F} \cdot d\vec{r} = 0.$$

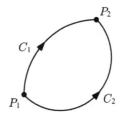

Figure 18.9

26. Since $\Delta \vec{r}$ points outward, in the opposite direction to $\vec{F}$, we expect the answer to be negative.

$$\int_C \vec{F} \cdot d\vec{r} = \int_C -\frac{GMm\vec{r}}{r^3} \cdot d\vec{r} = \int_{8000}^{10000} -\frac{GMm}{r^2} dr$$
$$= \frac{GMm}{r}\bigg|_{8000}^{10000} = GMm \left(\frac{1}{10000} - \frac{1}{8000} \right)$$
$$= -2.5 \cdot 10^{-5} GMm.$$

27. Since $\Delta \vec{r}$ points outward, in the opposite direction to $\vec{F}$, we expect a negative answer. We take the upper limit to be $r = \infty$, so the integral is improper.

$$\int_C \vec{F} \cdot d\vec{r} = \int_C -\frac{GMm\vec{r}}{r^3} \cdot d\vec{r} = \int_{8000}^{\infty} -\frac{GMm}{r^2} dr$$
$$= \lim_{b \to \infty} \int_{8000}^{b} -\frac{GMm}{r^2} dr = \lim_{b \to \infty} \frac{GMm}{r}\bigg|_{8000}^{b} = \lim_{b \to \infty} GMm \left(\frac{1}{b} - \frac{1}{8000} \right)$$
$$= -\frac{GMm}{8000}$$

28. We'll assume that the rod is positioned along the z-axis, and look at the magnetic field $\vec{B}$ in the xy-plane. If C is a circle of radius r in the plane, centered at the origin, then we are told that the magnetic field is tangent to the circle and has constant magnitude $\|\vec{B}\|$. We divide the curve C into little pieces C_i and then we sum $\vec{B} \cdot \Delta \vec{r}$ computed on each piece C_i. But $\Delta \vec{r}$ points nearly in the same direction as $\vec{B}$, that is, tangent to C,

and has magnitude nearly equal to the length of C_i. So the dot product is nearly equal to $\|\vec{B}\| \times$ length of C_i. When all of these dot products are summed and the limit is taken as $\|\Delta\vec{r}\| \to 0$, we get

$$\int_C \vec{B} \cdot d\vec{r} = \|\vec{B}\| \times \text{length of } C = \|\vec{B}\| \times 2\pi r$$

Now Ampère's Law also tells us that

$$\int_C \vec{B} \cdot d\vec{r} = kI$$

Setting these expressions for the line integral equal to each other and solving for $\|\vec{B}\|$ gives $kI = 2\pi r\|\vec{B}\|$, so

$$\|\vec{B}\| = \frac{kI}{2\pi r}.$$

Solutions for Section 18.2

1. The parameterization is given, so

$$\begin{aligned}
\int_C \vec{F} \cdot d\vec{r} &= \int_2^4 \vec{F}(2t, t^3) \cdot (2\vec{i} + 3t^2\vec{j}) \, dt \\
&= \int_2^4 [(\ln(t^3)\vec{i} + \ln(2t)\vec{j}] \cdot (2\vec{i} + 3t^2\vec{j}) \, dt \\
&= \int_2^4 (2\ln(t^3) + 3t^2 \ln(2t)) \, dt \\
&= \int_2^4 (6\ln(t) + 3t^2 \ln(2t)) \, dt \qquad \text{since } \ln(t^3) = 3\ln(t).
\end{aligned}$$

This integral can be computed numerically, or using integration by parts or the integral table, giving

$$\begin{aligned}
\int_C \vec{F} \cdot d\vec{r} &= \int_2^4 (6\ln(t) + 3t^2 \ln(2t)) \, dt \\
&= \left. (6(t\ln(t) - t) + t^3 \ln(2t) - t^3/3)\right|_2^4 \\
&= 240\ln 2 - \frac{136}{3} - (28\ln 2 - \frac{44}{3}) \\
&= 212\ln 2 - 92/3 \approx 116.28.
\end{aligned}$$

The expression containing $\ln 2$ was obtained using the properties of the natural log.

2. The curve C is parameterized by $(x, y) = (t, t)$ for $0 \le t \le 3$. Thus,

$$\int_C \vec{F} \cdot d\vec{r} = \int_0^3 (t\vec{i} + t\vec{j}) \cdot (\vec{i} + \vec{j}) \, dt = \int_0^3 2t \, dt = t^2 \Big|_0^3 = 9.$$

3. The line can be parameterized by $(1 + 2t, 2 + 2t)$, for $0 \leq t \leq 1$, so the integral looks like

$$
\begin{aligned}
\int_C \vec{F} \cdot d\vec{r} &= \int_0^1 \vec{F}(1 + 2t, 2 + 2t) \cdot (2\vec{i} + 2\vec{j}) \, dt \\
&= \int_0^1 [(1 + 2t)^2 \vec{i} + (2 + 2t)^2 \vec{j}] \cdot (2\vec{i} + 2\vec{j}) \, dt \\
&= \int_0^1 2(1 + 4t + 4t^2) + 2(4 + 8t + 4t^2) \, dt \\
&= \int_0^1 (10 + 24t + 16t^2) \, dt \\
&= (10t + 12t^2 + 16t^3/3) \Big|_0^1 \\
&= 10 + 12 + 16/3 - (0 + 0 + 0) = 82/3
\end{aligned}
$$

4. The curve C is parameterized by

$$
\vec{r} = \cos t \vec{i} + \sin t \vec{j}, \qquad \text{for } 0 \leq t \leq 2\pi,
$$

so,

$$
\vec{r}'(t) = -\sin t \vec{i} + \cos t \vec{j}.
$$

Thus,

$$
\begin{aligned}
\int_C \vec{F} \cdot d\vec{r} &= \int_0^{2\pi} (2 \sin t \vec{i} - \sin(\sin t) \vec{j}) \cdot (-\sin t \vec{i} + \cos t \vec{j}) \, dt \\
&= \int_0^{2\pi} (-2 \sin^2 t - \sin(\sin t) \cos t) \, dt \\
&= \sin t \cos t - t + \cos(\sin t) \Big|_0^{2\pi} \\
&= -2\pi.
\end{aligned}
$$

5. The portion of the ellipse can be parameterized by $(2 \cos t, \sin t)$, for $0 \leq t \leq \pi/2$, but this gives a *counterclockwise* orientation. Thus $t = \pi/2$ gives the beginning of the curve and $t = 0$ gives the end, so

$$
\begin{aligned}
\int_C \vec{F} \cdot d\vec{r} &= \int_{\pi/2}^0 \vec{F}(2 \cos t, \sin t) \cdot (-2 \sin t \vec{i} + \cos t \vec{j}) \, dt \\
&= -\int_0^{\pi/2} (e^{2\cos t} \vec{i} + e^{\sin t} \vec{j}) \cdot (-2 \sin t \vec{i} + \cos t \vec{j}) \, dt \\
&= -\int_0^{\pi/2} (-2e^{2\cos t} \sin t + e^{\sin t} \cos t) \, dt \\
&= -(e^{2\cos t} + e^{\sin t}) \Big|_0^{\pi/2} \\
&= -[e^0 + e^1 - (e^2 + e^0)] = e^2 - e.
\end{aligned}
$$

6. The path can be broken into three line segments: C_1, from $(1,0)$ to $(-1,0)$, and C_2, from $(-1,0)$ to $(0,1)$, and C_3, from $(0,1)$ to $(1,0)$. (See Figure 18.10.)

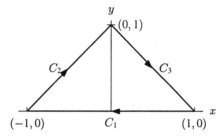

Figure 18.10

Along C_1 we have $y = 0$ so the vector field $xy\vec{i} + (x-y)\vec{j}$ is perpendicular to C_1; Thus, the line integral along C_1 is 0.

C_2 can be parameterized by $(-1+t, t)$, for $0 \le t \le 1$ so the integral is

$$\int_{C_2} \vec{F} \cdot d\vec{r} = \int_0^1 \vec{F}(-1+t, t) \cdot (\vec{i} + \vec{j}) \, dt$$
$$= \int_0^1 [t(-1+t)\vec{i} + (-1)\vec{j}] \cdot (\vec{i} + \vec{j}) \, dt$$
$$= \int_0^1 (-t + t^2 - 1) \, dt$$
$$= (-t^2/2 + t^3/3 - t)\big|_0^1$$
$$= -1/2 + 1/3 - 1 - (0 + 0 + 0) = -7/6$$

C_3 can be parameterized by $(t, 1-t)$, for $0 \le t \le 1$ so the integral is

$$\int_{C_3} \vec{F} \cdot d\vec{r} = \int_0^1 \vec{F}(t, 1-t) \cdot (\vec{i} - \vec{j}) \, dt$$
$$= \int_0^1 (t(1-t)\vec{i} + (2t - 1)\vec{j}) \cdot (\vec{i} - \vec{j}) \, dt$$
$$= \int_0^1 (-t^2 - t + 1) \, dt$$
$$= (-t^3/3 - t^2/2 + t)\big|_0^1$$
$$= -1/3 - 1/2 + 1 - (0 + 0 + 0) = 1/6$$

So the total line integral is

$$\int_C \vec{F} \cdot d\vec{r} = \int_{C_1} \vec{F} \cdot d\vec{r} + \int_{C_2} \vec{F} \cdot d\vec{r} + \int_{C_3} \vec{F} \cdot d\vec{r} = 0 + (-\frac{7}{6}) + \frac{1}{6} = -1$$

7. Since $\vec{r} = x(t)\vec{i} + y(t)\vec{j} + z(t)\vec{k} = t\vec{i} + t^2\vec{j} + t^3\vec{k}$, for $1 \le t \le 2$, we have $\vec{r}'(t) = x'(t)\vec{i} + y'(t)\vec{j} + z'(t)\vec{k} = \vec{i} + 2t\vec{j} + 3t^2\vec{k}$. Then

$$\int_C \vec{F} \cdot d\vec{r} = \int_1^2 (t\vec{i} + 2t^3 t^2\vec{j} + t\vec{k}) \cdot (\vec{i} + 2t\vec{j} + 3t^2\vec{k}) \, dt$$

$$= \int_1^2 (t + 4t^6 + 3t^3)\, dt$$

$$= \frac{t^2}{2} + \frac{4t^7}{7} + \frac{3t^4}{4} \Big|_1^2 = \frac{2389}{28}$$

$$\approx 85.32$$

8. We parameterize C by

$$\vec{r} = 2t\vec{i} + 3t\vec{j} + 4t\vec{k}, \qquad \text{for } 0 \le t \le 1.$$

Then $\vec{r}'(t) = 2\vec{i} + 3\vec{j} + 4\vec{k}$ and so

$$\int_C \vec{F} \cdot d\vec{r} = \int_0^1 \left((2t)^3\vec{i} + (3t)^2\vec{j} + (4t)\vec{k} \right) \cdot (2\vec{i} + 3\vec{j} + 4\vec{k})\, dt$$

$$= \int_0^1 (16t^3 + 27t^2 + 16t)\, dt$$

$$= 4t^4 + 9t^3 + 8t^2 \Big|_0^1 = 21.$$

9. Since C is given by $\vec{r} = \cos t\vec{i} + \sin t\vec{j} + t\vec{k}$, we have $\vec{r}'(t) = -\sin t\vec{i} + \cos t\vec{j} + \vec{k}$. Thus,

$$\int_C \vec{F} \cdot d\vec{r} = \int_0^{4\pi} (-\sin t\vec{i} + \cos t\vec{j} + 5\vec{k}) \cdot (-\sin t\vec{i} + \cos t\vec{j} + \vec{k})\, dt$$

$$= \int_0^{4\pi} (\sin^2 t + \cos^2 t + 5)\, dt = \int_0^{4\pi} 6\, dt = 24\pi.$$

10. The first step is to parameterize C by

$$(x(t), y(t), z(t)) = (0, 2\cos t, -2\sin t), \quad 0 \le t \le 2\pi.$$

Thus, we have

$$\vec{r}'(t) = x'(t)\vec{i} + y'(t)\vec{j} + z'(t)\vec{k} = -2\sin t\vec{j} - 2\cos t\vec{k}.$$

So we have

$$\int_C \vec{F} \cdot d\vec{r} = \int_0^{2\pi} (e^{2\cos t}\vec{i} + \vec{k}) \cdot ((-2\sin t)\vec{j} + (-2\cos t)\vec{k})\, dt$$

$$= \int_0^{2\pi} -2\cos t\, dt$$

$$= -2\sin t \Big|_0^{2\pi}$$

$$= 0$$

11. C_1: An equation for the circle is $(x - 1)^2 + y^2 = 1$ or $y = \sqrt{2x - x^2}$. Thus, a possible parameterization is given by

$$(x(t), y(t)) = (t, \sqrt{2t - t^2}), \quad 0 \le t \le 2.$$

C_2: An equation for the parabola must have the form $y = k(x - 1)^2$, since the vertex has been shifted left by 1. Substituting the point $(0, -2)$ yields $k = -2$. So a parameterization is

$$(x(t), y(t)) = (t, -2(t - 1)^2), \quad -1 \le t \le 2.$$

C_3: An equation for the curve is $y = \sin x$. A parameterization is given by

$$(x(t), y(t)) = (t, \sin t), \quad -2\pi \le t \le -\pi.$$

12. (a) The line integral $\int_C (xy\vec{i} + x\vec{j}) \cdot d\vec{r}$ is positive. This follows from the fact that all of the vectors of $xy\vec{i} + x\vec{j}$ at points along C point approximately in the same direction as C (meaning the angles between the vectors and the direction of C are less than $\pi/2$).

    (b) Using the parameterization $x(t) = t$, $y(t) = 3t$, with $x'(t) = 1$, $y'(t) = 3$, we have

$$\int_C \vec{F} \cdot d\vec{r} = \int_0^4 \vec{F}(t, 3t) \cdot (\vec{i} + 3\vec{j}) \, dt$$

$$= \int_0^4 (3t^2\vec{i} + t\vec{j}) \cdot (\vec{i} + 3\vec{j}) \, dt$$

$$= \int_0^4 (3t^2 + 3t) \, dt$$

$$= \left(t^3 + \frac{3}{2}t^2 \right) \Big|_0^4$$

$$= 88.$$

    (c) Figure 18.11 shows the oriented path C', with the "turn around" points P and Q. The particle first travels from the origin to the point P (call this path C_1), then backs up from P to Q (call this path C_2), then goes from Q to the point $(4, 12)$ in the original direction (call this path C_3). See Figure 18.12. Thus, $C' = C_1 + C_2 + C_3$. Along the parts of C_1 and C_2 that overlap, the line integrals cancel, so we are left with the line integral over the part of C_1 that does not overlap with C_2, followed by the line integral over C_3. Thus, the line integral over C' is the same as the line integral over the direct route from the point $(0, 0)$ to the point $(4, 12)$.

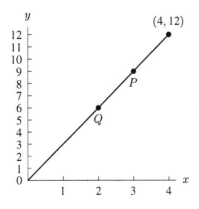

Figure 18.11

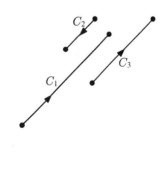

Figure 18.12

(d) The parameterization

$$(x(t), y(t)) = \left(\frac{1}{3}(t^3 - 6t^2 + 11t), (t^3 - 6t^2 + 11t) \right)$$

has $(x(0), y(0)) = (0,0)$ and $(x(4), y(4)) = (4, 12)$. The form of the parameterization we were given shows that the second coordinate is always three times the first. Thus all points on the parameterized curve lie on the line $y = 3x$.

We have to do a bit more work to guarantee that all points on the curve lie on the line *between* the point $(0,0)$ and the point $(4, 12)$; it is possible that they might shoot off to, say, $(100, 300)$ before returning to $(4, 12)$. Let's investigate the maximum and minimum values of $f(t) = t^3 - 6t^2 + 11t$ on the interval $0 \le t \le 4$. We can do this on a graphing calculator or computer, or use single-variable calculus. We already know the values of f at the endpoints, namely 0 and 12. We'll look for local extrema:

$$0 = f'(t) = 3t^2 - 12t + 11$$

which has roots at $t = 2 \pm \frac{1}{\sqrt{3}}$. These are the values of t where the particle changes direction: $t = 2 - \frac{1}{\sqrt{3}}$ corresponds to point P and $t = 2 + \frac{1}{\sqrt{3}}$ corresponds to point Q of C'. At these values of t we have $f(2 - \frac{1}{\sqrt{3}}) \approx 6.4$, and $f(2 + \frac{1}{\sqrt{3}}) \approx 5.6$. The fact that these values are between 0 and 12 shows that f takes on its maximum and minimum values at the endpoints of the interval and not in between.

(e) Using the parameterization given in part (d), we have

$$\vec{r}'(t) = x'(t)\vec{i} + y'(t)\vec{j} = \frac{1}{3}(3t^2 - 12t + 11)\vec{i} + (3t^2 - 12t + 11)\vec{j}.$$

Thus,

$$\int_{C'} \vec{F} \cdot d\vec{r}$$

$$= \int_0^4 \vec{F}\left(\frac{1}{3}(t^3 - 6t^2 + 11t), t^3 - 6t^2 + 11t\right) \cdot \left(\frac{1}{3}(3t^2 - 12t + 11)\vec{i} + (3t^2 - 12t + 11)\vec{j}\right) dt$$

$$= \int_0^4 \left(\frac{1}{3}(t^3 - 6t^2 + 11t)^2\vec{i} + \frac{1}{3}(t^3 - 6t^2 + 11t)\vec{j}\right) \cdot \left(\frac{1}{3}(3t^2 - 12t + 11)\vec{i} + (3t^2 - 12t + 11)\vec{j}\right) dt$$

$$= \int_0^4 \frac{1}{3}(t^3 - 6t^2 + 11t)(3t^2 - 12t + 11) \left\{((t^3 - 6t^2 + 11t)\vec{i} + \vec{j}) \cdot (\frac{1}{3}\vec{i} + \vec{j})\right\} dt$$

$$= \int_0^4 \frac{1}{3}(t^3 - 6t^2 + 11t)(3t^2 - 12t + 11) \left\{\frac{1}{3}(t^3 - 6t^2 + 11t) + 1\right\} dt$$

$$= \frac{1}{9} \int_0^4 (t^3 - 6t^2 + 11t)(3t^2 - 12t + 11)(t^3 - 6t^2 + 11t + 3) dt$$

Numerical integration yields an answer of 88, which agrees with the answer found in part b).

13. (a) Since $\vec{r}(t) = t\vec{i} + t^2\vec{j}$, we have $\vec{r}'(t) = \vec{i} + 2t\vec{j}$. Thus,

$$\int_C \vec{F} \cdot d\vec{r} = \int_0^1 \vec{F}(t, t^2) \cdot (\vec{i} + 2t\vec{j}) dt$$

$$= \int_0^1 [(3t - t^2)\vec{i} + t\vec{j}] \cdot (\vec{i} + 2t\vec{j}) dt$$

$$= \int_0^1 (3t + t^2) dt$$

$$= (\frac{3t^2}{2} + \frac{t^3}{3})\Big|_0^1$$

$$= \frac{3}{2} + \frac{1}{3} - (0 + 0) = \frac{11}{6}$$

(b) Since $\vec{r}(t) = t^2\vec{i} + t\vec{j}$, we have $\vec{r}'(t) = 2t\vec{i} + \vec{j}$. Thus,

$$\int_C \vec{F} \cdot d\vec{r} = \int_0^1 \vec{F}(t^2, t) \cdot (2t\vec{i} + \vec{j}) \, dt$$

$$= \int_0^1 [(3t^2 - t)\vec{i} + t^2\vec{j}] \cdot (2t\vec{i} + \vec{j}) \, dt$$

$$= \int_0^1 (6t^3 - t^2) \, dt$$

$$= (\frac{3t^4}{2} - \frac{t^3}{3})\Big|_0^1$$

$$= \frac{3}{2} - \frac{1}{3} - (0 - 0) = \frac{7}{6}$$

14. (a) The unit circle centered at the origin has equation $x^2 + y^2 = 1$. At any point in the plane, the magnitude of $\vec{F}$ is given by $\|\vec{F}\| = \sqrt{(-y)^2 + x^2}$. Along the unit circle, $\|\vec{F}\| = 1$.

(b) Suppose $\vec{r} = x\vec{i} + y\vec{j}$ is a radius vector to a point (x, y) on the unit circle centered at the origin. See Figure 18.13. Then

$$\vec{r} \cdot \vec{F} = (x\vec{i} + y\vec{j}) \cdot (-y\vec{i} + x\vec{j}) = -xy + xy = 0.$$

So the vector field is perpendicular to any corresponding radius vector, that is, the vector field is tangent to the circle at every point.

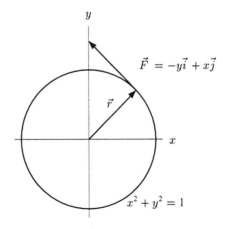

Figure 18.13

(c) We can parameterize C by $(\cos t, \sin t)$, for $0 \leq t \leq 2\pi$. Then

$$\int_C \vec{F} \cdot d\vec{r} = \int_0^{2\pi} \vec{F}(\cos t, \sin t) \cdot (-\sin t\vec{i} + \cos t\vec{j}) \, dt$$

$$= \int_0^{2\pi} (-\sin t \vec{i} + \cos t \vec{j}) \cdot (-\sin t \vec{i} + \cos t \vec{j}) \, dt$$

$$= \int_0^{2\pi} (\sin^2 t + \cos^2 t) \, dt$$

$$= \int_0^{2\pi} 1 \, dt$$

$$= 2\pi$$

Thus,

$$\int_C \vec{F} \cdot d\vec{r} = 2\pi = \text{Circumference of the unit circle.}$$

15. The line integral is defined by chopping the curve C into little pieces, C_i, and forming the sum

$$\sum_{C_i} \vec{F} \cdot \Delta \vec{r}.$$

When the pieces are small, $\Delta \vec{r}$ is approximately tangent to C_i, and its magnitude is approximately equal to the length of the little piece of curve C_i. This means that $\vec{F}$ and $\Delta \vec{r}$ are almost parallel, the dot product is approximately equal to the product of their magnitudes, i.e.,

$$\vec{F} \cdot \Delta \vec{r} \approx m \cdot (\text{Length of } C_i).$$

When we sum all the dot products, we get

$$\sum_{C_i} \vec{F} \cdot \Delta \vec{r} \approx \sum_{C_i} m \cdot (\text{Length of } C_i)$$

$$= m \cdot \sum_{C_i} (\text{Length of } C_i)$$

$$= m \cdot (\text{Length of } C)$$

16. First, check that each of these gives a parameterization of L: each has both coordinates equal (as do all points on L) and each begins at $(0,0)$ and ends at $(1,1)$. Now we calculate the line integral of the vector field $\vec{F} = (3x - y)\vec{i} + x\vec{j}$ using each parameterization.

(a) Using $B(t)$ gives

$$\int_L \vec{F} \cdot d\vec{r} = \int_0^{1/2} ((6t - 2t)\vec{i} + 2t\vec{j}) \cdot (2\vec{i} + 2\vec{j}) \, dt = \int_0^{1/2} 12t \, dt = 6t^2 \Big|_0^{1/2} = \frac{3}{2}.$$

(b) Now we use $C(t)$:

$$\int_L \vec{F} \cdot d\vec{r} = \int_1^2 \left(\left(\frac{3(t^2 - 1)}{3} - \frac{(t^2 - 1)}{3} \right) \vec{i} + \frac{t^2 - 1}{3} \vec{j} \right) \cdot \left(\frac{2t}{3} \vec{i} + \frac{2t}{3} \vec{j} \right) \, dt$$

$$= \int_1^2 \frac{2t}{3} (t^2 - 1) \, dt = \frac{2}{3} \int_1^2 (t^3 - t) \, dt$$

$$= \frac{2}{3} \left(\frac{t^4}{4} - \frac{t^2}{2} \right) \Big|_1^2 = \frac{3}{2}.$$

17. The integral corresponding to $A(t) = (t, t)$ is

$$\int_0^1 3t\, dt.$$

The integral corresponding to $B(t) = (2t, 2t)$ is

$$\int_0^{1/2} 12t\, dt.$$

The substitution $s = 2t$ has $ds = 2\, dt$ and $s = 0$ when $t = 0$ and $s = 1$ when $t = 1/2$. Thus, substituting $t = \dfrac{s}{2}$ into the integral corresponding to $B(t)$ gives

$$\int_0^{1/2} 12t\, dt = \int_0^1 12(\tfrac{s}{2})(\tfrac{1}{2}\, ds) = \int_0^1 3s\, ds.$$

The integral on the right-hand side is now the same as the integral corresponding to $A(t)$. Therefore we have

$$\int_0^{1/2} 12t\, dt = \int_0^1 3s\, ds = \int_0^1 3t\, dt.$$

Alternatively, a similar calculation shows that the substitution $t = 2w$ converts the integral corresponding to $A(t)$ into the integral corresponding to $B(t)$.

18. The integral corresponding to $A(t) = (t, t)$ is

$$\int_0^1 3t\, dt.$$

The integral corresponding to $C(t) = (\tfrac{t^2-1}{3}, \tfrac{t^2-1}{3})$ is

$$\frac{2}{3}\int_1^2 (t^3 - t)\, dt.$$

The substitution $s = \dfrac{t^2 - 1}{3}$ has $ds = \dfrac{2}{3}t\, dt$. Also $s = 0$ when $t = 1$ and $s = 1$ when $t = 2$. Thus, substituting into the integral corresponding to $C(t)$ gives

$$\frac{2}{3}\int_1^2 (t^3 - t)\, dt = \int_1^2 (t^2 - 1)\frac{2}{3}t\, dt = \int_0^1 3s\, ds.$$

The integral on the right-hand side is the same as the integral corresponding to $A(t)$. Therefore we have

$$\frac{2}{3}\int_1^2 (t^3 - t)\, dt = \int_0^1 3s\, ds = \int_0^1 3t\, dt.$$

Alternatively, the substitution $t = \dfrac{w^2 - 1}{3}$ converts the integral corresponding to $A(t)$ into the integral corresponding to $C(t)$.

19. The integral corresponding to $A(t) = (t, t)$ is

$$\int_0^1 3t \, dt.$$

The integral corresponding to $D(t) = (e^t - 1, e^t - 1)$ is

$$3 \int_0^{\ln 2} (e^{2t} - e^t) \, dt.$$

The substitution $s = e^t - 1$ has $ds = e^t \, dt$. Also $s = 0$ when $t = 0$ and $s = 1$ when $t = \ln 2$. Thus, substituting into the integral corresponding to $D(t)$ and using the fact that $e^{2t} = e^t \cdot e^t$ gives

$$3 \int_0^{\ln 2} (e^{2t} - e^t) \, dt = 3 \int_0^{\ln 2} (e^t - 1) e^t \, dt = \int_0^1 3s \, ds.$$

The integral on the right-hand side is the same as the integral corresponding to $A(t)$. Therefore we have

$$3 \int_0^{\ln 2} (e^{2t} - e^t) \, dt = \int_0^1 3s \, ds = \int_0^1 3t \, dt.$$

Alternatively, the substitution $t = e^w - 1$ converts the integral corresponding to $A(t)$ into the integral corresponding to $B(t)$.

20. We parameterize the helical staircase by observing that

$$x = 5 \cos t, \quad y = 5 \sin t, \quad z = t$$

has the correct radius, but climbs 2π in one revolution. To make it climb 4 meters in one revolution, we write:

$$x = 5 \cos t, \quad y = 5 \sin t, \quad z = \frac{4t}{2\pi} = \frac{2t}{\pi}.$$

Thus,

$$\vec{r}\,'(t) = -5 \sin t\vec{i} + 5 \cos t\vec{j} + \frac{2}{\pi}\vec{k}.$$

The gravitational force is given by $\vec{F} = -70g\vec{k}$, and we want to go around 2 turns of the staircase, so we take $0 \le t \le 4\pi$. Thus,

$$\text{Work done by gravity} = \int \vec{F} \cdot d\vec{r} = \int_0^{4\pi} -70g\vec{k} \cdot (-5 \sin t\vec{i} + 5 \cos t\vec{j} + \frac{2}{\pi}\vec{k}) dt$$

$$= \int_0^{4\pi} -\frac{140g}{\pi} dt = -\frac{140g}{\pi} t \Big|_0^{4\pi} = -560g.$$

Notice that the result can also be obtained by multiplying the force by the vertical distance:

$$\text{Gravitational force} \cdot \text{Vertical distance moved} = (-70g)8 = -560g.$$

Now

$$\text{Work done by person} = -\text{Work done by gravity} = 560 \text{ g}.$$

Solutions for Section 18.3

1. The vector field $\vec{F}$ points radially outward, and so is everywhere perpendicular to A; thus, $\int_A \vec{F} \cdot d\vec{r} = 0$.

    Along the first half of B, the terms $\vec{F} \cdot \Delta\vec{r}$ are negative; along the second half the terms $\vec{F} \cdot \Delta\vec{r}$ are positive. By symmetry the positive and negative contributions cancel out, giving a Riemann sum and a line integral of 0.

    The line integral is also 0 along C, by cancellation. Here the values of $\vec{F}$ along the x-axis have the same magnitude as those along the y-axis. On the first half of C the path is traversed in the opposite direction to $\vec{F}$; on the second half of C the path is traversed in the same direction as $\vec{F}$. So the two halves cancel.

2. We parameterize A by $x = t, y = t$, where $0 \le t \le 1$. Then

$$\int_A \vec{F} \cdot d\vec{r} = \int_0^1 (t\vec{i} + t\vec{j}) \cdot (\vec{i} + \vec{j}) \, dt$$
$$= \int_0^1 2t \, dt = t^2 \Big|_0^1 = 1.$$

    The path B has the parameterization $x = t, y = t^2$, where $0 \le t \le 1$. Then we have

$$\int_B \vec{F} \cdot d\vec{r} = \int_0^1 (t\vec{i} + t^2\vec{j}) \cdot (\vec{i} + 2t\vec{j}) \, dt$$
$$= \int_0^1 (t + 2t^3) \, dt = \frac{t^2}{2} + \frac{2t^4}{4} \Big|_0^1 = 1.$$

    We have to break the path C into two separate parameterizations: $x = t, y = 0$, where $0 \le t \le 1$ and $x = 1$, $y = t$, where $0 \le t \le 1$. Then

$$\int_C \vec{F} \cdot d\vec{r} = \int_0^1 (t\vec{i} \cdot \vec{i}) \, dt + \int_0^1 (\vec{i} + t\vec{j}) \cdot \vec{j} \, dt$$
$$= \int_0^1 t \, dt + \int_0^1 t \, dt = \frac{1}{2} + \frac{1}{2} = 1.$$

3. Yes. If $f(x, y) = \frac{1}{2}x^2$, then grad $f = x\vec{i}$.

4. Yes. If $f(x, y) = \frac{1}{3}x^3 - xy^2$, then grad $f = (x^2 - y^2)\vec{i} - 2xy\vec{j}$.

5. No. Suppose there were a function f such that grad $f = \vec{F}$. Then we would have

$$\frac{\partial f}{\partial x} = \frac{-z}{\sqrt{x^2 + z^2}}.$$

    Hence we would have

$$\frac{\partial^2 f}{\partial y \partial x} = \frac{\partial}{\partial y} \left(\frac{-z}{\sqrt{x^2 + z^2}} \right) = 0.$$

    In addition, since grad $f = \vec{F}$, we have that

$$\frac{\partial f}{\partial y} = \frac{y}{\sqrt{x^2 + z^2}}.$$

Thus we also know that

$$\frac{\partial^2 f}{\partial x \partial y} = \frac{\partial}{\partial x}\left(\frac{y}{\sqrt{x^2 + y^2}}\right) = -xy(x^2 + z^2)^{-3/2}.$$

Notice that

$$\frac{\partial^2 f}{\partial y \partial x} \neq \frac{\partial^2 f}{\partial x \partial y}.$$

Since we expect $\frac{\partial^2 f}{\partial y \partial x} = \frac{\partial^2 f}{\partial x \partial y}$, we have got a contradiction. The only way out of this contradiction is to conclude there is no function f with grad $f = \vec{F}$. Thus $\vec{F}$ is not a gradient vector field.

6. Yes. Let

$$f(\vec{r}) = -\frac{1}{r} = -(x^2 + y^2 + z^2)^{-1/2}$$

Then

$$\frac{\partial f}{\partial x} = x(x^2 + y^2 + z^2)^{-3/2}$$

$$\frac{\partial f}{\partial y} = y(x^2 + y^2 + z^2)^{-3/2}$$

$$\frac{\partial f}{\partial z} = z(x^2 + y^2 + z^2)^{-3/2}$$

So grad $f = (x^2 + y^2 + z^2)^{-3/2}(x\vec{i} + y\vec{j} + z\vec{k}) = \vec{r}/r^3$

7. Since $\vec{F} = \text{grad}\left(\dfrac{x^2 + y^2}{2}\right)$, the line integral can be calculated using the Fundamental Theorem of Line Integrals:

$$\int_c \vec{F} \cdot d\vec{r} = \left.\frac{x^2 + y^2}{2}\right|_{(0,0)}^{(3/\sqrt{2}, 3/\sqrt{2})} = \frac{9}{2}.$$

8. This vector field is not a gradient field, so we evaluate the line integral directly. Let C_1 be the path along the x-axis from $(0,0)$ to $(3,0)$ and let C_2 be the path from $(3,0)$ to $(3/\sqrt{2}, 3/\sqrt{2})$ along $x^2 + y^2 = 9$. Then

$$\int_C \vec{H} \cdot d\vec{r} = \int_{C_1} \vec{H} \cdot d\vec{r} + \int_{C_2} \vec{H} \cdot d\vec{r}.$$

On C_1, the vector field has only a $\vec{j}$ component (since $y = 0$), and $\vec{H}$ is therefore perpendicular to the path. Thus,

$$\int_{C_1} \vec{H} \cdot d\vec{r} = 0.$$

On C_2, the vector field is tangent to the path. The path is one eighth of a circle of radius 3 and so has length $2\pi(3/8) = 3\pi/4$.

$$\int_{C_2} \vec{H} \cdot d\vec{r} = \|\vec{H}\| \cdot \text{Length of path} = 3 \cdot \left(\frac{3\pi}{4}\right) = \frac{9\pi}{4}.$$

Thus,

$$\int_C \vec{H} \cdot d\vec{r} = \frac{9\pi}{4}.$$

9. Since $\vec{F} = \text{grad}(y \ln(x+1))$, we evaluate the line integral using the Fundamental Theorem of Line Integrals:

$$\int_C \vec{F} \cdot d\vec{r} = y \ln(x+1) \Big|_{(0,0)}^{(3/\sqrt{2},3/\sqrt{2})} = \frac{3}{\sqrt{2}} \ln\left(\frac{3}{\sqrt{2}} + 1\right) - 0 \ln 1 = \frac{3}{\sqrt{2}} \ln\left(\frac{3}{\sqrt{2}} + 1\right).$$

10. Since $\vec{G} = \text{grad}(e^{xy} + \sin(x+y))$, the line integral can be calculated using the Fundamental Theorem of Line Integrals:

$$\int_c \vec{F} \cdot d\vec{r} = e^{xy} + \sin(x+y) \Big|_{(0.0)}^{(3/\sqrt{2},3/\sqrt{2})} = e^{9/2} + \sin\left(\frac{6}{\sqrt{2}}\right) - e^0 = e^{9/2} + \sin(3\sqrt{2}) - 1.$$

11. (a) To find the change in f by computing a line integral, we first choose a path C between the points; the simplest is a line. We parameterize the line by $(x(t), y(t)) = (t, \pi t/2)$, with $0 \le t \le 1$. Then $(x'(t), y'(t)) = (1, \pi/2)$, so the Fundamental Theorem of Line Integrals tells us that

$$f\left(1, \frac{\pi}{2}\right) - f(0,0) = \int_C \text{grad} f \cdot d\vec{r}$$

$$= \int_0^1 \text{grad} f\left(t, \frac{\pi t}{2}\right) \cdot \left(\vec{i} + \frac{\pi}{2}\vec{j}\right) dt$$

$$= \int_0^1 \left(2te^{t^2} \sin\left(\frac{\pi t}{2}\right)\vec{i} + e^{t^2} \cos\left(\frac{\pi t}{2}\right)\vec{j}\right) \cdot \left(\vec{i} + \frac{\pi}{2}\vec{j}\right) dt$$

$$= \int_0^1 \left(2te^{t^2} \sin\left(\frac{\pi t}{2}\right) + \frac{\pi e^{t^2}}{2} \cos\left(\frac{\pi t}{2}\right)\right) dt$$

$$= \int_0^1 \frac{d}{dt}\left(e^{t^2} \sin\left(\frac{\pi t}{2}\right)\right) dt$$

$$= e^{t^2} \sin\left(\frac{\pi t}{2}\right)\Big|_0^1 = e = 2.718.$$

This integral can also be approximated numerically.

(b) The other way to find the change in f between these two points is to first find f. To do this, observe that

$$2xe^{x^2} \sin y\vec{i} + e^{x^2} \cos y\vec{j} = \frac{\partial}{\partial x}\left(e^{x^2} \sin y\right)\vec{i} + \frac{\partial}{\partial y}\left(e^{x^2} \sin y\right)\vec{j} = \text{grad}\left(e^{x^2} \sin y\right).$$

So one possibility for f is $f(x,y) = e^{x^2} \sin y$. Thus,

$$\text{Change in } f\Big|_{(0,0)}^{(1,\pi/2)} = e^{x^2} \sin y\Big|_{(0,0)}^{(1,\pi/2)} = e^1 \sin\left(\frac{\pi}{2}\right) - e^0 \sin 0 = e.$$

The exact answer confirms our calculations in part (a) which show that the answer is e.

12. (a) For path (i), we have $x(t) = t, y(t) = t^2$, so $x'(t) = 1, y'(t) = 2t$. Thus,

$$\int_C \vec{F} \cdot d\vec{r} = \int_0^1 \vec{F}(t, t^2) \cdot (\vec{i} + 2t\vec{j}) dt$$

$$= \int_0^1 [(t + t^2)\vec{i} + t\vec{j}] \cdot (\vec{i} + 2t\vec{j}) dt$$

$$= \int_0^1 (t + 3t^2)\, dt$$

$$= (\frac{t^2}{2} + t^3)\Big|_0^1$$

$$= \frac{1}{2} + 1 - (0 + 0) = \frac{3}{2}.$$

For path (ii), we have $x(t) = t^2, y(t) = t$, so $x'(t) = 2t, y'(t) = 1$. Thus,

$$\int_C \vec{F} \cdot d\vec{r} = \int_0^1 \vec{F}\,(t^2, t) \cdot (2t\vec{i} + \vec{j})\, dt$$

$$= \int_0^1 [(t^2 + t)\vec{i} + t^2\vec{j}] \cdot (2t\vec{i} + \vec{j})\, dt$$

$$= \int_0^1 (2t^3 + 3t^2)\, dt$$

$$= (\frac{t^4}{2} + t^3)\Big|_0^1$$

$$= \frac{1}{2} + 1 - (0 + 0) = \frac{3}{2}.$$

For path (iii), we have $x(t) = t, y(t) = t^n$, so $x'(t) = 1, y'(t) = nt^{n-1}$. Thus,

$$\int_C \vec{F} \cdot d\vec{r} = \int_0^1 \vec{F}\,(t, t^n) \cdot (\vec{i} + nt^{n-1}\vec{j})\, dt$$

$$= \int_0^1 [(t + t^n)\vec{i} + t\vec{j}] \cdot (\vec{i} + nt^{n-1}\vec{j})\, dt$$

$$= \int_0^1 (t + t^n + nt^n)\, dt$$

$$= \int_0^1 (t + (n + 1)t^n)\, dt$$

$$= \left(\frac{t^2}{2} + t^{n+1}\right)\Big|_0^1$$

$$= \frac{1}{2} + 1 - (0 + 0) = \frac{3}{2}.$$

(b)   If $f(x, y) = xy + x^2/2$, we have $\vec{F} = \operatorname{grad} f$. Each path goes from $(0, 0)$ to $(1, 1)$. Thus in each case

$$\int_C \vec{F} \cdot d\vec{r} = f(1, 1) - f(0, 0) = \frac{3}{2}.$$

13. (a) Three possible paths are shown in Figure 18.14.

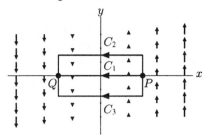

Figure 18.14

Since $\vec{F}$ is perpendicular to the horizontal axis everywhere, $\vec{F} \cdot d\vec{r} = 0$ along C_1.

Since C_2 starts out in the direction of $\vec{F}$, the first leg of C_2 will have a positive line integral. The second horizontal part of C_2 will have a 0 line integral, and the third leg that ends at Q will have a positive line integral. Thus the line integral along C_2 is positive.

A similar argument shows that the line integral along $C_3 < 0$.

(b) No, $\vec{F}$ is not a gradient field, since the line integrals along these three paths joining P and Q do not have the same value.

14. (a) The integral is positive, because the portion of the path that goes with the vector field is longer than the portion of the path that goes against it, and in addition the vectors are larger in magnitude along the former and smaller in magnitude along the latter.

(b) If it were true that $\vec{F} = \text{grad } f$ for some function f, then the integral around every closed path would be zero. But in part (a) we saw that the integral around one closed path was not zero, so $\vec{F}$ cannot be a gradient vector field.

(c) The region shown is in the first quadrant. In that quadrant, the vectors of $\vec{F_1}$ point away from the orgin, so $\vec{F_1}$ does not fit. The vectors of both $\vec{F_2}$ and $\vec{F_3}$ point up and to the left, so they are both possibilities; of these, $\vec{F_2}$ fits best because its vectors get larger in magnitude as you move away from the origin, which fits the diagram. The vectors in $\vec{F}_3$ shrink as you move away from the origin.

15. A path-independent vector field must have zero circulation around all closed paths. Consider a vector field like $\vec{F}(x,y) = |x|\vec{j}$, shown in Figure 18.15.

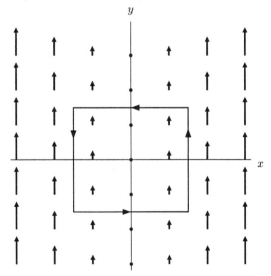

Figure 18.15

A rectangular path that is symmetric about the y-axis will have zero circulation: on the horizontal sides, the field is perpendicular, so the line integral is zero. The line integrals on the vertical sides are equal in magnitude and opposite in sign, so they cancel out, giving a line integral of zero. However, this field is not path-independent, because it is possible to find two paths with the same endpoints but different values of the line integral of $\vec{F}$. For example, consider the two points $(0, 0)$ and $(0, 1)$. The path C_1 in Figure 18.16 along the y axis gives zero for the line integral, because the field is 0 along the y axis, whereas a path like C_2 will have a nonzero line integral. Thus the line integral depends on the path between the points, so $\vec{F}$ is not path-independent.

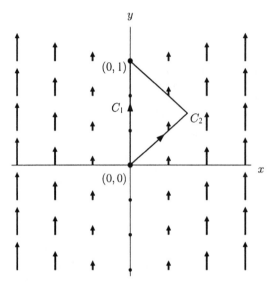

Figure 18.16

16. This is false, because the line integral yields a scalar whereas the total change in $\vec{F}$ would be a vector. In the *special* case when $\vec{F}$ happens to be the gradient of a scalar function f, the line integral does give the total change of the scalar function f along the path—but not of the vector function $\vec{F}$.

17. You can easily come up with counterexamples: suppose that $\vec{F} \neq \vec{G}$ but that both are gradient fields. For example, $\vec{F} = \vec{i}$ and $\vec{G} = \vec{j}$. Then, if C is a closed curve, the line integral around C of both $\vec{F}$ and $\vec{G}$ will equal to zero. But this does not mean that $\vec{F} = \vec{G}$.

18. The total change of f along C depends only on the endpoints of C. If f has the same values at each endpoint (or if C is closed, so that the endpoints coincide) then the total change will be zero. This in no way restricts the shape of the curve C. For example, take $f(x, y) = x^2 + y^2$ and C to be the straight line from the point $(1, 0)$ to the point $(0, 1)$. Then $f(1, 0) = f(0, 1) = 1$ so the change in f along C is zero, but C is not a contour of f.

19. (a) We parameterize the path by $(\cos t, \sin t)$ for $\pi/2 \leq t \leq \pi$. Since $t = \pi/2$ gives the end point, $(0, 1)$ and $t = \pi$ gives the starting point $(-1, 0)$, we have

$$\int_C \vec{F} \cdot d\vec{r} = \int_\pi^{\pi/2} \vec{F}(\cos t, \sin t) \cdot (-\sin t\vec{i} + \cos t\vec{j})\, dt$$

$$= -\int_{\pi/2}^{\pi} (\sin t\vec{i} - \cos t\vec{j}) \cdot (-\sin t\vec{i} + \cos t\vec{j})\, dt$$

$$= -\int_{\pi/2}^{\pi} (-\sin^2 t - \cos^2 t)\, dt$$

$$= \int_{\pi/2}^{\pi} 1\, dt$$

$$= t \Big|_{\pi/2}^{\pi}$$

$$= \pi/2.$$

The work done by the force is $+\pi/2$. The work is positive since the force is always in the direction of the path (in fact it is always tangent to C since $\vec{F} \cdot \vec{r} = 0$).

(b) If we redo our computations using the entire unit circle, the only change will be the limits of integration: they'll change to 0 to 2π. This yields an answer of 2π (or -2π, depending on orientation). Since the work around a closed path is not zero, the force is not path-independent.

20. Since $\|\vec{v}(t)\|^2 = \vec{v}(t) \cdot \vec{v}(t)$ and since

$$\vec{v}(t) = \vec{r}'(t) = x'(t)\vec{i} + y'(t)\vec{j} + z'(t)\vec{k},$$

we have

$$\frac{1}{2}\frac{d}{dt}\|\vec{v}(t)\|^2 = \frac{1}{2}\frac{d}{dt}(\vec{v}(t) \cdot \vec{v}(t))$$

$$= \frac{1}{2}\frac{d}{dt}(x'(t)^2 + y'(t)^2 + z'(t)^2)$$

$$= \frac{1}{2}(2x'(t)x''(t) + 2y'(t)y''(t) + 2z'(t)z''(t))$$

$$= x'(t)x''(t) + y'(t)y''(t) + z'(t)z''(t)$$

$$= (x'(t)\vec{i} + y'(t)\vec{j} + z'(t)\vec{k}) \cdot (x''(t)\vec{i} + y''(t)\vec{j} + z''(t)\vec{k})$$

$$= \vec{v}(t) \cdot \vec{a}(t) \quad \text{(Since } \vec{a}(t) = x''(t)\vec{i} + y''(t)\vec{j} + z''(t)\vec{k}.\text{)}$$

$$= \vec{a}(t) \cdot \vec{v}(t).$$

21. (a) The level surfaces are horizontal planes given by $gz = c$, so $z = c/g$. The potential energy increases with the height above the earth. This means that more energy is stored as "potential to fall" as height increases.

(b) The gradient of ϕ points upward (in the direction of increasing potential energy), so $\nabla\phi = g\vec{k}$. The gravitational force acts toward the earth in the direction of $-\vec{k}$. So, $\vec{F} = -g\vec{k}$. The negative sign represents the fact that the gravitational force acts in the direction of the decreasing potential energy.

22. (a) We use $\vec{F} = m\vec{a}$ and the parameterization of C given by $r(t)$ for $t_0 \leq t \leq t_1$. In addition, we need the fact that $\frac{1}{2}\frac{d}{dt}\|\vec{v}(t)\|^2 = \vec{a} \cdot \vec{v}$:

$$\int_C \vec{F} \cdot d\vec{r} = \int_C m\vec{a} \cdot d\vec{r}$$

$$= \int_{t_0}^{t_1} m\vec{a} \cdot \vec{r}'\, dt$$

$$= \int_{t_0}^{t_1} m(\vec{a} \cdot \vec{v})\, dt$$

$$= \int_{t_0}^{t_1} \frac{m}{2} \left(\frac{d}{dt} \|\vec{v}(t)\|^2 \right) dt$$

$$= \frac{m}{2} \|\vec{v}(t)\|^2 \Big|_{t_0}^{t_1}$$

$$= \frac{m}{2} \|\vec{v}(t_1)\|^2 - \frac{m}{2} \|\vec{v}(t_0)\|^2$$

$$= \text{Kinetic energy at } Q - \text{Kinetic energy at } P.$$

(b)   Since $\vec{F} = -\nabla f$ we use the Fundamental Theorem of Line Integrals:

$$\int_C \vec{F} \cdot d\vec{r} = \int_C -\nabla f \cdot d\vec{r}$$

$$= -\int_C \nabla f \cdot d\vec{r}$$

$$= -(f(\vec{r}(t_1)) - f(\vec{r}(t_0)))$$

$$= -(\text{Potential energy at } Q - \text{Potential energy at } P)$$

$$= \text{Potential energy at } P - \text{Potential energy at } Q.$$

(c)   In parts (a) and (b) we derived two expressions for the work done by $\vec{F}$ as the particle moves from P to Q. These two expressions must be equal, so

Kinetic energy at Q − Kinetic energy at P = Potential energy at P − Potential energy at Q.

Rewriting this equation we have,

(Potential energy + Kinetic energy) at P = (Potential energy + Kinetic energy) at Q.

This shows that the total energy is the same at P as at Q. Since P and Q are arbitrary points in space, the total energy of a particle moving in a force vector field $\vec{F} = -\nabla f$ is a constant.

Solutions for Section 18.4

1.   The drawing of the contour diagrams fitting this gradient field would look like Figure 18.17:

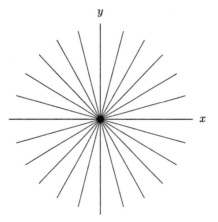

Figure 18.17

This diagram could not be the contour diagram because the origin is on all contours. This means that $f(0, 0)$ would have to take on more than one value, which is impossible. At a point P other than the origin, we have the same problem. The values on the contours increase as you go counterclockwise around, since the gradient vector points in the direction of greatest increase of a function. But, starting at P, and going all the way around the origin, you would eventually get back to P again, and with a larger value of f, which is impossible.

2. The drawing of the contour diagrams in Figure 18.18 fitting this gradient field would look like Figure 18.18. The values on the contours would increase both as y increases (for positive x) and as y decreases (for negative x), following the rule that the gradient vector points in the direction of greatest increase of a function. Therefore, it is impossible for this to be a contour diagram.

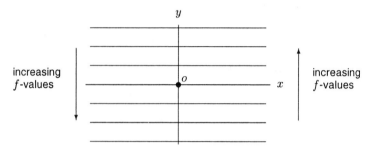

Figure 18.18

3. We know that
$$\frac{\partial f}{\partial x} = 2xy \quad \text{and} \quad \frac{\partial f}{\partial y} = x^2 + 8y^3$$

Now think of y as a constant in the equation for $\partial f/\partial x$ and integrate, giving
$$f(x, y) = x^2 y + C(y).$$

Since the constant of integration may depend on y, it is written $C(y)$. Differentiating this expression for $f(x, y)$ with respect to y and using the fact that $\partial f/\partial y = x^2 + 8y^3$, we get
$$\frac{\partial f}{\partial y} = x^2 + C'(y) = x^2 + 8y^3.$$

Therefore
$$C'(y) = 8y^3 \quad \text{so} \quad C(y) = 2y^4 + K.$$

for some constant K. Thus,
$$f(x, y) = x^2 y + 2y^4 + K.$$

4. Integrating
$$\frac{\partial f}{\partial x} = yze^{xyz} + z^2 \cos(xz^2)$$

with respect to x and thinking of y and z as constant gives
$$f(x, y, z) = e^{xyz} + \sin(xz^2) + C(y, z).$$

Differentiating with respect to y and using the fact that $\partial f/\partial y = xze^{xyz}$ gives

$$\frac{\partial f}{\partial y} = xze^{xyz} + \frac{\partial C}{\partial y} = xze^{xyz}.$$

Thus, $\partial C/\partial y = 0$. This means C does not depend on y and can be written $C(z)$, giving:

$$f(x,y,z) = e^{xyz} + \sin(xz^2) + C(z).$$

Differentiating with respect to z, we get

$$\frac{\partial f}{\partial z} = xye^{xyz} + 2zx\cos(xz^2) + C'(z).$$

The expression for grad f tells us that

$$\frac{\partial f}{\partial z} = xye^{xyz} + 2xz\cos(xz^2).$$

Thus, we have $C'(z) = 0$ so $C = $ constant, giving

$$f(x,y,z) = e^{xyz} + \sin(xz^2) + C.$$

5. The domain of the vector field $\vec{F}(x,y) = y\vec{i} + y\vec{j}$ is the whole xy-plane. In order to see if $\vec{F}$ is a gradient let us apply the curl test:

$$\frac{\partial F_1}{\partial y} = 1$$

and

$$\frac{\partial F_2}{\partial x} = 0$$

So $\vec{F}$ is not the gradient of any function.

6. The domain of the vector field $\vec{F}(x,y) = (x^2 + y^2)\vec{i} + 2xy\vec{j}$ is the whole xy-plane. Let us apply the curl test:

$$\frac{\partial F_1}{\partial y} = 2y = \frac{\partial F_2}{\partial x}$$

so $\vec{F}$ is the gradient of some function f. In order to compute f we first integrate

$$\frac{\partial f}{\partial x} = x^2 + y^2$$

with respect to x, thinking of y as a constant.
   We get

$$f(x,y) = \frac{x^3}{3} + xy^2 + C(y)$$

Differentiating with respect to y and using the fact that $\partial f/\partial y = 2xy$ gives

$$\frac{\partial f}{\partial y} = 2xy + C'(y) = 2xy$$

Thus $C'(y) = 0$ so C is a constant and

$$f(x,y) = \frac{x^3}{3} + xy^2 + C.$$

7. The domain of the vector field $\vec{F} = (2xy^3 + y)\vec{i} + (3x^2y^2 + x)\vec{j}$ is the whole xy-plane. We apply the curl test:

$$\frac{\partial F_1}{\partial y} = 6xy^2 + 1 = \frac{\partial F_2}{\partial x}$$

so $\vec{F}$ is the gradient of a function f. In order to compute f we first integrate

$$\frac{\partial f}{\partial x} = 2xy^3 + y$$

with respect to x thinking of y as a constant. We get

$$f(x, y) = x^2y^3 + xy + C(y)$$

Differentiating with respect to y and using the fact that $\partial f/\partial y = 3x^2y^2 + x$ gives

$$\frac{\partial f}{\partial y} = 3x^2y^2 + x + C'(y) = 3x^2y^2 + x$$

Thus $C'(y) = 0$ so C is constant and

$$f(x, y) = x^2y^3 + xy + C.$$

8. The domain of the vector field $\vec{F} = \dfrac{\vec{i}}{x} + \dfrac{\vec{j}}{y} + \dfrac{\vec{k}}{xy}$ is the set of points in the three space, (x, y, z) such that $x \neq 0$ and $y \neq 0$. This is the set of points in the three space left after removing the planes $x = 0$ and $y = 0$. This domain has the property that every closed curve is the boundary of a surface entirely contained in it, hence we can apply the curl test.

$$\text{curl } \vec{F} = \begin{vmatrix} \vec{i} & \vec{j} & \vec{k} \\ \frac{\partial}{\partial x} & \frac{\partial}{\partial y} & \frac{\partial}{\partial z} \\ \frac{1}{x} & \frac{1}{y} & \frac{1}{xy} \end{vmatrix}$$

$$= \vec{i} \left(\frac{\partial}{\partial y}(\frac{1}{xy}) - \frac{\partial}{\partial z}(\frac{1}{y}) \right) - \vec{j} \left(\frac{\partial}{\partial x}(\frac{1}{xy}) - \frac{\partial}{\partial z}(\frac{1}{x}) \right) + \vec{k} \left(\frac{\partial}{\partial x}(\frac{1}{y}) - \frac{\partial}{\partial y}(\frac{1}{x}) \right)$$

$$= -\frac{1}{xy^2}\vec{i} + \frac{1}{x^2y}\vec{j} \neq 0$$

Therefore $\vec{F}$ is not the gradient of any function.

9. The domain of the vector field $\vec{F} = \dfrac{\vec{i}}{x} + \dfrac{\vec{j}}{y} + \dfrac{\vec{k}}{z}$ is the set of points (x, y, z) in the three space such that $x \neq 0$, $y \neq 0$ and $z \neq 0$. This is what is left in the three space after removing the coordinate planes.

    This domain has the property that every closed curve is the boundary of a surface entirely contained in it, hence we can apply the curl test.

$$\text{curl } \vec{F} = \begin{vmatrix} \vec{i} & \vec{j} & \vec{k} \\ \frac{\partial}{\partial x} & \frac{\partial}{\partial y} & \frac{\partial}{\partial z} \\ \frac{1}{x} & \frac{1}{y} & \frac{1}{z} \end{vmatrix}$$

So curl $\vec{F} = \vec{0}$ and thus $\vec{F}$ is the gradient of a function f. In order to compute f we first integrate

$$\frac{\partial f}{\partial x} = \frac{1}{x}$$

with respect to x, thinking of y and z as constants.

We get
$$f(x, y, z) = \ln|x| + C(y, z)$$

Differentiating with respect to y and using the fact that $\partial f / \partial y = 1/y$ gives
$$\frac{\partial f}{\partial y} = \frac{\partial C}{\partial y} = \frac{1}{y}$$

We integrate this relation with respect to y thinking of z as a constant. We get
$$f(x, y, z) = \ln|xy| + K(z)$$

Differentiating with respect to z and using the fact that $\partial f / \partial z = 1/z$ gives
$$\frac{\partial f}{\partial z} = K'(z) = \frac{1}{z}$$

Now we integrate with respect to z and get
$$f(x, y, z) = \ln A|xyz|$$

where A is a positive constant.

10. The domain of the vector field $\vec{F} = 2x\cos(x^2 + z^2)\vec{i} + \sin(x^2 + z^2)\vec{j} + 2z\cos(x^2 + z^2)\vec{k}$ is the whole three space so we can apply the curl test.

$$\text{curl}\, \vec{F} = \begin{vmatrix} \vec{i} & \vec{j} & \vec{k} \\ \frac{\partial}{\partial x} & \frac{\partial}{\partial y} & \frac{\partial}{\partial z} \\ 2x\cos(x^2 + y^2) & \sin(x^2 + y^2) & 2z\cos(x^2 + y^2) \end{vmatrix}$$
$$= -4yz\sin(x^2 + y^2)\vec{i} + 4xz\sin(x^2 + y^2)\vec{j} + (2x\cos(x^2 + y^2) + 4xy\sin(x^2 + y^2))\vec{k}$$
$$\neq 0$$

As curl $\vec{F} \neq \vec{0}$, $\vec{F}$ is not the gradient of any function.

11. (a) The vector field points in the opposite direction to the orientation of the curve, hence the circulation is negative. See Figure 18.19.

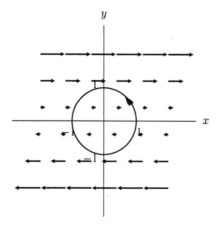

Figure 18.19

(b)   Since $\vec{F} = y\vec{i}$, we have $\partial F_1/\partial y = 1$ and $\partial F_2/\partial x = 0$ and $\partial F_1/\partial y = 1$. Thus, using Green's Theorem if R is the region enclosed by the closed curve C (the unit circle centered at the origin and traversed counterclockwise), we have

$$\int_C \vec{F} \cdot d\vec{r} = \int_R \left(\frac{\partial F_2}{\partial x} - \frac{\partial F_1}{\partial y} \right) dx\,dy = \int_R (-1)\,dx\,dy = -\text{Area of } R = -\pi.$$

12.   Since $\vec{F} = x\vec{j}$, we have $\partial F_2/\partial x = 1$ and $\partial F_1/\partial y = 0$. Thus, using Green's Theorem if R is the region enclosed by the closed curve C, we have

$$\int_C \vec{F} \cdot d\vec{r} = \int_R \left(\frac{\partial F_2}{\partial x} - \frac{\partial F_1}{\partial y} \right) dx\,dy = \int_R 1\,dx\,dy = \text{Area of } R$$

13.   Using $\vec{F} = x\vec{j} = a\cos t\vec{j}$ and $\vec{r}'(t) = -a\sin t\vec{i} + b\cos t\vec{j}$, we have

$$A = \int_C \vec{F} \cdot d\vec{r} = \int_0^{2\pi} (a\cos t)(b\cos t)\,dt$$

$$= ab \int_0^{2\pi} \cos^2 t\,dt$$

$$= ab \int_0^{2\pi} \frac{1 + \cos 2t}{2}\,dt$$

$$= \pi ab + \frac{ab}{4} \sin 2t \Big|_0^{2\pi}$$

$$= \pi ab$$

The ellipse is shown in Figure 18.20.

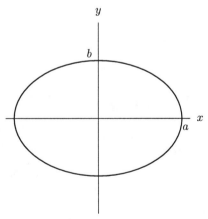

Figure 18.20: $\dfrac{x^2}{a^2} + \dfrac{y^2}{b^2} = 1$

14.   Using $\vec{F} = x\vec{j} = a\cos^3 t$ and $\vec{r}'(t) = -3a\cos^2 t \sin t\vec{i} + 3a\sin^2 t \cos t\vec{j}$, we have

$$A = \int_C \vec{F} \cdot d\vec{r} = \int_0^{2\pi} (a\cos^3 t)(3a\sin^2 t \cos t)\,dt$$

$$= 3a^2 \int_0^{2\pi} \cos^4 t \sin^2 t \, dt = 3a^2 \int_0^{2\pi} \cos^2 t (\sin t \cos t)^2 \, dt = 3a^2 \int_0^{2\pi} \cos^2 t \frac{\sin^2 2t}{4} \, dt$$

$$= \frac{3a^2}{16} \int_0^{2\pi} (1 + \cos 2t)(1 - \cos 4t) \, dt$$

$$= \frac{3a^2}{16} \int_0^{2\pi} (1 + \cos 2t - \cos 4t - \cos 2t \cos 4t) \, dt$$

$$= \frac{3a^2}{16} \int_0^{2\pi} \left(1 + \cos 2t - \cos 4t - \frac{1}{2}\cos 6t - \frac{1}{2}\cos 2t\right) dt$$

$$= \frac{3a^2}{16} \left(t - \frac{1}{2}\sin 2t - \frac{1}{4}\sin 4t + \frac{1}{12}\sin 6t + \frac{1}{4}\sin 2t\right)\Big|_0^{2\pi}$$

$$= \frac{3\pi a^2}{8}$$

For the last integral we use the trigonometric formula $\cos 2t \cos 4t = \frac{1}{2}(\cos 6t + \cos 2t)$. The hypocycloid is shown in Figure 18.21.

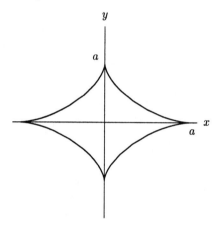

Figure 18.21: $x^{2/3} + y^{2/3} = a^{2/3}$

15. Using $\vec{F} = x\vec{j} = \dfrac{3t^2}{1+t^3}\vec{j}$ and $\vec{r}'(t) = \dfrac{1 - 2t^3}{(1+t^3)^2}\vec{i} + \dfrac{3t(2 - t^3)}{(1+t^3)^2}\vec{j}$, we have

$$A = \int_C \vec{F} \cdot d\vec{r} = \int_0^\infty \frac{3t}{1+t^3} \cdot \frac{3t(2 - t^3)}{(1+t^3)^2} \, dt$$

$$= 9 \int_0^\infty \frac{t^2(2 - t^3)}{(1+t^3)^3} \, dt$$

We make the change of variables $u = 1 + t^3$ so $du = 3t^2 dt$ and $2 - t^3 = 3 - u$. So

$$A = 3 \int_1^\infty \frac{3 - u}{u^3} \, du.$$

This is an improper integral, so it can be computed as follows

$$A = 3 \int_1^\infty \frac{3 - u}{u^3} \, du = \lim_{b \to \infty} 3 \int_1^b \left(\frac{3}{u^3} - \frac{1}{u^2}\right) du$$

$$= \lim_{b \to \infty} \left[9 \left(-\frac{1}{2} \right) u^{-2} \Big|_1^b + 3 \frac{1}{u} \Big|_1^b \right]$$

$$= \lim_{b \to \infty} \left[-\frac{9}{2} \left(\frac{1}{b^2} - 1 \right) + 3 \left(\frac{1}{b} - 1 \right) \right]$$

$$= -\frac{9}{2}(0 - 1) + 3(0 - 1) = \frac{3}{2}.$$

The Folium of Descartes is shown in Figure 18.22.

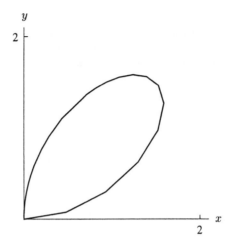

Figure 18.22: $x^3 + y^3 = 3xy$

16. (a) Suppose we try to integrate with respect to x first. Then using Figure 18.23, we see that

$$\int_R (2x - 2y)e^{x^2+y^2} \, dA = \int_{-1}^1 \int_0^{\sqrt{1-x^2}} (2x - 2y)e^{x^2+y^2} \, dy \, dx$$

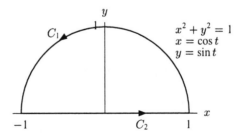

Figure 18.23

Since there is no elementary formula for the antiderivative of $(2x - 2y)e^{x^2+y^2}$ with respect to y, we cannot proceed. The same problem arises if we try integrating with respect to x first. Thus, we can not use iterated integration to evaluate this integral.

(b) To use Green's Theorem, we need to find F_1 and F_2 such that

$$(2x - 2y)e^{x^2+y^2} = \frac{\partial F_2}{\partial x} - \frac{\partial F_1}{\partial y}.$$

Let's take

$$2xe^{x^2+y^2} = \frac{\partial F_2}{\partial x} \quad \text{so} \quad F_2 = e^{x^2+y^2}$$

$$2ye^{x^2+y^2} = \frac{\partial F_1}{\partial y} \quad \text{so} \quad F_1 = e^{x^2+y^2}.$$

Thus,

$$\vec{F} = F_1\vec{i} + F_2\vec{j} = e^{x^2+y^2}(\vec{i} + \vec{j}).$$

We also need a parameterization of the curve C. This parameterization must trace out the curve once and once only, with the region R on the left. We split C into two parts, C_1 and C_2, as shown in Figure 18.23, and use the following parameterizations:

$$
\begin{aligned}
C_1: \quad & x(t) = \cos t, y(t) = \sin t, \quad 0 \le t \le \pi \\
& \text{so} \quad x'(t) = -\sin t, y'(t) = \cos t \\
C_2: \quad & x(t) = t - 1, y(t) = 0, \quad 0 \le t \le 2 \\
& \text{so} \quad x'(t) = 1, y'(t) = 0.
\end{aligned}
$$

Then, Green's Theorem tells us that

$$\int_R (2x - 2y)e^{x^2+y^2} dA$$

$$= \int_{C_1} \vec{F} \cdot d\vec{r} + \int_{C_2} \vec{F} \cdot d\vec{r}$$

$$= \int_0^\pi e^{\cos^2 t + \sin^2 t}(\vec{i} + \vec{j}) \cdot (-\sin t\vec{i} + \cos t\vec{j}) dt + \int_0^2 e^{(t-1)^2}(\vec{i} + \vec{j}) \cdot \vec{i} \, dt$$

$$= \int_0^\pi e^1(-\sin t + \cos t) dt + \int_0^2 e^{(t-1)^2} dt$$

$$= e(\cos t + \sin t)\Big|_0^\pi + \int_0^2 e^{(t-1)^2} dt$$

$$= e(-1 - 1) + 2.93 = -2.51.$$

The second integral, $\int_0^2 e^{(t-1)^2} dt$, can only be evaluated numerically.

Note: If we used polar coordinates on the original integral we got

$$\int_R (2x - 2y)e^{x^2+y^2} dA = \int_0^1 \int_0^\pi 2r^2(\cos \theta - \sin \theta)e^{r^2} d\theta dr =$$

$$= 2 \int_0^1 r^2 e^{r^2}(\sin \theta + \cos \theta)\Big|_0^\pi dr = -4 \int_0^1 r^2 e^{r^2} dr$$

The last integral cannot be computed, but by using numerical methods we get

$$\int_R (2x - 2y)e^{x^2+y^2} dA = -4(0.63) = -2.51$$

Solutions for Section 18.5

1. We use the polar coordinates r and θ as parameters:

    $$x = -1 + r\cos\theta, \quad y = 2 + r\sin\theta, \qquad 2 \le r \le 3, 0 \le \theta \le 2\pi.$$

    To obtain the annulus, the values of r and θ vary on a rectangle.

2. We let x be one of the parameters; we want the other parameter t to make y go from 0 to $\sin x$ as it goes from 0 to 1, so we set $y = t\sin x$, $0 \le t \le 1$. Thus the parameterization is

    $$x = x, \quad y = t\sin x, \qquad 0 \le x \le \pi, 0 \le t \le 1.$$

    Notice that the values of x and t vary on a rectangle. (In the special case that $f(x) = g(x)$, the rectangle is a line.)

3. (a)

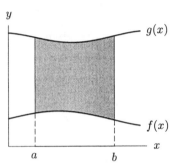

Figure 18.24

    (b) For $x = x_0$, we want to parameterize the line segment $f(x_0) \le y \le g(x_0)$. We write y as a linear function of t so that $y(0) = f(x_0)$ and $y(1) = g(x_0)$:

    $$x = x_0, \quad y = y(0) + t(y(1) - y(0)) = f(x_0) + t(g(x_0) - f(x_0)) = tg(x_0) + (1-t)f(x_0),$$

    for $0 \le t \le 1$.

    (c) We let x_0 be a parameter s that varies from a to b:

    $$x = s, \quad y = tg(s) + (1-t)f(s), \qquad a \le s \le b, 0 \le t \le 1.$$

    Notice that s and t vary on a rectangle. (In the special case that $f(x) = g(x)$, the rectangle is a line.)

4. (a)

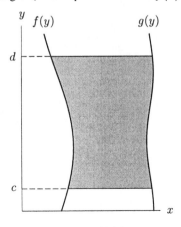

Figure 18.25

(b) For $y = y_0$, we want to parameterize the line segment $f(y_0) \leq x \leq g(y_0)$. We write x as a linear function of t so that $x(0) = f(y_0)$ and $x(1) = g(y_0)$:

$$x = x(0) + t(x(1) - x(0)) = f(y_0) + t(g(y_0) - f(y_0)) = tg(y_0) + (1 - t)f(y_0), \qquad 0 \leq t \leq 1.$$

(c) All we have to do now is let y_0 be a parameter s that varies from c to d:

$$x = tg(s) + (1 - t)f(s), \qquad y = s \qquad c \leq s \leq d, 0 \leq t \leq 1$$

Notice that s and t vary on a rectangle. (In the special case that $f(x) = g(x)$, the rectangle is a line.)

5. We prove this by breaking each path into four pieces, as shown in Figure 18.5 on page 377 of the text. For example, the path C_1 is parameterized by

$$\vec{r} = \vec{r}(s, c),$$

so

$$\int_{C_1} \vec{F} \cdot d\vec{r} = \int_a^b \vec{F} \cdot \frac{\partial \vec{r}}{\partial s} \, ds = \int_a^b G_1 \, ds = \int_{D_1} \vec{G} \cdot d\vec{u}.$$

The other pieces work the same way.

6. Since

$$G_1 = \vec{F} \cdot \frac{\partial \vec{r}}{\partial s}, \qquad G_2 = \vec{F} \cdot \frac{\partial \vec{r}}{\partial t},$$

we have

$$\frac{\partial G_2}{\partial s} - \frac{\partial G_1}{\partial t} = \frac{\partial \vec{F}}{\partial s} \cdot \frac{\partial \vec{r}}{\partial t} + \vec{F} \cdot \frac{\partial^2 \vec{r}}{\partial s \partial t} - \frac{\partial \vec{F}}{\partial t} \cdot \frac{\partial \vec{r}}{\partial s} - \vec{F} \cdot \frac{\partial^2 \vec{r}}{\partial t \partial s}$$

$$= \frac{\partial \vec{F}}{\partial s} \cdot \frac{\partial \vec{r}}{\partial t} - \frac{\partial \vec{F}}{\partial t} \cdot \frac{\partial \vec{r}}{\partial s}$$

where we assume

$$\vec{F} \cdot \left(\frac{\partial^2 \vec{r}}{\partial t \partial s} - \frac{\partial^2 \vec{r}}{\partial s \partial t} \right) = 0.$$

Now, by the chain rule,

$$\frac{\partial \vec{F}}{\partial s} = \frac{\partial F_1}{\partial s} \vec{i} + \frac{\partial F_2}{\partial s} \vec{j}$$

$$= \left(\frac{\partial F_1}{\partial x} \frac{\partial x}{\partial s} + \frac{\partial F_1}{\partial y} \frac{\partial y}{\partial s} \right) \vec{i} + \left(\frac{\partial F_2}{\partial x} \frac{\partial x}{\partial s} + \frac{\partial F_2}{\partial y} \frac{\partial y}{\partial s} \right) \vec{j}$$

and

$$\frac{\partial \vec{F}}{\partial t} = \left(\frac{\partial F_1}{\partial x} \frac{\partial x}{\partial t} + \frac{\partial F_1}{\partial y} \frac{\partial y}{\partial t} \right) \vec{i} + \left(\frac{\partial F_2}{\partial x} \frac{\partial x}{\partial t} + \frac{\partial F_2}{\partial y} \frac{\partial y}{\partial t} \right) \vec{j}.$$

So

$$\frac{\partial G_2}{\partial s} - \frac{\partial G_1}{\partial t} = \left(\frac{\partial F_1}{\partial x} \frac{\partial x}{\partial s} + \frac{\partial F_1}{\partial y} \frac{\partial y}{\partial s} \right) \frac{\partial x}{\partial t} + \left(\frac{\partial F_2}{\partial x} \frac{\partial x}{\partial s} + \frac{\partial F_2}{\partial y} \frac{\partial y}{\partial s} \right) \frac{\partial y}{\partial t}$$

$$- \left(\frac{\partial F_1}{\partial x} \frac{\partial x}{\partial t} + \frac{\partial F_1}{\partial y} \frac{\partial y}{\partial t} \right) \frac{\partial x}{\partial s} - \left(\frac{\partial F_2}{\partial x} \frac{\partial x}{\partial t} + \frac{\partial F_2}{\partial y} \frac{\partial y}{\partial t} \right) \frac{\partial y}{\partial s}.$$

Notice that in the expansion of this expression, the coefficients of $\frac{\partial F_1}{\partial x}$ and $\frac{\partial F_2}{\partial y}$ cancel, and the coefficients of $\frac{\partial F_2}{\partial x}$ and $\frac{\partial F_1}{\partial y}$ are, respectively, plus and minus $\frac{\partial x}{\partial s}\frac{\partial y}{\partial t} - \frac{\partial y}{\partial s}\frac{\partial x}{\partial t}$. So we have

$$\frac{\partial G_2}{\partial S} - \frac{\partial G_1}{\partial t} = \left(\frac{\partial F_2}{\partial x} - \frac{\partial F_1}{\partial y} \right) \begin{vmatrix} \frac{\partial x}{\partial s} & \frac{\partial y}{\partial s} \\ \frac{\partial x}{\partial t} & \frac{\partial y}{\partial t} \end{vmatrix}.$$

Solutions for Chapter 18 Review

1. (a) The line integral around A is negative, because the vectors of the field are all pointing in the opposite direction to the direction of the path.
    (b) Along C_1, the line integral is positive, since $\vec{F}$ points in the same direction as the curve. Along C_2 or C_4, the line integral is zero, since $\vec{F}$ is perpendicular to the curve everywhere. Along C_3, the line integral is negative, since $\vec{F}$ points in the opposite direction to the curve.
    (c) The line integral around C is negative because C_3 is longer than C_1 and the magnitude of the field is bigger along C_3 than C_1.

2. (a) The line integral around A is zero, because the curve is perpendicular to the field everywhere.
    (b) The line integral along C_1 or C_3 is zero because the curves are everywhere perpendicular to the vector field. Along C_2, the line integral is negative, since $\vec{F}$ points along the opposite direction to the curve. Along C_4, the line integral is positive, since $\vec{F}$ points in the same direction as the curve.
    (c) The line integral around C is zero because C_1 and C_3 are perpendicular to the field and the contributions from C_2 and C_4 cancel out.

3. We can parameterize the curve C by $(t, t^2 + 1)$, for $0 \le t \le 1$. Then

$$\int_C \vec{F} \cdot d\vec{r} = \int_0^1 \vec{F}(t, t^2 + 1) \cdot (\vec{i} + 2t\vec{j})dt = \int_0^1 ((-1)\vec{i} + (t^4 + 2t^2 + t + 1)\vec{j}) \cdot (\vec{i} + 2t\vec{j})dt$$

$$= \int_0^1 (-1 + 2t(t^4 + 2t^2 + t + 1))dt = \int_0^1 (-1 + 2t^5 + 4t^3 + 2t^2 + 2t)dt$$

$$= \left(-t + \frac{2t^6}{6} + \frac{4t^4}{4} + \frac{2t^3}{3} + t^2 \right) \Big|_0^1 = 2$$

4. We parameterize the path C by (t, t, t), for $0 \le t \le 1$. Then

$$\int_C \vec{F} \cdot d\vec{r} = \int_0^1 \vec{F}(t, t, t) \cdot (\vec{i} + \vec{j} + \vec{k})dt = \int_0^1 (t\vec{i} + 3t\vec{j} - t^2\vec{k}) \cdot (\vec{i} + \vec{j} + \vec{k})dt$$

$$= \int_0^1 (4t - t^2)dt = \left(2t^2 - \frac{t^3}{3} \right) \Big|_0^1 = \frac{5}{3}.$$

5.

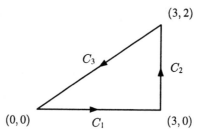

Figure 18.26

The triangle C consists of the three paths shown in Figure 18.26. Write $C = C_1 + C_2 + C_3$ where C_1, C_2, and C_3 are parameterized by

$$C_1 : (t, 0) \text{ for } 0 \le t \le 3; \quad C_2 : (3, t) \text{ for } 0 \le t \le 2; \quad C_3 : (3 - 3t, 2 - 2t) \text{ for } 0 \le t \le 1.$$

Then

$$\int_C \vec{F} \cdot d\vec{r} = \int_{C_1} \vec{F} \cdot d\vec{r} + \int_{C_2} \vec{F} \cdot d\vec{r} + \int_{C_3} \vec{F} \cdot d\vec{r}$$

where

$$\int_{C_1} \vec{F} \cdot d\vec{r} = \int_0^3 \vec{F}(t, 0) \cdot \vec{i} \, dt = \int_0^3 (2t + 4) dt = (t^2 + 4t)\big|_0^3 = 21$$

$$\int_{C_2} \vec{F} \cdot d\vec{r} = \int_0^2 \vec{F}(3, t) \cdot \vec{j} \, dt = \int_0^2 (5t + 3) dt = (5t^2/2 + 3t)\big|_0^2 = 16$$

$$\int_{C_3} \vec{F} \cdot d\vec{r} = \int_0^1 \vec{F}(3 - 3t, 2 - 2t) \cdot (-3\vec{i} - 2\vec{j}) dt$$

$$= \int_0^1 ((-4t + 8)\vec{i} + (-19t + 13)\vec{j}) \cdot (-3\vec{i} - 2\vec{j}) dt$$

$$= 50 \int_0^1 (t - 1) dt = -25.$$

So

$$\int_C \vec{F} \, d\vec{r} = 21 + 16 - 25 = 12.$$

6. False. Because $\vec{F} \cdot \Delta\vec{r}$ is a scalar quantity, $\int_C \vec{F} \cdot d\vec{r}$ is also a scalar quantity.

7. False. The left side is a scalar, the right side is a vector. A true statement is: If $\vec{F} = \text{grad } f$ and P and Q are the endpoints of C, then $\int_C \vec{F} \cdot d\vec{r} = f(Q) - f(P)$.

8. False. The line integral of a gradient vector field around this circle would be 0, but the converse is not necessarily true. That is, the fact that the line integral around this one circle is zero does not mean $\vec{F}$ is necessarily a gradient field.

9. True. You can trace out C_2 using the same subdivisions, but each $\Delta\vec{r}$ will have the opposite sign as before and will be traced out twice, so $\int_{C_2} \vec{F} \cdot d\vec{r} = -2 \int_{C_1} \vec{F} \cdot d\vec{r} = -6$.

10. Yes, the line integral over C_1 is the negative of the line integral over C_2. One way to see this is to observe that the vector field $x\vec{i} + y\vec{j}$ is symmetric in the y-axis and that C_1 and C_2 are reflections in the y axis (except for orientation). See Figure 18.27. Since the orientation of C_2 is the reverse of the orientation of a mirror image of C_1, the two line integrals are opposite in sign.

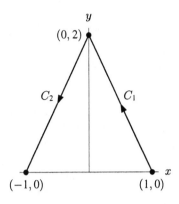

Figure 18.27

11. Since $\vec{F} = 6\vec{i} - 7\vec{j}$, consider the function f

$$f(x, y) = 6x - 7y.$$

Then we see that grad $f = 6\vec{i} - 7\vec{j}$, so we use the Fundamental Theorem of Calculus for Line Integrals:

$$\int_C \vec{F} \cdot d\vec{r} = \int_C \text{grad } f \cdot d\vec{r}$$
$$= f(4, 4) - f(2, -6)$$
$$= (-4) - (54)$$
$$= -58.$$

12. A contour of f is a set on which f doesn't change, so the total change of f from P to Q, $f(P) - f(Q)$, is zero. If C is a part of a contour of f, we know that grad f is perpendicular to C. This means that the line integral of grad f along C, which also computes the total change in f between its endpoints, must be zero, since the dot products in its definition are all zero.

13. (a) By the chain rule
$$\frac{dh}{dt} = \frac{\partial f}{\partial x}\frac{dx}{dt} + \frac{\partial f}{\partial y}\frac{dy}{dt} = f_x x'(t) + f_y y'(t),$$

which is the result we want.

(b) Using the parameterization of C that we were given,

$$\int_C \text{grad } f \cdot d\vec{r} = \int_a^b (f_x(x(t), y(t))\vec{i} + f_y(x(t), y(t))\vec{j}) \cdot (x'(t)\vec{i} + y'(t)\vec{j})dt$$
$$= \int_a^b (f_x(x(t), y(t))x'(t) + f_y(x(t), y(t))y'(t))dt.$$

Using the result of part (a), this gives us

$$\int_C \operatorname{grad} f \cdot d\vec{r} = \int_a^b h'(t)dt$$
$$= h(b) - h(a) = f(Q) - f(P).$$

14. (a) The curves C_1 and C_3 give line integrals which we expect to be zero because at every point, the curve looks perpendicular to the vector field.
     (b) The curve C_4 gives a negative line integral because the path is traversed in the direction opposite to the vector field.
     (c) The line integrals along C_2, C_5, C_6 and C_7 are all positive. The vector field is path-independent; it is the gradient of a function f whose contours appear to be equally spaced circles centered at the origin; the value of f increases going outward. By the Fundamental Theorem of Line Integrals, the value of a line integral is the difference between the values of f at the two endpoints. The difference between the radii of the circles containing the endpoints of C_2 and the difference between the radii of the circles containing the endpoints of C_6 look about the same, so the line integrals along C_2 and C_6 are approximately equal. Since C_6 and C_7 have the same endpoints, their line integrals are also equal. The difference between the radii of the circles containing the endpoints of C_2 is less than the difference between the radii of the circles containing the endpoints of C_5, so the line integral along C_2 is smaller than the line integral along C_5. Thus

$$C_2 = C_6 = C_7 < C_5.$$

15. (a) Since $-y\vec{i} + x\vec{j}$ is a counterclockwise rotation, both ω and K must be positive. In order to find the values of ω and K, we must look at the velocity field where we know the magnitude. At a radius of $100\,\mathrm{m}$ from the center, we know that $\sqrt{x^2 + y^2} = 100$, and that $\|\vec{v}\| = 3 \cdot 10^5$. Thus, using $\vec{v} = \omega(-y\vec{i} + x\vec{j})$ we get

$$\|\vec{v}\| = \omega\sqrt{(-y)^2 + x^2} = 100\omega = 3 \cdot 10^5 \text{ meters/hr},$$

so

$$\omega = 3000 \text{ rad/hr}.$$

Using $\vec{v} = K(x^2 + y^2)^{-1}(-y\vec{i} + x\vec{j})$ gives

$$\|\vec{v}\| = |K|(x^2 + y^2)^{-1}\sqrt{(-y)^2 + x^2} = \frac{K\,100}{100^2} = 3 \cdot 10^5 \text{ meters/hr},$$

so

$$K = 3 \cdot 10^7 \text{ meters}^2 \cdot \text{rad/hr}$$

     (b)

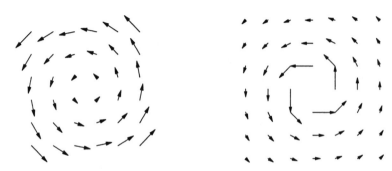

Figure 18.28   Figure 18.29

The vector field in Figure 18.28 shows the velocity vectors inside the tornado, (i.e. $r < 100$ meters). The vector field in Figure 18.29 shows the velocity vectors as seen from a great distance (i.e. $r >> 100$ meters) with the tornado at the origin.

(c)   Let C be the circle of radius r around the origin. If $r < 100$ meters, the velocity vectors at distance r from the origin have magnitude ωr. Since they are tangent, and point counterclockwise, the circulation is

$$\int_C \vec{v} \cdot d\vec{r} = \|\vec{v}\| \cdot \text{Length of } C = 2\omega\pi r^2.$$

If $r \geq 100$ meters, the vectors at distance r from the origin have magnitude K/r and are again tangent to the circle. The circulation here is

$$\int_C \vec{v} \cdot d\vec{r} = \|v\| \cdot \text{Length of } C = (\frac{K}{r})2\pi r = 2K\pi.$$

16.   The free vortex appears to starts at about $r = 200$ meters (that's where the graph changes its behavior) and the tangential velocity at this point is about 200 km/hr $= 2 \cdot 10^5$ meters/hr.
Since $\vec{v} = \omega(-y\vec{i} + x\vec{j})$ for $\sqrt{x^2 + y^2} \leq 200$, at $r = 200$ we have

$$\|\vec{v}\| = \omega\sqrt{(-y)^2 + x^2} = \omega(200) = 2 \cdot 10^5 \text{ meters/hr},$$

so

$$\omega = 10^3 \text{ rad/hr.}$$

Since $\vec{v} = K(x^2 + y^2)^{-1}(-y\vec{i} + x\vec{j})$ for $\sqrt{x^2 + y^2} \geq 200$, at $r = 200$ we have

$$\|\vec{v}\| = K(200^2)^{-1}(200) = \frac{K}{200} = 2 \cdot 10^5 \text{ meters/hr}$$

so

$$K = 4 \cdot 10^7 \text{ m}^2 \cdot \text{rad/hr.}$$

17.   (a)   An example of a central field is in Figure 18.30:

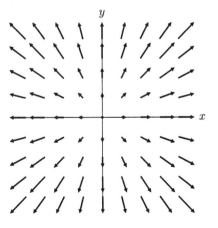

Figure 18.30

(b)   The vectors of $\vec{F}$ are radial and the contours of f must be perpendicular to the vectors. Therefore, every contour must be a circle centered at the origin. Sketching some contours results in a diagram like that in Figure 18.31:

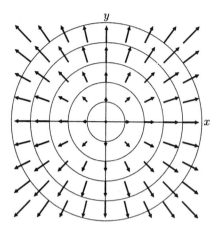

Figure 18.31

(c)   No, not every gradient field is a central field, because there are gradient fields which are not perpendicular to circles. An example is the gradient of $f(x, y) = y$, where grad $f = \vec{j}$, so the gradient is parallel to the y axis. Thus, $\vec{F} = \vec{j}$ is an example of a gradient field which is not a central field.

(d)   When a particle moves around a circle centered at O, no work is done, because $\vec{F}$ is tangent to the circle. Thus the only work done in moving from P to Q is in moving between the circles. Since $\vec{F}$ is central, the work done on any radial line between C_1 and C_3, for example, depends on only the radii of C_1 and C_3 ($\vec{F}$ is parallel to this path and its magnitude is a function of the distance to the center of the circle only). For that reason, on a path which goes from C_1 to C_2 and then from C_2 to C_3, the same amount of work will be done as on a path direct from C_1 to C_3.

(e)   Pick any two points P and Q. Any path between them can be well-approximated by a path which is partly radial and partly around a circle centered at O. By the answer to part d), the work along any such path depends only on the radii of the circles on which P and Q sit, not on the path. Thus, the work done is independent of the path. Hence, $\vec{F}$ must be path-independent and therefore a gradient field.

18.   Suppose C encloses a region R. Then, using Green's Theorem, we have

$$\int_C \vec{F} \cdot d\vec{r} = \int_R \left(\frac{\partial F_2}{\partial x} - \frac{\partial F_1}{\partial y} \right) dA$$

$$= \int_R \frac{\partial}{\partial x}(4x(1 - y^2) + x\sin(xy)) - \frac{\partial}{\partial y}(-y^3 + y\sin(xy)) \, dA$$

$$= \int_R 4(1 - y^2) + \sin(xy) + xy\cos(xy) + 3y^2 - \sin(xy) - xy\cos(xy) \, dA$$

$$= \int_R (4 - y^2) \, dA$$

This integral over R is largest if C encloses the maximum possible region where $4 - y^2 > 0$, that is, where $-2 \le y \le 2$. Therefore C should be the curve with two sides along the lines $y = -2$ and $y = 2$, as well as two arcs of the circle $x^2 + y^2 = 25$. See Figure 18.32.

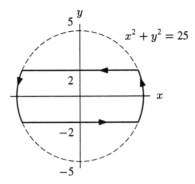

Figure 18.32

CHAPTER NINETEEN

Solutions for Section 19.1

1. (a) The flux is positive, since $\vec{F}$ points in direction of positive x-axis, the same direction as the normal vector.
    (b) The flux is negative, since below the xy-plane $\vec{F}$ points towards negative x-axis, which is opposite the orientation of the surface.
    (c) The flux is zero. Since $\vec{F}$ has only an x-component, there is no flow across the surface.
    (d) The flux is zero. Since $\vec{F}$ has only an x-component, there is no flow across the surface.
    (e) The flux is zero. Since $\vec{F}$ has only an x-component, there is no flow across the surface.

2. The vector field $\vec{F} = F_1\vec{i} + F_2\vec{j} + F_3\vec{k} = -z\vec{i} + x\vec{k}$ is a field parallel to the xz-plane that suggests swirling around the origin from the positive x-axis to the positive z-axis.
    (a) The flux going through this surface is negative, because $\vec{F} \cdot \vec{n} = (-z\vec{i} + x\vec{k}) \cdot \vec{i} = -z$, z is positive here.
    (b) The flux going through this surface is positive, because $\vec{F} \cdot \vec{n} = -z$, z is negative here.
    (c) The flux through this surface is negative, because $\vec{F} \cdot \vec{n} = (-z\vec{i} + x\vec{k}) \cdot (-\vec{k}) = -x$, x is positive.
    (d) The flux through this surface is negative, because $\vec{F} \cdot \vec{n} = -x$, x is positive.
    (e) The flux through this surface is zero, because it is in the xz-plane, which is parallel to the vector field.

3. The vector field $\vec{r}$ is a field that always points away from the origin.
    (a) The flux through this surface is zero, because the plane is parallel to the field.
    (b) The flux through this surface is zero also, for the same reason.
    (c) The flux through this surface is zero also, for the same reason.
    (d) The flux through this surface is negative, because the field in that quadrant is going up and away from the origin, and since the orientation is downward, the flux is negative.
    (e) The flux through this surface is zero also.

4.

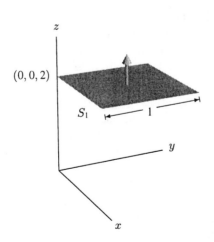

Figure 19.1

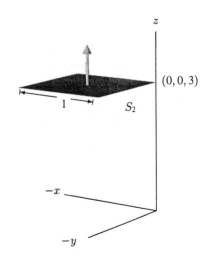

Figure 19.2

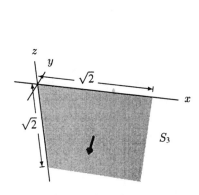

Figure 19.3

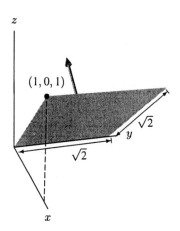

Figure 19.4

$$\text{Flux through } S_1 = \vec{F} \cdot \vec{A} = (-\vec{i} - \vec{j} + \vec{k}) \cdot (\vec{k}) = 1$$
$$\text{Flux through } S_2 = \vec{F} \cdot \vec{A} = (-\vec{i} - \vec{j} + \vec{k}) \cdot (\vec{k}) = 1$$
$$\text{Flux through } S_3 = \vec{F} \cdot \vec{A} = (-\vec{i} - \vec{j} + \vec{k}) \cdot (-2\vec{j}) = 2$$

For S_4, a normal is $-\vec{i} + \vec{k}$ and the area is 2, so $\vec{A} = -\sqrt{2}\vec{i} + \sqrt{2}\vec{k}$

$$\text{Flux through } S_4 = \vec{F} \cdot \vec{A} = (-\vec{i} - \vec{j} + \vec{k}) \cdot (-\sqrt{2}\vec{i} + \sqrt{2}\vec{k}) = 2\sqrt{2}$$

So,

Flux through S_1 = Flux through S_2 < Flux through S_3 < Flux through S_4.

5. (a) $\vec{v} \cdot \vec{A} = (2\vec{i} + 3\vec{j} + 5\vec{k}) \cdot \vec{k} = 5.$
    (b) $\vec{v} \cdot \vec{A} = (2\vec{i} + 3\vec{j} + 5\vec{k}) \cdot 2\vec{i} = 4.$
    (c) The rectangle lies in the plane $z + 2y = 2$. So a normal vector is $2\vec{j} + \vec{k}$ and a unit normal vector is $\frac{1}{\sqrt{5}}(2\vec{j} + \vec{k})$. Since this points in the positive z-direction it is indeed an orientation for the rectangle. Since the area of this rectangle is $\sqrt{5}$ we have $\vec{A} = 2\vec{j} + \vec{k}$,
    $\vec{v} \cdot \vec{A} = (2\vec{i} + 3\vec{j} + 5\vec{k}) \cdot (2\vec{j} + \vec{k}) = 6 + 5 = 11.$
    (d) The rectangle lies in the plane $z + 2x = 2$. So $2\vec{i} + \vec{k}$ is a normal vector and $\frac{1}{\sqrt{5}}(2\vec{i} + \vec{k})$ is a unit normal vector. Since this points in both the positive x-axis and the positive z it is an orientation for this surface. Since the area of the rectangle is $\sqrt{5}$, we have $\vec{A} = 2\vec{i} + \vec{k}$ and $\vec{v} \cdot \vec{A} = (2\vec{i} + 3\vec{j} + 5\vec{k}) \cdot (2\vec{i} + \vec{k}) = 4 + 5 = 9.$

6. (a) At the north pole, the area vector of the plate is upward (away from the center of the earth), and so is in the opposite direction to the magnetic field. Thus the magnetic flux is negative.
    (b) At the south pole, the area vector of the plate is again away from the center of the earth (because that is upward in the southern hemisphere), and so is in the same direction as the magnetic field. Thus, the magnetic flux is positive.
    (c) At the equator the magnetic field is parallel to the plate, so the flux is zero.

7. (a) The net electric flux through this surface is zero, because the surface is placed so that it is always parallel with the electric field, and there is no flow through the surface.
    (b) The net flux is zero, because the flow in through one half of the cylinder is cancelled by the flow out through the other half.

8.

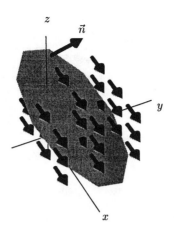

Figure 19.5

See Figure 19.5. Since the vector field is constant and the surface is flat, the flux integral

$$\int_S \vec{F} \cdot d\vec{A}$$

is $\vec{F} \cdot \vec{A}$, where $\vec{A}$ is the area vector of the circle. The vector $\vec{i} + \vec{j} + \vec{k}$ is normal to the plane and points up, so a unit normal is $\frac{1}{\sqrt{3}}\vec{i} + \frac{1}{\sqrt{3}}\vec{j} + \frac{1}{\sqrt{3}}\vec{k}$. The area of the circle is 4π, so the area vector of the circle is

$$\vec{A} = \frac{4\pi}{\sqrt{3}}(\vec{i} + \vec{j} + \vec{k})$$

Thus

$$\int_S \vec{F} \cdot d\vec{A} = \vec{F} \cdot \vec{A} = 2\vec{i} \cdot \frac{4\pi}{\sqrt{3}}(\vec{i} + \vec{j} + \vec{k}) = \frac{8\pi}{\sqrt{3}} \approx 14.5104.$$

9.

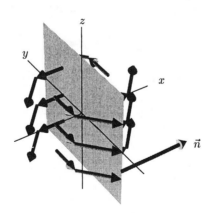

Figure 19.6

See Figure 19.6. The vector field is a vortex going around the z-axis, and the square is centered on the x-axis, so the flux going across one half of the square is balanced by the flux coming back across the other half. Thus, the net flux is zero, so

$$\int_S \vec{F} \cdot d\vec{A} = 0.$$

10.

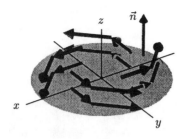

Figure 19.7

See Figure 19.7. Since $\vec{F}$ is parallel to the xy plane, there is no flux across the surface, so

$$\int_S \vec{F} \cdot d\vec{A} = 0.$$

11.

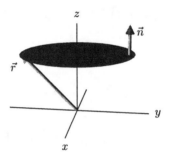

Figure 19.8

See Figure 19.8. The area vector of a small area element $\Delta\vec{A}$ is the vector pointing in the direction normal to the surface with magnitude ΔA. The unit vector normal to the surface is $\vec{k}$, so $\Delta\vec{A} = \Delta A\vec{k}$. Thus,

$$\int_S \vec{F} \cdot d\vec{A} = \lim_{\|\Delta\vec{A}\|\to 0} \sum \vec{r} \cdot \Delta\vec{A} = \lim_{\|\Delta A\|\to 0} \sum \vec{r} \cdot \vec{k}\,\Delta\vec{A} = \int_S \vec{r} \cdot \vec{k}\,dA.$$

Now $\vec{r} \cdot \vec{k} = 2$ for all the points on S because all such points have z-coordinate equal to 2. Thus, we have

$$\int_S \vec{F} \cdot d\vec{A} = \int_S \vec{r} \cdot \vec{k}\,dA = \int_S 2\,dA = 2 \cdot \text{Area of } S = 8\pi.$$

12.

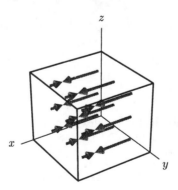

Figure 19.9

See Figure 19.9. Since the vector field is parallel to the x-axis, only the two sides perpendicular to the x-axis contribute to the flux integral. On the side where $x = 0$, the vector field is $2\vec{i}$, and hence the flux through that side is $-(2)(3^2) = -18$ (negative because the flow is inward and the normal vector is pointing out). The flow out the other side is at $x = 3$, so $\vec{F} = -\vec{i}$, so the flux out that side is $(-\vec{i}) \cdot (3^2\vec{i}) = -9$. So the net flux is $-18 - 9 = -27$. So

$$\int_S \vec{F} \cdot d\vec{A} = -27.$$

13. All the vectors in the vector field point horizontally (because their z-component is zero), and the surface is horizontal, so there is no flow through the surface and the flux is zero.

14. Since this vector field points radially out from the origin, it is everywhere parallel to the area vector, $\Delta\vec{A}$. Thus since $\|\vec{F}(\vec{r})\| = 1/R$ on the surface, S,

$$\vec{F}(\vec{r}) \cdot \Delta\vec{A} = \frac{1}{R}\Delta A$$

so

$$\int_S \vec{F}(\vec{r}) \cdot d\vec{A} = \frac{1}{R}\lim_{\Delta A \to 0}\sum \Delta A = \frac{1}{R} \cdot \text{Surface area of sphere} = \frac{1}{R}(4\pi R^2) = 4\pi R.$$

15. (a) For a flat surface, flux through $\vec{A}$ is $\vec{v} \cdot \vec{A}$. Therefore, the flux through each face of the cube is equal to $(-\vec{i} + 2\vec{j} + \vec{k}) \cdot (\vec{A} \text{ of the face})$.

First we shall find the flux through the two faces parallel to the xy-plane, beginning with the one with negative z. The unit vector normal to this face and pointing outward is $-\vec{k}$. The area of the face equals 4, so $\vec{A} = -4\vec{k}$. The flux through the face with negative z equals

$$(-\vec{i} + 2\vec{j} + \vec{k}) \cdot (-4\vec{k}) = 0 + 0 - 4 = -4$$

For the face with positive z, the unit normal vector that points outward is $\vec{k}$. Therefore $\vec{A} = 4\vec{k}$. The flux through this face is given by

$$(-\vec{i} + 2\vec{j} + \vec{k}) \cdot 4\vec{k} = 0 + 0 + 4 = 4$$

Next, we will find the flux through the two faces parallel to the xz-plane, beginning with the one with negative y. A unit vector normal to this face pointing outward is $-\vec{j}$. Therefore $\vec{A} = -4\vec{j}$. The flux then equals

$$(-\vec{i} + 2\vec{j} + \vec{k}) \cdot (-4\vec{j}) = 0 - 8 + 0 = -8$$

For the face with positive y, the unit normal vector pointing outward is $\vec{j}$. Therefore $\vec{A} = 4\vec{j}$. The flux then equals

$$(-\vec{i} + 2\vec{j} + \vec{k}) \cdot (4\vec{j}) = 0 + 8 + 0 = 8$$

Next, we will find the flux through the two faces parallel to the yz plane, beginning with the one with negative x. A unit vector normal to this plane pointing outward is $-\vec{i}$. Therefore $\vec{A} = -4\vec{i}$. The flux then equals

$$(-\vec{i} + 2\vec{j} + \vec{k}) \cdot (-4\vec{i}) = 4 + 0 + 0 = 4$$

For the face with positive x, the unit normal vector pointing outward is $\vec{i}$. Therefore $\vec{A} = 4\vec{i}$. The flux then equals

$$(-\vec{i} + 2\vec{j} + \vec{k}) \cdot (4\vec{i}) = -4 + 0 + 0 = -4$$

Adding up all of these fluxes to get the flux out of the entire cube, we get

$$\text{Total flux} = -4 + 4 - 8 + 8 + 4 - 4 = 0$$

(b) For any constant vector field $\vec{v} = a\vec{i} + b\vec{j} + c\vec{k}$, we can calculate the flux out of the cube by the same method.

First we shall find the flux out of the two faces parallel to the xy plane, beginning with the one with negative z. A unit vector normal to this plane, that points negative (because of the orientation of the face) is $-\vec{k}$. The area of the face equals 4, therefore $\vec{A} = -4\vec{k}$. The flux through $\vec{A}$ then equals

$$(a\vec{i} + b\vec{j} + c\vec{k}) \cdot (-4\vec{k}) = 0 + 0 - 4c = -4c$$

For the face with positive z, the unit normal vector pointing outward is $\vec{k}$. Therefore $\vec{A} = 4\vec{k}$. The flux then equals

$$(a\vec{i} + b\vec{j} + c\vec{k}) \cdot (4\vec{k}) = 0 + 0 + 4c = 4c.$$

Next, we will find the flux through the two faces parallel to the xz plane, beginning with the one with negative y. A unit vector normal to this plane pointing outward is $-\vec{j}$. Therefore $\vec{A} = -4\vec{j}$. The flux then equals

$$(a\vec{i} + b\vec{j} + c\vec{k}) \cdot (-4\vec{j}) = 0 - 4b + 0 = -4b$$

For the face with positive y, the unit normal vector pointing outward is $\vec{j}$. Therefore $\vec{A} == 4\vec{j}$. The flux then equals

$$(a\vec{i} + b\vec{j} + c\vec{k}) \cdot (4\vec{j}) = 0 + 4b + 0 = 4b$$

Next, we will find the flux through the two faces parallel to the yz plane, beginning with the one with negative x. A unit vector normal to this plane pointing outward is $-\vec{i}$. Therefore $\vec{A} = -4\vec{i}$. The flux then equals

$$(a\vec{i} + b\vec{j} + c\vec{k}) \cdot (-4\vec{i}) = -4a + 0 + 0 = -4a$$

For the face in the positive x, the unit normal vector pointing outward is $\vec{i}$. Therefore $\vec{A} = 4\vec{i}$. The flux then equals

$$(a\vec{i} + b\vec{j} + c\vec{k}) \cdot (4\vec{i}) = 4a + 0 + 0 = 4a$$

Adding up all of these fluxes to get the flux out of the entire cube, we get

$$\text{Total flux} = -4c + 4c - 4b + 4b + 4a - 4a = 0$$

(c) Yes, the answers in parts (a) and (b) make sense. The vector field is constant, and so it does not change as it comes in the one side of the cube, and exits the other side. Therefore the two fluxes cancel each other out, making the total flux zero.

16.

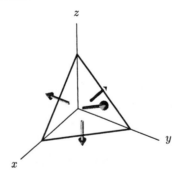

Figure 19.10

(a) The flux through $\vec{A}$ equals $\vec{v} \cdot \vec{A}$ because the vector field is constant. Therefore, the flux through each face of the tetrahedron is equal to $(-\vec{i} + 2\vec{j} + \vec{k}) \cdot \vec{A}$, where $\vec{A}$ is the area of that face.

First we shall find the flux out of the triangle in the xy plane. A unit vector normal to that plane, that points negative (because of the orientation of the face), is equal to $-\vec{k}$. The area of the face equals 0.5, therefore $\vec{A} = -0.5\vec{k}$. The flux through $\vec{A}$ then equals

$$(-\vec{i} + 2\vec{j} + \vec{k}) \cdot (-0.5\vec{k}) = 0 + 0 - 0.5 = -0.5.$$

Next, we will find the flux out of the triangle in the xz plane. A unit vector normal to that plane, that points negative, is equal to $-\vec{j}$. The area of the face equals 0.5, therefore $\vec{A} = -0.5\vec{j}$. The flux through $\vec{A}$ then equals

$$(-\vec{i} + 2\vec{j} + \vec{k}) \cdot (-0.5\vec{j}) = 0 - 1 + 0 = -1.$$

Next, we will find the flux out of the triangle in the yz plane. A unit vector normal to that plane, that points negative, is equal to $-\vec{i}$. The area of the face equals 0.5, therefore $\vec{A} = -0.5\vec{i}$. The flux through $\vec{A}$ then equals

$$(-\vec{i} + 2\vec{j} + \vec{k}) \cdot (-0.5\vec{i}) = 0.5 + 0 + 0 = 0.5.$$

Last, we will find the flux out of the triangle with vertices $(1,0,0), (0,1,0), (0,0,1)$. A unit vector normal to that plane, that points positive, is equal to $\frac{1}{\sqrt{3}}(\vec{i} + \vec{j} + \vec{k})$. The area of the face equals $\sqrt{3}/2$, since it is an equilateral triangle with side $\sqrt{2}$. Therefore:

$$\vec{A} = \frac{1}{\sqrt{3}}(\vec{i} + \vec{j} + \vec{k})(\sqrt{3}/2) = 0.5(\vec{i} + \vec{j} + \vec{k}).$$

The flux through $\vec{A}$ then equals

$$(-\vec{i} + 2\vec{j} + \vec{k}) \cdot (0.5\vec{i} + 0.5\vec{j} + 0.5\vec{k}) = -0.5 + 1 + 0.5 = 1.$$

The total flux out of the tetrahedron is $-0.5 - 1 + 0.5 + 1 = 0$. Therefore the flux equals zero.

(b) For any constant vector field $\vec{v} = a\vec{i} + b\vec{j} + c\vec{k}$, we can find the flux out of the tetrahedron.

First we shall find the flux out of the triangle in the xy plane. A unit vector normal to that plane, that points negative (because of the orientation of the face), is equal to $-\vec{k}$. The area of the face equals 0.5, therefore $\vec{A} = -0.5\vec{k}$. The flux through $\vec{A}$ then equals

$$(a\vec{i} + b\vec{j} + c\vec{k}) \cdot (-0.5\vec{k}) = 0 + 0 - 0.5 = -0.5c.$$

Next, we will find the flux out of the triangle in the xz plane. A unit vector normal to that plane, that points negative, is equal to $-\vec{j}$. The area of the face equals 0.5, therefore: $\vec{A} = -0.5\vec{j}$. The flux through $\vec{A}$ then equals

$$(a\vec{i} + b\vec{j} + c\vec{k}) \cdot (-0.5\vec{j}) = 0 - 0.5b + 0 = -0.5b.$$

Next, we will find the flux out of the triangle in the yz plane. A unit vector normal to that plane, that points negative, is equal to $(-\vec{i})$. The area of the face equals 0.5, therefore $\vec{A} = -0.5\vec{i}$. The flux through $\vec{A}$ then equals

$$(a\vec{i} + b\vec{j} + c\vec{k}) \cdot (-0.5\vec{i}) = -0.5a + 0 + 0 = -0.5a.$$

Last, we will find the flux out of the triangle with vertices $(1,0,0), (0,1,0), (0,0,1)$. A unit vector normal to that plane, that points positive, is equal to $\frac{1}{\sqrt{3}}(\vec{i} + \vec{j} + \vec{k})$. The area of the face equals $\sqrt{3}/2$, therefore:

$$\vec{A} = \frac{1}{\sqrt{3}}(\vec{i} + \vec{j} + \vec{k})(\sqrt{3}/2) = 0.5(\vec{i} + \vec{j} + \vec{k}).$$

The flux through $\vec{A}$ then equals

$$(a\vec{i} + b\vec{j} + c\vec{k}) \cdot (0.5\vec{i} + 0.5\vec{j} + 0.5\vec{k}) = 0.5a + 0.5b + 0.5c.$$

The total flux out of the tetrahedron is $-0.5c - 0.5b - 0.5a + 0.5a + 0.5b + 0.5c = 0$. Therefore, the flux is equal to zero.

(c)   Yes, the answers in (a) and (b) make sense. The vector field is constant, so it does not change as it enters through one side of the tetrahedron, and exits the other side. Therefore the two cancel each other out,causing the flux to be equal to zero.

17.   (a)   Figure 19.11 shows the electric field $\vec{E}$. Note that $\vec{E}$ points radially outward from the z-axis.

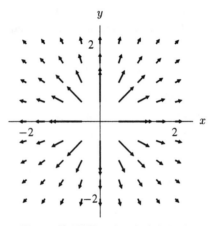

Figure 19.11: The electric field in the xy-plane due to a line of positive charge uniformly distributed along the z-axis:

$$\vec{E}(x, y, 0) = 2\lambda \frac{x\vec{i} + y\vec{j}}{x^2 + y^2}$$

(b)   On the cylinder $x^2 + y^2 = R^2$, the electric field $\vec{E}$ points in the same direction as the outward normal $\vec{n}$, and

$$\|\vec{E}\| = \frac{2\lambda}{R^2}\|x\vec{i} + y\vec{j}\| = \frac{2\lambda}{R}.$$

So

$$\int_S \vec{E} \cdot d\vec{A} = \int_S \vec{E} \cdot \vec{n}\, dA = \int_S \|\vec{E}\|\, dA = \int_S \frac{2\lambda}{R}\, dA$$

$$= \frac{2\lambda}{R} \int_S dA = \frac{2\lambda}{R} \cdot \text{Area of } S = \frac{2\lambda}{R} \cdot 2\pi Rh = 4\pi\lambda h,$$

which is positive, as we expected.

18.   Suppose $\vec{n}$ is the normal vector at a point on the surface, and $\vec{F}$ is the vector field at the same point, then

$$\vec{F} \cdot \vec{n} = \|\vec{F}\|\|\vec{n}\| \cos\theta.$$

By the assumption that $\vec{F}$ is normal to S, the angle $\theta = 0$ at all points on the surface. In addition $\|\vec{n}\| = 1$. Thus,

$$\vec{F} \cdot \vec{n} = \|\vec{F}\| = \text{constant}.$$

So,

$$\int_S \vec{F} \cdot d\vec{A} = \int_S \|\vec{F}\| dA = \|\vec{F}\| \int_S dA = \|\vec{F}\| \cdot \text{Area of } S.$$

19. Since Pressure = Force/Area, we have

$$\text{Force on a small patch with area } \Delta\vec{A} \text{ at point } (x, y, z) \approx P(x, y, z)\|\Delta\vec{A}\|.$$

This force is directed inward and normal to the surface, so the force is $P(x, y, z)\Delta\vec{A}$ (if S is oriented with the inward normal). For buoyancy, take the upward component of this force, so

$$\text{Buoyancy force} = P(x, y, z)\Delta\vec{A} \cdot \vec{k}.$$

Then:

$$\text{Total buoyancy} = \lim_{\|\Delta\vec{A}\|\to 0} \sum_S P(x, y, z)\Delta\vec{A} \cdot \vec{k}$$

$$= \int_S P(x, y, z)\vec{k} \cdot d\vec{A}$$

$$= \int_S \vec{F} \cdot d\vec{A}$$

20. (a) (i) The integral $\int_W \rho\, dV$ represents the total charge in the volume W.

   (ii) The integral $\int_S \vec{J} \cdot d\vec{A}$ represents the total current flowing out of the surface S.

   (b) The total current flowing out of the surface S is the rate at which the total charge inside the surface S (i.e., in the volume W) is decreasing. In other words,

$$\text{Rate current flowing out of } S = -\frac{\partial}{\partial t}(\text{charge in } W),$$

so

$$\int_S \vec{J} \cdot d\vec{A} = -\frac{\partial}{\partial t}\left(\int_W \rho\, dV\right).$$

21. (a) If we examine the equation for $\vec{v}$, we see that when $r = 0$, that is, at the center of the pipe, $\vec{v}(0)$ becomes $u\vec{i}$. So u is the speed at the center of the pipe; it is also the maximum speed since $u(1 - r^2/a^2)$ reaches its maximum at $r = 0$.

   (b) The flow rate at the wall of the pipe (where $r = a$) is

$$\vec{v}(a) = u(1 - a^2/a^2)\vec{i} = \vec{0}.$$

   (c) To find the flux through a circular cross-sectional area, we use polar coordinates in the plane perpendicular to the velocity. In these coordinates, an infinitesimal area, $d\vec{A}$ becomes $r\, dr\, d\theta\vec{i}$. So the flux is given by

$$\text{Flux} = \int_S \vec{v} \cdot d\vec{A} = \int_S u(1 - r^2/a^2)\vec{i} \cdot r\, dr\, d\theta\vec{i} = \int_0^{2\pi} \int_0^a u(1 - r^2/a^2)r\, dr\, d\theta$$

$$= 2\pi u \int_0^a (r - \frac{r^3}{a^2})\, dr = 2\pi u \left(\frac{a^2}{2} - \frac{a^2}{4}\right) = \frac{\pi u a^2}{2}.$$

22. (a) From Newton's law of cooling, we know that the temperature gradient will be proportional to the heat flow. If the constant of proportionality is k then we have the equation $\vec{F} = k \operatorname{grad} T$. Since $\operatorname{grad} T$ points in the direction of increasing T, but heat flows towards lower temperatures, the constant k must be negative.

(b) This form of Newton's law of cooling is saying that heat will be flowing in the direction in which temperature is decreasing most rapidly, in other words, in the direction exactly opposite to $\operatorname{grad} T$. This agrees with our intuition which tells us that a difference in temperature causes heat to flow from the higher temperature to the lower temperature, and the rate at which it flows depends on the temperature gradient.

(c) The rate of heat loss from W is given by the flux of the heat flow vector field through the surface of the body. Thus,

$$\begin{array}{ll} \text{Rate of heat} \\ \text{loss from } W \end{array} = \begin{array}{ll} \text{Flux of } \vec{F} \\ \text{out of } S \end{array} = \int_S \vec{F} \cdot d\vec{A} = k \int_S (\operatorname{grad} T) \cdot d\vec{A}$$

23. (a) Since the direction of the field is normal to the surface of any cylinder with the wire as an axis, it is parallel to the surfaces of the two washers. Consequently, there is no flux through the washers.

(b) Gauss's Law tells us that the total flux through the surfaces must be zero, since no charge is contained therein. Since the flux through the washers is zero, then the flux into the inner surface must equal the flux out of the outer surface in order for the net flux through the surfaces to be zero.

(c) Since the surface area of a cylinder is given by $A = 2\pi R L$ where R is the radius of the cylinder and L is its length, and we know that $E_a A_a = E_b A_b$ (equality of the fluxes), we know that:

$$E_b(2\pi b L) = E_a(2\pi a L)$$
$$\frac{E_b}{E_a} = \frac{2\pi a L}{2\pi b L}$$
$$\frac{E_b}{E_a} = \frac{a}{b}$$

(d) The equations in part (c) also imply that:

$$a E_a = b E_b.$$

Since a, b are arbitrary radii we can say:

$$r E_r = \text{Constant},$$

$$E_r = \text{Constant} \left(\frac{1}{r} \right),$$

for any radius r. This statement tells us that the strength of the electric field at r is proportional to $1/r$.

24. Since the field points perpendicular to the sheet, it is parallel to all sides of the box, except the two sides parallel to the sheet. Additionally, since there is no charge contained in the box, Gauss's Law tells us the net flux through the surface of the box must be zero, which implies that the flux into the near face must equal the flux out of the far face. Since the faces have the same area, the field must have equal strengths at the two faces in order for their fluxes to be equal. Since we did not use the values of a or b, we see that for all points in space on the same side of the sheet, the field has the same magnitude.

Solutions for Section 19.2

1. Using $z = f(x, y) = x + y$, we have $d\vec{A} = (-\vec{i} - \vec{j} + \vec{k}) \, dx \, dy$. As S is oriented upward, we have

$$\int_S \vec{F} \cdot d\vec{A} = \int_0^3 \int_0^2 ((x - y)\vec{i} + (x + y)\vec{j} + 3x\vec{k}) \cdot (-\vec{i} - \vec{j} + \vec{k}) \, dx \, dy$$

$$= \int_0^3 \int_0^2 (-x + y - x - y + 3x) \, dx \, dy = \int_0^3 \int_0^2 x \, dx \, dy = 6.$$

2. Using $z = 1 - x - y$, the upward pointing area element is $d\vec{A} = (\vec{i} + \vec{j} + \vec{k}) \, dx \, dy$, so the downward one is $d\vec{A} = (-\vec{i} - \vec{j} - \vec{k}) \, dx \, dy$. Since S is oriented downward, we have

$$\int_S \vec{F} \cdot d\vec{A} = \int_S (x\vec{i} + y\vec{j} + z\vec{k}) \cdot d\vec{A}$$

$$= \int_0^3 \int_0^2 (x\vec{i} + y\vec{j} + (1 - x - y)\vec{k}) \cdot (-\vec{i} - \vec{j} - \vec{k}) \, dx \, dy$$

$$= \int_0^3 \int_0^2 (-x - y - 1 + x + y) \, dx \, dy = -6.$$

3. Using $z = x^2 + y^2$, we find that the upward pointing area element is $d\vec{A} = (-2x\vec{i} - 2y\vec{j} + \vec{k}) \, dx \, dy$. Since S is oriented downward, we have $d\vec{A} = (2x\vec{i} + 2y\vec{j} - \vec{k}) \, dx \, dy$, so

$$\int_S \vec{F} \cdot d\vec{A} = \int_S (x\vec{i} + y\vec{j} + z\vec{k}) \cdot d\vec{A}$$

$$= \int_{\text{Disk}} (x\vec{i} + y\vec{j} + (x^2 + y^2)\vec{k}) \cdot (2x\vec{i} + 2y\vec{j} - \vec{k}) \, dx \, dy$$

$$= \int_{\text{Disk}} (2x^2 + 2y^2 - x^2 - y^2) \, dx \, dy = \int_{\text{Disk}} (x^2 + y^2) \, dx \, dy$$

$$= \int_0^{2\pi} \int_0^1 r^2 r \, dr \, d\theta = 2\pi \cdot \frac{r^4}{4} \Big|_0^1 = \frac{\pi}{2}.$$

4. Writing the surface S as $z = f(x, y) = -y + 1$, we have

$$d\vec{A} = (-f_x\vec{i} - f_y\vec{j} + \vec{k}) dx \, dy.$$

Thus,

$$\int_S \vec{F} \cdot d\vec{A} = \int_R \vec{F}(x, y, f(x, y)) \cdot (-f_x\vec{i} - f_y\vec{j} + \vec{k}) \, dx \, dy$$

$$= \int_0^1 \int_0^1 (2x\vec{j} + y\vec{k}) \cdot (\vec{j} + \vec{k}) \, dx \, dy$$

$$= \int_0^1 \int_0^1 (2x + y) \, dx \, dy = \int_0^1 (x^2 + xy) \Big|_0^1 dy$$

$$= \int_0^1 (1 + y) \, dy = (y + \frac{y^2}{2}) \Big|_0^1 = \frac{3}{2}.$$

5.

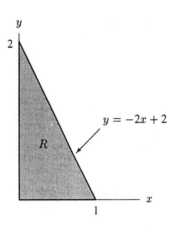

Figure 19.12

Writing the surface S as $z = f(x, y) = -2x - 4y + 1$, we have

$$d\vec{A} = (-f_x\vec{i} - f_y\vec{j} + \vec{k})dxdy.$$

With R as shown in Figure 19.12, we have

$$\int_S \vec{F} \cdot d\vec{A} = \int_R \vec{F}(x, y, f(x, y)) \cdot (-f_x\vec{i} - f_y\vec{j} + \vec{k}) \, dxdy$$

$$= \int_R (3x\vec{i} + y\vec{j} + (-2x - 4y + 1)\vec{k}) \cdot (2\vec{i} + 4\vec{j} + \vec{k}) \, dxdy$$

$$= \int_R (4x + 1) \, dxdy = \int_0^1 \int_0^{-2x+2} (4x + 1) \, dydx$$

$$= \int_0^1 (4x + 1)(-2x + 2) \, dx$$

$$= \int_0^1 (-8x^2 + 6x + 2) \, dx = \left. \left(-\frac{8x^3}{3} + 3x^2 + 2x \right) \right|_0^1 = \frac{7}{3}.$$

6.   Writing the surface S as $z = f(x, y) = 25 - x^2 - y^2$, we have

$$d\vec{A} = (-f_x\vec{i} - f_y\vec{j} + \vec{k})dxdy.$$

Thus,

$$\int_S \vec{F} \cdot d\vec{A} = \int_R \vec{F}(x, y, f(x, y)) \cdot (-f_x\vec{i} - f_y\vec{j} + \vec{k}) \, dxdy$$

$$= \int_R (x\vec{i} + y\vec{j}) \cdot (2x\vec{i} + 2y\vec{j} + \vec{k}) \, dxdy$$

$$= \int_R 2(x^2 + y^2) \, dxdy = \int_0^{2\pi} \int_0^5 2r^2 r \, drd\theta$$

$$= \int_0^{2\pi} \left. \frac{r^4}{2} \right|_0^5 d\theta = \frac{625}{2}(2\pi) = 625\pi.$$

7. Writing the surface S as $z = f(x,y) = 25 - x^2 - y^2$, we have

$$d\vec{A} = (-f_x\vec{i} - f_y\vec{j} + \vec{k})dxdy = (2x\vec{i} + 2y\vec{j} + \vec{k})dxdy.$$

Thus

$$\int_S \vec{F} \cdot d\vec{A} = \int_R \cos(x^2 + y^2)\vec{k} \cdot (2x\vec{i} + 2y\vec{j} + \vec{k})\,dxdy$$

$$= \int_R \cos(x^2 + y^2)\,dxdy = \int_0^{2\pi} \int_0^5 \cos r^2 \cdot r\,drd\theta$$

$$= \int_0^{2\pi} \frac{\sin r^2}{2}\Big|_0^5 \, d\theta = \pi\sin 25.$$

8. Writing the surface S as $z = f(x,y) = y^2 + 5$, we have

$$d\vec{A} = (-f_x\vec{i} + -f_y\vec{j} + \vec{k})dxdy.$$

Thus,

$$\int_S \vec{F} \cdot d\vec{A} = \int_R \vec{F}(x,y,f(x,y)) \cdot (-f_x\vec{i} - f_y\vec{j} + \vec{k})\,dxdy$$

$$= \int_R (-y\vec{j} + (y^2+5)\vec{k}) \cdot (-2y\vec{j} + \vec{k})\,dxdy$$

$$= \int_0^1 \int_{-2}^1 (3y^2 + 5)\,dxdy = \int_0^1 (9y^2 + 15)\,dy$$

$$= (3y^3 + 15y)\Big|_0^1 = 18.$$

9. We have $0 \le z \le 6$ so $0 \le x^2 + y^2 \le 36$. Let R be the disk of radius 6 in the xy-plane centered at the origin. Because of the cone's point, the flux integral is improper; however, it does converge. We have

$$\int_S \vec{F} \cdot d\vec{A} = \int_R \vec{F}(x,y,f(x,y)) \cdot (-f_x\vec{i} - f_y\vec{j} + \vec{k})\,dxdy$$

$$= \int_R (-x\sqrt{x^2+y^2}\vec{i} - y\sqrt{x^2+y^2}\vec{j} + (x^2+y^2)\vec{k})$$

$$\cdot \left(-\frac{x}{\sqrt{x^2+y^2}}\vec{i} - \frac{y}{\sqrt{x^2+y^2}}\vec{j} + \vec{k}\right)\,dxdy$$

$$= \int_R 2(x^2+y^2)\,dxdy$$

$$= 2\int_0^6 \int_0^{2\pi} r^3\,d\theta dr$$

$$= 4\pi \int_0^6 r^3\,dr = 1296\pi.$$

10. Since $y = f(x, z) = x^2 + z^2$, we have

$$d\vec{A} = (-f_x\vec{i} + \vec{j} - f_z\vec{k})\, dx\, dz = (-2x\vec{i} + \vec{j} - 2z\vec{k})\, dx\, dz.$$

Thus, substituting $y = x^2 + z^2$ into $\vec{F}$, we have

$$\int_S \vec{F} \cdot d\vec{A} = \int_{x^2+z^2\le 1} ((x^2 + z^2)\vec{i} + \vec{j} - xz\vec{k}) \cdot (-2x\vec{i} + \vec{j} - 2z\vec{k})\, dx\, dz$$

$$= \int_{x^2+z^2\le 1} (-2x^3 - 2xz^2 + 1 + 2xz^2)\, dx\, dz$$

$$= \int_{-1}^{1} \int_{-\sqrt{1-z^2}}^{\sqrt{1-z^2}} (1 - 2x^3)\, dx\, dz$$

$$= \int_{-1}^{1} \int_{-\sqrt{1-z^2}}^{\sqrt{1-z^2}} dx\, dz - \int_{-1}^{1} \int_{-\sqrt{1-z^2}}^{\sqrt{1-z^2}} 2x^3\, dx\, dz$$

$$= \text{Area of disk} - \int_{-1}^{1} \left(\frac{x^4}{2} \Big|_{-\sqrt{1-z^2}}^{\sqrt{1-z^2}} \right) dz = \pi - 0 = \pi$$

11. Here $z = \sqrt{9 - x^2 - y^2}$, so

$$z_x = -\frac{x}{\sqrt{9 - x^2 - y^2}} \qquad z_y = -\frac{y}{\sqrt{9 - x^2 - y^2}}.$$

The flux integral is given by

$$\int_S \vec{F} \cdot d\vec{A}$$

$$= \int_S \left(x\sqrt{9 - x^2 - y^2}\vec{i} + y\vec{k} \right) \cdot \left(-\frac{x}{\sqrt{9 - x^2 - y^2}}\vec{i} - \frac{y}{\sqrt{9 - x^2 - y^2}}\vec{j} + \vec{k} \right) dx\, dy$$

$$= \int_{-3}^{3} \int_{-\sqrt{9-x^2}}^{\sqrt{9-x^2}} (-x^2 + y)\, dy\, dx$$

Changing to polar coordinates gives

$$\int_S \vec{F} \cdot d\vec{A} = \int_0^{2\pi} \int_0^3 (-r^2 \cos^2 \theta + r \sin \theta)\, r\, dr\, d\theta$$

$$= \int_0^{2\pi} (-\frac{81}{4} \cos^2 \theta + \frac{27}{3} \sin \theta)\, d\theta$$

$$= -\frac{81}{4}\pi.$$

12. The plane through the points $(1, 0, 0)$, $(0, 1, 0)$, and $(0, 0, 1)$ is given by $x + y + z = 1$, so S is the part of the graph of $z = f(x, y) = 1 - x - y$ above the region R in the xy-plane where $x \ge 0$, $y \ge 0$, and $x + y \le 1$. Thus

$$\int_S \vec{F} \cdot d\vec{A} = \int_R \vec{F}(x, y, f(x, y)) \cdot (-f_x\vec{i} - f_y\vec{j} + \vec{k})\, dx\, dy$$

$$= \int_R (x^2\vec{i} + y^2\vec{j} + (1-x-y)^2\vec{k}) \cdot (\vec{i}+\vec{j}+\vec{k})\,dx\,dy$$

$$= \int_R (x^2 + y^2 + (1-x-y)^2)\,dx\,dy$$

$$= \int_0^1 \int_0^{1-x} (1 + 2x^2 + 2y^2 - 2x - 2y + 2xy)\,dy\,dx$$

$$= \int_0^1 [(1-x) + 2x^2(1-x) + \frac{2}{3}(1-x)^3 - 2x(1-x)$$

$$-(1-x)^2 + x(1-x)^2]\,dx$$

$$= \frac{1}{4}.$$

13. Since the radius of the cylinder is 1, using cylindrical coordinates we have

$$d\vec{A} = (\cos\theta\vec{i} + \sin\theta\vec{j})d\theta dz.$$

Thus,

$$\int_S \vec{F} \cdot d\vec{A} = \int_0^6 \int_0^{2\pi} (\cos\theta\vec{i} + \sin\theta\vec{j}) \cdot (\cos\theta\vec{i} + \sin\theta\vec{j})\,d\theta\,dz$$

$$= \int_0^6 \int_0^{2\pi} 1\,d\theta\,dz = 12\pi.$$

14. Since the radius of the cylinder is 1, using cylindrical coordinates we have $d\vec{A} = (\cos\theta\vec{i} + \sin\theta\vec{j})d\theta dz$.
Thus,

$$\int_S \vec{F} \cdot d\vec{A} = \int_0^6 \int_0^{2\pi} (z\cos\theta\vec{i} + z\sin\theta\vec{j} + z^3\vec{k}) \cdot (\cos\theta\vec{i} + \sin\theta\vec{j})\,d\theta\,dz$$

$$= \int_0^6 \int_0^{2\pi} z\,d\theta\,dz = 2\pi \left(\frac{z^2}{2}\right)\Big|_0^6 = 36\pi.$$

15. Since the radius of the sphere is 5, using spherical coordinates we have

$$d\vec{A} = (\sin\phi\cos\theta\vec{i} + \sin\phi\sin\theta\vec{j} + \cos\phi\vec{k})25\sin\phi\,d\theta\,d\phi.$$

Thus,

$$\int_S \vec{F} \cdot d\vec{A} = \int_0^{\frac{\pi}{2}} \int_0^{2\pi} (25\cos^2\phi\vec{k}) \cdot (\sin\phi\cos\theta\vec{i} + \sin\phi\sin\theta\vec{j} + \cos\phi\vec{k})25\sin\phi\,d\theta\,d\phi$$

$$= 625 \int_0^{\frac{\pi}{2}} \int_0^{2\pi} \cos^3\phi\sin\phi\,d\theta d\phi$$

$$= -1250\pi\frac{(\cos\phi)^4}{4}\Big|_0^{\frac{\pi}{2}} = \frac{625}{2}\pi.$$

16. Since the radius of the sphere is a, using spherical coordinates we have

$$d\vec{A} = (\sin\phi\cos\theta\vec{i} + \sin\phi\sin\theta\vec{j} + \cos\phi\vec{k})a^2\sin\phi\,d\phi\,d\theta.$$

Thus,

$$\int_S \vec{F}\cdot d\vec{A} = \int_0^{2\pi}\int_0^\pi (a\sin\phi\cos\theta\vec{i} + a\sin\phi\sin\theta\vec{j} + a\cos\phi\vec{k})\cdot$$

$$(\sin\phi\cos\theta\vec{i} + \sin\phi\sin\theta\vec{j} + \cos\phi\vec{k})a^2\sin\phi\,d\phi\,d\theta$$

$$= a^3\int_0^{2\pi}\int_0^\pi \sin\phi\,d\phi\,d\theta$$

$$= 2\pi a^3\int_0^\pi \sin\phi\,d\phi = (2\pi a^3)(2) = 4\pi a^3.$$

17. (a) The position vector of a point (R, θ, z) (in cylindrical coordinates) on the cylinder is given by

$$\vec{r} = R\cos\theta\vec{i} + R\sin\theta\vec{j} + z\vec{k}.$$

So $\|\vec{r}\| = \sqrt{R^2 + z^2}$. Furthermore, for an area element on the cylinder we have the following

$$d\vec{A} = (\cos\theta\vec{i} + \sin\theta\vec{j})R\,dz\,d\theta,$$

and the integral is:

$$\int_S \vec{E}\cdot d\vec{A} = \int_0^{2\pi}\int_{-H}^H q\frac{\vec{r}}{\|\vec{r}\|^3}\cdot(\cos\theta\vec{i} + \sin\theta\vec{j})R\,dz\,d\theta$$

$$= q\int_0^{2\pi}\int_{-H}^H \frac{R\cos^2\theta + R\sin^2\theta}{(R^2 + z^2)^{3/2}}R\,dz\,d\theta = 2\pi q\int_{-H}^H \frac{R^2\,dz}{(R^2 + z^2)^{3/2}}.$$

In order to compute this one variable integral, we write:

$$\int_{-H}^H \frac{R^2 dz}{(R^2 + z^2)^{3/2}} = \int_{-H}^H \frac{(R^2 + z^2)dz}{(R^2 + z^2)^{3/2}} - \int_{-H}^H \frac{z^2 dz}{(R^2 + z^2)^{3/2}}$$

$$= \int_{-H}^H \frac{dz}{(R^2 + z^2)^{1/2}} - \left(\int_{-H}^H \frac{dz}{(R^2 + z^2)^{1/2}} - \frac{z}{(R^2 + z^2)}\bigg|_{-H}^H\right)$$

$$= \int_{-H}^H \frac{dz}{(R^2 + z^2)^{1/2}} - \int_{-H}^H \frac{dz}{(R^2 + z^2)^{1/2}} + \frac{z}{(R^2 + z^2)}\bigg|_{-H}^H$$

$$= \frac{2H}{\sqrt{R^2 + H^2}}.$$

(The integral $\displaystyle\int_{-H}^H \frac{z^2 dz}{(R^2 + z^2)^{3/2}} = \int_{-H}^H z\frac{z\,dz}{(R^2 + z^2)^{3/2}}$ was computed using integration by parts).

Therefore

$$\int_S \vec{E}\cdot d\vec{A} = 4\pi q\frac{H}{\sqrt{R^2 + H^2}}.$$

(b) (i) Let R be fixed. We have

$$\lim_{H \to 0} \int_S \vec{E} \cdot d\vec{A} = \lim_{H \to 0} 4\pi q \frac{H}{\sqrt{H^2 + R^2}} = 0.$$

$$\lim_{H \to \infty} \int_S \vec{E} \cdot d\vec{A} = \lim_{H \to \infty} 4\pi q \frac{H}{\sqrt{H^2 + R^2}} = 4\pi q.$$

(ii) Now let H be fixed. We have

$$\lim_{R \to 0} \int_S \vec{E} \cdot d\vec{A} = \lim_{R \to 0} 4\pi q \frac{H}{\sqrt{H^2 + R^2}} = 4\pi q.$$

$$\lim_{R \to \infty} \int_S \vec{E} \cdot d\vec{A} = \lim_{R \to \infty} 4\pi q \frac{H}{\sqrt{H^2 + R^2}} = 0.$$

18. Let L denote the curved surface of the cylinder and T and B its top and bottom respectively. We have:

$$\int_S \vec{E} \cdot d\vec{A} = \int_L \vec{E} \cdot d\vec{A} + \int_T \vec{E} \cdot d\vec{A} + \int_B \vec{E} \cdot d\vec{A}.$$

It is shown in Problem 17 that

$$\int_L \vec{E} \cdot d\vec{A} = 4\pi q \frac{H}{\sqrt{R^2 + H^2}}.$$

Let's compute $\int_T \vec{E} \cdot d\vec{A}$. The normal at any point of T is $\vec{n} = \vec{k}$ and so $d\vec{A} = \vec{k} \, r \, dr d\theta$. Therefore we have:

$$\int_T \vec{E} \cdot d\vec{A} = \int_0^{2\pi} \int_0^R q \frac{\vec{r}}{\|\vec{r}\|^3} \cdot \vec{k} \, r \, dr \, d\theta$$

$$= q \int_0^{2\pi} \int_0^R \frac{(r\cos\theta\vec{i} + r\sin\theta\vec{j} + H\vec{k}) \cdot \vec{k}}{(r^2 + H^2)^{3/2}} r \, dr \, d\theta$$

$$= qH \int_0^{2\pi} \int_0^R \frac{r \, dr \, d\theta}{(r^2 + H^2)^{3/2}} = -2\pi q H \frac{1}{\sqrt{r^2 + H^2}} \Big|_0^R$$

$$= -2\pi q \frac{H}{\sqrt{R^2 + H^2}} + 2\pi q.$$

Similarly, or using a symmetry argument, we find that the flux through the bottom is given by

$$\int_B \vec{E} \cdot d\vec{A} = -2\pi q \frac{H}{\sqrt{R^2 + H^2}} + 2\pi q.$$

Thus the total flux is given by

$$\int_S \vec{E} \cdot d\vec{A} = 4\pi q \frac{H}{\sqrt{R^2 + H^2}} - 2\pi q \frac{H}{\sqrt{R^2 + H^2}} + 2\pi q - 2\pi q \frac{H}{\sqrt{R^2 + H^2}} + 2\pi q = 4\pi q.$$

19. On the disk, $z = 0$ and $d\vec{A} = \vec{k} \, dx \, dy$, so

$$\int_S \vec{F} \cdot d\vec{A} = \int_{x^2+y^2 \le 1} (xze^{yz}\vec{i} + x\vec{j} + (5 + x^2 + y^2)\vec{k}) \cdot \vec{k} \, dx \, dy$$

$$= \int_{x^2+y^2 \le 1} (5 + x^2 + y^2) \, dx \, dy = \int_0^{2\pi} \int_0^1 (5 + r^2) r \, dr \, d\theta$$

$$= 2\pi \left(\frac{5r^2}{2} + \frac{r^4}{4} \right) \Big|_0^1 = \frac{11\pi}{2}.$$

20.   The plane is $x - z = 0$ over region $0 \leq x \leq \sqrt{2}$, $0 \leq y \leq 2$. See Figure 19.13.

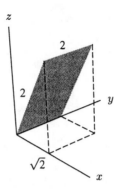

Figure 19.13

$$\text{Flux} = \int_0^2 \int_0^{\sqrt{2}} \left((e^{xy} + 3z + 5)\vec{i} + (e^{xy} + 5z + 3)\vec{j} + (3z + e^{xy})\vec{k} \right) \cdot (\vec{i} - \vec{k}) \, dx \, dy$$

$$= \int_0^2 \int_0^{\sqrt{2}} (e^{xy} + 3z + 5 - 3z - e^{xy}) \, dx \, dy = 5(2)(\sqrt{2}) = 10\sqrt{2}$$

Alternatively, since a unit normal to the surface is $\vec{n}/\sqrt{2} = (\vec{i} - \vec{j})/\sqrt{2}$, writing $dA = ||d\vec{A}||$, we have

$$\text{Flux} = \int_S \vec{H} \cdot d\vec{A} = \int \vec{H} \cdot \frac{\vec{i} - \vec{k}}{\sqrt{2}} \, dA = \int \frac{5}{\sqrt{2}} \, dA$$

$$= \frac{5}{\sqrt{2}}(\text{Area of slanted square}) = \frac{5}{\sqrt{2}} 4 = 10\sqrt{2}.$$

Solutions for Section 19.3

1.   Since S is given by

$$\vec{r}(s, t) = (s + t)\vec{i} + (s - t)\vec{j} + (s^2 + t^2)\vec{k},$$

we have

$$\frac{\partial \vec{r}}{\partial s} = \vec{i} + \vec{j} + 2s\vec{k} \quad \text{and} \quad \frac{\partial \vec{r}}{\partial t} = \vec{i} - \vec{j} + 2t\vec{k},$$

and

$$\frac{\partial \vec{r}}{\partial s} \times \frac{\partial \vec{r}}{\partial t} = \begin{vmatrix} \vec{i} & \vec{j} & \vec{k} \\ 1 & 1 & 2s \\ 1 & -1 & 2t \end{vmatrix} = (2s + 2t)\vec{i} + (2s - 2t)\vec{j} - 2\vec{k}.$$

Since the $\vec{i}$ component of this vector is positive for $0 < s < 1$, $0 < t < 1$, it points away from the z-axis, and so has the opposite orientation to the one specified. Thus, we use

$$d\vec{A} = -\frac{\partial \vec{r}}{\partial s} \times \frac{\partial \vec{r}}{\partial t} \, ds \, dt,$$

and so we have

$$\int_S \vec{F} \cdot d\vec{A} = -\int_0^1 \int_0^1 (s^2 + t^2)\vec{k} \cdot \left((2s + 2t)\vec{i} + (2s - 2t)\vec{j} - 2\vec{k}\right) ds\, dt$$

$$= 2\int_0^1 \int_0^1 (s^2 + t^2) \, ds\, dt = 2\int_0^1 \left(\frac{s^3}{3} + st^2\right)\Bigg|_{s=0}^{s=1} dt$$

$$= 2\int_0^1 (\frac{1}{3} + t^2) \, dt = 2(\frac{1}{3}t + \frac{t^3}{3})\Bigg|_0^1 = 2(\frac{1}{3} + \frac{1}{3}) = \frac{4}{3}.$$

2. Since S is parameterized by

$$\vec{r}(s, t) = 3\sin s\vec{i} + 3\cos s\vec{j} + (t + 1)\vec{k},$$

we have

$$\frac{\partial \vec{r}}{\partial s} = 3\cos s\vec{i} - 3\sin s\vec{j} \quad \text{and} \quad \frac{\partial \vec{r}}{\partial t} = \vec{k}.$$

So

$$\frac{\partial \vec{r}}{\partial s} \times \frac{\partial \vec{r}}{\partial t} = \begin{vmatrix} \vec{i} & \vec{j} & \vec{k} \\ 3\cos s & -3\sin s & 0 \\ 0 & 0 & 1 \end{vmatrix} = -3\sin s\vec{i} - 3\cos s\vec{j},$$

which points towards the z-axis and thus opposite to the orientation we were given. Hence, we use

$$d\vec{A} = -\frac{\partial \vec{r}}{\partial s} \times \frac{\partial \vec{r}}{\partial t} \, ds\, dt,$$

and so we have

$$\int_S \vec{F} \cdot d\vec{A} = -\int_0^1 \int_0^\pi (3\cos s\vec{i} + 3\sin s\vec{j}) \cdot (-3\sin s\vec{i} - 3\cos s\vec{j}) \, ds\, dt$$

$$= 9\int_0^1 \int_0^\pi 2\sin s\cos s \, ds\, dt = 9\int_0^1 \int_0^\pi \sin 2s \, ds\, dt$$

$$= 9\int_0^1 \left(-\frac{\cos 2s}{2}\Bigg|_{s=0}^{s=\pi}\right) dt = 0.$$

3. The cross product $\partial \vec{r}/\partial s \times \partial \vec{r}/\partial t$ is given by

$$\frac{\partial \vec{r}}{\partial s} \times \frac{\partial \vec{r}}{\partial t} = \begin{vmatrix} \vec{i} & \vec{j} & \vec{k} \\ 2s & 2 & 0 \\ 0 & 2t & 5 \end{vmatrix} = 10\vec{i} - 10s\vec{j} + 4st\vec{k}.$$

Since the z-component, $4st$, of the vector $\partial \vec{r}/\partial s \times \partial \vec{r}/\partial t$ is positive for $0 < s \leq 1, 1 \leq t \leq 3$, we see that $\partial \vec{r}/\partial s \times \partial \vec{r}/\partial t$ points upward, in the direction of the orientation of S we were given. Thus, we use

$$d\vec{A} = \left(\frac{\partial \vec{r}}{\partial s} \times \frac{\partial \vec{r}}{\partial t}\right) ds\, dt,$$

and so we have

$$
\int_S \vec{F} \cdot d\vec{A} = \int_0^1 \int_1^3 (5t\vec{i} + s^2\vec{j}) \cdot (10\vec{i} - 10s\vec{j} + 4st\vec{k}) \, dt \, ds
$$

$$
= \int_0^1 \int_1^3 (50t - 10s^3) \, dt \, ds = \int_0^1 (25t^2 - 10s^3 t)\Big|_{t=1}^{t=3} ds
$$

$$
= \int_0^1 (200 - 20s^3) \, ds = (200s - 5s^4)\Big|_0^1
$$

$$
= 200 - 5 = 195.
$$

4. Since

$$
\vec{r}(a, \theta) = a\cos\theta\vec{i} + a\sin\theta\vec{j} + \sin a^2\vec{k},
$$

we have

$$
\frac{\partial\vec{r}}{\partial a} \times \frac{\partial\vec{r}}{\partial \theta} = \begin{vmatrix} \vec{i} & \vec{j} & \vec{k} \\ \cos\theta & \sin\theta & 2a\cos a^2 \\ -a\sin\theta & a\cos\theta & 0 \end{vmatrix} = -2a^2\cos\theta\cos a^2\,\vec{i} - 2a^2\sin\theta\cos a^2\,\vec{j} + a\vec{k}.
$$

The z-component, a, of the vector $\partial\vec{r}/\partial a \times \partial\vec{r}/\partial\theta$ is positive for $1 \le a \le 3$, $0 \le \theta \le \pi$, so $\partial\vec{r}/\partial a \times \partial\vec{r}/\partial\theta$ points upward, in the direction of the orientation of S we were given. Thus, we use $d\vec{A} = (\partial\vec{r}/\partial a \times \partial\vec{r}/\partial\theta) \, da \, d\theta$, giving

$$
\int_S \vec{F} \cdot d\vec{A} = \int_1^3 \int_0^\pi \left(\left(-\frac{2}{a\cos\theta}\right)\vec{i} + \left(\frac{2}{a\sin\theta}\right)\vec{j} \right) \cdot \frac{\partial\vec{r}}{\partial a} \times \frac{\partial\vec{r}}{\partial\theta} \, d\theta \, da
$$

$$
= \int_1^3 \int_0^\pi (4a\cos a^2 - 4a\cos a^2) \, d\theta \, da = 0.
$$

5. Using cylindrical coordinates, we see that the surface S is parameterized by

$$
\vec{r}(r, \theta) = r\cos\theta\vec{i} + r\sin\theta\vec{j} + r\vec{k}.
$$

We have

$$
\frac{\partial\vec{r}}{\partial r} \times \frac{\partial\vec{r}}{\partial\theta} = \begin{vmatrix} \vec{i} & \vec{j} & \vec{k} \\ \cos\theta & \sin\theta & 1 \\ -r\sin\theta & r\cos\theta & 0 \end{vmatrix} = -r\cos\theta\vec{i} - r\sin\theta\vec{j} + r\vec{k}.
$$

Since the vector $\partial\vec{r}/\partial r \times \partial\vec{r}/\partial\theta$ points upward, in the direction opposite to the specified orientation, we use $d\vec{A} = -(\partial\vec{r}/\partial r \times \partial\vec{r}/\partial\theta) \, dr \, d\theta$. Hence

$$
\int_S \vec{F} \cdot d\vec{A} = -\int_0^{2\pi} \int_0^R (r^5\cos^2\theta\sin^2\theta\vec{k}) \cdot (-r\cos\theta\vec{i} - r\sin\theta\vec{j} + r\vec{k}) \, dr \, d\theta
$$

$$
= -\int_0^{2\pi} \int_0^R r^6\cos^2\theta\sin^2\theta \, dr \, d\theta
$$

$$= -\frac{R^7}{7} \int_0^{2\pi} \sin^2 \theta \cos^2 \theta \, d\theta$$

$$= -\frac{R^7}{7} \int_0^{2\pi} \sin^2 \theta (1 - \sin^2 \theta) \, d\theta$$

$$= -\frac{R^7}{7} \int_0^{2\pi} (\sin^2 \theta - \sin^4 \theta) \, d\theta$$

$$= -(\frac{R^7}{7})(\frac{\pi}{4}) = \frac{-\pi}{28} R^7.$$

The cone is not differentiable at the point $(0,0)$. However the flux integral, which is improper, converges.

6.

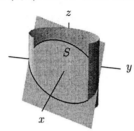

Figure 19.14

The plane is parameterized by

$$\vec{r} = x\vec{i} + y\vec{j} + z\vec{k} = x\vec{i} + y\vec{j} + (2 - 2x - y)\vec{k},$$

where (x, y) is in the disk R lying inside the circle $x^2 + y^2 = 2x$. By completing the square, this circle can be rewritten as $(x - 1)^2 + y^2 = 1$ and so the disk has area π.

We have $dA = \|\frac{\partial \vec{r}}{\partial x} \times \frac{\partial \vec{r}}{\partial y}\| \, dxdy$, where

$$\frac{\partial \vec{r}}{\partial x} \times \frac{\partial \vec{r}}{\partial y} = \begin{vmatrix} \vec{i} & \vec{j} & \vec{k} \\ 1 & 0 & -2 \\ 0 & 1 & -1 \end{vmatrix} = 2\vec{i} + \vec{j} + \vec{k}$$

and so

$$\left\| \frac{\partial \vec{r}}{\partial x} \times \frac{\partial \vec{r}}{\partial y} \right\| = \sqrt{6}.$$

Thus, the surface area of the ellipse S is given by

$$\text{Surface area} = \int_S 1 \, dA = \int_R \sqrt{6} \, dxdy$$

$$= \sqrt{6} \times (\text{Area of disk } x^2 + y^2 = 2x)$$

$$= \sqrt{6}\pi.$$

7. The elliptic cylindrical surface is parameterized by

$$\vec{r} = x\vec{i} + y\vec{j} + z\vec{k} = a\cos\theta\vec{i} + b\sin\theta\vec{j} + z\vec{k} \qquad \text{where } 0 \le \theta \le 2\pi, -c \le z \le c.$$

We have

$$\frac{\partial \vec{r}}{\partial \theta} \times \frac{\partial \vec{r}}{\partial z} = \begin{vmatrix} \vec{i} & \vec{j} & \vec{k} \\ -a\sin\theta & b\cos\theta & 0 \\ 0 & 0 & 1 \end{vmatrix} = b\cos\theta\vec{i} + a\sin\theta\vec{j}.$$

This vector points away from the z-axis, so we use $d\vec{A} = (b\cos\theta\vec{i} + a\sin\theta\vec{j})\,d\theta dz$, giving

$$\int_S \vec{F} \cdot d\vec{A} = \int_{-c}^{c} \int_0^{2\pi} \left(\frac{b}{a}(a\cos\theta)\vec{i} + \frac{a}{b}(b\sin\theta\vec{j})\right) \cdot (b\cos\theta\vec{i} + a\sin\theta\vec{j})\,d\theta\,dz$$

$$= \int_{-c}^{c} \int_0^{2\pi} (b^2\cos^2\theta + a^2\sin^2\theta)\,d\theta dz$$

$$= 2\pi c(a^2 + b^2).$$

8. The surface S is parameterized by

$$\vec{r} = x\vec{i} + f(x)\cos\theta\vec{j} + f(x)\sin\theta\vec{k}, \qquad a \le x \le b, 0 \le \theta \le 2\pi.$$

The area element on A is

$$dA = \left\| \frac{\partial \vec{r}}{\partial x} \times \frac{\partial \vec{r}}{\partial \theta} \right\| dx\,d\theta$$

$$= \|(\vec{i} + f'(x)\cos\theta\vec{j} + f'(x)\sin\theta\vec{k}) \times (-f(x)\sin\theta\vec{j} + f(x)\cos\theta\vec{k})\|\,dx\,d\theta$$

$$= \|f(x)f'(x)\vec{i} - f(x)\cos\theta\vec{j} - f(x)\sin\theta\vec{k}\|\,dx\,d\theta$$

$$= f(x)\sqrt{f'(x)^2 + \cos^2\theta + \sin^2\theta}\,dx\,d\theta$$

$$= f(x)\sqrt{1 + f'(x)^2}\,dx\,d\theta.$$

So

$$\text{Surface area} = \int_S dA = \int_0^{2\pi}\int_a^b f(x)\sqrt{1 + f'(x)^2}\,dx\,d\theta = 2\pi\int_a^b f(x)\sqrt{1 + f'(x)^2}\,dx.$$

9. If S is the part of the graph of $z = f(x,y)$ lying over a region R in the xy-plane, then S is parameterized by

$$\vec{r}(x,y) = x\vec{i} + y\vec{j} + f(x,y)\vec{k}, \qquad (x,y) \text{ in } R.$$

So

$$\frac{\partial \vec{r}}{\partial x} \times \frac{\partial \vec{r}}{\partial y} = (\vec{i} + f_x\vec{k}) \times (\vec{j} + f_y\vec{k}) = -f_x\vec{i} - f_y\vec{j} + \vec{k}.$$

Since the $\vec{k}$ component is positive, this points upward, so if S is oriented upward

$$d\vec{A} = (-f_x\vec{i} - f_y\vec{j} + \vec{k})\,dx\,dy$$

and therefore we have the expression for the flux integral obtained on page 399:

$$\int_S \vec{F} \cdot d\vec{A} = \int_R \vec{F}(x,y,f(x,y)) \cdot (-f_x\vec{i} - f_y\vec{k} + \vec{k})\,dx\,dy.$$

10. If S is the part of the cylinder of radius R corresponding to the region T in θz-space, then S is parameterized in cylindrical coordinates by

$$\vec{r}(\theta, z) = R\cos\theta\vec{i} + R\sin\theta\vec{j} + z\vec{k}, \qquad (\theta, z) \text{ in } T.$$

So

$$\frac{\partial\vec{r}}{\partial\theta} \times \frac{\partial\vec{r}}{\partial z} = (-R\sin\theta\vec{i} + R\cos\theta\vec{j}) \times \vec{k} = R\cos\theta\vec{i} + R\sin\theta\vec{j}.$$

This points outward, so

$$d\vec{A} = (R\cos\theta\vec{i} + R\sin\theta\vec{j})\, d\theta\, dz = (\cos\theta\vec{i} + \sin\theta\vec{j})R\, d\theta\, dz$$

and therefore we obtain the expression for the flux integral in cylindrical coordinates on page 400:

$$\int_S \vec{F} \cdot d\vec{A} = \int_T \vec{F}(R\cos\theta, R\sin\theta, z) \cdot (\cos\theta\vec{i} + \sin\theta\vec{j})R\, d\theta\, dz.$$

11. If S is the part of the sphere of radius R corresponding to the region T in $\theta\phi$-space, then S is parameterized in spherical coordinates by

$$\vec{r}(\theta, \phi) = R\sin\phi\cos\theta\vec{i} + R\sin\phi\sin\theta\vec{j} + R\cos\phi\vec{k}, \qquad (\theta, \phi) \text{ in } T.$$

So

$$\frac{\partial\vec{r}}{\partial\theta} \times \frac{\partial\vec{r}}{\partial\phi} = \begin{vmatrix} \vec{i} & \vec{j} & \vec{k} \\ -R\sin\phi\sin\theta & R\sin\phi\cos\theta & 0 \\ R\cos\phi\cos\theta & R\cos\phi\sin\theta & -R\sin\phi \end{vmatrix}$$

$$= -R^2\sin^2\phi\cos\theta\vec{i} - R^2\sin^2\phi\sin\theta\vec{j} - R^2\sin\phi\cos\phi\vec{k}$$

$$= -R^2\sin\phi(\sin\phi\cos\theta\vec{i} + \sin\phi\sin\theta\vec{j} + \cos\phi\vec{k}).$$

This points inward, so the outward area alement is

$$d\vec{A} = (\sin\phi\cos\theta\vec{i} + \sin\phi\sin\theta\vec{j} + \cos\phi\vec{k})R^2\sin\phi\, d\theta\, d\phi,$$

and therefore we obtain the expression for the flux integral in spherical coordinates on page 402:

$$\int_S \vec{F} \cdot d\vec{A}$$
$$= \int_T \vec{F}(R\sin\phi\cos\theta, R\sin\phi\sin\theta, R\cos\phi) \cdot (\sin\phi\cos\theta\vec{i} + \sin\phi\sin\theta\vec{j} + \cos\phi\vec{k})R^2\sin\phi\, d\theta\, d\phi.$$

12. In terms of the st-parameterization,

$$d\vec{A} = \frac{\partial\vec{r}}{\partial s} \times \frac{\partial\vec{r}}{\partial t}\, ds\, dt.$$

By the chain rule, we have

$$\frac{\partial\vec{r}}{\partial s} = \frac{\partial\vec{r}}{\partial u}\frac{\partial u}{\partial s} + \frac{\partial\vec{r}}{\partial v}\frac{\partial v}{\partial s}$$
$$\frac{\partial\vec{r}}{\partial t} = \frac{\partial\vec{r}}{\partial u}\frac{\partial u}{\partial t} + \frac{\partial\vec{r}}{\partial v}\frac{\partial v}{\partial t}.$$

So taking the cross product, we get

$$\frac{\partial \vec{r}}{\partial s} \times \frac{\partial \vec{r}}{\partial t} = \left(\frac{\partial \vec{r}}{\partial u} \frac{\partial u}{\partial s} + \frac{\partial \vec{r}}{\partial v} \frac{\partial v}{\partial s} \right) \times \left(\frac{\partial \vec{r}}{\partial u} \frac{\partial u}{\partial t} + \frac{\partial \vec{r}}{\partial v} \frac{\partial v}{\partial t} \right)$$

$$= \left(\frac{\partial u}{\partial s} \frac{\partial v}{\partial t} - \frac{\partial u}{\partial t} \frac{\partial v}{\partial s} \right) \frac{\partial \vec{r}}{\partial u} \times \frac{\partial \vec{r}}{\partial v}.$$

Now suppose we are going to change variables in a double integral from uv-coordinates to st-coordinates. The Jacobian is

$$\frac{\partial(u, v)}{\partial(s, t)} = \begin{vmatrix} \frac{\partial u}{\partial s} & \frac{\partial v}{\partial s} \\ \frac{\partial u}{\partial t} & \frac{\partial v}{\partial t} \end{vmatrix} = \frac{\partial u}{\partial s} \frac{\partial v}{\partial t} - \frac{\partial u}{\partial t} \frac{\partial v}{\partial s}.$$

Since the Jacobian is assumed to be positive, converting from a uv-integral to an st-integral gives:

$$\int_T \vec{F} \cdot \frac{\partial \vec{r}}{\partial u} \times \frac{\partial \vec{r}}{\partial v} \, du dv = \int_R \vec{F} \cdot \frac{\partial \vec{r}}{\partial u} \times \frac{\partial \vec{r}}{\partial v} \frac{\partial(u, v)}{\partial(s, t)} \, ds dt$$

$$= \int_R \vec{F} \cdot \frac{\partial \vec{r}}{\partial u} \times \frac{\partial \vec{r}}{\partial v} \left(\frac{\partial u}{\partial s} \frac{\partial v}{\partial t} - \frac{\partial u}{\partial t} \frac{\partial v}{\partial s} \right) \, ds dt.$$

However, we know that this gives us

$$\int_T \vec{F} \cdot \frac{\partial \vec{r}}{\partial u} \times \frac{\partial \vec{r}}{\partial v} \, du \, dv = \int_R \vec{F} \cdot \frac{\partial \vec{r}}{\partial u} \times \frac{\partial \vec{r}}{\partial v} \left(\frac{\partial u}{\partial s} \frac{\partial v}{\partial t} - \frac{\partial u}{\partial t} \frac{\partial v}{\partial s} \right) \, ds dt = \int_R \vec{F} \cdot \frac{\partial \vec{r}}{\partial s} \times \frac{\partial \vec{r}}{\partial t} \, ds dt.$$

Thus, the flux integral in uv-coordinates equals the flux integral in st-coordinates.

Solutions for Chapter 19 Review

1. For convention, orient the square so that the positive direction of flow is from down to up. Then when S is far up the positive z-axis, the flux is positive and large, because $\|\vec{r}\|$ is large. As S moves down, the flux gets smaller. When S reaches the xy-plane, the flux is zero, because $\vec{r}$ and $\Delta \vec{A}$ are perpendicular. As S moves down more, the flux becomes more and more negative.

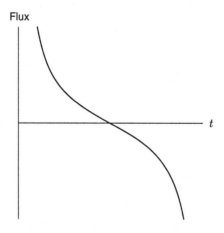

Figure 19.15

2. Suppose S starts off above the z-axis and that the orientation of S is upward. (See Figure 19.16.) If $r = \|\vec{r}\|$ and $dA = \|d\vec{A}\|$, the term $\vec{r} \cdot d\vec{A} = r \, dA \cos\theta$ approximates the flux. The flux starts off positive, as θ equals zero. As S rotates, θ increases. So the flux behaves like the cosine function. (See Figure 19.17.) Notice that r is not really constant as the surface rotates, but if S is far up the z-axis, and small, r can be considered to be constant.

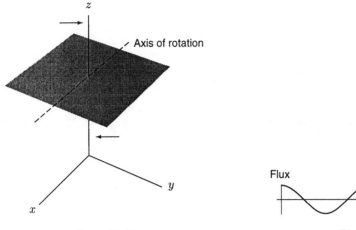

Figure 19.16   *Figure 19.17*

3. (a) Problem 1:
        We have $\vec{F}(\vec{r}) = \vec{r}/r^3$. The magnitude of the vector is $1/r^2$. In other words, as the vectors get farther from the origin, their lengths decrease. So when the square S is far up on the z-axis, the flux is small. As S moves down, the flux increases. When S reaches the xy-plane, the flux is undefined because S contains the point where $r = 0$. When S is below the xy-plane, it starts off negative and large, and it increases towards 0. (See Figure 19.18.)

    (b) Problem 2:
        As before, the term $(\vec{r}/r^3) \cdot \Delta\vec{A} = (1/r^2)\Delta A \cos\theta$ approximates the flux. The flux starts off positive, as θ initially equals zero. As S rotates, θ increases. So the flux behaves like the cosine function.

        Notice that r is not really constant as the surface rotates, but since S is far up the z-axis and small, r can be considered to be constant. (See Figure 19.19.)

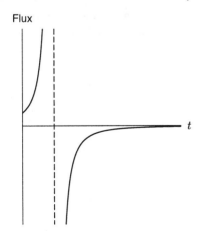

Figure 19.18: Flux for Problem 1 where
$\vec{F} = \vec{r}/r^3$

Figure 19.19: Flux for Problem 2 when
$\vec{F} = \vec{r}/r^3$

4. The flux through the surface equals

$$\vec{v} \cdot \vec{A}$$

where $\vec{A}$ is the area vector of the surface. The vector $\vec{k}$ is normal to the xy-plane. The area of the disk is equal to 4π, so the area vector of the disk equals

$$\vec{A} = 4\pi\vec{k}.$$

Thus

$$\vec{v} \cdot \vec{A} = 12\pi$$

5. The flux through the surface equals

$$\vec{v} \cdot \vec{A}$$

where $\vec{A}$ is the area vector of the surface. The vector $\vec{i}$ is normal to the yz-plane. The area of the triangle is equal to 4, so the area vector of the triangle equals

$$\vec{A} = 4\vec{i}.$$

Thus

$$\vec{v} \cdot \vec{A} = 4$$

6. The flux through the surface equals

$$\vec{v} \cdot \vec{A}$$

where $\vec{A}$ is the area vector of the surface. The vector $\vec{i}$ is normal to the yz-plane. The area of the square is equal to 4, so the area vector of the square equals

$$\vec{A} = 4\vec{i}.$$

Thus

$$\vec{v} \cdot \vec{A} = 4$$

7. The flux through the surface equals

$$\vec{v} \cdot \vec{A}$$

where $\vec{A}$ is the area vector of the surface. The vector $\vec{i} + \vec{j} + \vec{k}$ is normal to the triangle with vertices $(1,0,0), (0,1,0),$ and $(0,0,1)$, so a unit normal is $(1/\sqrt{3})\vec{i} + (1/\sqrt{3})\vec{j} + (1/\sqrt{3})\vec{k}$. The area of the triangle is equal to $\sqrt{3}/2$, so the area vector of the triangle is

$$\vec{A} = \frac{1}{2}\vec{i} + \frac{1}{2}\vec{j} + \frac{1}{2}\vec{k}.$$

Thus

$$\vec{v} \cdot \vec{A} = \frac{1}{2} - \frac{1}{2} + \frac{3}{2} = 1.5$$

8. We have $d\vec{A} = \vec{k}\,dA$, and $z = 4$, so,

$$\int_S \vec{F} \cdot d\vec{A} = \int_S (x\vec{i} + y\vec{j} + (4^2 + 3)\vec{k}) \cdot \vec{k}\,dA = \int_S 19\,dA$$
$$= 19(\text{Area of rectangle}) = 19(6) = 114.$$

9. We have $d\vec{A} = \vec{k}\, dA$, so

$$\int_S \vec{F} \cdot d\vec{A} = \int_S (z\vec{i} + y\vec{j} + 2x\vec{k}) \cdot \vec{k}\, dA = \int_S 2x\, dA$$

$$= \int_0^3 \int_0^2 2x\, dx dy = 12.$$

10. We have $d\vec{A} = \vec{i}\, dA$, so

$$\int_S \vec{F} \cdot d\vec{A} = \int_S ((2 + \cos z)\vec{i} + y\vec{j} + 2x\vec{k}) \cdot \vec{i}\, dA = \int_S (2 + \cos z)\, dA$$

$$= \int_0^4 \int_0^3 (2 + \cos z)\, dy dz = 3(8 + \sin 4)$$

11. On the surface S, y is constant, $y = -1$, and $d\vec{A} = -\vec{j}\, dA$, so,

$$\int_S \vec{F} \cdot d\vec{A} = \int_S (x^2\vec{i} + (x + e^{-1})\vec{j} - \vec{k}) \cdot (-\vec{j})\, dA = -\int_S (x + e^{-1})\, dA$$

$$= -\int_0^4 \int_0^2 (x + e^{-1})\, dx\, dz = -4(2 + 2e^{-1}) = -8(1 + e^{-1}).$$

12. Observe that the $\vec{j}$ and $\vec{k}$ components of $\vec{F}$ are parallel to the surface S, so they contribute nothing to the flux integral. On the surface S, the $\vec{i}$ component of $\vec{F}$ equals $5\vec{i}$, because $x = 0$ on S. Since $5\vec{i}$ is normal to S and in the direction of the orientation of S, $\int_S \vec{F} \cdot d\vec{A} = \int_S 5\vec{i} \cdot d\vec{A} = \|5\vec{i}\|(\text{Area of } S) = 20$.

13. There is no flux through the base or top of the cylinder because the vector field is parallel to these faces. For the curved surface, consider a small patch with area $\Delta\vec{A}$. The vector field is pointing radially outward from the z-axis and so is parallel to $\Delta\vec{A}$. Since $\|\vec{F}\| = \sqrt{x^2 + y^2} = 2$ on the curved surface of the cylinder, we have $\vec{F} \cdot \Delta\vec{A} = \|\vec{F}\|\|\Delta\vec{A}\| = 2\Delta A$. Replacing ΔA with dA, we get

$$\int_S \vec{F} \cdot d\vec{A} = \int_{\substack{\text{Curved} \\ \text{surface}}} 2\, dA = 2(\text{Area of curved surface}) = 2(2\pi \cdot 2 \cdot 3) = 24\pi.$$

14. The vector field $\vec{F} = -y\vec{i} + x\vec{j} + z\vec{k}$ is tangent to the curved surface of the cylinder. (The area vector is parallel to the vector pointing radially outward from the z-axis, namely $x\vec{i} + y\vec{j}$ and $(-y\vec{i} + x\vec{j} + z\vec{k}) \cdot (x\vec{i} + y\vec{j}) = 0$.) Thus the only contributions to the flux integral are from the top and the bottom. On the top, $z = 1$ and $d\vec{A} = dA\vec{k}$, so

$$\vec{F} \cdot d\vec{A} = (-y\vec{i} + x\vec{j} + \vec{k}) \cdot dA\vec{k} = dA.$$

Thus

$$\int_{\text{Top}} \vec{F} \cdot d\vec{A} = \int_{\text{Top}} dA = \text{Area of top} = \pi(1)^2 = \pi.$$

Similarly, on the base, $z = -1$ and $d\vec{A} = (-dA\vec{k})$, so

$$\vec{F} \cdot d\vec{A} = (-y\vec{i} + x\vec{j} - \vec{k}) \cdot (-dA\vec{k}) = dA.$$

$$\int_{\text{Base}} \vec{F} \cdot d\vec{A} = \int_{\text{Base}} dA = \text{Area of base} = \pi.$$

Therefore,

$$\text{Total flux through cylinder} = \text{Flux through top} + \text{Flux through base} = 2\pi.$$

15. First we have

$$z_x = \frac{x}{\sqrt{x^2 + y^2}} \qquad z_y = \frac{y}{\sqrt{x^2 + y^2}}.$$

Although z is not a smooth function of x and y at $(0,0)$, the improper integral that we get converges:

$$\int_S \vec{F} \cdot d\vec{A} = \int_S (x^2\vec{i} + y^2\vec{j} + \sqrt{x^2 + y^2}\vec{k}) \cdot (-\frac{x}{\sqrt{x^2 + y^2}}\vec{i} - \frac{y}{\sqrt{x^2 + y^2}}\vec{j} + \vec{k}) \, dA$$

$$= \int_S \left(-\frac{x^3 + y^3}{\sqrt{x^2 + y^2}} + \sqrt{x^2 + y^2}\right) dA$$

Changing to polar coordinates we have

$$\int_S \vec{F} \cdot d\vec{A} = \int_0^{\pi/2} \int_0^1 (-r^2\cos^3\theta - r^2\sin^3\theta + r)r \, dr d\theta$$

$$= \int_0^{\pi/2} \left(-\frac{r^4}{4}(\cos^3\theta + \sin^3\theta) + \frac{1}{3}r^3\Big|_{r=0}^{r=1}\right) d\theta$$

$$= \int_0^{\pi/2} \left(-\frac{1}{4}(\cos^3\theta + \sin^3\theta) + \frac{1}{3}\right) d\theta$$

$$= \int_0^{\pi/2} \left(-\frac{1}{4}(\cos\theta - \cos\theta\sin^2\theta + \sin\theta - \sin\theta\cos^2\theta) + \frac{1}{3}\right) d\theta$$

$$= -\frac{1}{4}(\sin\theta - \frac{1}{3}\sin^3\theta - \cos\theta + \frac{1}{3}\cos^3\theta) + \frac{\theta}{3}\Big|_0^{\pi/2}$$

$$= \frac{\pi}{6} - \frac{1}{3}.$$

16. Notice that the speed is 3 cm/sec at the center of the pipe and 0 cm/sec at the sides. Suppose $\vec{i}$ is the unit vector parallel to the direction of flow. Then, at a distance r from the center of the pipe, the velocity is given by

$$\vec{v} = \left(3 - \frac{3}{4}r^2\right)\vec{i} \text{ cm/sec.}$$

Divide the circular cross-section into concentric rings of width Δr, so that the velocity is approximately constant on each one. The area of a typical ring is $\Delta A \approx 2\pi r\Delta r$. Then since $\vec{v}$ and $\Delta\vec{A}$ are parallel (see Figure 19.20), we have

$$\text{Flux through ring} \approx \vec{v}\Delta\vec{A} = \|\vec{v}\|\|\Delta\vec{A}\| \approx \left(3 - \frac{3}{4}r^2\right)\frac{\text{cm}}{\text{sec}} \cdot (2\pi r\Delta r)\,\text{cm}^2.$$

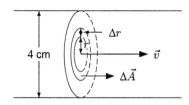

Figure 19.20: Flux through pipe when
velocity varies with distance from the center

Thus, the flux through the circular cross-section of the pipe is given by

$$\text{Flux} = \lim_{\|\Delta\vec{A}\|\to 0} \sum \vec{v} \cdot \Delta\vec{A}$$

$$= \lim_{\Delta r\to 0} \sum \left(3 - \frac{3}{4}r^2\right) 2\pi r \Delta r$$

$$= \int_{r=0}^{r=2} \left(3 - \frac{3}{4}r^2\right) 2\pi r \, dr = 6\pi \int_0^2 \left(r - \frac{r^3}{4}\right) dr = 6\pi \text{ cm}^3/\text{sec}.$$

17. (a) Consider two opposite faces of the cube, S_1 and S_2. The corresponding area vectors are $\vec{A}_1 = 4\vec{i}$ and $\vec{A}_2 = -4\vec{i}$ (since the side of the cube has length 2). Since $\vec{E}$ is constant, we find the flux by taking the dot product, giving

$$\text{Flux through } S_1 = \vec{E} \cdot \vec{A}_1 = (a\vec{i} + b\vec{j} + c\vec{k}) \cdot 4\vec{i} = 4a.$$

$$\text{Flux through } S_2 = \vec{E} \cdot \vec{A}_2 = (a\vec{i} + b\vec{j} + c\vec{k}) \cdot (-4\vec{i}) = -4a.$$

Thus the fluxes through S_1 and S_2 cancel. Arguing similarly, we conclude that, for any pair of opposite faces, the sum of the fluxes of $\vec{E}$ through these faces is zero. Hence, by addition, $\int_S \vec{E} \cdot d\vec{A} = 0$.

(b) The basic idea is the same as in part (a), except that we now need to use Riemann sums. First divide S into two hemispheres H_1 and H_2 by the equator C located in a plane perpendicular to $\vec{E}$. For a tiny patch S_1 in the hemisphere H_1, consider the patch S_2 in the opposite hemisphere which is symmetric to S_1 with respect to the center O of the sphere. The area vectors $\Delta\vec{A}_1$ and $\Delta\vec{A}_2$ satisfy $\Delta\vec{A}_2 = -\Delta\vec{A}_1$, so if we consider S_1 and S_2 to be approximately flat, then $\vec{E} \cdot \Delta\vec{A}_1 = -\vec{E} \cdot \Delta\vec{A}_2$. By decomposing H_1 and H_2 into small patches as above and using Riemann sums, we get

$$\int_{H_1} \vec{E} \cdot d\vec{A} = -\int_{H_2} \vec{E} \cdot d\vec{A}, \quad \text{so} \quad \int_S \vec{E} \cdot d\vec{A} = 0.$$

(c) The reasoning in part (b) can be used to prove that the flux of $\vec{E}$ through any surface with a center of symmetry is zero. For instance, in the case of the cylinder, cut it in half with a plane $z = 1$ and denote the two halves by H_1 and H_2. Just as before, take patches in H_1 and H_2 with $\Delta A_1 = -\Delta A_2$, so that $\vec{E} \cdot \Delta A_1 = -\vec{E} \cdot \Delta\vec{A}_2$. Thus, we get

$$\int_{H_1} \vec{E} \cdot d\vec{A} = -\int_{H_2} \vec{E} \cdot d\vec{A},$$

which shows that

$$\int_S \vec{E} \cdot d\vec{A} = 0.$$

18.   There are two possible methods:

(a)   The flux of $\vec{E}$ through S_a is given by

$$\int_{S_a} \vec{E} \cdot d\vec{A} = \int_{S_a} q \frac{\vec{r}}{\|\vec{r}\|^3} \cdot \frac{\vec{r}}{\|\vec{r}\|} \, dA,$$

since $d\vec{A} = \vec{n} \, dA$, where $\vec{n}$ is the outward pointing unit normal vector field on S and $\vec{n} = \vec{r}/\|\vec{r}\|$, and dA is the scalar area element. On the sphere S_a, we have $\|\vec{r}\| = a$ and therefore

$$\int_{S_a} \vec{E} \cdot d\vec{A} = \int_{S_a} \frac{q}{\|\vec{r}\|^2} \, dA$$

$$= \frac{q}{a^2} \int_{S_a} dA$$

$$= \frac{q}{a^2} \cdot 4\pi a^2$$

$$= 4\pi q.$$

(b)   Alternatively, we may compute the flux of $\vec{E}$ through S_a by using the definition of the flux integral as a limit of Riemann sums. Divide the sphere into approximately flat patches P_i. The vector area $\Delta \vec{A}_i$ of the patch P_i has the direction of the outward normal vector to the sphere at the point with position vector $\vec{r}_i$ and the magnitude area(P_i). The outward normal vector at any point of the sphere is proportional to the position vector of the point, hence we must have

$$\Delta \vec{A}_i = \text{Area}(P_i) \frac{\vec{r}_i}{\|\vec{r}_i\|}.$$

The Riemann sum corresponding to the above division of the sphere is therefore

$$\sum_i \vec{E}(\vec{r}_i) \cdot \Delta \vec{A}_i = \sum_i q \frac{\vec{r}_i}{\|\vec{r}_i\|^3} \, \text{Area}(P_i) \frac{\vec{r}_i}{\|\vec{r}_i\|} = \sum_i q \frac{\text{Area}(P_i)}{a^2}$$

since $\|\vec{r}_i\| = a$ for any i. But $\sum_i \text{Area}(P_i) = \text{Area}(S_a) = 4\pi a^2$, so we get

$$\sum_i \vec{E}(\vec{r}_i) \cdot \Delta \vec{A}_i = 4\pi q.$$

The right-hand side is independent of the way the sphere is divided, therefore in the limit, as $\|\Delta \vec{A}_i\| \to 0$, we get:

$$\int_{S_a} \vec{E} \cdot d\vec{A} = \lim_{\|\Delta \vec{A}_i\| \to 0} \sum_i \vec{E}(\vec{r}_i) \cdot \Delta \vec{A}_i = 4\pi q.$$

19.   (a)   The vector field $\vec{B}$ is sketched in Figure 19.21 for $I > 0$.

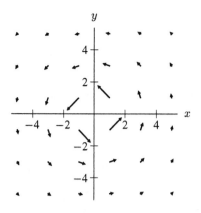

Figure 19.21

(b)   The disk S can be parameterized as $z = h$ (viewed as a constant function of x and y), for x, y in the region $\{x^2 + y^2 \leq a^2\}$. Hence

$$\int_S \vec{B} \cdot d\vec{A} = \int_{\{x^2+y^2\leq a^2\}} \frac{I}{2\pi} \cdot \frac{-y\vec{i} + x\vec{j}}{x^2 + y^2} \cdot \vec{k} \, dx \, dy = 0,$$

since $\vec{i} \cdot \vec{k} = 0$ and $\vec{j} \cdot \vec{k} = 0$. The answer is as we would expect, since the vector field $\vec{B}$ is tangent to the surface S, hence there is no flux through S.

(c)   The flux of $\vec{B}$ through S_2 is given by $\int_{S_2} \vec{B} \cdot d\vec{A}$. On S_2 we have

$$\vec{B}\,(x, y, z) = \frac{I}{2\pi} \cdot \frac{-y\vec{i}}{y^2} = -\frac{I}{2\pi y}\vec{i},$$

and

$$d\vec{A} = \vec{n} \, dA = -\vec{i} \, dy \, dz$$

Hence,

$$\int_{S_2} \vec{B} \cdot d\vec{A} = \int_0^h \int_a^b \frac{I}{2\pi} \cdot \frac{(-\vec{i})}{y} \cdot (-\vec{i}) \, dy \, dz$$

$$= \frac{I}{2\pi} \int_0^h \int_a^b \frac{1}{y} \, dy \, dz$$

$$= \frac{I}{2\pi} \int_0^h \left[\ln |y|\right]_a^b \, dz$$

$$= \frac{I}{2\pi} \int_0^h \left(\ln |b| - \ln |a|\right) \, dz$$

$$= \frac{I}{2\pi} h \left(\ln \left|\frac{b}{a}\right|\right).$$

This time we get a non-zero flux since the direction of $\vec{B}$ is everywhere parallel to the orientation of S_2. For $0 < a < b$ the flux is positive since $|b/a| > 1$ and increases as the area S_2 increases. This is as Figure 19.21 would lead us to expect. For $a < b < 0$, the flux is negative since $|b/a| < 1$. If $a < 0 < b$, the flux can be either positive or negative.

20. (a)   If $\vec{r} = x\vec{i} + y\vec{j} + z\vec{k}$ is the position vector of a point on the sphere, then

$$\vec{D}(\vec{r}) = \frac{3zp}{a^5}\vec{r} - \frac{p}{a^3}\vec{k}.$$

The second term is a constant vector field. Hence, by symmetry,

$$\int_S \left(-\frac{p}{a^3}\vec{k}\right) \cdot d\vec{A} = 0.$$

(See the solution to Problem 17 on page 479). Let us also apply a symmetry argument to

$$\int_S (\frac{3zp}{a^5}\vec{r}) \cdot d\vec{A}.$$

We will show that the flux of $\vec{D}$ through the upper hemisphere H_1 equals minus the flux of $\vec{D}$ through the lower hemisphere H_2. The flux of $\vec{D}$ through H_1 and H_2 will be computed as limits of Riemann sums.

Consider a small patch P_1 in H_1 and call its reflection about the xy-plane P_2. The contribution of P_1 to the flux of $\vec{D}$ through S is

$$\frac{3z_1p}{a^5}\vec{r} \cdot d\vec{A}_1 = \frac{3z_1p}{a^5}\vec{r} \cdot \text{Area}(P_1)\frac{\vec{r}}{\|\vec{r}\|} = \frac{3z_1p}{a^4} \cdot \text{Area}(P_1),$$

whereas the contribution from P_2 is

$$\frac{3z_2p}{a^5}\vec{r} \cdot d\vec{A}_1 = \frac{3z_2p}{a^4}\text{Area}(P_2).$$

But $\text{Area}(P_1) = \text{Area}(P_2)$ and $z_2 = -z_1$, so the contributions from P_1 and P_2 cancel each other. Dividing H_1 and H_2 into symmetric patches as above, and taking the limit as the areas of the patches become smaller and smaller, one gets

$$\int_{H_1} \left(\frac{3zp}{a^5}\vec{r}\right) \cdot d\vec{A} = -\int_{H_2} \left(\frac{3zp}{a^5}\vec{r}\right) \cdot d\vec{A}.$$

i.e.

$$\int_S \left(\frac{3zp}{a^5}\vec{r}\right) \cdot d\vec{A} = 0.$$

Since we also know that

$$\int_S \left(-\frac{p}{a^3}\vec{k}\right) \cdot d\vec{A} = 0,$$

we can conclude that

$$\int_S \vec{D} \cdot d\vec{A} = 0.$$

(b)   By Gauss's law,

$$\int_S \vec{E} \cdot d\vec{A} = 4\pi(q - q) = 0,$$

which is the same as the flux of $\vec{D}$ through S.

CHAPTER TWENTY

Solutions for Section 20.1

1. Two vector fields that have positive divergence everywhere are as follows:

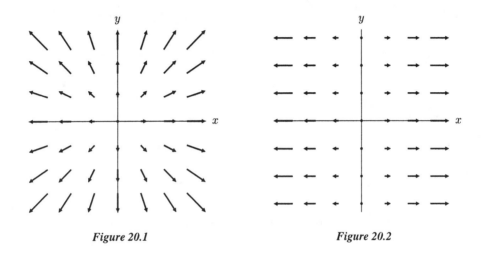

Figure 20.1   Figure 20.2

2. Two vector fields that have negative divergence everywhere are as follows:

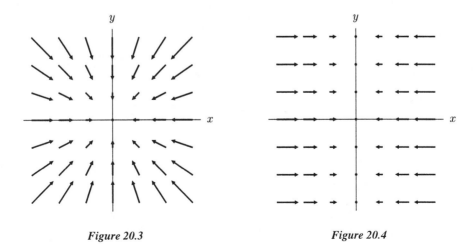

Figure 20.3   Figure 20.4

3. Two vector fields that have zero divergence everywhere are as follows:

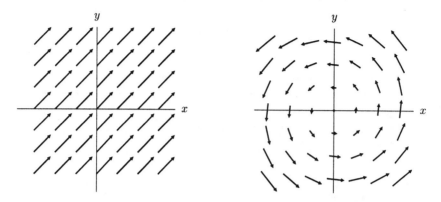

Figure 20.5   Figure 20.6

4. $\operatorname{div}\vec{F} = \dfrac{\partial}{\partial x}(-x) + \dfrac{\partial}{\partial y}(y) = -1 + 1 = 0$

5. $\operatorname{div}\vec{F} = \dfrac{\partial}{\partial x}(-y) + \dfrac{\partial}{\partial y}(x) = 0$

6. $\operatorname{div}\vec{F} = \dfrac{\partial}{\partial x}(x^2 - y^2) + \dfrac{\partial}{\partial y}(2xy) = 2x + 2x = 4x$

7. Using the formula for $\vec{a} \times \vec{r}$ in Cartesian coordinates, we get

$$\operatorname{div}\vec{F} = \frac{\partial}{\partial x}(a_2 z - a_3 y) + \frac{\partial}{\partial y}(a_3 x - a_1 z) + \frac{\partial}{\partial z}(a_1 y - a_2 x) = 0$$

8. Taking partial derivatives, we get

$$\operatorname{div}\vec{F} = \frac{\partial}{\partial x}\left(\frac{-y}{(x^2 + y^2)}\right) + \frac{\partial}{\partial y}\left(\frac{x}{(x^2 + y^2)}\right) = \frac{2xy}{(x^2 + y^2)^2} - \frac{2yx}{(x^2 + y^2)^2} = 0.$$

9. In coordinates, we have

$$\vec{F}(x, y, z) = \frac{(x - x_0)}{\sqrt{(x - x_0)^2 + (y - y_0)^2 + (z - z_0)^2}}\vec{i} + \frac{(y - y_0)}{\sqrt{(x - x_0)^2 + (y - y_0)^2 + (z - z_0)^2}}\vec{j}$$

$$+ \frac{(z - z_0)}{\sqrt{(x - x_0)^2 + (y - y_0)^2 + (z - z_0)^2}}\vec{k}.$$

So if $(x, y, z) \neq (x_0, y_0, z_0)$, then

$$\operatorname{div}\vec{F} = \left(\frac{1}{\sqrt{(x - x_0)^2 + (y - y_0)^2 + (z - z_0)^2}} - \frac{(x - x_0)^2}{((x - x_0)^2 + (y - y_0)^2 + (z - z_0)^2)^{3/2}}\right)$$

$$+ \left(\frac{1}{\sqrt{(x - x_0)^2 + (y - y_0)^2 + (z - z_0)^2}} - \frac{(y - y_0)^2}{((x - x_0)^2 + (y - y_0)^2 + (z - z_0)^2)^{3/2}}\right)$$

$$+ \left(\frac{1}{\sqrt{(x - x_0)^2 + (y - y_0)^2 + (z - z_0)^2}} - \frac{(z - z_0)^2}{((x - x_0)^2 + (y - y_0)^2 + (z - z_0)^2)^{3/2}} \right)$$

$$= \left(\frac{(x - x_0)^2 + (y - y_0)^2 + (z - z_0)^2}{((x - x_0)^2 + (y - y_0)^2 + (z - z_0)^2)^{3/2}} - \frac{(x - x_0)^2}{((x - x_0)^2 + (y - y_0)^2 + (z - z_0)^2)^{3/2}} \right)$$

$$+ \left(\frac{(x - x_0)^2 + (y - y_0)^2 + (z - z_0)^2}{((x - x_0)^2 + (y - y_0)^2 + (z - z_0)^2)^{3/2}} - \frac{(y - y_0)^2}{((x - x_0)^2 + (y - y_0)^2 + (z - z_0)^2)^{3/2}} \right)$$

$$+ \left(\frac{(x - x_0)^2 + (y - y_0)^2 + (z - z_0)^2}{((x - x_0)^2 + (y - y_0)^2 + (z - z_0)^2)^{3/2}} - \frac{(z - z_0)^2}{((x - x_0)^2 + (y - y_0)^2 + (z - z_0)^2)^{3/2}} \right)$$

$$= \frac{3((x - x_0)^2 + (y - y_0)^2 + (z - z_0)^2) - ((x - x_0)^2 + (y - y_0)^2 + (z - z_0)^2)}{((x - x_0)^2 + (y - y_0)^2 + (z - z_0)^2)^{3/2}}$$

$$= \frac{2}{\sqrt{(x - x_0)^2 + (y - y_0)^2 + (z - z_0)^2}} = \frac{2}{\|\vec{r} - \vec{r}_0\|}.$$

10. $\operatorname{div} \vec{F} = \dfrac{\partial}{\partial x}(-x + y) + \dfrac{\partial}{\partial y}(y + z) + \dfrac{\partial}{\partial z}(-z + x) = -1 + 1 - 1 = -1$

11. Let $\vec{a} = a_1 \vec{i} + a_2 \vec{j} + a_3 \vec{k}$ with a_1, a_2, and a_3 constant. Then $f\vec{a} = f(x, y, z)(a_1 \vec{i} + a_2 \vec{j} + a_3 \vec{k}) = f(x, y, z)a_1 \vec{i} + f(x, y, z)a_2 \vec{j} + f(x, y, z)a_3 \vec{k} = fa_1 \vec{i} + fa_2 \vec{j} + fa_3 \vec{k}$. So

$$\operatorname{div}(f\vec{a}) = \frac{\partial(fa_1)}{\partial x} + \frac{\partial(fa_2)}{\partial y} + \frac{\partial(fa_3)}{\partial z}$$

$$= a_1 \frac{\partial f}{\partial x} + a_2 \frac{\partial f}{\partial y} + a_3 \frac{\partial f}{\partial z} \quad \text{since } a_1, a_2, a_3 \text{ are constants}$$

$$= \left(\frac{\partial f}{\partial x} \vec{i} + \frac{\partial f}{\partial y} \vec{j} + \frac{\partial f}{\partial z} \vec{k} \right) \cdot (a_1 \vec{i} + a_2 \vec{j} + a_3 \vec{k})$$

$$= (\operatorname{grad} f) \cdot \vec{a}.$$

12. Let $\vec{F} = F_1 \vec{i} + F_2 \vec{j} + F_3 \vec{k}$. Then

$$\operatorname{div}(g\vec{F}) = \operatorname{div}(gF_1 \vec{i} + gF_2 \vec{j} + gF_3 \vec{k})$$

$$= \frac{\partial}{\partial x}(gF_1) + \frac{\partial}{\partial y}(gF_2) + \frac{\partial}{\partial z}(gF_3)$$

$$= \frac{\partial g}{\partial x} F_1 + g \frac{\partial F_1}{\partial x} + \frac{\partial g}{\partial y} F_2 + g \frac{\partial F_2}{\partial y} + \frac{\partial g}{\partial z} F_3 + g \frac{\partial F_3}{\partial z}$$

$$= \frac{\partial g}{\partial x} F_1 + \frac{\partial g}{\partial y} F_2 + \frac{\partial g}{\partial z} F_3 + g \left(\frac{\partial F_1}{\partial x} + \frac{\partial F_2}{\partial y} + \frac{\partial F_3}{\partial z} \right)$$

$$= (\operatorname{grad} g) \cdot \vec{F} + g \operatorname{div} \vec{F}.$$

13. Using $\operatorname{div}(g\vec{F}) = (\operatorname{grad} g) \cdot \vec{F} + g \operatorname{div} \vec{F}$, we have

$$\operatorname{div} \vec{F} = \frac{1}{\|\vec{r}\|^p} \operatorname{div}(\vec{a} \times \vec{r}) + \operatorname{grad}\left(\frac{1}{\|\vec{r}\|^p} \right) \cdot \vec{a} \times \vec{r}$$

$$= \frac{1}{\|\vec{r}\|^p} 0 + \frac{-p}{\|\vec{r}\|^{p+2}} \vec{r} \cdot (\vec{a} \times \vec{r})$$

$$= 0 \quad \text{since } \vec{r} \text{ and } \vec{a} \times \vec{r} \text{ are perpendicular.}$$

14. Using $\mathrm{div}(g\vec{F}) = (\mathrm{grad}\, g) \cdot \vec{F} + g\,\mathrm{div}\,\vec{F}$, we have

$$\mathrm{div}\,\vec{B} = \mathrm{grad}\left(\frac{1}{x^a}\right) \cdot \vec{r} + \frac{1}{x^a}\,\mathrm{div}\,\vec{r} = -ax^{-(a+1)}\vec{i} \cdot \vec{r} + x^{-a}(3) = (3-a)x^{-a}.$$

15. Using $\mathrm{div}(g\vec{F}) = (\mathrm{grad}\, g) \cdot \vec{F} + g\,\mathrm{div}\,\vec{F}$, we have

$$\mathrm{div}\,\vec{G} = \mathrm{grad}(\vec{b} \cdot \vec{r}) \cdot (\vec{a} \times \vec{r}) + \vec{b} \cdot \vec{r}\,\mathrm{div}(\vec{a} \times \vec{r}) = \vec{b} \cdot (\vec{a} \times \vec{r}) + \vec{v} \cdot \vec{r}\,0 = \vec{b} \cdot (\vec{a} \times \vec{r}).$$

16. The first vector field appears to be diverging more at the origin, since both fields are zero at the origin and the vectors near the origin are larger in the first field than they are in the second field.

17. (a) Positive. The inflow from the lower left is less than the outflow from the upper right. The net outflow is positive.
    (b) Zero. The inflow on the right side is equal to outflow on the left.
    (c) Negative. The inflow from above is greater than the outflow below. The net outflow is negative.

18. (a) We have

$$\mathrm{div}\,\vec{F} = \frac{\partial z}{\partial z} = 1.$$

   (b) Above the xy plane, the vector field consists of vectors pointing vertically upwards, getting longer as you go up. Below the xy plane, it consists of vectors pointing vertically downwards, getting longer as you go down. You can clearly see the divergence on the xy plane, since vectors on either side of it point in opposite directions, but it is not so clear elsewhere. However, the fact that the vectors are getting longer as you go up means that the flux through a cube situated above the xy plane will be non-zero, since the flux out of its top face will be greater than the flux into the bottom face.

19. (a) In Cartesian coordinates,

$$\vec{F}(x,y,z) = \frac{x}{(x^2+y^2+z^2)^{3/2}}\vec{i} + \frac{y}{(x^2+y^2+z^2)^{3/2}}\vec{j} + \frac{z}{(x^2+y^2+z^2)^{3/2}}\vec{k}.$$

   So if $(x,y,z) \neq (0,0,0)$, then

$$\begin{aligned}
\mathrm{div}\,\vec{F}(x,y,z) &= \left(\frac{1}{(x^2+y^2+z^2)^{3/2}} - \frac{3x^2}{(x^2+y^2+z^2)^{5/2}}\right) \\
&\quad + \left(\frac{1}{(x^2+y^2+z^2)^{3/2}} - \frac{3y^2}{(x^2+y^2+z^2)^{5/2}}\right) \\
&\quad + \left(\frac{1}{(x^2+y^2+z^2)^{3/2}} - \frac{3z^2}{(x^2+y^2+z^2)^{5/2}}\right) \\
&= \left(\frac{x^2+y^2+z^2}{(x^2+y^2+z^2)^{5/2}} - \frac{3x^2}{(x^2+y^2+z^2)^{5/2}}\right) \\
&\quad + \left(\frac{x^2+y^2+z^2}{(x^2+y^2+z^2)^{5/2}} - \frac{3y^2}{(x^2+y^2+z^2)^{5/2}}\right) \\
&\quad + \left(\frac{x^2+y^2+z^2}{(x^2+y^2+z^2)^{5/2}} - \frac{3z^2}{(x^2+y^2+z^2)^{5/2}}\right) \\
&= \frac{3(x^2+y^2+z^2) - 3(x^2+y^2+z^2)}{(x^2+y^2+z^2)^{5/2}} \\
&= 0.
\end{aligned}$$

(b) The vector field is radial (all the arrows point out), so you might think that it is has non-zero divergence. (See Figure 20.7.) However the fact that the divergence is 0 at every point shows that flux density out of any small volume around a point must be 0. This is possible because the arrows also get shorter as you go out.

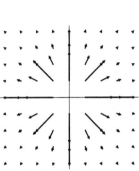

Figure 20.7: The vector field $\vec{F}(\vec{r}) = \dfrac{\vec{r}}{\|\vec{r}\|^3}$ shown in the xz-plane

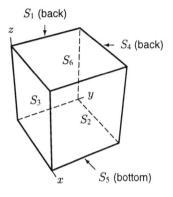

Figure 20.8

20. (a) The vector field is parallel to the x-axis and zero on the yz-plane. Thus the only contribution to the flux is from S_2. On S_2, $x = c$, the normal is outward. Since $\vec{F}$ is constant on S_2, the flux through face S_2 is

$$\int_{S_2} \vec{F} \cdot d\vec{A} = \vec{F} \cdot \vec{A}\, S_2$$
$$= c\vec{i} \cdot c^2 \vec{i}$$
$$= c^3.$$

Thus, total flux through box $= c^3$.

(b) Using the geometric definition of convergence

$$\operatorname{div} \vec{F} = \lim_{c \to 0} \left(\frac{\text{Flux through box}}{\text{Volume of box}} \right)$$
$$= \lim_{c \to 0} \left(\frac{c^3}{c^3} \right)$$
$$= 1$$

(c)

$$\frac{\partial}{\partial x}(x) + \frac{\partial}{\partial y}(0) + \frac{\partial}{\partial z}(0) = 1 + 0 + 0 = 1.$$

21. See Figure 20.8.

(a) Since $2\vec{i} + 3\vec{k}$ is a constant field, its contribution to the flux is zero (flux in cancels flux out). Therefore $\int \vec{F} \cdot d\vec{A} = \int (y\vec{j}) \cdot d\vec{A} = \int_{S_3} y\vec{j} \cdot d\vec{A} + \int_{S_4} y\vec{j} \cdot d\vec{A}$ since only S_3 and S_4 are perpendicular to $y\vec{j}$. On S_3, $y = 0$ so $\int_{S_3} y\vec{j} \cdot d\vec{A} = 0$. On S_4, $y = c$ and normal is in the positive y-direction, so $\int_{S_4} y\vec{j} \cdot d\vec{A} = c(\text{Area of } S_4) = c \cdot c^2 = c^3$. Thus, total flux $= c^3$.

(b)   Using the geometric definition of convergence

$$\text{div } \vec{F} = \lim_{c \to 0} \left(\frac{\text{Flux through box}}{\text{Volume of box}} \right)$$

$$= \lim_{c \to 0} \left(\frac{c^3}{c^3} \right) = 1.$$

(c)

$$\frac{\partial}{\partial x}(2) + \frac{\partial}{\partial y}(y) + \frac{\partial}{\partial z}(3) = 0 + 1 + 0 = 1.$$

22.   See Figure 20.8.

(a)   This vector field points radially outward from the z-axis. Thus, the vector field is parallel to the surface on S_1, S_3, S_5 and S_6, so the only contributions to the flux integral are from S_2 and S_4. On $S_2, x = c$ and normal is in the positive x-direction, so the flux is

$$\int_{S_2} \vec{F} \cdot d\vec{A} = \int_{S_2} (c\vec{i} + y\vec{j}) \cdot (dA\vec{i}) = \int_{S_2} c\, dA = c(\text{Area of } S_2) = c^3.$$

Similarly, the flux through S_4 is

$$\int_{S_4} \vec{F} \cdot d\vec{A} = \int_{S_4} (x\vec{i} + c\vec{j}) \cdot (dA\vec{j}) = \int_{S_4} c\, dA = c(\text{Area of } S_4) = c^3.$$

Thus, the total flux through the box $= 2c^3$.

(b)   Using the geometric definition of convergence, we have

$$\text{div } \vec{F} = \lim_{c \to 0} \left(\frac{\text{Flux through surface of box}}{\text{Volume of box}} \right)$$

$$= \lim_{c \to 0} \left(\frac{2c^3}{c^3} \right) = 2.$$

(c)

$$\frac{\partial}{\partial x}(x) + \frac{\partial}{\partial y}(y) + \frac{\partial}{\partial z}(0) = 1 + 1 + 0 = 2.$$

23.   (a)   Since $\vec{F}$ is parallel to the top and bottom of the cylinder, the only contribution to the flux is through the curved surface. Since the area vector of a small patch points directly outward parallel to $\vec{F}$, and $\|\vec{F}\| = \sqrt{x^2 + y^2} = c$ on the curved surface of the cylinder:

$$\vec{F} \cdot \Delta\vec{A} = \|\vec{F}\|\|\Delta\vec{A}\| = c\Delta A$$

Replacing ΔA with dA, we get

$$\int_S \vec{F} \cdot d\vec{A} = \int_{\substack{\text{curved} \\ \text{surface}}} c\, dA$$

$$= c \int_{\substack{\text{curved} \\ \text{surface}}} dA$$

$$= c(\text{Area of the curved surface of cylinder})$$

$$= c(2\pi c \cdot c)$$

$$= 2\pi c^3.$$

(b)   At the origin,

$$\operatorname{div}\vec{F} = \lim_{c\to0}\left(\frac{\text{Flux through cylinder}}{\text{Volume inside cylinder}}\right)$$

$$= \lim_{c\to0}\left(\frac{2\pi c^3}{\pi c^2 \cdot c}\right)$$

$$= 2.$$

(c)

$$\operatorname{div}\vec{F} = \frac{\partial}{\partial x}(x) + \frac{\partial}{\partial y}(y) = 1 + 1 = 2.$$

24. The charges that produce this electric field are concentrated along two vertical lines, one near $x = -1$ and the another one near $x = 1$. This is seen by the change in direction of the field at those lines. Near $x = -1$ the field is being repulsed by the line (seen by the field going away from the line), and the charge is therefore positive. Near $x = 1$ the field is being attracted to the line (seen by the field going towards the line), and the charge is therefore negative.

25. (a)   Translating the vector field into rectangular coordinates gives, if $(x, y, z) \neq (0, 0, 0)$

$$\vec{E}(x, y, z) = \frac{kx}{(x^2 + y^2 + z^2)^{3/2}}\vec{i} + \frac{ky}{(x^2 + y^2 + z^2)^{3/2}}\vec{j} + \frac{kz}{(x^2 + y^2 + z^2)^{3/2}}\vec{k}.$$

We now take the divergence of this to get

$$\operatorname{div}\vec{E} = k\left(-3\frac{x^2 + y^2 + z^2}{(x^2 + y^2 + z^2)^{5/2}} + \frac{3}{(x^2 + y^2 + z^2)^{3/2}}\right)$$

$$= 0.$$

(b)   Let S be the surface of a sphere centered at the origin. We have seen that for this field, the flux $\int \vec{E} \cdot d\vec{A}$ is the same for all such spheres, regardless of their radii. So let the constant c stand for $\int \vec{E} \cdot d\vec{A}$. Then

$$\operatorname{div}\vec{E}(0, 0, 0) = \lim_{\text{vol}\to0}\frac{\int \vec{E} \cdot d\vec{A}}{\text{Volume inside } S} = \lim_{\text{vol}\to0}\frac{c}{\text{Volume}}.$$

(c)   For a point charge, the charge density is not defined. The charge density is 0 everywhere else.

26. (a) $\operatorname{div}\vec{B} = \frac{\partial}{\partial x}(-y) + \frac{\partial}{\partial y}(x) + \frac{\partial}{\partial z}(x + y) = 0$, so this could be a magnetic field.

(b) $\operatorname{div}\vec{B} = \frac{\partial}{\partial x}(-z) + \frac{\partial}{\partial y}(y) + \frac{\partial}{\partial z}(x) = 0 + 1 + 0 = 1$, so this could not be a magnetic field.

(c) $\operatorname{div}\vec{B} = \frac{\partial}{\partial x}(x^2 - y^2 - x) + \frac{\partial}{\partial y}(y - 2xy) + \frac{\partial}{\partial z}(0) = 2x - 1 + 1 - 2x + 0 = 0$, so this could be a magnetic field.

27. Now $\operatorname{grad} f = f_x\vec{i} + f_y\vec{j} + f_z\vec{k}$ and $\operatorname{grad} g$ is similar. Thus

$$\operatorname{grad} f \times \operatorname{grad} g = \begin{vmatrix} \vec{i} & \vec{j} & \vec{k} \\ f_x & f_y & f_z \\ g_x & g_y & g_z \end{vmatrix} = (f_y g_z - f_z g_y)\vec{i} - (f_x g_z - f_z g_x)\vec{j} + (f_x g_y - f_y g_x)\vec{k}.$$

Therefore

$$\operatorname{div}(\operatorname{grad} f \times \operatorname{grad} g) = \frac{\partial}{\partial x}(f_y g_z - f_z g_y) + \frac{\partial}{\partial y}(f_z g_x - f_x g_z) + \frac{\partial}{\partial z}(f_x g_y - f_y g_x).$$

Expanding using the product rule gives

$$\text{div}(\text{grad } f \times \text{grad } g) = f_{yx}g_z + f_y g_{zx} - f_{zx}g_y - f_z g_{yx} + f_{zy}g_x + f_z g_{xy}$$
$$- f_{xy}g_z - f_x g_{zy} + f_{xz}g_y + f_x g_{yz} - f_{yz}g_x - f_y g_{xz}.$$

Now consider pairs of terms such as $f_{yx}g_z - f_{xy}g_z$. Since $f_{yx} = f_{xy}$ provided the second derivatives are continuous, these two terms cancel out. All the other terms cancel in pairs, showing that

$$\text{div}(\text{grad } f \times \text{grad } g) = 0.$$

28.   We have now our temperature a function depending on t, x, y, z, hence $T = T(t, x, y, z)$. For a fixed moment, say t_0, T is a function of only x, y, z. For this moment, $t = t_0$, we have:

$$\begin{array}{c}\text{Rate of heat loss} \\ \text{from volume } V\end{array} = k \int_S (\text{grad } T) \cdot d\vec{A}.$$

where $\text{grad } T = \left(\frac{\partial T}{\partial x}\vec{i} + \frac{\partial T}{\partial y}\vec{j} + \frac{\partial T}{\partial z}\vec{k} \right)\Big|_{t=t_0}$. Now the rate of change, with respect to time, in the average temperature in the region, at $t = t_0$, is proportional to the average rate at which heat is being lost per unit volume at $t = t_0$, so

$$\frac{\partial T_{avg}}{\partial t}\Big|_{t=t_0} = -c\left(\frac{\text{Rate heat lost}}{\text{Volume } V} \right)_{t=t_0} = \frac{-ck \int_S (\text{grad } T) \cdot d\vec{A}}{\text{Volume } V}$$

Taking the limit as V shrinks around the point, the average temperature through the region becomes the temperature at that point. Thus using the definition of the divergence (with respect to x, y, z), we have

$$\frac{\partial T}{\partial t}\Big|_{t=t_0} = -ck \lim_{V \to 0} \left(\frac{\int_S (\text{grad } T) \cdot d\vec{A}}{\text{Volume } V} \right)$$
$$= (-ck \, \text{div grad } T)_{t=t_0}$$

As this holds at every moment t_0, one has:

$$\frac{\partial T}{\partial t} = B \cdot \text{div grad } T,$$

where $B = -ck$ is a function of time only, and the gradient and divergence are taken with respect to the variables x, y, z.

29.   (a)   At any point $\vec{r} = x\vec{i} + y\vec{j}$, the direction of the vector field $\vec{v}$ is pointing away from the origin, which means it is of the form $\vec{v} = f\vec{r}$ for some positive function f, whose value can vary depending on $\vec{r}$. The magnitude of $\vec{v}$ depends only on the distance r, thus f must be a function depending only on r, which is equivalent to depending only on r^2 since $r \geq 0$. So $\vec{v} = f(r^2)\vec{r} = \left(f(x^2 + y^2) \right)(x\vec{i} + y\vec{j})$.

(b)   At $(x, y) \neq (0, 0)$ the divergence of $\vec{v}$ is

$$\text{div } \vec{v} = \frac{\partial (K(x^2 + y^2)^{-1}x)}{\partial x} + \frac{\partial (K(x^2 + y^2)^{-1}y)}{\partial y} = \frac{Ky^2 - Kx^2}{(x^2 + y^2)^2} + \frac{Kx^2 - Ky^2}{(x^2 + y^2)^2} = 0.$$

Therefore, $\vec{v}$ is a point source at the origin.

(c)   The magnitude of $\vec{v}$ is

$$\|\vec{v}\| = K(x^2 + y^2)^{-1}|x\vec{i} + y\vec{j}| = K(x^2 + y^2)^{-1}(x^2 + y^2)^{1/2} = K(x^2 + y^2)^{-1/2} = \frac{K}{r}.$$

(d) The vector field looks like that in Figure 20.9:

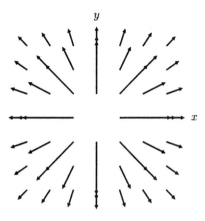

Figure 20.9

(e) We need to show that grad $\phi = \vec{v}$.

$$\text{grad } \phi = \frac{\partial}{\partial x}(\frac{K}{2}\log(x^2 + y^2))\vec{i} + \frac{\partial}{\partial y}(\frac{K}{2}\log(x^2 + y^2))\vec{j}$$
$$= \frac{Kx}{x^2 + y^2}\vec{i} + \frac{Ky}{x^2 + y^2}\vec{j}$$
$$= K(x^2 + y^2)^{-1}(x\vec{i} + y\vec{j})$$
$$= \vec{v}$$

30. (a) At any point $\vec{r} = x\vec{i} + y\vec{j}$, the direction of the vector field $\vec{v}$ is pointing toward the origin, which means it is of the form $\vec{v} = f\vec{r}$ for some negative function f whose value can vary depending on $\vec{r}$. The magnitude of $\vec{v}$ depends only on the distance r, thus f must be a function depending only on r, which is equivalent to depending only on r^2 since $r \geq 0$. So $\vec{v} = f(r^2)\vec{r} = (f(x^2 + y^2))(x\vec{i} + y\vec{j})$.

    (b) At $(x, y) \neq (0, 0)$ the divergence of $\vec{v}$ is

$$\text{div } \vec{v} = \frac{\partial(K(x^2 + y^2)^{-1}x)}{\partial x} + \frac{\partial(K(x^2 + y^2)^{-1}y)}{\partial y} = \frac{Ky^2 - Kx^2}{(x^2 + y^2)^2} + \frac{Kx^2 - Ky^2}{(x^2 + y^2)^2} = 0.$$

    Therefore, $\vec{v}$ is a point sink at the origin.

    (c) The magnitude of $\vec{v}$ is

$$\|\vec{v}\| = |K|(x^2 + y^2)^{-1}|x\vec{i} + y\vec{j}| = |K|(x^2 + y^2)^{-1}(x^2 + y^2)^{1/2} = |K|(x^2 + y^2)^{-1/2} = \frac{|K|}{r}.$$

    (remember, $K < 0$)

    (d) The vector field looks like the following:

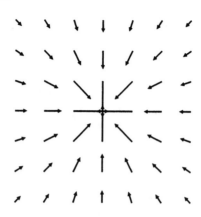

Figure 20.10

(e)   We need to show that grad $\phi = \vec{v}$.

$$\text{grad } \phi = \frac{\partial}{\partial x}(\frac{K}{2}\log(x^2+y^2))\vec{i} + \frac{\partial}{\partial y}(\frac{K}{2}\log(x^2+y^2))\vec{j}$$

$$= \frac{Kx}{x^2+y^2}\vec{i} + \frac{Ky}{x^2+y^2}\vec{j}$$

$$= K(x^2+y^2)^{-1}(x\vec{i}+y\vec{j})$$

$$= \vec{v}$$

Solutions for Section 20.2

1. First directly: On the faces $x=0, y=0, z=0$, the flux is zero. On the face $x=2$, a unit normal is $\vec{i}$ and $d\vec{A} = dA\vec{i}$. So

$$\int_{S_{x=2}} \vec{r} \cdot d\vec{A} = \int_{S_{x=2}} (2\vec{i}+y\vec{j}+z\vec{k}) \cdot (dA\vec{i})$$

(since on that face, $x=2$)

$$= \int_{S_{x=2}} 2dA = 2 \cdot (\text{Area of face}) = 2 \cdot 4 = 8.$$

In exactly the same way, you get

$$\int_{S_{y=2}} \vec{r} \cdot d\vec{A} = \int_{S_{z=2}} \vec{r} \cdot d\vec{A} = 8,$$

so

$$\int_{S} \vec{r} \cdot d\vec{A} = 3 \cdot 8 = 24.$$

Now using divergence:

$$\text{div } \vec{F} = \frac{\partial x}{\partial x} + \frac{\partial y}{\partial y} + \frac{\partial z}{\partial z} = 3,$$

so

$$\text{Flux} = \int_0^2 \int_0^2 \int_0^2 3\,dx\,dy\,dz = 3 \cdot (\text{Volume of Cube}) = 3 \cdot 8 = 24$$

2. First directly, since the vector field is totally in the $\vec{j}$ direction, there is no flux through the ends. On the side of the cylinder, a normal vector at (x, y, z) is $x\vec{i} + y\vec{j}$. This is in fact a unit normal, since $x^2 + y^2 = 1$ (the cylinder has radius 1). Also, using $x = \cos\theta, y = \sin\theta$, in this case, the element of area dA equals $1 d\theta dz$. So

$$\text{Flux} = \int \vec{F} \cdot d\vec{A} = \int_0^2 \int_0^{2\pi} (y\vec{j}) \cdot (x\vec{i} + y\vec{j})\, d\theta\, dz = \int_0^2 \int_0^{2\pi} y^2\, d\theta\, dz = \int_0^2 \int_0^{2\pi} \sin^2\theta\, d\theta\, dz = \int_0^2 \pi dz = 2\pi.$$

Now we calculate the flux using the divergence theorem. The divergence of the field is given by the sum of the respective partials of the components, so the divergence is simply $\dfrac{\partial y}{\partial y} = 1$. Since the divergence is constant, we can simply calculate the volume of the cylinder and multiply by the divergence

$$\text{Flux} = 1\pi r^2 h = 2\pi$$

3. The location of the pyramid has not been completely specified. For instance, where is it centered on the xy plane? How is base oriented with respect to the axes? Thus, we cannot compute the flux by direct integration with the information we have. However, we can calculate it using the divergence theorem. First we calculate the divergence of $\vec{F}$.

$$\text{div}\,\vec{F} = \frac{\partial(-z)}{\partial x} + \frac{\partial 0}{\partial y} + \frac{\partial x}{\partial z} = 0 + 0 + 0 = 0$$

Thus for any closed surface the flux will be zero, so the flux through our pyramid, regardless of its location or orientation, is zero.

4. (a) True. Flux that goes in one face goes out the other, since the vector field is constant and the surface is closed.

   (b) True. The flux out of S_1 along the face shared with S_2 cancels with the flux out of S_2 over the same face. (The normals are in opposite directions.) The other five faces of S_1 and the other five faces of S_2 are each faces of S.

5. Finding flux directly:
   1) On bottom face, $z = 0$ so $\vec{F} = x^2\vec{i} + 2y^2\vec{j}$ is parallel to face so flux is zero.
   2) On front face, $y = 0$ so $\vec{F} = x^2\vec{i} + 3z^2\vec{k}$ is parallel to face so flux is zero.
   3) On back face, $y = 1$ so $\vec{F} = x^2\vec{i} + 2\vec{j} + 3z^2\vec{k}$ and $\vec{A} = \vec{j}$ so flux is 2.
   4) On top face, $z = 1$ so $\vec{F} = x^2\vec{i} + 2y^2\vec{j} + 3\vec{k}$ and $\vec{A} = \vec{k}$ so flux is 3.
   5) On side $x = 1$, $\vec{F} = \vec{i} + 2y^2\vec{j} + 3z^2\vec{k}$ and $\vec{A} = -\vec{i}$ so flux is -1.
   6) On side $x = 2$, $\vec{F} = 4\vec{i} + 2y^2\vec{j} + 3z^2\vec{k}$ and $\vec{A} = \vec{i}$ so flux is 4.
   Total flux is thus 8.

   By the Divergence Theorem:

$$\text{div}\,\vec{F} = 2x + 4y + 6z$$

So

$$\int_S \vec{F} \cdot d\vec{A} = \int_V (2x + 4y + 6z)dV = 2\int_1^2 \int_0^1 \int_0^1 (x + 2y + 3z)dz\, dy\, dx$$

$$= 2\int_1^2 \int_0^1 \left[xz + 2yz + \frac{3z^2}{2}\right]_0^1 dy\, dx = 2\int_1^2 \int_0^1 \left(x + 2y + \frac{3}{2}\right) dy\, dx$$

$$= 2\int_1^2 \left[xy + y^2 + \frac{3y}{2}\right]_0^1 dx = 2\int_1^2 \left(x + 1 + \frac{3}{2}\right) dx$$

$$= \int_1^2 (2x + 5)dx = (x^2 + 5x)\Big|_1^2 = 8$$

6. First compute $\text{div}\vec{F} = 11$ at origin, where $\vec{F} = x^2\vec{i} + (y - 2xy)\vec{j} + 10z\vec{k}$. By the Divergence Theorem,
    $\int_S \vec{F} \cdot d\vec{A} = \int_R 11\,dV = 11(\text{volume of } R) = 11(4\pi 5^3/3) = 5500\pi/3 \approx 5759.6$

7. The divergence of the field is
    $$\text{div }\vec{F} = \frac{\partial(-z)}{\partial x} + \frac{\partial(0)}{\partial y} + \frac{\partial x}{\partial z} = 0.$$

    Hence,
    $$\int_{\text{Sphere}} \vec{F} \cdot d\vec{A} = \int_{\text{Ball}} \text{div }\vec{F}\,dV = \int_{\text{Ball}} 0\,dV = 0,$$

    and so the flux through the sphere is zero. This makes sense, because, as Figure 20.11 shows, the vector field is flowing around the y-axis and is always tangent to the sphere.

Figure 20.11

8. By the Divergence Theorem, if W is the cylinder and S is its surface:
    $$\int_S \vec{F} \cdot d\vec{A} = \int_W \text{div }\vec{F}\,dV = \int_W 10\,dV = 10 \cdot \text{Volume of cylinder} = 10\pi a^3.$$

9. (a) Taking partial derivatives, we have for $\vec{r} \neq \vec{0}$,
    $$\begin{aligned}
    \text{div }\vec{F} &= \text{div}\left(\frac{x\vec{i} + y\vec{j} + z\vec{k}}{(x^2 + y^2 + z^2)^{3/2}}\right) \\
    &= -\frac{3(x^2 + y^2 + z^2)}{(x^2 + y^2 + z^2)^{5/2}} + \frac{3}{(x^2 + y^2 + z^2)^{3/2}} \\
    &= 0.
    \end{aligned}$$

    (b) We could compute the flux out of the box S by computing the flux out of each side separately. However, since $\text{div }\vec{F} = \vec{0}$ everywhere except the origin, we instead consider a region W between the box S and a sphere S_b of radius b centered at the origin and which fits inside the box. If the sphere is oriented inward, since $\text{div }\vec{F} = \vec{0}$ throughout W, the Divergence Theorem says
    $$0 = \int_W \text{div }\vec{F}\,dV = \int_S \vec{F} \cdot d\vec{A} + \int_{S_b} \vec{F} \cdot d\vec{A}.$$

The flux of $\vec{F}$ through S_b is easier to calculate than the flux through the box. Since S_b is oriented inward,

$$\int_{S_b} \vec{F} \cdot d\vec{A} = -\int_{S_b} \|\vec{F}\| \|d\vec{A}\| = -\int_{S_b} \frac{1}{b^2} \|dA\|$$

$$= -\frac{1}{b^2} \cdot \text{Surface area of } S_b = -\frac{1}{b^2} \cdot 4\pi b^2 = -4\pi.$$

Thus,

$$\int_{S} \vec{F} \cdot d\vec{A} = -\int_{S_b} \vec{F} \cdot d\vec{A} = 4\pi.$$

10. Let S_3 the sphere of radius 3 and let S_5 be the sphere of radius 5. Then we know that

$$\int_{S_3} \vec{G} \cdot d\vec{A} = 8\pi.$$

We consider the region W between S_3 and S_5. For $2 \leq \|\vec{r}\| \leq 7$, we have $\operatorname{div} \vec{G} = 3$. We apply the Divergence Theorem to the region W between S_3 and S_5, with S_5 oriented outward and S_3 oriented inward:

$$\int_{S_5} \vec{G} \cdot d\vec{A} - \int_{S_3} \vec{G} \cdot d\vec{A} = \int_{W} \operatorname{div} \vec{G} \, dV = 3 \cdot \text{Volume of } W$$

$$= 3\left(\frac{4}{3}\pi 5^3 - \frac{4}{3}\pi 3^3\right) = 392\pi.$$

Thus,

$$\int_{S_5} \vec{G} \cdot d\vec{A} = \int_{S_3} \vec{G} \cdot d\vec{A} + \int_{W} \operatorname{div} \vec{G} \, dV = 8\pi + 392\pi = 400\pi$$

11. By the divergence theorem $\int_{S} \vec{F} \cdot d\vec{A} = \int_{R} \operatorname{div}\vec{F} \, dV = \int_{R} 0 \, dV = 0$ for a closed surface S, where R is the region enclosed by S.

12. We can rewrite $\vec{F}$ in terms of (x, y, z) as

$$\vec{F} = \frac{-Gmx}{\sqrt{(x^2 + y^2 + z^2)^3}}\vec{i} + \frac{-Gmy}{\sqrt{(x^2 + y^2 + z^2)^3}}\vec{j} + \frac{-Gmz}{\sqrt{(x^2 + y^2 + z^2)^3}}\vec{k}$$

Now we find the divergence of $\vec{F}$ in the usual manner

$$\operatorname{div} \vec{F} = \frac{\partial}{\partial x} \frac{-Gmx}{\sqrt{(x^2 + y^2 + z^2)^3}} + \frac{\partial}{\partial y} \frac{-Gmy}{\sqrt{(x^2 + y^2 + z^2)^3}} + \frac{\partial}{\partial z} \frac{-Gmz}{\sqrt{(x^2 + y^2 + z^2)^3}}$$

$$= -3Gm\frac{x^2 + y^2 + z^2}{(x^2 + y^2 + z^2)^{5/2}} + \frac{3Gm}{(x^2 + y^2 + z^2)^{3/2}}$$

$$= 0.$$

Thus, $\operatorname{div} \vec{F} = 0$ for all points except the origin. We consider a region enclosed by two concentric spheres. Since the divergence of the field is zero at all points except the origin, the volume enclosed contains only points with zero divergence. Consequently, the flux through the surface of the enclosed volume must be zero. Since the field is always inward pointing, this is equivalent to saying that the flux into the outer sphere must equal the flux out of the inner sphere, and so we see that for any two spheres, the flux must be equal, which shows that the flux is independent of the radius of the spheres.

13. Since the divergence is zero at all points not containing the charge, the flux must be zero through any closed surface containing no charge. We imagine a surface composed of two concentric cylinders and their end-caps, where the axis of both cylinders is the z-axis. Then, since no charge is contained in the region enclosed, the flux through the surface must be zero. Now, we know that the field points away from the axis, which means it is parallel to the end-caps. Consequently, there must be no flux through the end-caps. This implies that the flux through the inner cylinder must equal the flux out of the outer cylinder. Since the strength of the field only depends upon the distance from the z axis, the flux through each cylinder is a constant. This implies that the following equation must hold

$$\text{Flux through each cylinder} = E_a 2\pi r_a L = E_b 2\pi r_b L$$

where E_a and E_b are the strengths of the field at r_a and r_b respectively, and L is the length of the cylinders. Dividing through, we can arrive at the following relationship:

$$E_a / E_b = r_b / r_a$$

If we take E_b to be a constant at a fixed value of r_b, then the equation can be simplified to

$$E_a = k / r_a$$

where $k = E_b r_b$. Thus we see that the strength of the field is proportional to $1/r$.

14. (a)

$$\text{Force in } \vec{i} \text{ direction} = \vec{F} \cdot \vec{i} = -\int_S \rho g z \vec{i} \; d\vec{A}$$

$$= -\rho g \int_S z \vec{i} \; d\vec{A}.$$

Now apply the Divergence Theorem to this integral. (Notice that in order to do this, you need to orient S outwardly, hence the minus sign dispears.)

$$\vec{F} \cdot \vec{i} = \rho g \int_R \frac{\partial z}{\partial x} dV = 0.$$

Similarly:

$$\text{Force in } \vec{j} \text{ direction} = \vec{F} \cdot \vec{j} = -\rho g \int_S z \vec{j} \cdot d\vec{A}$$

$$= \rho g \int_R \frac{\partial z}{\partial y} dV = 0$$

(b)

$$\text{Force in } \vec{k} \text{ direction} = \vec{F} \cdot \vec{k} = -\rho g \int_S z \vec{k} \cdot d\vec{A}$$

$$= \rho g \int_R \frac{\partial z}{\partial z} dV = \rho g \int_R dV = \rho g V.$$

15. (a) The rate at which heat is generated is div $\vec{F} = 30$ watts/km^3.
    (b) $\text{div}(\alpha(x\vec{i} + y\vec{j} + z\vec{k})) = \alpha(1 + 1 + 1) = 3\alpha$ so $\alpha = 30/3 = 10$ watts/km^3. Thus, $\vec{F} = \alpha\vec{r}$ has constant divergence, and $\vec{F} = \alpha\vec{r}$ has flow lines going radially outward, and symmetric about the origin.

(c) The vector grad T gives the direction of greatest increase in temperature. Thus, $-\text{grad } T$ gives the direction of greatest decrease in temperature. The equation $\vec{F} = -k \text{ grad } T$ says that heat will flow in the direction of greatest decrease in temperature (i.e. from hot regions to cold), and at a rate proportional to the temperature gradient.

(d) Since

$$\vec{F} = 10(x\vec{i} + y\vec{j} + z\vec{k}) = -30{,}000 \text{ grad } T,$$

so

$$\text{grad } T = -\frac{10}{30{,}000}(x\vec{i} + y\vec{j} + z\vec{k}).$$

Integrating we get

$$T = \frac{-10}{2(30{,}000)}(x^2 + y^2 + z^2) + C.$$

At the surface of the earth, $x^2 + y^2 + z^2 = 6400^2$, and $T = 20°C$, so

$$T = \frac{-1}{6000}(6400^2) + C = 20.$$

Thus,

$$C = 20 + \frac{6400^2}{6000} = 6847.$$

At the center of the earth, $x^2 + y^2 + z^2 = 0$, so

$$T = 6847°C.$$

16. Writing out the partial derivatives

$$\nabla^2 \phi(x, y, z) = \nabla(\nabla \phi(x, y, z)) = \nabla\left(\frac{\partial \phi}{\partial x}\vec{i} + \frac{\partial \phi}{\partial y}\vec{j} + \frac{\partial \phi}{\partial z}\vec{k}\right)$$

$$= \frac{\partial}{\partial x}\left(\frac{\partial \phi}{\partial x}\right) + \frac{\partial}{\partial y}\left(\frac{\partial \phi}{\partial y}\right) + \frac{\partial}{\partial z}\left(\frac{\partial \phi}{\partial z}\right) = \frac{\partial^2 \phi}{\partial x^2} + \frac{\partial^2 \phi}{\partial y^2} + \frac{\partial^2 \phi}{\partial z^2}.$$

17. Suppose $\phi(x, y, z) = ax + by + cz + d$ is linear (a, b, c, d are constants). We have

$$\nabla^2 \phi(x, y, z) = \frac{\partial^2}{\partial x^2}(ax + by + cz + d) + \frac{\partial^2}{\partial y^2}(ax + by + cz + d)$$

$$+ \frac{\partial^2}{\partial z^2}(ax + by + cz + d) = 0$$

Hence ϕ is harmonic.

18. Let $\phi(x, y, z) = ax^2 + by^2 + cz^2 + dxy + exz + fyz$, where a, b, c, d, e, f, are constants. Then

$$\nabla^2 \phi(x, y, z) = \frac{\partial^2 \phi}{\partial x^2} + \frac{\partial^2 \phi}{\partial y^2} + \frac{\partial^2 \phi}{\partial z^2} = 2a + 2b + 2c.$$

Hence ϕ is harmonic if and only if $a + b + c = 0$.

19. Apply the Divergence Theorem to the integral: $\int_S (\nabla \phi) \cdot d\vec{A}$ and get:

$$\int_S (\nabla \phi) \, d\vec{A} = \int_{R'} \text{div}(\nabla \phi) \cdot d\vec{A},$$

where R' is the domain enclosed by S.

But $\text{div}(\nabla \phi) = \nabla^2 \phi = 0$ as ϕ is harmonic in R, so $\int_S (\nabla \phi) \cdot d\vec{A} = 0$.

20. (a) Let $(x, y, z) \neq (0, 0, 0)$ and compute $\nabla^2\phi(x, y, z)$.

$$\nabla^2\phi(x, y, z) = \frac{\partial^2(x^2 + y^2 + z^2)^{-1/2}}{\partial x^2} + \frac{\partial^2(x^2 + y^2 + z^2)^{-1/2}}{\partial y^2} + \frac{\partial^2(x^2 + y^2 + z^2)^{-1/2}}{\partial z^2}$$

$$= -\frac{3}{(x^2 + y^2 + z^2)^{3/2}} + \frac{3(x^2 + y^2 + z^2)}{(x^2 + y^2 + z^2)^{3/2}} = 0.$$

So ϕ is harmonic in any region which does not contain the origin.

(b) Let $P \neq (0, 0, 0)$ be a point, then $\phi(P) = 1/r$. So if Q_1 is a point further away from the origin, and Q_2 is a point closer to the origin than P, we have $\phi(Q_1) < \phi(P) < \phi(Q_2)$. Hence P can be neither a maximum nor a minimum.

(c) Calculating the gradient, we get

$$\nabla\phi = -\frac{x}{(x^2 + y^2 + z^2)^{3/2}}\vec{i} - \frac{y}{(x^2 + y^2 + z^2)^{3/2}}\vec{j} - \frac{z}{(x^2 + y^2 + z^2)^{3/2}}\vec{k} = -\frac{\vec{r}}{\|\vec{r}\|^3}.$$

Thus, $\nabla\phi = -\vec{r}$ on S, a sphere of radius 1. Therefore,

$$\int_S \nabla\phi \cdot d\vec{A} = \int_S -\vec{r} \cdot d\vec{A} = -\int_S \|d\vec{A}\| = -(\text{Surface area of } S) = -4\pi.$$

This doesn't contradict the assertion of Problem 19, as the region contains the origin, and ϕ is not defined at the origin.

21. Let ϕ be our nonconstant harmonic function on the region R. Denote $\psi = -\phi$. Then ψ is nonconstant, and

$$\nabla^2\psi = -\nabla^2\phi = 0.$$

Hence, ψ is also harmonic.

Moreover, any minimum value for ϕ in R is a maximum value for ψ in the same region. Now use Example 5 on page 424 to derive that if ϕ has a minimum value at a point, ψ has a maximum one at the same point. So the point must be on the boundary of R.

22. Writing out the partial derivatives and using the product rule:

$$\text{div}(\phi \, \text{grad} \, \phi) = \text{div}\left(\phi\frac{\partial\phi}{\partial x}\vec{i} + \phi\frac{\partial\phi}{\partial y}\vec{j} + \phi\frac{\partial\phi}{\partial z}\vec{k}\right)$$

$$= \frac{\partial}{\partial x}\left(\phi\frac{\partial\phi}{\partial x}\right) + \frac{\partial}{\partial y}\left(\phi\frac{\partial\phi}{\partial y}\right) + \frac{\partial}{\partial z}\left(\phi\frac{\partial\phi}{\partial z}\right)$$

$$= \left(\frac{\partial\phi}{\partial x}\right)^2 + \phi\frac{\partial^2\phi}{\partial x^2} + \left(\frac{\partial\phi}{\partial y}\right)^2 + \phi\frac{\partial^2\phi}{\partial y^2} + \left(\frac{\partial\phi}{\partial z}\right)^2 + \phi\frac{\partial^2\phi}{\partial z^2}$$

$$= \|\text{grad}\,\phi\|^2 + \phi \cdot \nabla^2\phi = \|\text{grad}\,\phi\|^2, \quad \text{because } \nabla^2\phi = 0.$$

23. We apply the Divergence Theorem to the integral:

$$\int_S \phi \, \text{grad}\, \phi \cdot d\vec{A} = \int_W \text{div}(\phi \, \text{grad}\, \phi) \, dV$$

$$= \int_W \|\text{grad}\,\phi\|^2 \, dV. \quad (\text{see Problem 22}).$$

Since ϕ is zero on S,

$$\int_S \phi \, \text{grad}\, \phi \, d\vec{A} = 0,$$

and

$$\int_W \| \operatorname{grad} \phi \|^2 \, dV = 0.$$

But $\| \operatorname{grad} \phi \|^2 \geq 0$, and $\| \operatorname{grad} \phi(P) \|^2 > 0$, for a point P, would imply

$$\int_W \| \operatorname{grad} \phi \|^2 \, dV > 0.$$

Hence $\| \operatorname{grad} \phi \|^2 = 0$ everywhere. So grad $\phi = \vec{0}$ everywhere and ϕ is a constant on W. But ϕ is zero on S, so it must be zero on W.

24. Let $\phi = \phi_1 - \phi_2$, then

$$\nabla^2 \phi = \nabla^2 (\phi_1 - \phi_2) = \nabla^2 \phi_1 - \nabla^2 \phi_2 = 0.$$

Hence, ϕ is harmonic. Moreover, for a point P on S,

$$\phi(P) = \phi_1(P) - \phi_2(P) = 0.$$

Applying Problem 23, we get that $\phi = 0$ on R, hence $\phi_1 = \phi_2$ on the whole of R.

25. Apply Divergence Theorem to the integral:

$$\int_S u \operatorname{grad} v \cdot d\vec{A} = \int_R \operatorname{div}(u \operatorname{grad} v) \, dV.$$

Now:

$$\operatorname{div}(u \operatorname{grad} v) = \operatorname{div}(u \frac{\partial v}{\partial x} \vec{i} + u \frac{\partial v}{\partial y} \vec{j} + u \frac{\partial v}{\partial z} \vec{k})$$
$$= \frac{\partial}{\partial x}(u \frac{\partial v}{\partial x}) + \frac{\partial}{\partial y}(u \frac{\partial v}{\partial y}) + \frac{\partial}{\partial z}(u \frac{\partial v}{\partial z})$$
$$= \frac{\partial u}{\partial x}\frac{\partial v}{\partial x} + \frac{\partial u}{\partial y}\frac{\partial v}{\partial y} + \frac{\partial u}{\partial z}\frac{\partial v}{\partial z} + u \cdot (\frac{\partial^2 v}{\partial x^2} + \frac{\partial^2 v}{\partial y^2} + \frac{\partial^2 v}{\partial z^2})$$
$$= \operatorname{grad} u \cdot \operatorname{grad} v.$$

Hence

$$\int_S u \operatorname{grad} v \cdot d\vec{A} = \int_R (\operatorname{grad} u \cdot \operatorname{grad} v) \, dV.$$

Similarly,

$$\int_S v \operatorname{grad} v \cdot d\vec{A} = \int_R \operatorname{div}(v \operatorname{grad} u) \, dV = \int_R (\operatorname{grad} v \cdot \operatorname{grad} u) \, dV.$$

Hence

$$\int_S u \operatorname{grad} v \cdot d\vec{A} = \int_S v \operatorname{grad} u \cdot d\vec{A}.$$

Solutions for Section 20.3

1. Using the definition in Cartesian coordinates, we have

$$\text{curl } \vec{F} = \begin{vmatrix} \vec{i} & \vec{j} & \vec{k} \\ \frac{\partial}{\partial x} & \frac{\partial}{\partial y} & \frac{\partial}{\partial z} \\ x^2 - y^2 & 2xy & 0 \end{vmatrix}$$

$$= \left(\frac{\partial}{\partial y}(0) - \frac{\partial}{\partial z}(2xy) \right) \vec{i} + \left(-\frac{\partial}{\partial x}(0) + \frac{\partial}{\partial z}(x^2 - y^2) \right) \vec{j} + \left(\frac{\partial}{\partial x}(2xy) - \frac{\partial}{\partial y}(x^2 - y^2) \right) \vec{k}$$

$$= 4y\vec{k} .$$

2. This vector field points radically outward and has unit length everywhere (except the origin). Thus, we would expect its curl to be $\vec{0}$. Computing the curl directly we get

$$\text{curl} \left(\frac{\vec{r}}{\|\vec{r}\|} \right) = \begin{vmatrix} \vec{i} & \vec{j} & \vec{k} \\ \frac{\partial}{\partial x} & \frac{\partial}{\partial y} & \frac{\partial}{\partial z} \\ \frac{x}{(x^2+y^2+z^2)^{\frac{1}{2}}} & \frac{y}{(x^2+y^2+z^2)^{\frac{1}{2}}} & \frac{z}{(x^2+y^2+z^2)^{\frac{1}{2}}} \end{vmatrix}$$

The $\vec{i}$-component is given by $= \left(-\frac{1}{2} \cdot \frac{2yz}{(x^2 + y^2 + z^2)^{\frac{3}{2}}} - \left(-\frac{1}{2} \cdot \frac{2yz}{(x^2 + y^2 + z^2)^{\frac{1}{2}}} \right) \right) \vec{i}$

$$= \vec{0}$$

Similarly, the $\vec{j}$ and $\vec{k}$ components are also both $\vec{0}$.

3. Using the definition of Cartesian coordinates,

$$\text{curl } \vec{F} = \begin{vmatrix} \vec{i} & \vec{j} & \vec{k} \\ \frac{\partial}{\partial x} & \frac{\partial}{\partial y} & \frac{\partial}{\partial z} \\ x^2 & y^3 & z^4 \end{vmatrix}$$

$$= \left(\frac{\partial}{\partial y}(z^4) - \frac{\partial}{\partial z}(y^3) \right) \vec{i} + \left(-\frac{\partial}{\partial x}(z^4) + \frac{\partial}{\partial z}(x^2) \right) \vec{j} + \left(\frac{\partial}{\partial x}(y^3) - \frac{\partial}{\partial y}(x^2) \right) \vec{k}$$

$$= \vec{0} .$$

4. Using the definition of Cartesian coordinates,

$$\text{curl } \vec{F} = \begin{vmatrix} \vec{i} & \vec{j} & \vec{k} \\ \frac{\partial}{\partial x} & \frac{\partial}{\partial y} & \frac{\partial}{\partial z} \\ e^x & \cos y & e^{z^2} \end{vmatrix}$$

$$= \left(\frac{\partial}{\partial y}(e^{z^2}) - \frac{\partial}{\partial z}(\cos y) \right) \vec{i} + \left(-\frac{\partial}{\partial x}(e^{z^2}) + \frac{\partial}{\partial z}(e^x) \right) \vec{j} + \left(\frac{\partial}{\partial x}(\cos y) - \frac{\partial}{\partial y}(e^x) \right) \vec{k}$$

$$= \vec{0} .$$

5. Using the definition of Cartesian coordinates,

$$\text{curl } \vec{F} = \begin{vmatrix} \vec{i} & \vec{j} & \vec{k} \\ \frac{\partial}{\partial x} & \frac{\partial}{\partial y} & \frac{\partial}{\partial z} \\ 2yz & 3xz & 7xy \end{vmatrix}$$

$$= (7x - 3x)\vec{i} - (7y - 2y)\vec{j} + (3z - 2z)\vec{k}$$

$$= 4x\vec{i} - 5y\vec{j} + z\vec{k}.$$

6. Using the definition of Cartesian coordinates,

$$\text{curl } \vec{F} = \begin{vmatrix} \vec{i} & \vec{j} & \vec{k} \\ \frac{\partial}{\partial x} & \frac{\partial}{\partial y} & \frac{\partial}{\partial z} \\ (-x + y) & (y + z) & (-z + x) \end{vmatrix}$$

$$= \left(\frac{\partial}{\partial y}(-z + x) - \frac{\partial}{\partial z}(y + z) \right) \vec{i} + \left(-\frac{\partial}{\partial x}(-z + x) + \frac{\partial}{\partial z}(-x + y) \right) \vec{j}$$

$$+ \left(\frac{\partial}{\partial x}(y + z) - \frac{\partial}{\partial y}(-x + y) \right) \vec{k}$$

$$= -\vec{i} - \vec{j} - \vec{k}.$$

7. Using the definition in Cartesian coordinates

$$\text{curl } \vec{F} = \begin{vmatrix} \vec{i} & \vec{j} & \vec{k} \\ \frac{\partial}{\partial x} & \frac{\partial}{\partial y} & \frac{\partial}{\partial z} \\ x + yz & y^2 + xzy & zx^3y^2 + x^7y^6 \end{vmatrix}$$

$$= (2x^3yz + 6x^7y^5 - xy)\vec{i} + (-3x^2y^2z - 7x^6y^6 + y)\vec{j} + (yz - z)\vec{k}$$

8. The part of this vector field in the xy-plane looks like Figure 20.1 on page 412, and shows no rotational tendency. Thus we expect the curl to be $\vec{0}$. In fact it is, because the circulation around *every* closed curve C is zero, since

$$\vec{F} = x\vec{i} + y\vec{j} + z\vec{k} = \text{grad}(x^2/2 + y^2/2 + z^2/2),$$

so $\vec{F}$ is a gradient field. Thus the circulation density is zero in any direction, and hence curl $\vec{F}(P) = \vec{0}$ for every point P. Using the formula, we see that

$$\text{curl } \vec{F} = \nabla \times \vec{F} = \begin{vmatrix} \vec{i} & \vec{j} & \vec{k} \\ \frac{\partial}{\partial x} & \frac{\partial}{\partial y} & \frac{\partial}{\partial z} \\ x & y & z \end{vmatrix}$$

$$= \left(\frac{\partial z}{\partial y} - \frac{\partial y}{\partial z}\right)\vec{i} + \left(\frac{\partial x}{\partial z} - \frac{\partial z}{\partial x}\right)\vec{j} + \left(\frac{\partial y}{\partial x} - \frac{\partial x}{\partial y}\right)\vec{k} = 0\vec{i} + 0\vec{j} + 0\vec{k} = \vec{0}.$$

9. The conjecture is that when the first component of $\vec{F}$ depends only on x, the second component depends only on y, and the third component depends only on z, that is, if

$$\vec{F} = F_1(x)\vec{i} + F_2(y)\vec{j} + F_3(z)\vec{k}$$

then

$$\text{curl } \vec{F} = \vec{0}$$

The reason for this is that if $\vec{F} = F_1(x)\vec{i} + F_2(y)\vec{j} + F_3(z)\vec{k}$, then

$$\text{curl } \vec{F} = \begin{vmatrix} \vec{i} & \vec{j} & \vec{k} \\ \frac{\partial}{\partial x} & \frac{\partial}{\partial y} & \frac{\partial}{\partial z} \\ F_1(x) & F_2(y) & F_3(z) \end{vmatrix}$$

$$= \left(\frac{\partial}{\partial y}F_3(z) - \frac{\partial}{\partial z}F_2(y)\right)\vec{i} + \left(-\frac{\partial}{\partial x}F_3(z) + \frac{\partial}{\partial z}F_1(x)\right)\vec{j} + \left(\frac{\partial}{\partial x}F_2(y) - \frac{\partial}{\partial y}F_1(x)\right)\vec{k}$$

$$= \vec{0}.$$

10. (a) The vector field always points perpendicularly to both $\vec{k}$ and $\vec{r}$, in the direction determined by the right hand rule, and its magnitude is twice the magnitude of $\vec{r}$. Thus

$$\vec{F}(\vec{r}) = 2\vec{k} \times \vec{r} = -2y\vec{i} + 2x\vec{j}.$$

(b)   Using the definition in Cartesian coordinates

$$\text{curl } \vec{v} = \begin{vmatrix} \vec{i} & \vec{j} & \vec{k} \\ \frac{\partial}{\partial x} & \frac{\partial}{\partial y} & \frac{\partial}{\partial z} \\ -2y & 2x & 0 \end{vmatrix}$$

$$= (\frac{\partial}{\partial y}(0) - \frac{\partial}{\partial z}(2x))\vec{i} + (\frac{\partial}{\partial z}(-2y) - \frac{\partial}{\partial x}(0))\vec{j} + (\frac{\partial}{\partial x}(2x) - \frac{\partial}{\partial y}(-2y))\vec{k} = 4\vec{k}.$$

This makes sense, because we computed the circulation density of this vector field in the z-direction and found it was 4, and we would expect the z-direction to give the maximum circulation density from the symmetry of the vector field.

11.   (a)   This vector field shows no rotation, and the circulation around any closed curve appears to be zero, so we will suspect a zero curl here.

(b)   This vector field is definitely swirling, so we suspect a non-zero curl here.

(c)   This vector field appears to have zero curl, but if you find the circulation around the boundary of a square that is in quadrant one, you will discover that there is a positive curl, for the vectors on the top of the square are larger than those at the bottom. This is therefore a non-zero curl also.

12.   Investigate the velocity vector field of the atmosphere near the fire. If the curl of this vector field is non-zero, there is circulatory motion. Consequently, if the magnitude of the curl of this vector field is large near the fire, a fire storm has probably developed.

13.   Let $\vec{C} = a\vec{i} + b\vec{j} + c\vec{k}$. Then

$$\text{curl}(\vec{F} + \vec{C}) = \left(\frac{\partial}{\partial y}(F_3 + c) - \frac{\partial}{\partial z}(F_2 + b) \right)\vec{i} + \left(\frac{\partial}{\partial z}(F_1 + a) - \frac{\partial}{\partial x}(F_3 + c) \right)\vec{j}$$

$$+ \left(\frac{\partial}{\partial x}(F_2 + b) - \frac{\partial}{\partial y}(F_1 + a) \right)\vec{k}$$

$$= \left(\frac{\partial F_3}{\partial y} - \frac{\partial F_2}{\partial z} \right)\vec{i} + \left(\frac{\partial F_1}{\partial z} - \frac{\partial F_3}{\partial x} \right)\vec{j} + \left(\frac{\partial F_2}{\partial x} - \frac{\partial F_1}{\partial y} \right)\vec{k}$$

$$= \text{curl } \vec{F}.$$

14.   Let $\vec{c} = c_1\vec{i} + c_2\vec{j} + c_3\vec{k}$ and $\vec{F} = F_1\vec{i} + F_2\vec{j} + F_3\vec{k}$. We then show the desired result as follows:

$$\text{div}(\vec{F} \times \vec{c}) = \text{div}((F_1\vec{i} + F_2\vec{j} + F_3\vec{k}) \times (c_1\vec{i} + c_2\vec{j} + c_3\vec{k}))$$

$$= \text{div}((F_2 c_3 - F_3 c_2)\vec{i} + (F_3 c_1 - F_1 c_3)\vec{j} + (F_1 c_2 - F_2 c_1)\vec{k})$$

$$= \frac{\partial}{\partial x}(F_2 c_3 - F_3 c_2) + \frac{\partial}{\partial y}(F_3 c_1 - F_1 c_3) + \frac{\partial}{\partial z}(F_1 c_2 - F_2 c_1)$$

$$= c_3 \frac{\partial F_2}{\partial x} - c_2 \frac{\partial F_3}{\partial x} + c_1 \frac{\partial F_3}{\partial y} - c_3 \frac{\partial F_1}{\partial y} + c_2 \frac{\partial F_1}{\partial z} - c_1 \frac{\partial F_2}{\partial z}$$

$$= c_1 (\frac{\partial F_3}{\partial y} - \frac{\partial F_2}{\partial z}) + c_2 (\frac{\partial F_1}{\partial z} - \frac{\partial F_3}{\partial x}) + c_3 (\frac{\partial F_2}{\partial x} - \frac{\partial F_1}{\partial y})$$

$$= (c_1\vec{i} + c_2\vec{j} + c_3\vec{k}) \cdot ((\frac{\partial F_3}{\partial y} - \frac{\partial F_2}{\partial z})\vec{i} + (\frac{\partial F_1}{\partial z} - \frac{\partial F_3}{\partial x})\vec{j} + (\frac{\partial F_2}{\partial x} - \frac{\partial F_1}{\partial y})\vec{k})$$

$$= \vec{c} \cdot \text{curl } \vec{F}.$$

15. The Fundamental Theorem of Calculus for Line Integrals states that if C is a path from P to Q, then

$$\int_C \text{grad } f \cdot d\vec{r} = f(Q) - f(P).$$

Since C is a closed path we have

$$\int_c \text{grad } f \cdot d\vec{r} = f(P) - f(P) = 0$$

(a) For any unit vector $\vec{n}$

$$\text{circ}_{\vec{n}} \text{ grad } f = \lim_{\text{Area} \to 0} \left(\frac{\int \text{grad } f \cdot d\vec{r}}{\text{Area of C}} \right) = \lim_{\text{Area} \to 0} \left(\frac{0}{\text{Area}} \right) = 0$$

where the limit is taken over curves C in a plane perpendicular to $\vec{n}$, and oriented by the right hand rule. Thus the circulation density of grad f is zero in every direction, and hence curl grad $f = \vec{0}$.

(b) Using the Cartesian coordinate definition

$$\text{curl grad } f = \text{curl}(\frac{\partial f}{\partial x}\vec{i} + \frac{\partial f}{\partial y}\vec{j} + \frac{\partial f}{\partial z}\vec{k})$$

$$= \left(\frac{\partial^2 f}{\partial y \partial z} - \frac{\partial^2 f}{\partial z \partial y} \right) \vec{i} + \left(\frac{\partial^2 f}{\partial z \partial x} - \frac{\partial^2 f}{\partial x \partial z} \right) \vec{j} + \left(\frac{\partial^2 f}{\partial x \partial y} - \frac{\partial^2 f}{\partial y \partial x} \right) \vec{k}$$

$$= 0\vec{i} + 0\vec{j} + 0\vec{k} = \vec{0}.$$

16.

$$\text{curl } \vec{F} = (\frac{\partial F_3}{\partial y} - \frac{\partial F_2}{\partial z})\vec{i} + (\frac{\partial F_1}{\partial z} - \frac{\partial F_3}{\partial x})\vec{j} + (\frac{\partial F_2}{\partial x} - \frac{\partial F_1}{\partial y}\vec{k})$$

$$\text{div curl } \vec{F} = \frac{\partial}{\partial x}(\frac{\partial F_3}{\partial y} - \frac{\partial F_2}{\partial z}) + \frac{\partial}{\partial y}(\frac{\partial F_1}{\partial z} - \frac{\partial F_3}{\partial x}) + \frac{\partial}{\partial z}(\frac{\partial F_2}{\partial x} - \frac{\partial F_1}{\partial y})$$

$$= \frac{\partial^2 F_3}{\partial x \partial y} - \frac{\partial^2 F_2}{\partial x \partial z} + \frac{\partial^2 F_1}{\partial y \partial z} - \frac{\partial^2 F_3}{\partial y \partial x} + \frac{\partial^2 F_2}{\partial z \partial x} - \frac{\partial^2 F_1}{\partial z \partial y}$$

Since, if $\vec{F}$ has continuous second partial derivatives,

$$\frac{\partial^2 F_3}{\partial x \partial y} = \frac{\partial^2 F_3}{\partial y \partial x}, \quad \frac{\partial^2 F_2}{\partial x \partial z} = \frac{\partial^2 F_2}{\partial z \partial x}, \quad \text{and} \quad \frac{\partial^2 F_1}{\partial y \partial z} = \frac{\partial^2 F_1}{\partial z \partial y}$$

everything cancels out and we get div curl $\vec{F} = \vec{0}$.

17.

$$\text{curl}(\phi\vec{F})$$

$$= \left(\frac{\partial}{\partial y}(\phi F_3) - \frac{\partial}{\partial z}(\phi F_2) \right) \vec{i} + \left(\frac{\partial}{\partial z}(\phi F_1) - \frac{\partial}{\partial x}(\phi F_3) \right) \vec{j} + \left(\frac{\partial}{\partial x}(\phi F_2) - \frac{\partial}{\partial y}(\phi F_1) \right) \vec{k}$$

$$= \left(\phi\frac{\partial F_3}{\partial y} + \frac{\partial\phi}{\partial y}F_3 - \phi\frac{\partial F_2}{\partial z} - \frac{\partial\phi}{\partial z}F_2 \right) \vec{i} + \left(\phi\frac{\partial F_1}{\partial z} + \frac{\partial\phi}{\partial z}F_1 - \phi\frac{\partial F_3}{\partial x} - \frac{\partial\phi}{\partial x}F_3 \right) \vec{j}$$

$$+ \left(\phi \frac{\partial F_2}{\partial x} + \frac{\partial \phi}{\partial x} F_2 - \phi \frac{\partial F_1}{\partial y} - \frac{\partial \phi}{\partial y} F_1 \right) \vec{k}$$

$$= \phi \left(\left(\frac{\partial F_3}{\partial y} - \frac{\partial F_2}{\partial z} \right) \vec{i} + \left(\frac{\partial F_1}{\partial z} - \frac{\partial F_3}{\partial x} \right) \vec{j} + \left(\frac{\partial F_2}{\partial x} - \frac{\partial F_1}{\partial y} \right) \vec{k} \right)$$

$$+ \left(\left(\frac{\partial \phi}{\partial y} F_3 - \frac{\partial \phi}{\partial z} F_2 \right) \vec{i} + \left(\frac{\partial \phi}{\partial z} F_1 - \frac{\partial \phi}{\partial x} F_3 \right) \vec{j} + \left(\frac{\partial \phi}{\partial x} F_2 - \frac{\partial \phi}{\partial y} F_1 \right) \vec{k} \right)$$

$$= \phi \operatorname{curl} \vec{F} + \left(\frac{\partial \phi}{\partial x} \vec{i} + \frac{\partial \phi}{\partial y} \vec{j} + \frac{\partial \phi}{\partial z} \vec{k} \right) \times (F_1 \vec{i} + F_2 \vec{j} + F_3 \vec{k})$$

$$= \phi \operatorname{curl} \vec{F} + (\operatorname{grad} \phi) \times \vec{F}.$$

18. (a) Cross-section of the vector field in xy-plane with $\omega = 1$, so $\vec{v} = -y\vec{i} + x\vec{j}$.

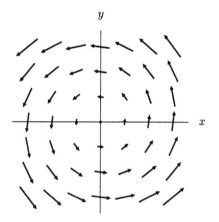

Figure 20.12: $\vec{v} = -y\vec{i} + x\vec{j}$

Cross-section of vector field in xy-plane with $\omega = -1$, so $\vec{v} = y\vec{i} - x\vec{j}$.

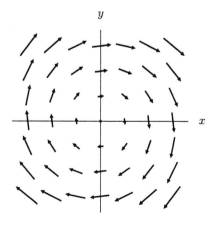

Figure 20.13: $\vec{v} = y\vec{i} - x\vec{j}$

(b) The distance from the center of the vortex is given by $r = \sqrt{x^2 + y^2}$. The velocity of the vortex at any point is $-\omega y\vec{i} + \omega x\vec{j}$, and the speed of the vortex at any point is the magnitude of the velocity, or
$$s = \|\vec{v}\| = \sqrt{(-\omega y)^2 + (\omega x)^2} = |\omega|\sqrt{x^2 + y^2} = |\omega|\, r.$$

(c) The divergence of the velocity field is given by:
$$\operatorname{div}\vec{v} = \frac{\partial(-\omega y)}{\partial x} + \frac{\partial(\omega x)}{\partial y} = 0$$

The curl of the field is:
$$\operatorname{curl}\vec{v} = \operatorname{curl}(-\omega y\vec{i} + \omega x\vec{j}) = \left(\frac{\partial}{\partial x}(\omega x) - \frac{\partial}{\partial y}(-\omega y)\right)\vec{k} = 2\omega\vec{k}$$

(d) We know that $\vec{v}$ has constant magnitude $|\omega|R$ everywhere on the circle and is everywhere tangential to the circle. In addition, if $\omega > 0$, the vector field rotates counterclockwise; if $\omega < 0$, the vector field rotates clockwise. Thus if $\omega > 0$, $\vec{v}$ and $\Delta\vec{r}$ are parallel and in the same direction, so
$$\int_C \vec{v} \cdot d\vec{r} = |\vec{v}| \cdot (\text{Length of } C) = \omega R \cdot 2\pi R = 2\pi\omega R^2$$

If $\omega < 0$, then $|\omega| = -\omega$ and $\vec{v}$ and $\Delta\vec{r}$ are in opposite directions, so
$$\int_C \vec{v} \cdot d\vec{r} = -|\vec{v}| \cdot (\text{Length of } C) = -|\omega|\, R \cdot (2\pi R) = 2\pi\omega R^2.$$

19. Fields with zero curl: (c), (d), (f) because these don't appear to be swirling.

20. Fields (a), (c), (e) because they do not appear to be exploding or collapsing.

21. C_2, C_3, C_4, C_6, since line integrals around C_1 and C_5 are clearly nonzero. You can see directly that $\int_{C_2} \vec{F} \cdot d\vec{r}$ and $\int_{C_6} \vec{F} \cdot d\vec{r}$ are zero, because C_2 and C_6 are perpendicular to their fields at every point.

22. Since ϕ is harmonic, $\operatorname{div}(\operatorname{grad}\phi) = 0$, and so $\operatorname{grad}\phi$ is divergence free. On the other hand, every gradient field is curl free (see Problem 15), and so $\operatorname{grad}\phi$ is also curl free.

23. $\vec{F} = \nabla\phi + (\vec{F} - \nabla\phi)$ does the job. The vector field $\nabla\phi$ is curl free, because every gradient field is curl free, and $\vec{F} - \nabla\phi$ is divergence free, because $\operatorname{div}(\vec{F} - \nabla\phi) = \operatorname{div}\vec{F} - \nabla^2\phi = 0$.

24. Note that
$$\operatorname{div}(2y\vec{i} + 4x\vec{j}) = \frac{\partial}{\partial x}(2y) + \frac{\partial}{\partial y}(4x) = 0$$

and
$$\operatorname{curl}(3x\vec{i} + 9y\vec{j}) = \begin{vmatrix} \vec{i} & \vec{j} & \vec{k} \\ \frac{\partial}{\partial x} & \frac{\partial}{\partial y} & \frac{\partial}{\partial z} \\ 3x & 9y & 0 \end{vmatrix} = \vec{0},$$

so $(3x\vec{i} + 9y\vec{j}) + (2y\vec{i} + 4x\vec{j})$ is the required decomposition.

25. Let $\vec{v} = a\vec{i} + b\vec{j} + c\vec{k}$ and try
$$\vec{F} = \vec{v} \times \vec{r} = (a\vec{i} + b\vec{j} + c\vec{k}) \times (x\vec{i} + y\vec{j} + z\vec{k}) = (bz - cy)\vec{i} + (cx - az)\vec{j} + (ay - bx)\vec{k}.$$

Then
$$\operatorname{curl}\vec{F} = \begin{vmatrix} \vec{i} & \vec{j} & \vec{k} \\ \frac{\partial}{\partial x} & \frac{\partial}{\partial y} & \frac{\partial}{\partial z} \\ bz - cy & cx - az & ay - bx \end{vmatrix} = 2a\vec{i} + 2b\vec{j} + 2c\vec{k}.$$

Taking $a = 1$, $b = -\frac{3}{2}$, $c = 2$ gives curl $\vec{F} = 2\vec{i} - 3\vec{j} + 4\vec{k}$, so the desired vector field is $\vec{F} = (-\frac{3}{2}z - 2y)\vec{i} + (2x - z)\vec{j} + (y + \frac{3}{2}x)\vec{k}$.

26. **(a)** A thin twig at the origin along the x-axis would only feel the velocity along that axis, and thus go counterclockwise.

   (b) Clockwise.

   (c) Using the Cartesian coordinate definition, we get

$$\text{curl}\,\vec{F} = \begin{vmatrix} \vec{i} & \vec{j} & \vec{k} \\ \frac{\partial}{\partial x} & \frac{\partial}{\partial y} & \frac{\partial}{\partial z} \\ y & x & 0 \end{vmatrix} = 0\vec{i} + 0\vec{j} + (1-1)\vec{k} = \vec{0}.$$

   This is as expected, since a paddle-wheel (instead of a twig) placed in the field would not rotate at all.

27. If $\vec{F} = F_1\vec{i} + F_2\vec{j} + F_3\vec{k}$, $\vec{u} = u_1\vec{i} + u_2\vec{j} + u_3\vec{k}$, $\vec{v} = v_1\vec{i} + v_2\vec{j} + v_3\vec{k}$, then

$$\text{grad}(\vec{F} \cdot \vec{v}) \cdot \vec{u} - \text{grad}(\vec{F} \cdot \vec{u}) \cdot \vec{v}$$
$$= \text{grad}(F_1v_1 + F_2v_2 + F_3v_3) \cdot (u_1\vec{i} + u_2\vec{j} + u_3\vec{k}) - \text{grad}(F_1u_1 + F_2u_2 + F_3u_3) \cdot (v_1\vec{i} + v_2\vec{j} + v_3\vec{k})$$
$$= \frac{\partial F_1}{\partial x}v_1u_1 + \frac{\partial F_2}{\partial x}v_2u_1 + \frac{\partial F_3}{\partial x}v_3u_1 + \frac{\partial F_1}{\partial y}v_1u_2 + \frac{\partial F_2}{\partial y}v_2u_2 + \frac{\partial F_3}{\partial y}v_3u_2 +$$
$$\frac{\partial F_1}{\partial z}v_1u_3 + \frac{\partial F_2}{\partial z}v_2u_3 + \frac{\partial F_3}{\partial z}v_3u_3 -$$
$$\left(\frac{\partial F_1}{\partial x}u_1v_1 + \frac{\partial F_2}{\partial x}u_2v_1 + \frac{\partial F_3}{\partial x}u_3v_1 + \frac{\partial F_1}{\partial y}u_1v_2 + \frac{\partial F_2}{\partial y}u_2v_2 + \frac{\partial F_3}{\partial y}u_3v_2 + \right.$$
$$\left. \frac{\partial F_1}{\partial z}u_1v_3 + \frac{\partial F_2}{\partial z}u_2v_3 + \frac{\partial F_3}{\partial z}u_3v_3 \right)$$
$$= \left(\frac{\partial F_3}{\partial y} - \frac{\partial F_2}{\partial z} \right)(u_2v_3 - u_3v_2) + \left(\frac{\partial F_1}{\partial z} - \frac{\partial F_3}{\partial x} \right)(u_3v_1 - u_1v_3) + \left(\frac{\partial F_2}{\partial x} - \frac{\partial F_1}{\partial y} \right)(u_1v_2 - u_2v_1)$$
$$= (\text{curl}\,\vec{F}) \cdot \vec{u} \times \vec{v}.$$

28. **(a)** We can write $\vec{F} = \vec{G} + \vec{F}_{\text{perp}}$, where $\vec{F}_{\text{perp}}$ is the component of $\vec{F}$ perpendicular to L. Since $\vec{u}$ and $\vec{v}$ are parallel to L, $\vec{F}_{\text{perp}} \cdot \vec{u} = \vec{F}_{\text{perp}} \cdot \vec{v} = 0$. So

$$\vec{F} \cdot \vec{u} = (\vec{G} + \vec{F}_{\text{perp}}) \cdot \vec{u} = \vec{G} \cdot \vec{u}$$
$$\vec{F} \cdot \vec{v} = (\vec{G} + \vec{F}_{\text{perp}}) \cdot \vec{b} = \vec{G} \cdot \vec{v}.$$

Since $\vec{u}$ and $\vec{v}$ are orthogonal unit vectors,

$$\vec{G} \cdot \vec{u} = (G_1\vec{u} + G_2\vec{v}) \cdot \vec{u} = G_1(\vec{u} \cdot \vec{u}) + G_2(\vec{u} \cdot \vec{v}) = G_1,$$

and

$$\vec{G} \cdot \vec{v} = (G_1\vec{u} + G_2\vec{v}) \cdot \vec{v} = G_1(\vec{u} \cdot \vec{v}) + G_2(\vec{v} \cdot \vec{v}) = G_2.$$

Thus $\vec{F} \cdot \vec{u} = G_1$, and $\vec{F} \cdot \vec{v} = G_2$, which is what we were supposed to show.

(b)   By the chain rule

$$\frac{\partial G_2}{\partial s} = \frac{\partial}{\partial s}(\vec{F} \cdot \vec{v})$$

$$= \text{grad}(\vec{F} \cdot \vec{v}) \cdot \frac{\partial \vec{r}}{\partial s}.$$

From the parametric equation

$$\vec{r} = \vec{r}_0 + s\vec{u} + t\vec{v},$$

we find that $\frac{\partial \vec{r}}{\partial s} = \vec{u}$. Hence

$$\frac{\partial G_2}{\partial s} = \text{grad}(\vec{F} \cdot b) \cdot \vec{u}.$$

Similarly,

$$\frac{\partial G_1}{\partial t} = \frac{\partial}{\partial t}(\vec{F} \cdot \vec{u})$$

$$= \text{grad}(\vec{F} \cdot \vec{u}) \cdot \frac{\partial \vec{r}}{\partial t}$$

$$= \text{grad}(\vec{F} \cdot \vec{u}) \cdot \vec{v}.$$

Thus

$$\frac{\partial G_2}{\partial s} - \frac{\partial G_1}{\partial t} = \text{grad}(\vec{F} \cdot \vec{v}) \cdot \vec{u} - \text{grad}(\vec{F} \cdot \vec{u}) \cdot \vec{v}.$$

(c)   Problem 27 says that

$$\text{curl}\,\vec{F} \cdot \vec{n} = \text{curl}\,\vec{F} \cdot \vec{u} \times \vec{v} = \text{grad}(\vec{F} \cdot \vec{v}) \cdot \vec{u} - \text{grad}(\vec{F} \cdot \vec{u}) \cdot \vec{v}.$$

Part (b) says that the right hand side of this equation is $\frac{\partial G_2}{\partial s} - \frac{\partial G_1}{\partial t}$.

(d)   By definition,

$$\text{circ}_{\vec{n}}\,\vec{F}(\vec{r}_0) = \lim_{R \to 0} \frac{\int_C \vec{F} \cdot d\vec{r}}{\pi R^2},$$

where C is a circle of radius R in L, centered at the point with position vector $\vec{r}_0$. Since $d\vec{r}$ is parallel to L, we have $\vec{F} \cdot d\vec{r} = \vec{G} \cdot d\vec{r}$, so by Green's Theorem in the st-plane,

$$\int_C \vec{F} \cdot d\vec{r} = \int_C \vec{G} \cdot d\vec{r} = \int_T \left(\frac{\partial G_2}{\partial s} - \frac{\partial G_1}{\partial t} \right) dA,$$

where T is the disk in L enclosed by C. So

$$\text{circ}_{\vec{n}}\,\vec{F}(\vec{r}_0) = \lim_{R \to 0} \frac{\int_T \left(\frac{\partial G_2}{\partial s} - \frac{\partial G_1}{\partial t} \right) dA}{\text{Area of } T}$$

$$= \frac{\partial G_2}{\partial s}(\vec{r}_0) - \frac{\partial G_1}{\partial t}(\vec{r}_0)$$

$$= \left. (\text{grad}(\vec{F} \cdot \vec{v}) \cdot \vec{u} - \text{grad}(\vec{F} \cdot \vec{a}) \cdot \vec{v}) \right|_{\vec{r} = \vec{r}_0}$$

$$= \text{curl}\,\vec{F}(\vec{r}_0) \cdot \vec{u} \times \vec{v}.$$

Solutions for Section 20.4

1. No, because the curve C over which the integral is taken is not a closed curve, and so it is not the boundary of a surface.

2. First C is parameterized by
    $$\vec{r}(\theta) = 2\cos\theta\vec{i} + 2\sin\theta\vec{j} + \vec{k}.$$
    Note that C bounds the disk S given by $x^2 + y^2 \le 4$, $z = 1$. Then
    $$\vec{r}'(\theta) = -2\sin\theta\vec{i} + 2\cos\theta\vec{j}.$$

    Now,
    $$\operatorname{curl}\vec{F} = \begin{vmatrix} \vec{i} & \vec{j} & \vec{k} \\ \frac{\partial}{\partial x} & \frac{\partial}{\partial y} & \frac{\partial}{\partial z} \\ z - 2y & 3x - 4y & z + 3y \end{vmatrix} = 3\vec{i} + \vec{j} + 5\vec{k},$$

    and $d\vec{A} = \vec{k}\,dA$. Using Stokes' Theorem we get
    $$\int_C \vec{F}\cdot d\vec{r} = \int_S \operatorname{curl}\vec{F}\cdot d\vec{A}$$
    $$= \int_S (3\vec{i} + \vec{j} + 5\vec{k})\cdot\vec{k}\,dA = \int_S 5\,dA$$
    $$= 5(\text{Area of circle}) = 5(4\pi) = 20\pi.$$

3. First note that $\operatorname{curl} = 2\vec{k}$.
    (a) By Stokes' Theorem, $\int_C \vec{F}\cdot d\vec{r} = \int_S 2\vec{k}\cdot d\vec{A}$ where S is the disk of radius 10 in the xy-plane centered at the origin, oriented downwards. Since this orientation is opposite to $2\vec{k}$, $\int_S 2\vec{k}\cdot d\vec{A} = -\|2\vec{k}\|(\text{area of } S) = -200\pi$.
    (b) By Stokes' Theorem, $\int_C \vec{F}\cdot d\vec{r} = \int_S 2\vec{k}\cdot d\vec{A}$ where S is the disk of radius 10 in the yz-plane centered at the origin, oriented in the negative x direction. Since the vector field $2\vec{k}$ is parallel to the surface S, its flux through the surface is zero.

4. The graph of $\vec{F} = \vec{r}/\|\vec{r}\|^3$ consists of vectors pointing radially outward. There is no swirl, so $\operatorname{curl}\vec{F} = \vec{0}$. From Stokes' Theorem,
    $$\int_C \vec{F}\cdot d\vec{r} = \int_S \operatorname{curl}\vec{F}\cdot d\vec{A} = \int_S \vec{0}\cdot d\vec{A} = 0$$

5. The circulation is the line integral $\int_C \vec{F}\cdot d\vec{r}$ which can be evaluated directly by parameterizing the circle, C. Or, since C is the boundary of a flat disk S, we can use Stokes' Theorem:
    $$\int_C \vec{F}\cdot d\vec{r} = \int_S \operatorname{curl}\vec{F}\cdot d\vec{A}$$

    where S is the disk $x^2 + y^2 \le 1$, $z = 2$ and is oriented upward (using the right hand rule). Then $\operatorname{curl}\vec{F} = -y\vec{i} - x\vec{j} + \vec{k}$ and the unit normal to S is $\vec{k}$. So
    $$\int_S \operatorname{curl}\vec{F}\cdot d\vec{A} = \int_S (-y\vec{i} - x\vec{j} + \vec{k})\cdot\vec{k}\,dx\,dy$$

$$= \int_S 1 \, dxdy$$
$$= \text{Area of } S$$
$$= \pi$$

6. (a) Let us parameterize the curve C by $\vec{r}(t) = 3 \cos t \vec{i} + 3 \sin t \vec{j}, \quad 0 \le t \le 2\pi$.
        Then $d\vec{r} = (-3 \sin t \vec{i} + 3 \cos t \vec{j}) dt$ and so

$$\int_C ((yz^2 - y)\vec{i} + (xz^2 + x)\vec{j} + 2xyz\vec{k}) \cdot d\vec{r} = \int_C (-3 \sin t \vec{i} + 3 \cos t \vec{j}) \cdot d\vec{r}$$
$$= \int_0^{2\pi} 9dt = 18\pi.$$

   (b) Since C is a closed curve, Stokes' Theorem applies. We choose the surface S to be the disk in the
        xy-plane bounded by C, and it must be oriented upwards. Since $\text{curl}\vec{F} = 2\vec{k}$,

$$\int_C \vec{F} \cdot d\vec{r} = \int_S 2\vec{k} \cdot d\vec{A} = \|2\vec{k}\|(\text{area of } S) = 2(\pi 3^2) = 18\pi.$$

7. A sketch of the surface S and curve C which is the union of four curves C_1, C_2, C_3, and C_4, and the region
    R is shown in Figure 20.14.

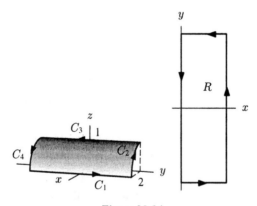

Figure 20.14

To compute the flux integral, we find $d\vec{A}$, oriented upward.

$$d\vec{A} = (2x\vec{i} + \vec{k})dxdy \quad \text{and} \quad \text{curl } \vec{F} = -y\vec{i} - z\vec{j} - x\vec{k}.$$

Thus,

$$\int_S \text{curl } \vec{F} \cdot d\vec{A} = \int_S (-y\vec{i} - z\vec{j} - x\vec{k}) \cdot (2x\vec{i} + \vec{k})dxdy$$
$$= \int_0^1 \int_{-2}^2 (-2xy - x)dydx = -2.$$

The line integral $\int_S \vec{F} \cdot d\vec{r}$ is the sum of four integrals along C_1, C_2, C_3, and C_4.

On C_1: $x = 1, z = 0, dx = 0, dz = 0$, so

$$\int_S \vec{F} \cdot d\vec{r} = \int_{-2}^{2} 0 \, dy = 0.$$

On C_2: $y = 2, z = 1 - x^2, dy = 0, dz = -2x dx$, so

$$\int_{C_2} \vec{F} \cdot d\vec{r} = \int_{C_2} 2x dx + 2(1 - x^2)0 + x(1 - x^2)(-2x)dx$$
$$= \int_1^0 (2x - 2x^2 + 2x^4)dx = -\frac{11}{15}.$$

On C_3: $x = 0, z = 1, dx = 0, dz = 0$, so

$$\int_{C_3} \vec{F} \cdot d\vec{r} = \int_{C_3} 0 + y dy + 0 = \int_2^{-2} y dy = 0.$$

On C_4: $y = -2, z = 1 - x^2, dy = 0, dz = -2x dx$, so

$$\int_{C_4} \vec{F} \cdot d\vec{r} = \int_{C_4} -2x dx - 2(1 - x^2)0 + x(1 - x^2)(-2x)dx$$
$$= \int_0^1 (-2x - 2x^2 + 2x^4)dx = -\frac{19}{15}.$$

Hence

$$\int_C \vec{F} \cdot d\vec{r} = 0 - \frac{11}{15} + 0 - \frac{19}{15} = -2.$$

Thus,

$$\int_C \vec{F} \cdot d\vec{r} = \int_S \text{curl}\,\vec{F} \cdot d\vec{A}.$$

8. The boundary of S is C, the circle $x^2 + y^2 = 1$, $z = 0$, oriented counterclockwise and parameterized in polar coordinates by
$$\vec{r}(\theta) = \cos\theta \vec{i} + \sin\theta \vec{j}, \quad 0 \le \theta \le 2\pi,$$

so,
$$\vec{r}'(\theta) = -\sin\theta \vec{i} + \cos\theta \vec{j}.$$

Hence

$$\int_C \vec{F} \cdot d\vec{r} = \int_0^{2\pi} (\sin\theta \vec{i} + 0\vec{j} + \cos\theta \vec{k}) \cdot (-\sin\theta \vec{i} + \cos\theta \vec{j} + 0\vec{k})d\theta$$
$$= \int_0^{2\pi} -\sin^2\theta d\theta = -\pi.$$

Now consider the integral $\int_S \text{curl}\,\vec{F} \cdot d\vec{A}$. Here $\text{curl}\,\vec{F} = -\vec{i} - \vec{j} - \vec{k}$ and the area vector $d\vec{A}$, oriented upward, is given by
$$d\vec{A} = 2x\vec{i} + 2y\vec{j} + \vec{k} \, dx dy.$$

If R is the disk $x^2 + y^2 \le 1$, then we have

$$\int_S \text{curl}\,\vec{F} \cdot d\vec{A} = \int_R (-\vec{i} - \vec{j} - \vec{k}) \cdot (2x\vec{i} + 2y\vec{j} + \vec{k})dx dy.$$

Converting to polar coordinates gives:

$$
\begin{aligned}
\int_S \operatorname{curl} \vec{F} \cdot d\vec{A} &= \int_0^{2\pi} \int_0^1 (-\vec{i} - \vec{j} - \vec{k}) \cdot (2r \cos\theta \vec{i} + 2r \sin\theta \vec{j} + \vec{k}) r \, dr \, d\theta \\
&= \int_0^{2\pi} \int_0^1 (-2r \cos\theta - 2r \sin\theta - 1) r \, dr \, d\theta \\
&= \int_0^{2\pi} \left(\frac{2}{3}(-\cos\theta - \sin\theta) - \frac{1}{2} \right) d\theta \\
&= -\pi.
\end{aligned}
$$

Thus, we confirm that

$$
\int_C \vec{F} \cdot d\vec{r} = \int_S \operatorname{curl} \vec{F} \cdot d\vec{A}.
$$

9. Use Stokes' theorem, applied to the surface R, oriented upwards. Since $\operatorname{curl}\vec{F} = \vec{k}$ for $\vec{F} = \frac{1}{2}(-y\vec{i} + x\vec{j})$, we have $\frac{1}{2} \int_C (-y\vec{i} + x\vec{j}) \cdot d\vec{r} = \int_R \vec{k} \cdot d\vec{A} = \|\vec{k}\| (\text{area of } R) = \text{area of } R$.

10. (a) It appears that $\operatorname{div} \vec{F} < 0$, and $\operatorname{div} \vec{G} < 0$; $\operatorname{div} \vec{G}$ is larger in magnitude (more negative) if the scales are the same.
    (b) $\operatorname{curl} \vec{F}$ and $\operatorname{curl} \vec{G}$ both appear to be zero at the origin (and elsewhere).
    (c) Yes, the cylinder with axis along the z-axis will have negative flux through it (ends parallel to xy-plane).
    (d) Same as part(c).
    (e) No, you cannot draw a closed curve around the origin such that $\vec{F}$ has a non-zero circulation around it because curl is zero. By Stokes' theorem, circulation equals the integral of the curl over the surface bounded by the curve.
    (f) Same as part(e)

11. Suppose we have a surface S which completely bounds a solid region W. Pick a fixed point P on the surface. Imagine a closed curve C which lies on S and encloses P. Consider the surface S' which is S with the patch around P cut out. Then C can be thought of as the boundary of S', so from Stokes' Theorem,

$$
\int_{S'} \operatorname{curl} \vec{F} \cdot d\vec{A} = \int_C \vec{F} \cdot d\vec{r}
$$

Now let the length of C shrink to zero. The surface S' becomes S, and C becomes the point P, the line integral of $\vec{F}$ around C becomes zero, and so

$$
\int_S \operatorname{curl} \vec{F} \cdot d\vec{A} = 0
$$

Thus, if Q is a point in the domain of $\operatorname{curl} \vec{F}$ and S denotes closed surfaces around Q,

$$
\operatorname{div} \operatorname{curl} \vec{F}(Q) = \lim_{\text{Vol} \to 0} \frac{\int_S \operatorname{curl} \vec{F} \cdot d\vec{A}}{\text{Vol enclosed by } S} = \lim_{\text{Vol} \to 0} \frac{0}{\text{Vol}} = 0
$$

12. Let $\vec{F} = F_1(x, y)\vec{i} + F_2(x, y)\vec{j}$ be a continuously differentiable vector function in a domain in the xy-plane which contains a closed curve C surrounding a region R. Assume that the orientation of C is chosen so that R is on the left as we move around C. Then, R is oriented upward, so the area element $d\vec{A}$ is given by

$$
d\vec{A} = \vec{k} \, dx \, dy.
$$

Thus,

$$\text{curl}\,\vec{F}\cdot d\vec{A} = \text{curl}\,\vec{F}\cdot\vec{k}\,dxdy = \left(\frac{\partial F_2}{\partial x} - \frac{\partial F_1}{\partial y}\right)dxdy.$$

Thus the formula in Stokes' Theorem takes the form

$$\int_C \vec{F}\cdot d\vec{r} = \int_C \left(\frac{\partial F_2}{\partial x} - \frac{\partial F_1}{\partial y}\right)dxdy.$$

This shows that Green's Theorem in the plane is a special case of Stokes' Theorem.

13. (a) You can't say anything, because any surface bounded by the circle must intersect the z-axis. Since curl $\vec{F}$ is not defined on the z-axis, the surface integral in Stokes' Theorem is not defined.

(b) In this case curl $\vec{F}$ is defined and equal to 0 on a surface S bounded by the the circle, so Stokes' Theorem says that

$$\int_C \vec{F}\cdot d\vec{r} = \int_S \text{curl}\,\vec{F}\cdot d\vec{A} = 0.$$

14. The curl $\vec{F}(x,y,z)$ of this vector field is equal to $-2\vec{j}$. Notice, as a check, that this field rotates in a direction opposite to the direction of C. Therefore we expect a negative line integral. The surface S is the disk parallel to the xz plane with radius 2. The curl points in the opposite direction to the normal vector to the surface with this orientation, so by Stokes' Theorem,

$$\text{Circulation around } S = \int_S \text{curl}\,\vec{F}\cdot d\vec{A} = \int_S (-2\vec{j})\cdot d\vec{A} = \int_S -2\,dA$$
$$= -2(\text{Area of circle}) = -2\pi 2^2 = -8\pi.$$

15. (a) At every point of C the vector field $\vec{F}$ is tangent to the curve C and of magnitude a. Since $\vec{F}$ is in the direction opposite to the orientation of C, we have $\int_C \vec{F}\cdot d\vec{r} = -\|\vec{F}\|(\text{length of } C) = -a(2\pi a) = -2\pi a^2$.

(b) Since curl$\vec{F} = 2\vec{i}$, which is normal to S and in the direction in which S is oriented, $\int_S \text{curl}\vec{F}\cdot d\vec{A} = \|2\vec{i}\|(\text{area of } S) = 2\pi a^2$.

(c) The orientations of C and S are not related by the right-hand rule, so Stokes' Theorem does not apply.

16. (a) Computing curl $\vec{G}$, we get

$$\text{curl}\,\vec{G} = \left(\frac{\partial(xz)}{\partial y} - \frac{\partial(3x)}{\partial z}\right)\vec{i} + \left(\frac{\partial(4yz^2)}{\partial z} - \frac{\partial(xz)}{\partial x}\right)\vec{j} + \left(\frac{\partial(3x)}{\partial x} - \frac{\partial(4yz^2)}{\partial y}\right)\vec{k}$$
$$= 0\vec{i} + (8yz - z)\vec{j} + (3 - 4z^2)\vec{k}$$
$$= \vec{F}$$

(b) Because $\vec{F}$ is a curl field, the flux of $\vec{F}$ through S is the same as its flux through any surface with the same boundary as S. Let us replace S by the circular disk S_1 of radius 5 in the xy-plane. On S_1, $z = 0$, and so the vector field reduces to $\vec{F} = 3\vec{k}$. Since the vector $3\vec{k}$ is normal to S_1 and in the direction of the orientation of S_1, we have

$$\int_S \vec{F}\cdot d\vec{A} = \int_{S_1} \vec{F}\cdot d\vec{A} = \int_{S_1} 3\vec{k}\cdot d\vec{A} = \|3\vec{k}\|(\text{area of } S_1) = 75\pi.$$

17. Using Stokes' Theorem, the flux integral $\int_S \text{curl } \vec{F} \cdot d\vec{A}$ has the same value as the line integral $\int_C \vec{F} \cdot d\vec{r}$, where C is the boundary curve of S with the appropriate orientation. Here C is the unit circle $x^2 + y^2 = 1, z = 0$ oriented clockwise when viewed from above. We can parameterize C by $\vec{r}(t) = \cos t\vec{i} - \sin t\vec{j}$ with $0 \le t \le 2\pi$.

Then $\vec{r}'(t) = -\sin t\vec{i} - \cos t\vec{j}$, so

$$\int_C \vec{F} \cdot d\vec{r} = \int_0^{2\pi} (\sin t\vec{i} + \cos t\vec{j}) \cdot (-\sin t\vec{i} - \cos t\vec{j}) dt$$

$$= \int_0^{2\pi} (-\sin^2 t - \cos^2 t)\, dt$$

$$= \int_0^{2\pi} -1 dt$$

$$= -2\pi$$

18. (a) Let D be the disk representing the drain, oriented downward. Then the rate at which the water is leaving the bathtub is the flux of water flowing out of the drain:

$$\int_D \vec{F} \cdot d\vec{A} = \int_D \vec{F} \cdot (-dA\vec{k}) = \int_D \frac{1}{z^2 + 1} dA.$$

Because D is in the xy-plane, $z = 0$, so

$$\text{Flux out of } D = \int_D dA = \pi \text{ cm}^3/\text{sec}.$$

(b) We have, in units/sec,

$$\text{div } \vec{F} = -\frac{z}{(z^2 + 1)^2} - \frac{z}{(z^2 + 1)^2} + \frac{2z}{(z^2 + 1)^2} = 0.$$

(c) Let W be the closed region bounded by the hemisphere S of radius 1 and the disk, D, in the xy-plane representing the drain. By the Divergence Theorem, we have:

$$0 = \int_W \text{div } \vec{F}\, dV = \text{Flux out of } W - \text{Flux into } W$$

$$= \int_S \vec{F} \cdot d\vec{A} - \int_D \vec{F} \cdot d\vec{A}$$

$$= \int_S \vec{F} \cdot d\vec{A} - \pi.$$

Thus,

$$\int_S \vec{F} \cdot d\vec{A} = \pi \text{ cm}^3/\text{sec}.$$

(d) Since the orientation is clockwise, we parameterize the circle by $\vec{r}(t) = \sin t\vec{i} + \cos t\vec{j}$. In addition, on the drain, $z = 0$. Thus,

$$\int_C \vec{G} \cdot d\vec{r} = \int_C \frac{1}{2}(y\vec{i} - x\vec{j} - (x^2 + y^2)\vec{k}) \cdot d\vec{r}$$

$$= \frac{1}{2} \int_0^{2\pi} (\cos t\vec{i} - \sin t\vec{j} - \vec{k}) \cdot (\cos t\vec{i} - \sin t\vec{j})\, dt$$

$$= \frac{1}{2} \int_0^{2\pi} (\sin^2 t + \cos^2 t)dt = \pi.$$

So,

$$\int_C \vec{G} \cdot d\vec{r} = \pi \text{ cm}^3/\text{sec}.$$

(e) Computing the partial derivatives, we find that

$$\text{curl } \vec{G} = -\frac{y + xz}{(z^2 + 1)^2}\vec{i} - \frac{yz - x}{(z^2 + 1)^2}\vec{j} - \frac{1}{z^2 + 1}\vec{k} = \vec{F}.$$

(f) By Stokes' Theorem, we have:

$$\int_C \vec{G} \cdot d\vec{r} = \int_D \text{curl } \vec{G} \cdot d\vec{A} = \int_D \vec{F} \cdot d\vec{A}.$$

Solutions for Section 20.5

1. Since curl $\vec{F} = \vec{0}$ and $\vec{F}$ is defined everywhere, we know by the curl test that $\vec{F}$ is a gradient field. In fact, $\vec{F} = \text{grad} f$, where $f(x, y, z) = xyz + yz^2$, so f is a potential function for $\vec{F}$.

2. Since curl $\vec{G} = 2\vec{k} \neq \vec{0}$, the vector field $\vec{G}$ is not a gradient field.

3. (a) We calculate the curl of each of these vector fields.

$$\text{curl } \vec{A} = \text{curl}(-by\vec{i}) = \begin{vmatrix} \vec{i} & \vec{j} & \vec{k} \\ \frac{\partial}{\partial x} & \frac{\partial}{\partial y} & \frac{\partial}{\partial z} \\ -by & 0 & 0 \end{vmatrix} = -\frac{\partial}{\partial y}(-by)\vec{k} = b\vec{k}.$$

(b)

$$\text{curl } \vec{A} = \text{curl}(bx\vec{j}) = \begin{vmatrix} \vec{i} & \vec{j} & \vec{k} \\ \frac{\partial}{\partial x} & \frac{\partial}{\partial y} & \frac{\partial}{\partial z} \\ 0 & bx & 0 \end{vmatrix} = \frac{\partial}{\partial x}(bx)\vec{k} = b\vec{k}.$$

(c)

$$-\frac{1}{2}\vec{r} \times \vec{B} = -\frac{1}{2}yb\vec{i} + \frac{1}{2}xb\vec{j}$$

$$\text{curl } \vec{A} = \text{curl}\left(\frac{1}{2}\vec{B} \times \vec{r}\right)$$

$$= \begin{vmatrix} \vec{i} & \vec{j} & \vec{k} \\ \frac{\partial}{\partial x} & \frac{\partial}{\partial y} & \frac{\partial}{\partial z} \\ -(1/2)by & (1/2)bx & 0 \end{vmatrix}$$

$$= \left(\frac{\partial}{\partial x}\left(\left(\frac{1}{2}\right)bx\right) - \frac{\partial}{\partial y}\left(-\left(\frac{1}{2}\right)by\right)\right)\vec{k} = b\vec{k}.$$

4. In Example 3 on page 431 we showed that $\text{curl}(\vec{b} \times \vec{r}) = 2\vec{b}$. Thus $(1/2)\vec{b} \times \vec{r}$ is a vector potential for $\vec{B}$.

5. We must show curl $\vec{A} = \vec{B}$.

$$\text{curl}\,\vec{A} = \frac{\partial}{\partial y}\left(\frac{-I}{c}\ln(x^2+y^2)\right)\vec{i} - \frac{\partial}{\partial x}\left(\frac{-I}{c}\ln(x^2+y^2)\right)\vec{j}$$

$$= \frac{-I}{c}\left(\frac{2y}{x^2+y^2}\right)\vec{i} + \frac{I}{c}\left(\frac{2x}{(x^2+y^2)}\right)\vec{j}$$

$$= \frac{2I}{c}\left(\frac{-y\vec{i}+x\vec{j}}{x^2+y^2}\right)$$

$$= \vec{B}.$$

6. Since $\text{div}(y\vec{i}+x\vec{j}) = 0$, there is such a vector field $\vec{G}$. In fact, $\vec{G} = (1/2)(y^2-x^2)\vec{k}$ is one possibility, though there are others.

7. Since $\text{div}\,\vec{F} = 2+3-5 = 0$, a vector potential does exist. One such is the vector field $\vec{H} = (-xy + 5yz)\vec{i} + (2xy+xz^2)\vec{k}$, but there are many others.

8. Since $\text{div}\,\vec{G} = 2x+2y+2z \neq 0$, there is not a vector potential for $\vec{G}$.

9. (a) Yes. To show this, we use a version of the product rule for curl (Problem 17 on page 504):

    $$\text{curl}(\phi\vec{F}) = \phi\,\text{curl}\,\vec{F} + (\text{grad}\,\phi) \times \vec{F},$$

    where ϕ is a scalar function and $\vec{F}$ is a vector field. So

    $$\text{curl}\left(q\frac{\vec{r}}{\|\vec{r}\|^3}\right) = \text{curl}\left(\frac{q}{\|\vec{r}\|^3}\vec{r}\right) = \frac{q}{\|\vec{r}\|^3}\text{curl}\,\vec{r} + \text{grad}\left(\frac{q}{\|\vec{r}\|^3}\right) \times \vec{r}$$

    $$= \vec{0} + q\,\text{grad}\left(\frac{1}{\|\vec{r}\|^3}\right) \times \vec{r}$$

    Since the level surfaces of $1/\|\vec{r}\|^3$ are spheres centered at the origin, $\text{grad}(1/\|\vec{r}\|^3)$ is parallel to $\vec{r}$, so $\text{grad}(1/\|\vec{r}\|^3) \times \vec{r} = \vec{0}$. Thus, $\text{curl}\,\vec{E} = \vec{0}$.

    (b) Yes. The domain of $\vec{E}$ is 3-space minus $(0,0,0)$. Any closed curve in this region is the boundary of a surface contained entirely in the region. (If the first surface you pick happens to contain $(0,0,0)$, change its shape slightly to avoid it.)

    (c) Yes. Since $\vec{E}$ satisfies both conditions of the curl test, it must be a gradient field. In fact,

    $$\vec{E} = \text{grad}\left(-q\frac{1}{\|\vec{r}\|}\right).$$

10. (a) Yes. This is the case $p = 2$ of Example 5 on page 432.

    (b) No. The domain of $\vec{B}$ is 3-space minus the z-axis. A closed curve C which surrounds the z-axis cannot be contracted to a point without hitting the z-axis, so it cannot remain at all times within the domain.

    (c) No. In Example 2 on page 439 we found that if C is a circle around the origin,

    $$\int_C \vec{B} \cdot d\vec{r} = \frac{4\pi I}{c}.$$

    Thus $\vec{B}$ has non-zero circulation around C, and hence cannot be a gradient field.

11. (a) Using the product rule from Problem 17 on page 504, we find

$$\operatorname{curl} \vec{E} = \operatorname{curl}\left(\frac{\vec{r}}{\|\vec{r}\|^p}\right) = \frac{1}{\|\vec{r}\|^p}\operatorname{curl}\vec{r} + \operatorname{grad}\left(\frac{1}{\|\vec{r}\|^p}\right) \times \vec{r}.$$

Now $\operatorname{curl}\vec{r} = \vec{0}$ and $\operatorname{grad}\left(\frac{1}{\|\vec{r}\|^p}\right)$ is parallel to $\vec{r}$, so both terms are zero. Thus $\operatorname{curl}\vec{E} = \vec{0}$.

(b) The domain of $\vec{E}$ is 3-space minus the origin if $p > 0$, and it is all of 3-space if $p \leq 0$.

(c) Both domains have the property that any closed curve can be contracted to a point without hitting the origin, so $\vec{E}$ satisfies the curl test for all p. Since $\vec{E}$ has constant magnitude r^{1-p} on the sphere of radius r centered at the origin, and is parallel to the outward normal at every point of the sphere, the sphere must be a level surface of the potential function ϕ, that is, ϕ is a function of r alone. Further, since $\|\vec{E}\| = r^{1-p}$, a good guess is

$$\phi(r) = \int r^{1-p}\,dr,$$

that is,

$$\phi(r) = \begin{cases} \frac{r^{2-p}}{2-p} & \text{if } p \neq 2 \\ \ln r & \text{if } p = 2. \end{cases}$$

You can check that this is indeed a potential function for $\vec{E}$ by checking that $\operatorname{grad}\phi = \vec{E}$.

12. (a) Although $\operatorname{div}\vec{B} = 0$, $\vec{B}$ does not satisfy the divergence test because its domain is 3-space minus the origin, which does not have the required property that every closed surface is the boundary of a solid region which is entirely contained within the domain. For example, the solid region inside a sphere centered at the origin contains the origin, hence is not in the domain of $\vec{B}$.

(b) Using the product rule from Problem 17 on page 504, we find

$$\operatorname{curl}\vec{A} = \left(\frac{1}{\|\vec{r}\|^3}\right)\operatorname{curl}(\mu \times \vec{r}) + \operatorname{grad}\left(\frac{1}{\|\vec{r}\|^3}\right) \times (\vec{\mu} \times \vec{r}).$$

By Example 3 on page 431,

$$\operatorname{curl}(\vec{\mu} \times \vec{r}) = 2\vec{\mu}$$

and by Problem 19 on page 134

$$\operatorname{grad}\left(\frac{1}{\|\vec{r}\|^3}\right) = -3\frac{1}{\|\vec{r}\|^5}\vec{r}.$$

So

$$\operatorname{curl}\vec{A} = 2\frac{\vec{\mu}}{\|\vec{r}\|^3} - 3\left(\frac{1}{\|\vec{r}\|^5}\right)\vec{r} \times (\vec{\mu} \times \vec{r}).$$

From Problem 24 on page 92, we have

$$\vec{r} \times (\vec{\mu} \times \vec{r}) = \|\vec{r}\|^2\vec{\mu} - (\vec{\mu} \cdot \vec{r})\vec{r}.$$

So

$$\operatorname{curl}\vec{A} = 2\frac{\vec{\mu}}{\|\vec{r}\|^3} - 3\left(\frac{1}{\|\vec{r}\|^5}\right)(\|\vec{r}\|^2\vec{\mu} - (\vec{\mu} \cdot \vec{r})\vec{r}) = -\frac{\vec{\mu}}{\|\vec{r}\|^3} + \frac{3(\vec{\mu} \cdot \vec{r})\vec{r}}{\|\vec{r}\|^5}.$$

(c) No. The divergence test says a vector field must be a curl field if it satisfies the conditions of the test; it does not say the vector field cannot be a curl field if the vector field fails to satisfy the test.

13. (a)   Since curl grad $\psi = 0$ for any function ψ, curl$(\vec{A} + \text{grad}\,\psi) = \text{curl}\,\vec{A} + \text{curl grad}\,\psi = \text{curl}\,\vec{A} = \vec{B}$.
     (b)   We have

$$\text{div}(\vec{A} + \text{grad}\,\psi) = \text{div}\,\vec{A} + \text{div grad}\,\psi = \text{div}\,\vec{A} + \nabla^2 \psi.$$

Thus ψ should be chosen to satisfy the partial differential equation

$$\nabla^2 \psi = -\,\text{div}\,\vec{A}.$$

14. The idea is to let C_s be the circle of radius $1 - s$, so that C_0 is C and C_1 is the origin. In parameters,

$$x = (1 - s)\cos t, \quad y = (1 - s)\sin t, \quad z = 0.$$

15. Suppose that C has a parameterization

$$\vec{r} = \vec{r}(t), \qquad a \le t \le b.$$

Suppose that P has position vector $\vec{r}_0$. The straight line segment from $\vec{r}(t)$ to $\vec{r}_0$, parameterized by s, is

$$\vec{r} = (1 - s)\vec{r}(t) + s\vec{r}_0, \qquad 0 \le s \le 1.$$

Notice that this gives $\vec{r}(t)$ when $s = 0$, and $\vec{r}_0$ when $s = 1$. Thus the family of curves we want is

$$C_s : \vec{r} = (1 - s)\vec{r}(t) + s\vec{r}_0, \qquad a \le t \le b.$$

16. There is a family of curves C_s parameterized by

$$\vec{r} = \vec{r}_s(t), \qquad 0 \le s \le 1, a \le t \le b.$$

Define a parametric surface S by

$$\vec{r}(s, t) = \vec{r}_s(t), \qquad 0 \le s \le 1, a \le t \le b.$$

As the curves C_s move from $s = 0$ to $s = 1$, they trace out the surface S, which is smoothly parameterized by the rectangle $0 \le s \le 1, a \le t \le b$. The sides $t = a$ and $t = b$ of the rectangle are glued together, because $\vec{r}(s, a) = \vec{r}(s, b)$. The side $s = 0$ is the original curve C, and the side $s = 1$ is collapsed to the point P. Thus the boundary of S is C.

Solutions for Section 20.6

1. Suppose the cylinder goes from a to b along the z-axis, and has inner radius r_1 and outer radius r_2. The in cylindrical coordinates the region is parameterized by the rectangular region $a \le z \le b, r_1 \le r \le r_2$, and $0 \le \theta \le 2\pi$. The faces $\theta = 0$ and $\theta = 2\pi$ are pasted together.

2. (a)   By the Divergence Theorem

$$\frac{\int_{S_R} \vec{F} \cdot d\vec{A}}{\text{Volume of } U_R} = \frac{\int_{U_R} \text{div}\,\vec{F}\ dV}{\text{Volume of } U_R}.$$

Now, $m_R \le \text{div}\,\vec{F}(x, y, z) \le M_R$ for all (x, y, z) in U_R, so

$$\int_{U_R} m_R\ dV \le \int_{U_R} \text{div}\,\vec{F}(x, y, z)\ dV \le \int_{U_R} M_R\ dV,$$

that is,

$$m_R(\text{Volume of } U_R) \leq \int_{U_R} \text{div } \vec{F} \, dV \leq m_R(\text{Volume of } U_R).$$

Dividing through by the volume, we get the required inequalities.

(b) Taking the limit as R approaches zero we get

$$\lim_{R \to 0} m_R \leq \lim_{R \to 0} \frac{\int_{S_R} \vec{F} \cdot d\vec{A}}{\text{Volume of } U_R} \leq \lim_{R \to 0} M_R.$$

Since div $\vec{F}$ is continuous, the minimum and maximum values of div $\vec{F}$ on U_R approach its value at (x_0, y_0, z_0) as the sphere shrinks around (x_0, y_0, z_0), that is,

$$\lim_{R \to 0} m_R = \lim_{R \to 0} M_R = \text{div } \vec{F}(x_0, y_0, z_0).$$

So we must have

$$\lim_{R \to 0} \frac{\int_{S_R} \vec{F} \cdot d\vec{A}}{\text{Volume of } U_R} = \text{div } \vec{F}(x_0, y_0, z_0).$$

(c) The Divergence Theorem is true for cubes as well, so the argument in part (a) is the same as before, where m_R and M_R are now the minimum and maximum values of div $\vec{F}$ on the cube. Furthermore, by making R small enough we can ensure that all points in the cube are as close as we like to (x_0, y_0, z_0), so again $\lim_{R \to 0} m_R = \lim_{R \to 0} M_R = \text{div } \vec{F}(x_0, y_0, z_0)$. Thus the argument in part (b) works the same way as well.

3. (a) At a fixed point on A_1 with parameters (a, t_0, u_0), the curve

$$\vec{r} = \vec{r}(s, t_0, u_0),$$

parameterized by s, is a curve that starts outside V (when $s < a$), passes through A_1 (when $s = a$), and ends up in V (when $s > a$). Thus its tangent vector, $\frac{\partial \vec{r}}{\partial s}$, points into V.

(b) The surface A_1 is parameterized by

$$\vec{r} = \vec{r}(a, t, u), \qquad c \leq t \leq d, e \leq u \leq f.$$

Thus the vector $\frac{\partial \vec{r}}{\partial t} \times \frac{\partial \vec{r}}{\partial u}$ is normal to A_1. Since $\frac{\partial \vec{r}}{\partial s} \cdot \left(\frac{\partial \vec{r}}{\partial t} \times \frac{\partial \vec{r}}{\partial u} \right)$ is positive, the angle between the vectors $\frac{\partial \vec{r}}{\partial t}$ and $\frac{\partial \vec{r}}{\partial t} \times \frac{\partial \vec{r}}{\partial u}$ is an acute angle. Since $\frac{\partial \vec{r}}{\partial s}$ points into the surface, this means that $\frac{\partial \vec{r}}{\partial t} \times \frac{\partial \vec{r}}{\partial u}$ is the inward pointing normal. Thus its negative is the outward pointing normal.

(c) On A_2, the surface where $s = b$, the vector $\frac{\partial \vec{r}}{\partial s}$ points outward (because as s increases from values less than b to values greater than b, we move from points inside V to points outside V). Since $\frac{\partial \vec{r}}{\partial s} \cdot \left(\frac{\partial \vec{r}}{\partial t} \times \frac{\partial \vec{r}}{\partial u} \right)$ is positive, this means that $\frac{\partial \vec{r}}{\partial t} \times \frac{\partial \vec{r}}{\partial u}$ is an outward pointing normal.

4. In Problem 3 we computed the area elements $d\vec{A}$ on each face:

$$\text{On } A_2: \quad d\vec{A} = \frac{\partial \vec{r}}{\partial t} \times \frac{\partial \vec{r}}{\partial u} \, dt \, du$$

$$\text{On } A_3: \quad d\vec{A} = -\frac{\partial \vec{r}}{\partial u} \times \frac{\partial \vec{r}}{\partial s} \, du \, ds$$

$$\text{On } A_4: \quad d\vec{A} = \frac{\partial \vec{r}}{\partial u} \times \frac{\partial \vec{r}}{\partial s} \, du \, ds$$

$$\text{On } A_5: \quad d\vec{A} = -\frac{\partial \vec{r}}{\partial s} \times \frac{\partial \vec{r}}{\partial t} \, ds \, dt$$

$$\text{On } A_6: \quad d\vec{A} = \frac{\partial \vec{r}}{\partial s} \times \frac{\partial \vec{r}}{\partial t} \, ds \, dt.$$

Also,

$$\text{On } S_2: \quad d\vec{S} = \vec{i} \, dt \, du$$
$$\text{On } S_3: \quad d\vec{S} = -\vec{j} \, du \, ds$$
$$\text{On } S_4: \quad d\vec{S} = \vec{j} \, du \, ds$$
$$\text{On } S_5: \quad d\vec{S} = -\vec{k} \, ds \, dt$$
$$\text{On } S_6: \quad d\vec{S} = \vec{k} \, ds \, dt.$$

So

$$\text{On } A_2: \quad \vec{F} \cdot d\vec{A} = \vec{F} \cdot \frac{\partial \vec{r}}{\partial t} \times \frac{\partial \vec{r}}{\partial u} \, dt \, du = G_1 \, dt \, du = \vec{G} \cdot d\vec{S}$$

$$\text{On } A_3: \quad \vec{F} \cdot d\vec{A} = -\vec{F} \cdot \frac{\partial \vec{r}}{\partial u} \times \frac{\partial \vec{r}}{\partial s} \, du \, ds = -G_2 \, du \, ds = \vec{G} \cdot d\vec{S}$$

$$\text{On } A_4: \quad \vec{F} \cdot d\vec{A} = \vec{F} \cdot \frac{\partial \vec{r}}{\partial u} \times \frac{\partial \vec{r}}{\partial s} \, du \, ds = G_2 \, du \, ds = \vec{G} \cdot d\vec{S}$$

$$\text{On } A_5: \quad \vec{F} \cdot d\vec{A} = -\vec{F} \cdot \frac{\partial \vec{r}}{\partial s} \times \frac{\partial \vec{r}}{\partial t} \, ds \, dt = -G_3 \, ds \, dt = \vec{G} \cdot d\vec{S}$$

$$\text{On } A_6: \quad \vec{F} \cdot d\vec{A} = \vec{F} \cdot \frac{\partial \vec{r}}{\partial s} \times \frac{\partial \vec{r}}{\partial t} \, ds \, dt = G_3 \, ds \, dt = \vec{G} \cdot d\vec{S}.$$

Thus the integrals of boths sides are equal.

5. (a) Consider a small parallelepiped with one corner at the point $\vec{r}_0 = (x_0, y_0, z_0)$ and with edges $\vec{a} \, \Delta x$, $\vec{b} \, \Delta y$, and $\vec{c} \Delta z$, as in Figure 20.15.

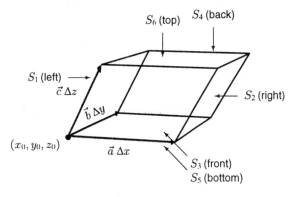

Figure 20.15

On S_1 (the left face of the parallelepiped shown in Figure 20.15) the outward area vector is $\Delta \vec{A} = -(\vec{b} \, \Delta y) \times (\vec{c} \, \Delta z) = -\vec{b} \times \vec{c} \, \Delta y \, \Delta z$. Assuming $\vec{F}$ is approximately constant on S_1, we have

$$\int_{S_1} \vec{F} \cdot d\vec{A} \approx \vec{F}(\vec{r}_0) \cdot \Delta \vec{A} = -\vec{F}(\vec{r}_0) \cdot (\vec{b} \times \vec{c}) \, \Delta y \, \Delta z.$$

On S_2, the outward normal points in the other direction, so

$$\int_{S_1} \vec{F} \cdot d\vec{A} \approx \vec{F}(\vec{r}_0 + \vec{a} \, \Delta x) \cdot (\vec{b} \times \vec{c}) \, \Delta y \, \Delta z.$$

Thus

$$\int_{S_1} \vec{F} \cdot d\vec{A} + \int_{S_2} \vec{F} \cdot d\vec{A} \approx (\vec{F}(\vec{r}_0 + \vec{a}\,\Delta x) \cdot (\vec{b} \times \vec{c}) - \vec{F}(\vec{r}_0) \cdot (\vec{b} \times \vec{c}))\,\Delta y\,\Delta z.$$

In other words, if f is the function $f(\vec{r}) = \vec{F}(\vec{r}) \cdot (\vec{b} \times \vec{c})$, then

$$\int_{S_1} \vec{F} \cdot d\vec{A} + \int_{S_2} \vec{F} \cdot d\vec{A} \approx (f(\vec{r}_0 + \Delta x \vec{a}) - f(\vec{r}_0)).$$

Now, by local linearity

$$f(\vec{r}_0 + \Delta x \vec{a}) \approx f(\vec{r}_0) + \operatorname{grad} f \cdot (\Delta x \vec{a}),$$

so

$$\int_{S_1} \vec{F} \cdot d\vec{A} + \int_{S_2} \vec{F} \cdot d\vec{A} \approx \operatorname{grad} f \cdot \vec{a}\,\Delta x\,\Delta y\,\Delta z = \operatorname{grad}(\vec{F} \cdot \vec{b} \times \vec{c}) \cdot \vec{a}\,\Delta x\,\Delta y\,\Delta z.$$

By an analogous argument, the contribution to the flux from S_3 and S_4 is approximately

$$\operatorname{grad}(\vec{F} \cdot \vec{c} \times \vec{a}) \cdot \vec{b}\,\Delta x\,\Delta y\,\Delta z$$

and the contribution to the flux from S_5 and S_6 is approximately

$$\operatorname{grad}(\vec{F} \cdot \vec{a} \times \vec{b}) \cdot \vec{c}\,\Delta x\,\Delta y\,\Delta z.$$

Thus, adding these contributions we have

Total flux through $S \approx (\operatorname{grad}(\vec{F} \cdot \vec{b} \times \vec{c}) \cdot \vec{a} + \operatorname{grad}(\vec{F} \cdot \vec{c} \times \vec{a}) \cdot \vec{b} + \operatorname{grad}(\vec{F} \cdot \vec{a} \times \vec{b}) \cdot \vec{c})\,\Delta x\,\Delta y\,\Delta z.$

The volume of the parallelepiped is $(\Delta x \vec{a}) \cdot ((\Delta y \vec{b}) \times (\Delta z \vec{c})) = \vec{a} \cdot (\vec{b} \times \vec{c})\,\Delta x\,\Delta y\,\Delta z$, so

Total flux through $S \approx \operatorname{div} \vec{F}\,$(Volume of S),

that is

$$(\operatorname{grad}(\vec{F} \cdot \vec{b} \times \vec{c}) \cdot \vec{a} + \operatorname{grad}(\vec{F} \cdot \vec{c} \times \vec{a}) \cdot \vec{b} + \operatorname{grad}(\vec{F} \cdot \vec{a} \times \vec{b}) \cdot \vec{c})\,\Delta x\,\Delta y\,\Delta z$$
$$\approx \operatorname{div} \vec{F}\,(\vec{a} \cdot (\vec{b} \times \vec{c}))\,\Delta x\,\Delta y\,\Delta z.$$

This is the formula which we are trying to prove holds, multiplied on both sides by $\Delta x\,\Delta y\,\Delta z$.

(b) The triple product $\vec{F} \cdot \vec{b} \times \vec{c}$ expands into 6 terms; therefore each component of its gradient has 6 terms, so $\operatorname{grad}(\vec{F} \cdot \vec{b} \times \vec{c})$ has 18 terms; there are three terms like this on the left-hand side, so the left-hand side has 54 terms.

(c) Since we are only interested in F_1, we only need to look at the first term in each of the dot products $\vec{F} \cdot \vec{b} \times \vec{c}$, $\vec{F} \cdot \vec{c} \times \vec{a}$, and $\vec{F} \cdot \vec{a} \times \vec{b}$. These terms are

$$F_1(b_2 c_3 - b_3 c_2), \quad F_1(c_2 a_3 - c_3 a_2), \quad F_1(a_2 b_3 - a_3 b_2).$$

Then, since we are only interested in the x partial derivative, we only need to look at the first term in each of the dot products $\operatorname{grad}(\cdots) \cdot \vec{a}$, $\operatorname{grad}(\cdots) \cdot \vec{b}$, and $\operatorname{grad}(\cdots) \cdot \vec{c}$. So we get 6 terms,

$$\frac{\partial F_1}{\partial x}((b_2 c_3 - b_3 c_2)a_1 + (c_2 a_3 - c_3 a_2)b_1 + (a_2 b_3 - a_3 b_2)c_1)$$
$$= \frac{\partial F_1}{\partial x}(b_2 c_3 a_1 - b_3 c_2 a_1 + c_2 a_3 b_1 - c_3 a_2 b a_1 + a_2 b_3 c_1 - a_2 b_3 c_1) = \frac{\partial F_1}{\partial x}\vec{a} \cdot \vec{b} \times \vec{c}.$$

(d)   This starts out the same way, but this time we look at the second component in each gradient, so we get

$$\frac{\partial F_1}{\partial y}((b_2c_3 - b_3c_2)a_2 + (c_2a_3 - c_3a_2)b_2 + (a_2b_3 - a_3b_2)c_2)$$

$$= \frac{\partial F_1}{\partial y}(b_2c_3a_2 - b_3c_2a_2 + c_2a_3b_2 - c_3a_2b_2 + a_2b_3c_2 - a_2b_3c_2) = 0.$$

(e)   The terms involving $\frac{\partial F_2}{\partial y}$ add up to $\frac{\partial F_2}{\partial y}(\vec{a} \cdot \vec{b} \times \vec{c})$ and those involving $\frac{\partial F_3}{\partial z}$ add up to $\frac{\partial F_3}{\partial z}(\vec{a} \cdot \vec{b} \times \vec{c})$. All the rest have some cancellation which makes them zero. So the whole thing works out to

$$\left(\frac{\partial F_1}{\partial x} + \frac{\partial F_2}{\partial y} + \frac{\partial F_3}{\partial z} \right) \vec{a} \cdot \vec{b} \times \vec{c} = \text{div}\,\vec{F}\,(\vec{a} \cdot \vec{b} \times \vec{c}).$$

6.   (a)   Differentiating the expressions for G_1, G_2 and G_3 we have

$$\frac{\partial G_1}{\partial s} = \frac{\partial \vec{F}}{\partial s} \cdot \frac{\partial \vec{r}}{\partial t} \times \frac{\partial \vec{r}}{\partial u} + \vec{F} \cdot \frac{\partial^2 \vec{r}}{\partial s \partial t} \times \frac{\partial \vec{r}}{\partial u} + \vec{F} \cdot \frac{\partial \vec{r}}{\partial t} \times \frac{\partial^2 \vec{r}}{\partial s \partial u}$$

$$\frac{\partial G_2}{\partial t} = \frac{\partial \vec{F}}{\partial t} \cdot \frac{\partial \vec{r}}{\partial u} \times \frac{\partial \vec{r}}{\partial s} + \vec{F} \cdot \frac{\partial^2 \vec{r}}{\partial t \partial u} \times \frac{\partial \vec{r}}{\partial s} + \vec{F} \cdot \frac{\partial \vec{r}}{\partial u} \times \frac{\partial^2 \vec{r}}{\partial t \partial s}$$

$$\frac{\partial G_3}{\partial u} = \frac{\partial \vec{F}}{\partial u} \cdot \frac{\partial \vec{r}}{\partial s} \times \frac{\partial \vec{r}}{\partial t} + \vec{F} \cdot \frac{\partial^2 \vec{r}}{\partial u \partial s} \times \frac{\partial \vec{r}}{\partial t} + \vec{F} \cdot \frac{\partial \vec{r}}{\partial s} \times \frac{\partial^2 \vec{r}}{\partial u \partial t}.$$

When we add up these equations the terms involving the second order mixed partial derivatives all cancel (since $\vec{r} = \vec{r}\,(s, t, u)$, is a smooth change of variables), and we get

$$\frac{\partial G_1}{\partial s} + \frac{\partial G_2}{\partial t} + \frac{\partial G_3}{\partial u} = \frac{\partial \vec{F}}{\partial s} \cdot \frac{\partial \vec{r}}{\partial t} \times \frac{\partial \vec{r}}{\partial u} + \frac{\partial \vec{F}}{\partial t} \cdot \frac{\partial \vec{r}}{\partial u} \times \frac{\partial \vec{r}}{\partial s} + \frac{\partial \vec{F}}{\partial u} \cdot \frac{\partial \vec{r}}{\partial s} \times \frac{\partial \vec{r}}{\partial t}.$$

(b)   Evaluating the expression in part (a) at $\vec{r} = \vec{r}_0$, we get

$$\left. \frac{\partial G_1}{\partial s} + \frac{\partial G_2}{\partial t} + \frac{\partial G_3}{\partial u} \right|_{\vec{r} = \vec{r}_0} =$$

$$\frac{\partial \vec{F}}{\partial s} \cdot \vec{b} \times \vec{c} + \frac{\partial \vec{F}}{\partial t} \cdot \vec{c} \times \vec{a} + \frac{\partial \vec{F}}{\partial u} \cdot \vec{a} \times \vec{b}.$$

Now

$$\frac{\partial \vec{F}}{\partial s} \cdot \vec{b} \times \vec{c} = \frac{\partial}{\partial s}(\vec{F} \cdot \vec{b} \times \vec{c}) = \text{grad}(\vec{F} \cdot \vec{b} \times \vec{c}) \cdot \frac{\partial \vec{r}}{\partial s}$$

$$\frac{\partial \vec{F}}{\partial t} \cdot \vec{c} \times \vec{a} = \frac{\partial}{\partial t}(\vec{F} \cdot \vec{c} \times \vec{a}) = \text{grad}(\vec{F} \cdot \vec{c} \times \vec{a}) \cdot \frac{\partial \vec{r}}{\partial t}$$

$$\frac{\partial \vec{F}}{\partial u} \cdot \vec{a} \times \vec{b} = \frac{\partial}{\partial u}(\vec{F} \cdot \vec{a} \times \vec{b}) = \text{grad}(\vec{F} \cdot \vec{a} \times \vec{b}) \cdot \frac{\partial \vec{r}}{\partial u}.$$

Adding these all up and evaluating at $\vec{r} = \vec{r}_0$, we get

$$\left. \left(\frac{\partial G_1}{\partial s} + \frac{\partial G_2}{\partial t} + \frac{\partial G_3}{\partial u} \right) \right|_{\vec{r} = \vec{r}_0} = \text{grad}(\vec{F} \cdot \vec{b} \times \vec{c}) \cdot \vec{a} + \text{grad}(\vec{F} \cdot \vec{c} \times \vec{a}) \cdot \vec{b} + \text{grad}(\vec{F} \cdot \vec{a} \times \vec{b}) \cdot \vec{c}.$$

(c) By Problem 5, the right hand side of the previous equation is $(\vec{a} \cdot \vec{b} \times \vec{c}) \operatorname{div} \vec{F}$. Now

$$\vec{a} \cdot \vec{b} \times \vec{c} = \frac{\partial \vec{r}}{\partial s} \cdot \frac{\partial \vec{r}}{\partial t} \times \frac{\partial \vec{r}}{\partial u}\bigg|_{\vec{r}=\vec{r}_0} = \left|\frac{\partial(x,y,z)}{\partial(s,t,u)}\right|_{\vec{r}=\vec{r}_0}.$$

So we have

$$\left(\frac{\partial G_1}{\partial s} + \frac{\partial G_2}{\partial t} + \frac{\partial G_3}{\partial u}\right)\bigg|_{\vec{r}=\vec{r}_0} = \left|\frac{\partial(x,y,z)}{\partial(s,t,u)}\right|_{\vec{r}=\vec{r}_0} \operatorname{div} \vec{F}.$$

Since $\vec{r}_0$ is arbitrary, this equation is true in general.

7. (a) Differentiating the expressions for G_1 and G_2, we have

$$\frac{\partial G_1}{\partial t} = \frac{\partial \vec{F}}{\partial t} \cdot \frac{\partial \vec{r}}{\partial s} + \vec{F} \cdot \frac{\partial^2 \vec{r}}{\partial t \partial s}$$

$$\frac{\partial G_2}{\partial s} = \frac{\partial \vec{F}}{\partial s} \cdot \frac{\partial \vec{r}}{\partial t} + \vec{F} \cdot \frac{\partial^2 \vec{r}}{\partial s \partial t}$$

When we subtract these two equations, the mixed partials cancel, so

$$\frac{\partial G_1}{\partial t} - \frac{\partial G_2}{\partial s} = \frac{\partial \vec{F}}{\partial t} \cdot \frac{\partial \vec{r}}{\partial s} - \frac{\partial \vec{F}}{\partial s} \cdot \frac{\partial \vec{r}}{\partial t}.$$

Now we evaluate this expression at $\vec{r}_0$:

$$\frac{\partial G_1}{\partial t}(\vec{r}_0) - \frac{\partial G_2}{\partial s}(\vec{r}_0) = \frac{\partial \vec{F}}{\partial t}(\vec{r}_0) \cdot \vec{a} - \frac{\partial \vec{F}}{\partial s}(\vec{r}_0) \cdot \vec{b}.$$

Now,

$$\frac{\partial \vec{F}}{\partial t} \cdot \vec{a} = \frac{\partial}{\partial t}(\vec{F} \cdot \vec{a}) = \operatorname{grad}(\vec{F} \cdot \vec{a}) \cdot \frac{\partial \vec{r}}{\partial t}$$

$$\frac{\partial \vec{F}}{\partial s} \cdot \vec{b} = \frac{\partial}{\partial s}(\vec{F} \cdot \vec{b}) = \operatorname{grad}(\vec{F} \cdot \vec{b}) \cdot \frac{\partial \vec{r}}{\partial s}.$$

Evaluating both sides at $\vec{r} = \vec{r}_0$, we get

$$\frac{\partial \vec{F}}{\partial t}(\vec{r}_0) \cdot \vec{a} = \operatorname{grad}(\vec{F} \cdot \vec{a}) \cdot \vec{b}$$

$$\frac{\partial \vec{F}}{\partial s}(\vec{r}_0) \cdot \vec{b} = \operatorname{grad}(\vec{F} \cdot \vec{b}) \cdot \vec{a}$$

Hence

$$\frac{\partial G_1}{\partial t}(\vec{r}_0) - \frac{\partial G_2}{\partial s}(\vec{r}_0) = \operatorname{grad}(\vec{F} \cdot \vec{a}) \cdot \vec{b} - \operatorname{grad}(\vec{F} \cdot \vec{b}) \cdot \vec{a}.$$

(b) Problem 27 on page 507 says

$$\operatorname{grad}(\vec{F} \cdot \vec{a}) \cdot \vec{b} - \operatorname{grad}(\vec{F} \cdot \vec{b}) \cdot \vec{a} = \operatorname{curl} \vec{F} \cdot (\vec{b} \times \vec{a}) = \operatorname{curl} \vec{F}(\vec{r}_0) \cdot \left(\frac{\partial \vec{r}}{\partial t}(\vec{r}_0) \times \frac{\partial \vec{r}}{\partial s}(\vec{r}_0)\right).$$

Thus, from part (b)

$$\frac{\partial G_1}{\partial t}(\vec{r}_0) - \frac{\partial G_2}{\partial s}(\vec{r}_0) = \text{curl}\,\vec{F}\,(\vec{r}_0) \cdot \left(\frac{\partial \vec{r}}{\partial t}(\vec{r}_0) \times \frac{\partial \vec{r}}{\partial s}(\vec{r}_0)\right).$$

Since $\vec{r}_0$ can be any point, this means

$$\frac{\partial G_1}{\partial t} - \frac{\partial G_2}{\partial s} = \text{curl}\,\vec{F} \cdot \left(\frac{\partial \vec{r}}{\partial t} \times \frac{\partial \vec{r}}{\partial s}\right).$$

Solutions for Chapter 20 Review

1. Figure 20.16 shows a two dimensional cross-section of the vector field $\vec{v} = -2\vec{r}$. The vector field points radially inwards, so if we take S to be a sphere of radius R centered at the origin, oriented outward, we have

$$\vec{v} \cdot \Delta\vec{A} = -2R\,\|\Delta\vec{A}\|,$$

for a small area vector $\Delta\vec{A}$ on the sphere. Therefore,

$$\int_S \vec{v} \cdot d\vec{A} = \int_S -2R\,\|d\vec{A}\| = -2R(\text{Surface area of sphere}) = -2R(4\pi R^2) = -8\pi R^3.$$

Thus, we find that

$$\text{div}\,\vec{v}\,(0,0,0) = \lim_{\text{vol}\to 0}\left(\frac{\int_S \vec{v} \cdot d\vec{A}}{\text{Volume of sphere}}\right) = \lim_{R\to 0}\left(\frac{-8\pi R^3}{\frac{4}{3}\pi R^3}\right) = -6.$$

Notice that the divergence is negative. This is what you would expect, since the vector field represents an inward flow at the origin.

Since $\vec{v} = -2\vec{r} = -2x\vec{i} - 2y\vec{j} - 2z\vec{k}$, the coordinate definition give

$$\text{div}\,\vec{v} = \frac{\partial}{\partial x}(-2x) + \frac{\partial}{\partial y}(-2y) + \frac{\partial}{\partial z}(-2z) = -2 - 2 - 2 = -6.$$

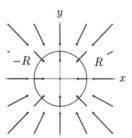

Figure 20.16: The vector field $\vec{v} = -2\vec{r}$

2. No, because the surface S is not a closed surface.

3. Since $\operatorname{div}\vec{r} = \operatorname{div}(x\vec{i} + y\vec{j} + z\vec{k}) = 3$, applying the Divergence Theorem to the vector field $\vec{F} = \vec{r}$ gives

$$\int_S \vec{r} \cdot d\vec{A} = \int_V 3\,dV = 3\int_V dV = 3V.$$

Thus $\frac{1}{3}\int_S \vec{r} \cdot d\vec{A} = V$.

4. We consider a sphere of radius R centered at the origin and compute the flux of $\vec{r}$ through its surface. Since $\vec{r}$ and $d\vec{A}$ both point radially outward, $\vec{r} \cdot d\vec{A} = \|\vec{r}\|\|d\vec{A}\| = R\,dA$ on the surface of the sphere, so

$$\int_S \vec{F} \cdot d\vec{A} = \int_S R\,dA = R\int_S dA = R(\text{Surface area of a sphere}) = R(4\pi R^2) = 4\pi R^3.$$

Therefore, volume of sphere $= \frac{1}{3}\int_S \vec{F} \cdot d\vec{A} = \frac{4}{3}\pi R^3$.

5. Since $\vec{r}$ is parallel to the slanted edges of the cone, the flux of $\vec{r}$ through the surface is all through the base (See Figure 20.17). On the base, $z = h$, and the normal is upward, so

$$\int_S \vec{r} \cdot d\vec{A} = \int_{\text{base}} (x\vec{i} + y\vec{j} + z\vec{k}) \cdot (dA\vec{k})$$
$$= \int_{\text{base}} h\,dA$$
$$= h(\text{Area of base})$$
$$= h(\pi b^2)$$

Thus

$$V = \frac{1}{3}\int_S \vec{r} \cdot d\vec{A} = \frac{\pi}{3}b^2 h.$$

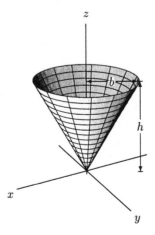

Figure 20.17

6. (a) On S_1, $x = a$ and normal is in negative x-direction, so

$$\vec{F} \cdot \Delta\vec{A} = (2a\vec{i} - 3y\vec{j} + 5z\vec{k}) \cdot (-\Delta A\vec{i}) = -2a\Delta A.$$

Thus

$$\int_{S_1} \vec{F} \cdot d\vec{A} = \int_{S_1} -2a \, dA = -2a(\text{Area of } S_1) = -2aw^2.$$

On S_2, $x = a + w$ and normal is in positive x-direction, so

$$\vec{F} \cdot \Delta\vec{A} = (2(a+w)\vec{i} - 3y\vec{j} + 5z\vec{k}) \cdot (\Delta A \vec{i}) = 2(a+w)\Delta A.$$

Thus

$$\int_{S_2} \vec{F} \cdot d\vec{A} = \int_{S_2} 2(a+w) \, dA = 2(a+w)(\text{Area of } S_2) = 2(a+w)w^2$$

Calculating the flux through the other sides similarly, we get

Total flux

$$= \int_{S_1} \vec{F} \cdot d\vec{A} + \int_{S_2} \vec{F} \cdot d\vec{A} + \int_{S_3} \vec{F} \cdot d\vec{A} + \int_{S_4} \vec{F} \cdot d\vec{A} + \int_{S_5} \vec{F} \cdot d\vec{A} + \int_{S_6} \vec{F} \cdot d\vec{A}$$

$$= -2aw^2 + 2(a+w)w^2 + 3bw^2 - 3(b+w)w^2 - 5cw^2 + 5(c+w)w^2$$

$$= (2w - 3w + 5w)w^2 = 4w^3.$$

(b)   To find div $\vec{F}$ at the point (a, b, c), let the box shrink to the point by letting $w \to 0$. Then

$$\text{div } \vec{F} = \lim_{w \to 0} \left(\frac{\text{Flux through box}}{\text{Volume of box}} \right)$$

$$= \lim_{w \to 0} \left(\frac{4w^3}{w^3} \right) = 4.$$

(c)

$$\text{div } \vec{F} = \frac{\partial}{\partial x}(2x) + \frac{\partial}{\partial y}(-3y) + \frac{\partial}{\partial z}(5z) = 2 - 3 + 5 = 4.$$

7.   Using flux: On S_1, $x = a$ and normal is in negative x-direction, so

$$\vec{F} \cdot \Delta\vec{A} = ((3a+2)\vec{i} + 4a\vec{j} + (5a+1)\vec{k}) \cdot (-\Delta A \vec{i}) = -(3a+2)\Delta A$$

Thus

$$\int_{S_1} \vec{F} \cdot d\vec{A} = \int_{S_1} -(3a+2) \, dA = -(3a+2)(\text{Area of } S_1) = -(3a+2)w^2.$$

On S_2, $x = a + w$ and normal is in the positive x-direction, so

$$\vec{F} \cdot \Delta\vec{A} = [(3(a+w)+2)\vec{i} + 4(a+w)\vec{j} + (5(a+w)+1)\vec{k}] \cdot (\Delta A \vec{i}) = (3a + 3w + 2)\Delta A$$

Thus

$$\int_{S_2} \vec{F} \cdot d\vec{A} = \int_{S_2} (3a + 3w + 2) \, dA = (3a + 3w + 2)(\text{Area of } S_2) = (3a + 3w + 2)w^2.$$

Next, we have $\int_{S_3} \vec{F} \cdot d\vec{A} = \int_{S_3} -4x \, dA$ and $\int_{S_4} \vec{F} \cdot d\vec{A} = \int_{S_4} 4x \, dA$. Since these two are integrated over the same region in the xz-plane, the two integrals cancel. Similarly, $\int_{S_5} \vec{F} \cdot d\vec{A} = \int_{S_5} -(5x+1) \, d\vec{A}$ cancels out $\int_{S_6} \vec{F} \cdot d\vec{A} = \int_{S_6} (5x + 1)dA$. Therefore,

$$
\begin{aligned}
\text{Total flux} \\
&= \int_{S_1} \vec{F} \cdot d\vec{A} + \int_{S_2} \vec{F} \cdot d\vec{A} + \int_{S_3} \vec{F} \cdot d\vec{A} + \int_{S_4} \vec{F} \cdot d\vec{A} + \int_{S_5} \vec{F} \cdot d\vec{A} + \int_{S_6} \vec{F} \cdot d\vec{A} \\
&= -(3a + 2)w^2 + (3a + 3w + 2)w^2 + \int_{S_3} -4x\,dA + \int_{S_4} 4x\,dA \\
&\quad + \int_{S_5} -(5x + 1)dA + \int_{S_6} (5x + 1)dA \\
&= 3w^3.
\end{aligned}
$$

To find div $\vec{F}$ at the point (a, b, c), let the box shrink to the point by letting $w \to 0$. Then

$$
\begin{aligned}
\text{div}\,\vec{F} &= \lim_{w \to 0} \left(\frac{\text{Flux through box}}{\text{Volume of box}} \right) \\
&= \lim_{w \to 0} \left(\frac{3w^3}{w^3} \right) = 3.
\end{aligned}
$$

Using partial derivatives:

$$
\text{div}\,\vec{F} = \frac{\partial}{\partial x}(3x + 2) + \frac{\partial}{\partial y}(4x) + \frac{\partial}{\partial z}(5x + 1) = 3
$$

8. (a) We will compute separately the flux of the vector field $\vec{F} = x^3\vec{i} + 2y\vec{j} + 3\vec{k}$ through each of the six faces of the cube.

The face S_I where $x = 1$, which has normal vector $\vec{i}$. Only the $\vec{i}$ component $x^3\vec{i} = \vec{i}$ of $\vec{F}$ has flux through S_I.

$$
\int_{S_I} \vec{F} \cdot d\vec{A} = \int_{S_I} \vec{i} \cdot d\vec{A} = \|\vec{i}\|(\text{area of } S_I) = 4.
$$

The face S_{II} where $x = -1$, which has normal vector $-\vec{i}$. Only the $\vec{i}$ component $x^3\vec{i} = -\vec{i}$ of $\vec{F}$ has flux through S_{II}.

$$
\int_{S_{II}} \vec{F} \cdot d\vec{A} = \int_{S_{II}} -\vec{i} \cdot d\vec{A} = \| -\vec{i}\|(\text{area of } S_{II}) = 4.
$$

The face S_{III} where $y = 1$, which has normal vector $\vec{j}$. Only the $\vec{j}$ component $2y\vec{j} = 2\vec{j}$ of $\vec{F}$ has flux through S_{III}.

$$
\int_{S_{III}} \vec{F} \cdot d\vec{A} = \int_{S_{III}} 2\vec{j} \cdot d\vec{A} = \|2\vec{j}\|(\text{area of } S_{III}) = 8.
$$

The face S_{IV} where $y = -1$, which has normal vector $-\vec{j}$. Only the $\vec{j}$ component $2y\vec{j} = -2\vec{j}$ of $\vec{F}$ has flux through S_{IV}.

$$
\int_{S_{IV}} \vec{F} \cdot d\vec{A} = \int_{S_{IV}} -2\vec{j} \cdot d\vec{A} = \| -2\vec{j}\|(\text{area of } S_{IV}) = 8.
$$

The face S_V where $z = 1$, which has normal vector $\vec{k}$. Only the $\vec{k}$ component $3\vec{k}$ of $\vec{F}$ has flux through S_V.

$$\int_{S_V} \vec{F} \cdot d\vec{A} = \int_{S_V} 3\vec{k} \cdot d\vec{A} = \|3\vec{k}\|(\text{area of } S_V) = 12.$$

The face S_{VI} where $z = -1$, which has normal vector $-\vec{k}$. Only the $\vec{k}$ component $3\vec{k}$ of $\vec{F}$ has flux through S_{VI}.

$$\int_{S_{VI}} \vec{F} \cdot d\vec{A} = \int_{S_{VI}} 3\vec{k} \cdot d\vec{A} = -\|3\vec{k}\|(\text{area of } S_{VI}) = -12.$$

$$(\text{Total flux through } S) = 4 + 4 + 8 + 8 + 12 - 12 = 24.$$

(b)   Since S is a closed surface the Divergence Theorem applies. Since $\text{div}\vec{F} = 3x^2 + 2$,

$$\int_S \vec{F} \cdot d\vec{A} = \int_{x=-1}^{1} \int_{y=-1}^{1} \int_{z=-1}^{1} (3x^2 + 2)dzdydx = 24.$$

9.   True. $\text{curl}\,\vec{F}$ is a vector whose value depends on the point at which it is calculated.

10. False. Left side is a vector, right side is a scalar.

11. True.

$$\text{div}(\vec{F} + \vec{G}) = \frac{\partial(F_1 + G_1)}{\partial x} + \frac{\partial(F_2 + G_2)}{\partial y} + \frac{\partial(F_3 + G_3)}{\partial z}$$
$$= \frac{\partial F_1}{\partial x} + \frac{\partial F_2}{\partial y} + \frac{\partial F_3}{\partial z} + \frac{\partial G_1}{\partial x} + \frac{\partial G_2}{\partial y} + \frac{\partial G_3}{\partial z}$$
$$= \text{div}\,\vec{F} + \text{div}\,\vec{G}.$$

12. False. Let's compare the x–components of each side of the equation. The x–component of $\text{grad}(\vec{F} \cdot \vec{G})$ is given by

$$(\text{grad}(\vec{F} \cdot \vec{G}))_1 = \frac{\partial(F_1 G_1 + F_2 G_2 + F_3 G_3)}{\partial x}$$
$$= \frac{\partial F_1}{\partial x}G_1 + F_1\frac{\partial G_1}{\partial x} + \frac{\partial F_2}{\partial x}G_2 + F_2\frac{\partial G_2}{\partial x} + \frac{\partial F_3}{\partial x}G_3 + F_3\frac{\partial G_3}{\partial x}.$$

However, the x–component of $\vec{F} \cdot (\text{div}\,\vec{G}) + (\text{div}\,\vec{F}) \cdot \vec{G}$ is

$$(\vec{F}(\text{div}\,\vec{G}) + (\text{div}\,\vec{F})\vec{G})_1 = F_1(\text{div}\,\vec{G}) + (\text{div}\,\vec{F})G_1$$
$$= F_1(\frac{\partial G_1}{\partial x} + \frac{\partial G_2}{\partial y} + \frac{\partial G_3}{\partial z}) + (\frac{\partial F_1}{\partial x} + \frac{\partial F_2}{\partial y} + \frac{\partial F_3}{\partial z})G_1.$$

These two x–components are different and therefore

$$\text{grad}(\vec{F} \cdot \vec{G}) \neq \vec{F}(\text{div}\,\vec{G}) + (\text{div}\,\vec{F})\vec{G}.$$

13. True. We calculate the x–components for each side of the equation:

$$(\text{curl}(f\vec{G}))_1 = \frac{\partial(fG_3)}{\partial y} - \frac{\partial(fG_2)}{\partial z}$$

$$= \frac{\partial f}{\partial y}G_3 + f\frac{\partial G_3}{\partial y} - \frac{\partial f}{\partial z}G_2 - f\frac{\partial G_2}{\partial z}$$

$$= (\frac{\partial f}{\partial y}G_3 - \frac{\partial f}{\partial z}G_2) + f(\frac{\partial G_3}{\partial y} - \frac{\partial G_2}{\partial z})$$

$$= ((\text{grad } f) \times \vec{G})_1 + (f(\text{curl } \vec{G}))_1.$$

Computations for the other two components are similar, so

$$\text{curl}(f\vec{G}) = (\text{grad } f) \times \vec{G} + f \cdot (\text{curl } \vec{G}).$$

14. True. div $\vec{F}$ is a scalar whose value depends on the point at which it is calculated.

15. False. You need a closed surface to use the divergence theorem.

16. True. $\vec{F}$ is rotating around the y axis, so by the right hand rule curl $\vec{F}$ has a positive y component. Therefore taking the dot product of curl $\vec{F}$ and $\vec{j}$ will give a positive number.

17. The flux of $\vec{E}$ through a small sphere of radius R around the point marked P is negative, because all the arrows are pointing into the sphere. The divergence at P is

$$\text{div } \vec{E}(P) = \lim_{\text{vol}\to 0} \left(\frac{\int_S \vec{E} \cdot d\vec{A}}{\text{Volume of sphere}} \right) = \lim_{R\to 0} \left(\frac{\text{Negative number}}{\frac{4}{3}\pi R^3} \right) \le 0.$$

By a similar argument, the divergence at Q must be positive or zero.

18. We have:

$$\vec{F}(x, y, z) = \frac{x\vec{i} + y\vec{j} + z\vec{k}}{(x^2 + y^2 + z^2)^{3/2}}$$

Calculating the flux of $\vec{F}$ through the ellipsoid directly would be difficult. However, since div $\vec{F} = 0$, we can replace the ellipsoid by a sphere. Except at the origin, we have div $\vec{F} = 0$. Let T be the surface of a sphere centered at the origin inside the ellipsoid S, and let W be the region between S and T. Suppose both S and T are oriented away from the origin. By the Divergence Theorem, we have

$$\text{Flux out of } W = \int_W \text{div } \vec{F} \cdot dV = 0,$$

and therefore

$$\text{Flux out of } W = (\text{Flux out} - \text{Flux in}) = \int_S \vec{F} \cdot d\vec{A} - \int_T \vec{F} \cdot d\vec{A} = 0.$$

Thus, we have

$$\int_S \vec{F} \cdot d\vec{A} = \int_T \vec{F} \cdot d\vec{A} = (\text{Magnitude of } \vec{F} \text{ on sphere}) \cdot (\text{Surface area})$$

$$= (\frac{1}{\text{radius}^2}) \cdot (4\pi \cdot \text{radius})$$

$$= 4\pi.$$

19. To show that the force field is irrotational we must show that its curl is zero. Let us do this in Cartesian coordinates:

$$\vec{F} = f(r)\vec{r} = f(\sqrt{x^2 + y^2 + z^2})(x\vec{i} + y\vec{j} + z\vec{k})$$

The third component of curl $\vec{F}$ is

$$\frac{\partial}{\partial x}(f(\sqrt{x^2 + y^2 + z^2})y) - \frac{\partial}{\partial y}(f(\sqrt{x^2 + y^2 + z^2})x)$$

$$= f'(\sqrt{x^2 + y^2 + z^2}) \cdot \frac{2xy}{2\sqrt{x^2 + y^2 + z^2}} - f'(\sqrt{x^2 + y^2 + z^2}) \cdot \frac{2xy}{2\sqrt{x^2 + y^2 + z^2}}$$

$$= 0.$$

A similar computation shows that the other components of curl $\vec{F}$ are 0 too.

20. (a) Taking partial derivatives of $\vec{E}$ gives

$$\frac{\partial E_1}{\partial x} = \frac{\partial}{\partial x}[qx(x^2 + y^2 + z^2)^{-3/2}] = q[(x^2 + y^2 + z^2)^{-3/2} + x(-3/2)(2x)(x^2 + y^2 + z^2)^{-5/2}]$$

$$= q(y^2 + z^2 - 2x^2)(x^2 + y^2 + z^2)^{-5/2}.$$

Similarly,

$$\frac{\partial E_2}{\partial x} = q(x^2 + z^2 - 2y^2)(x^2 + y^2 + z^2)^{-5/2}$$

$$\frac{\partial E_3}{\partial x} = q(x^2 + y^2 - 2z^2)(x^2 + y^2 + z^2)^{-5/2}.$$

Summing, we obtain div $\vec{E} = 0$.

(b) Since on the surface of the sphere, the vector field $\vec{E}$ and the area vector $\Delta\vec{A}$ are parallel,

$$\vec{E} \cdot \Delta\vec{A} = \|\vec{E}\|\|\Delta\vec{A}\|.$$

Now, on the surface of a sphere of radius a,

$$\|\vec{E}\| = \frac{q\|\vec{r}\|}{\|\vec{r}\|^3} = \frac{q}{a^2}.$$

Thus,

$$\int_{S_a} \vec{E} \cdot d\vec{A} = \int \frac{q}{a^2}\|d\vec{A}\| = \frac{q}{a^2} \cdot \text{Surface area of sphere} = \frac{q}{a^2} \cdot 4\pi a^2 = 4\pi q.$$

(c) It is not possible to apply the Divergence Theorem in part (b) since $\vec{E}$ is not defined at the origin (which lies inside the region of space bounded by S_a), and the Divergence Theorem requires that the vector field be defined everywhere inside S.

(d) Let R be the solid region lying between a small sphere S_a, centered at the origin, and the surface S. Applying the Divergence Theorem and the result of part (a), we get:

$$0 = \int_R \text{div} \vec{E} \, dV = \int_{S_a} \vec{E} \cdot d\vec{A} + \int_S \vec{E} \cdot d\vec{A},$$

where S is oriented with the outward normal vector, and S_a with the inward normal vector (since this is "outward" with respect to the region R). Since

$$\int_{S_a, \text{ inward}} \vec{E} \cdot d\vec{A} = -\int_{S_a, \text{ outward}} \vec{E} \cdot d\vec{A},$$

the result of part (b) yields

$$\int_S \vec{E} \cdot d\vec{A} = 4\pi q.$$

[Note: It is legitimate to apply the Divergence Theorem to the region R since the vector field $\vec{E}$ is defined everywhere in R.]

21. Check that div $\vec{E} = 0$ by taking partial derivatives. For instance,

$$\frac{\partial E_1}{\partial x} = \frac{\partial}{\partial x}[q(x - x_0)[(x - x_0)^2 + (y - y_0)^2 + (z - z_0)^2]^{-3/2}]$$
$$= q[(y - y_0)^2 + (z - z_0)^2 - 2(x - x_0)^2][(x - x_0)^2 + (y - y_0)^2 + (z - z_0)^2]^{-5/2}$$

and similarly,

$$\frac{\partial E_2}{\partial y} = q[(x - x_0)^2 + (z - z_0)^2 - 2(y - y_0)^2][(x - x_0)^2 + (y - y_0)^2 + (z - z_0)^2]^{-5/2}$$
$$\frac{\partial E_3}{\partial z} = q[(x - x_0)^2 + (y - y_0)^2 - 2(z - z_0)^2][(x - x_0)^2 + (y - y_0)^2 + (z - z_0)^2]^{-5/2}.$$

Therefore,

$$\frac{\partial E_1}{\partial x} + \frac{\partial E_2}{\partial y} + \frac{\partial E_3}{\partial z} = 0.$$

The vector field $\vec{E}$ is defined everywhere but at the point with position vector $\vec{r}_0$. If this point lies outside the surface S, the Divergence Theorem can be applied to the region R enclosed by S, yielding:

$$\int_S \vec{E} \cdot d\vec{A} = \int_R \text{div}\,\vec{E}\, dV = 0.$$

If the charge q is located inside S, consider a small sphere S_a centered at q and contained in R. The Divergence Theorem for the region R' between the two spheres yields:

$$\int_S \vec{E} \cdot d\vec{A} + \int_{S_a} \vec{E} \cdot d\vec{A} = \int_{R'} \text{div}\,\vec{E}\, dV = 0.$$

In this formula, the Divergence Theorem requires S to be given the outward orientation, and S_a the inward orientation. To compute $\int_{S_a} \vec{E} \cdot d\vec{A}$, we use the fact that on the surface of the sphere, $\vec{E}$ and $\Delta\vec{A}$ are parallel and in opposite directions, so

$$\vec{E} \cdot \Delta\vec{A} = -\|\vec{E}\|\|\Delta\vec{A}\|$$

since on the surface of a sphere of radius a,

$$\|\vec{E}\| = q\frac{\|\vec{r} - \vec{r}_0\|}{\|\vec{r} - \vec{r}_0\|^3} = \frac{q}{a^2}.$$

Then,

$$\int_{S_a} \vec{E} \cdot d\vec{A} = \int -\frac{q}{a^2}\|d\vec{A}\| = \frac{-q}{a^2} \cdot \text{Surface area of sphere} = -\frac{q}{a^2} \cdot 4\pi a^2 = -4\pi q.$$

$$\int_{S_a} \vec{E} \cdot d\vec{A} = -4\pi q.$$

$$\int_S \vec{E} \cdot d\vec{A} - \int_{S_a} \vec{E} \cdot d\vec{A} = 4\pi q.$$

22. (a) If $\vec{p} = p_1\vec{i} + p_2\vec{j} + p_3\vec{k}$ we have

$$\vec{E}(\vec{r}) = 3\frac{(xp_1 + yp_2 + zp_3)\vec{r}}{(x^2 + y^2 + z^2)^{5/2}} - \frac{\vec{p}}{(x^2 + y^2 + z^2)^{3/2}}$$

Thus taking partial derivatives

$$\frac{\partial E_1}{\partial x} = \frac{\partial}{\partial x}\frac{3x(xp_1 + yp_2 + zp_3) - p_1(x^2 + y^2 + z^2)}{(x^2 + y^2 + z^2)^{3/2}}$$

$$= \frac{\partial}{\partial x}\frac{2p_1 x^2 + 3(p_2 y + p_3 z)x - p_1(y^2 + z^2)}{(x^2 + y^2 + z^2)^{3/2}}$$

$$= \frac{(4p_1 x + 3p_2 y + 3p_3 z)(x^2 + y^2 + z^2)^{3/2}}{(x^2 + y^2 + z^2)^3}$$

$$- \frac{[2p_1 x^2 + 3(p_2 y + p_3 z)x - p_1(y^2 + z^2)](3/2)(2x)(x^2 + y^2 + z^2)^{1/2}}{(x^2 + y^2 + z^2)^3}$$

$$= \frac{p_1 x(-2x^2 + y^2 + z^2) + p_2 y(-6x^2 + 3y^2 + 3z^2) + p_3 z(-6x^2 + 3y^2 + 3z^2)}{(x^2 + y^2 + z^2)^{5/2}}$$

The expressions for $\partial E_2/\partial y$ and $\partial E_3/\partial z$ are obtained similarly; Summing, we obtain

$$\operatorname{div}\vec{E} = \frac{\partial E_1}{\partial x} + \frac{\partial E_2}{\partial y} + \frac{\partial E_3}{\partial z} = 0.$$

(b) The Divergence Theorem cannot be applied directly since $\vec{E}$ is not defined at the origin (which lies inside S) and the theorem requires that the vector field be defined everywhere inside S. However, it can be applied to the region R bounded by S and a small sphere S_a with center at the origin and contained inside S. We conclude that

$$\int_R \operatorname{div}\vec{E}\, dV = \int_S \vec{E}\cdot d\vec{A} - \int_{S_a} \vec{E}\cdot d\vec{A},$$

where S and S_a are outward oriented. Since $\operatorname{div}\vec{E} = 0$ inside R, it follows that

$$\int_S \vec{E}\cdot d\vec{A} = \int_{S_a} \vec{E}\cdot d\vec{A}.$$

The integral on the right-hand side was computed in Problem 20 on page 410, and is equal to zero no matter what the radius a is. Therefore,

$$\int_S \vec{E}\cdot d\vec{A} = 0.$$

23. (a) The velocity vector for the traffic flow would look like:

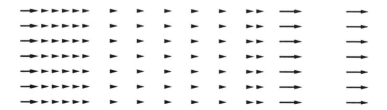

(b) When $0 \leq x < 2000$, the velocity is decreasing linearly from 55 to 15, so its formula is $(55 - x/50)\vec{i}$ mph. Then, when $2000 \leq x < 7000$, the speed is constant, so $\vec{v}(x) = 15\vec{i}$ mph. Next, when $7000 \leq x < 8000$, the velocity is increasing linearly from 15 to 55, so $\vec{v}(x) = (15 + (x - 7000)/25)\vec{i}$ mph. Finally, when $x \geq 8000$, the speed is constant, so $\vec{v}(x) = 55\vec{i}$ mph.

(c) $\operatorname{div}\vec{v} = dv(x)/dx$.

At $x = 1000$, $v(x) = 55 - x/50$, so $\operatorname{div}\vec{v} = -1/50$.
At $x = 5000$, $v(x) = 15$, so $\operatorname{div}\vec{v} = 0$.
At $x = 7500$, $v(x) = 15 + (x - 7000)/25$, so $\operatorname{div}\vec{v} = 1/25$.
At $x = 10,000$, $v(x) = 55$, so $\operatorname{div}\vec{v} = 0$.
In each case the units of $\operatorname{div}\vec{v}$ are $\dfrac{\text{miles/hour}}{\text{feet}}$.

24. (a) Usually, the distance between cars is more at higher speeds and less at lower speeds. The cars are traveling the fastest at $x = 0$, so at that point, the traffic should be the least dense. Thus,

$$\rho(0) < \rho(1000) < \rho(5000)$$

(b) Since ρ is in cars/mile, $\vec{v}$ is in miles/hour, $\rho\vec{v}$ is in cars/hour. The vector quantity $\rho\vec{v}$ gives the number of cars passing through a fixed point in a time interval.

(c) Pick any two points on the highway, $x = a$ and $x = b$ ($a < b$). We expect $\rho\vec{v}$ to be the same at both places. This is because if more cars pass through a than b, that would mean cars are disappearing (or at least stopping, which we know is not the case since the velocity field is not 0) between a and b. On the other hand, if more cars pass through b than a, that would mean cars are being created between a and b. So we expect $\rho\vec{v}$ to be the same at a and b. Since a and b were chosen arbitrarily, we can say that $\rho\vec{v}$ is constant at all x. This means $\operatorname{div}(\rho\vec{v}) = 0$.

(d) At $x = 0$, $\vec{v}(0) = 55\vec{i}$ and $\rho(0) = 75$. We have $\rho\vec{v}(0) = 4125\vec{i} = \text{constant}$. So $\rho(x) = \|\rho\vec{v}\|/\|\vec{v}(x)\| = 4125/v$.

$$\rho(x) = \frac{4125}{55 - \frac{x}{50}} \text{ if } 0 \leq x < 2000$$

$$\rho(x) = \frac{4125}{15} = 275 \text{ if } 2000 \leq x < 7000$$

$$\rho(x) = \frac{4125}{15 + \frac{x-7000}{25}} \text{ if } 7000 \leq x < 8000$$

$$\rho(x) = \frac{4125}{55} = 75 \text{ if } x \geq 8000$$

(e) We have $\rho(0) = 75$, $\rho(1000) = 118$, $\rho(5000) = 275$, where ρ is given in cars/mile. At $x = 0$, there are 75 cars in a mile stretch of highway. Since there are two lanes, there are about 38 cars in a mile in one lane. A mile is 5280 feet, so that says on average, one car occupies 139 feet. So at $x = 0$, the distance between two cars is 139 feet.

Similarly, we find that at $x = 1000$, the distance is 89 feet, and at $x = 5000$, the distance is 38 feet.

25. (a) Since $\vec{v} = \operatorname{grad}\phi$ we have

$$\vec{v} = \left(1 + \frac{y^2 - x^2}{(x^2 + y^2)^2}\right)\vec{i} + \frac{-2xy}{(x^2 + y^2)^2}\vec{j}$$

(b) Differentiating the components of $\vec{v}$, we have

$$\operatorname{div}\vec{v} = \frac{\partial}{\partial x}\left(1 + \frac{y^2 - x^2}{(x^2 + y^2)^2}\right) + \frac{\partial}{\partial y}\left(\frac{-2xy}{(x^2 + y^2)^2}\right) = \frac{2x(x^2 - 3y^2)}{(x^2 + y^2)^3} + \frac{2x(3y^2 - x^2)}{(x^2 + y^2)^3} = 0$$

(c) The vector $\vec{v}$ is tangent to the circle $x^2+y^2=1$, if and only if the dot product of the field on the circle with any radius vector of that circle is zero. Let (x,y) be a point on the circle. We want to show: $\vec{v}\cdot\vec{r}=\vec{v}(x,y)\cdot(x\vec{i}+y\vec{j})=0$. We have:

$$\begin{aligned}\vec{v}(x,y)\cdot(x\vec{i}+y\vec{j}) &= ((1+\frac{y^2-x^2}{(x^2+y^2)^2})\vec{i}+\frac{-2xy}{(x^2+y^2)^2}\vec{j})\cdot(x\vec{i}+y\vec{j})\\ &= x+x\frac{y^2-x^2}{(x^2+y^2)^2}-\frac{2xy^2}{(x^2+y^2)^2}\\ &= \frac{x(x^2+y^2-1)}{x^2+y^2},\end{aligned}$$

but we know that for any point on the circle, $x^2+y^2=1$, thus we have $\vec{v}\cdot\vec{r}=0$. Therefore, the velocity field is tangent to the circle. Consequently, there is no flow through the circle and any water on the outside of the circle must flow around it.

(d)

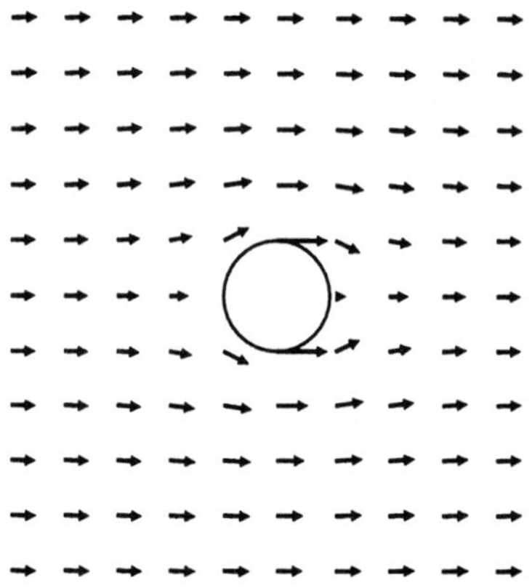

Figure 20.18

26.

$$\begin{aligned}\operatorname{grad}\phi &= \left(a_1x+\frac{1}{2}(a_2+b_1)y+\frac{1}{2}(a_3+c_1)z\right)\vec{i}+\left(b_2y+\frac{1}{2}(a_2+b_1)x+\frac{1}{2}(b_3+c_2)z\right)\vec{j}\\ &+\left(c_3z+\frac{1}{2}(a_3+c_1)x+\frac{1}{2}(b_3+c_2)y\right)\vec{k}.\end{aligned}$$

$$\begin{aligned}\vec{v}\times\vec{r} &= \left(\frac{1}{2}(a_3-c_1)z-\frac{1}{2}(b_1-a_2)y\right)\vec{i}+\left(\frac{1}{2}(b_1-a_2)x-\frac{1}{2}(c_2-b_3)z\right)\vec{j}\\ &+\left(\frac{1}{2}(c_2-b_3)y-\frac{1}{2}(a_3-c_1)x\right)\vec{k}.\end{aligned}$$

$$\operatorname{grad}\phi+\vec{v}\times\vec{r}=(a_1x+a_2y+a_3z)\vec{i}+(b_1x+b_2y+b_3z)\vec{j}+(c_1x+c_2y+c_3z)\vec{k}.$$

Since this is the general form for any linear vector field $\vec{F}$, any such $\vec{F}$ can be written as $\vec{F}=\operatorname{grad}\phi+\vec{v}\times\vec{r}$, if ϕ and $\vec{v}$ are as given.

27. Calculating the gradient gives us

$$\operatorname{grad}\vec{\phi}=\begin{cases}2x\vec{i}+2y\vec{j}+2z\vec{k} & \text{for } x^2+y^2+z^2\le\frac{b^2}{4}\\ \frac{b^3x}{4(x^2+y^2+z^2)^{3/2}}\vec{i}+\frac{b^3y}{4(x^2+y^2+z^2)^{3/2}}\vec{j}+\frac{b^3z}{4(x^2+y^2+z^2)^{3/2}}\vec{k} & \text{for } \frac{b^2}{4}\le x^2+y^2+z^2\end{cases}$$

The electric field, $\vec{E}$, is the negative of grad ϕ. Both vector fields are radial. The magnitude of the vector field in the region $x^2 + y^2 + z^2 \leq b^2/4$ is proportional to $\|\vec{r}\|$ while the magnitude of the vector field in the region $b^2/4 \leq x^2 + y^2 + z^2$ is proportional to $1/\|\vec{r}\|^2$, or the inverse square. Thus, for $x^2 + y^2 + z^2 \leq b^2/4$,

$$\text{Charge distribution} = \rho = \frac{1}{4\pi} \operatorname{div} \vec{E} = -\frac{2+2+2}{4\pi} = -\frac{3}{2\pi}$$

For $b^2/4 \leq x^2 + y^2 + z^2$,

$$\text{Charge distribution} = \rho = \frac{1}{4\pi} \operatorname{div} \vec{E}$$

$$= -\frac{b^3}{4(x^2+y^2+z^2)^{3/2}} + \frac{b^3}{4}\frac{3}{2}\frac{2x \cdot x}{(x^2+y^2+z^2)^{5/2}}$$

$$\quad - \frac{b^3}{4(x^2+y^2+z^2)^{3/2}} + \frac{b^3}{4}\frac{3}{2}\frac{2y \cdot y}{(x^2+y^2+z^2)^{5/2}}$$

$$\quad - \frac{b^3}{4(x^2+y^2+z^2)^{3/2}} + \frac{b^3}{4}\frac{3}{2}\frac{2z \cdot z}{(x^2+y^2+z^2)^{5/2}}$$

$$= -\frac{b^3}{4}\frac{x^2+y^2+z^2-3x^2+x^2+y^2+z^2-3y^2+x^2+y^2+z^2-3z^2}{(x^2+y^2z^2)^{5/2}}$$

$$= 0$$

The charge density is constant: $-3/(2\pi)$ for $x^2 + y^2 + z^2 \leq b^2/4$ and zero further away.

28. (a) The path along which we integrate is shown in Figure 20.19.

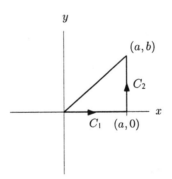

Figure 20.19

The path C_1 is given by

$$C_1 : \begin{cases} x = t & 0 \leq t \leq a \\ y = 0 \end{cases}$$

and path C_2 is given by

$$C_2 : \begin{cases} x = a \\ y = t & 0 \leq t \leq b \end{cases}.$$

We integrate along the path $C = C_1 + C_2$. Then,

$$\int_C \vec{E} \cdot d\vec{r} = \int_{C_1} \vec{E} \cdot d\vec{r} + \int_{C_2} \vec{E} \cdot d\vec{r}$$

$$= \int_0^a 5t^2 \vec{j} \cdot \vec{i} \; dt + \int_0^b (10at\vec{i} + (5a^2 - 5t^2)\vec{j}) \cdot \vec{j} \; dt$$

$$= 5a^2b - \frac{5}{3}b^3$$

(b) The path along which we integrate is shown in Figure 20.20.

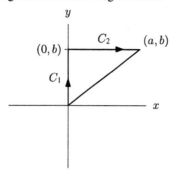

Figure 20.20

The path C_1 is given by

$$C_1 : \begin{cases} x = 0 \\ y = t \quad 0 \le t \le b \end{cases}$$

and Path C_2 is given by

$$C_2 : \begin{cases} x = t \quad 0 \le t \le a \\ y = b \end{cases}.$$

We integrate along the path $C = C_1 + C_2$. Then,

$$\int_C \vec{E} \cdot d\vec{r} = \int_{C_1} \vec{E} \cdot d\vec{r} + \int_{C_2} \vec{E} \cdot d\vec{r}$$

$$= \int_0^b -5t^2 \vec{j} \cdot \vec{j} \; dt + \int_0^a (10bt\vec{i} + (5t^2 - 5b^2)\vec{j}) \cdot \vec{i} \; dt$$

$$= -\frac{5}{3}b^3 + 5a^2b$$

(c) Notice that the line integrals along both paths in part (a) and (b) are equal. Thus $\vec{E}$ could be path independent. This property is a property of electric fields.

(d) Using the calculations of the parts (a) and (b), with the point (a, b) replaced by (x, y), we see that the electric potential, ϕ, at (x, y) is given by

$$\phi = \frac{5}{3}y^3 - 5x^2y.$$

Then, taking the gradient gives

$$\text{grad } \phi = \phi_x \vec{i} + \phi_y \vec{j}$$

$$= -10xy\vec{i} + (\frac{5}{3}(3y^2) - 5x^2)\vec{j}$$

$$= -10xy\vec{i} + (5y^2 - 5x^2)\vec{j}$$

$$= -\vec{E}.$$

Thus, we have confirmed that $\vec{E} = -\text{grad } \phi$.

29. (a) Differentiating div $\vec{E} = 4\pi\rho$ with respect to time gives

$$\frac{\partial}{\partial t}\left(\text{div}\,\vec{E}\right) = \frac{\partial}{\partial t}\left(\frac{\partial E_1}{\partial x} + \frac{\partial E_2}{\partial y} + \frac{\partial E_3}{\partial z}\right) = 4\pi\frac{\partial\rho}{\partial t}.$$

Since, for example, $\frac{\partial}{\partial t}\left(\frac{\partial E_1}{\partial x}\right) = \frac{\partial}{\partial x}\left(\frac{\partial E_1}{\partial t}\right)$, we can rewrite this as

$$\frac{\partial}{\partial x}\left(\frac{\partial E_1}{\partial t}\right) + \frac{\partial}{\partial y}\left(\frac{\partial E_2}{\partial t}\right) + \frac{\partial}{\partial z}\left(\frac{\partial E_3}{\partial t}\right) = 4\pi\frac{\partial\rho}{\partial t}.$$

So we have shown that

$$\text{div}\left(\frac{\partial\vec{E}}{\partial t}\right) = 4\pi\frac{\partial\rho}{\partial t}.$$

Now consider the equation

$$\text{curl}\,\vec{B} - \frac{1}{c}\frac{\partial\vec{E}}{\partial t} = \frac{4\pi}{c}\vec{J}$$

and take the divergence of both sides:

$$\text{div curl}\,\vec{B} - \frac{1}{c}\,\text{div}\left(\frac{\partial\vec{E}}{\partial t}\right) = \frac{4\pi}{c}\,\text{div}\,\vec{J}.$$

Since div curl $\vec{B} = 0$, by Problem 16 on page 504, we have

$$-\frac{1}{c}\,\text{div}\left(\frac{\partial\vec{E}}{\partial t}\right) = \frac{4\pi}{c}\,\text{div}\,\vec{J}.$$

Thus

$$-\frac{1}{c}\left(4\pi\frac{\partial\rho}{\partial t}\right) = \frac{4\pi}{c}\,\text{div}\,\vec{J},$$

so

$$-\frac{\partial\rho}{\partial t} = \text{div}\,\vec{J},$$

or

$$\frac{\partial\rho}{\partial t} + \text{div}\,\vec{J} = 0.$$

(b) The equation derived in part (a) says that the rate of change with time of the charge density at a point is the negative of the divergence of the current density at that point.

Why is this reasonable? Suppose div $\vec{J} < 0$ at some point, that is, there is a current "sink" there. This means that current is "piling up" at this point – in other words, the charge is "piling up" there. Thus we would expect $\partial\rho/\partial t > 0$ there. Similarly, if div $\vec{J} > 0$, at some point, there is a current source there. This means that current is being "created" near that point, which means that charge density is decreasing there. Thus we would expect $\partial\rho/\partial t < 0$ there.

(c) The equation is called the charge conservation equation because it reflects the fact that charge is neither created nor destroyed. If div $\vec{J}$ is negative at some point, there is a net influx of current into a small surface around the point, so the charge density must be increasing there. If div $\vec{J}$ is positive, there is a net outflow of current through a small surface around the point, so the charge density must be decreasing there.

30.   (a)   The distance from any point to the origin is given by $\sqrt{x^2 + y^2 + z^2}$, so the denominator is simply r^3. The field can then be rewritten in components as $(\frac{kx}{r^3}\vec{i} + \frac{ky}{r^3}\vec{j} + \frac{kz}{r^3}\vec{k})$. Its magnitude is thus:

$$\|\vec{v}\| = \sqrt{K^2 \frac{x^2 + y^2 + z^2}{r^6}} = \sqrt{\frac{K^2}{r^4}} = \frac{K}{r^2}$$

which is only a function of the distance from the origin. It is clear that the vector field points away from the origin for all points (x, y, z), because it is the radius vector $x\vec{i} + y\vec{j} + z\vec{k}$, multiplied by the positive scalar K/r^3. Suppose $(x, y, z) \neq (0, 0, 0)$, then

$$\text{div}\,\vec{v}\,(x, y, z) = \frac{\partial}{\partial x}\left(\frac{Kx}{r^3}\right) + \frac{\partial}{\partial y}\left(\frac{Ky}{r^3}\right) + \frac{\partial}{\partial z}\left(\frac{Kz}{r^3}\right)$$
$$= K\left(\frac{-2x^2 + y^2 + z^2}{r^5} + \frac{x^2 - 2y^2 + z^2}{r^5} + \frac{x^2 + y^2 - 2z^2}{r^5}\right) = 0.$$

Hence, indeed $\vec{v}$ is a point source at the origin.

(b)   The dependence of $\vec{v}$ on r, the distance from the origin, is shown in part (a).

(c)   The flux through a sphere centered at the origin is calculated as:

$$\text{Flux} = \int_S \vec{v} \cdot d\vec{A}$$

Since the vector field's magnitude is a function only of the distance from the origin, it will be constant over the surface of a sphere centered at the origin. Furthermore, since it is pointed away from the origin, $\vec{v} \cdot d\vec{A}$ will be simply $\|\vec{v}\| \cdot \|d\vec{A}\|$. Thus

$$\text{Total flux out of a sphere of radius } r = \|\vec{v}_r\| \cdot \|\vec{A}_{\text{sphere}}\| = \frac{K}{r^2}\frac{4}{3}\pi r^2 = \frac{4}{3}\pi K.$$

So the flux does not even depend upon r since the rate at which the area of the sphere is increasing is exactly equal to the rate at which the magnitude of the field is decreasing.

(d)   This is best handled by observing (as in part (a)) that the divergence of the vector field at any point besides the origin is zero. Since the divergence anywhere but the origin is zero, the net flux through any closed surface not enclosing the origin must also be zero.

31.   Set up the wire as the z-axis with the current flowing upward, and let $\vec{B}$ be the magnetic field due to the current in the wire. Then by the right-hand rule, in the xy-plane $\vec{B}$ points counterclockwise around a circle centered around the z-axis. (See Figure 20.21.) Let us take the circulation on the boundary of a ring of outer radius R and inner radius r around the z-axis. We choose the upward orientation for S. (This is arbitrary.) The boundary of this region, C, is a curve in two pieces, C_P and C_Q. Given the orientation of S, we must have C_P oriented clockwise and C_Q oriented counterclockwise when viewed from above. (See Figure 20.21.) The boundary of S is $C_P + C_Q$. From Stokes' Theorem,

$$\int_S \text{curl}\,\vec{B} \cdot d\vec{A} = \int_C \vec{B} \cdot d\vec{r}$$

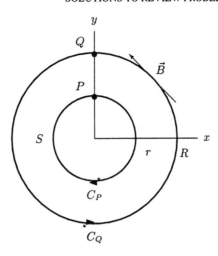

Figure 20.21: Current along positive z-axis

Thus, since curl $\vec{B} = \vec{0}$, we have

$$0 = \int_S \text{curl}\,\vec{B} \cdot d\vec{A} = \int_{C_Q} \vec{B} \cdot d\vec{r} + \int_{C_P} \vec{B} \cdot d\vec{r}$$

so

$$\int_{C_Q} \vec{B} \cdot d\vec{r} = -\int_{C_P} \vec{B} \cdot d\vec{r}$$

Since the vector field has constant magnitude on each circle and is parallel to the circle

$$\int_{C_Q} \vec{B} \cdot d\vec{r} = \|B(Q)\| \cdot \text{Length of } C_Q = 2\pi R \|B(Q)\|$$

and because C_P is oriented in the opposite direction

$$\int_{C_P} \vec{B} \cdot d\vec{r} = -\|B(P)\| \cdot \text{Length of } C_P = -2\pi r \|B(P)\|$$

Thus,

$$2\pi R \|B(Q)\| = -(-2\pi r \|B(P)\|)$$

which simplifies to:

$$\frac{\|\vec{B}(Q)\|}{\|\vec{B}(P)\|} = \frac{R}{r} = \frac{1/r}{1/R}.$$

This relationship shows that $\|\vec{B}\|$ varies inversely with the radial distance, r.

32. (a) Cross-section in the xy-plane of the vector field with $K = 1$ is in Figure 20.22.

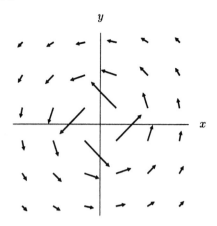

Figure 20.22

Cross-section in the xy-plane of the vector field with $K = -1$ is in Figure 20.23.

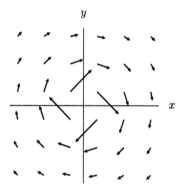

Figure 20.23

(b) $\|\vec{v}\| = |K| \cdot \| - y\vec{i} + x\vec{j} \|/(x^2 + y^2) = |K| \cdot r/r^2 = |K|/r.$

(c) For $(x, y, z) \neq (0, 0, z)$,

$$\operatorname{div} \vec{v} = \frac{\partial}{\partial x}\left(\frac{-Ky}{x^2 + y^2}\right) + \frac{\partial}{\partial y}\left(\frac{Kx}{x^2 + y^2}\right) = K\frac{(-y)(-2x)}{(x^2 + y^2)^2} + K\frac{x(-2y)}{(x^2 + y^2)^2} = 0$$

(d) In order for curl $\vec{v}$ to make sense, $\vec{v}$ needs to have 3 components. Write

$$\vec{v} = \frac{K}{x^2 + y^2}(-y\vec{i} + x\vec{j} + 0\vec{k}),$$

Hence, for $(x, y, z) \neq (0, 0, z)$,

$$\begin{aligned}
\operatorname{curl} \vec{v} &= \left(\frac{\partial}{\partial x}\left(\frac{Kx}{x^2 + y^2}\right) - \frac{\partial}{\partial y}\left(\frac{Ky}{x^2 + y^2}\right)\right)\vec{k} \\
&= \left(K\frac{y^2 - x^2}{(x^2 + y^2)^2} + K\frac{x^2 - y^2}{(x^2 + y^2)^2}\right)\vec{k} \\
&= \vec{0}.
\end{aligned}$$

(e) On the circle of radius R, the magnitude of $\vec{v}$ is constant; $\|\vec{v}\| = |K|/R$. If $K > 0, \vec{v}$ and $\Delta\vec{r}$ are parallel and in the same direction so that

$$\int_C \vec{v} \cdot d\vec{r} = \frac{K}{R} \cdot (\text{Length of circle}) = \frac{K}{R} \cdot 2\pi R = 2\pi K.$$

If $K < 0, |K| = -K$, and $\vec{v}$ and $\Delta\vec{r}$ are in opposite directions so that

$$\int_C \vec{v} \cdot d\vec{r} = \frac{-|K|}{R} \cdot (\text{Length of circle}) = \frac{K}{R} \cdot 2\pi R = 2\pi K.$$

(f) Since $\vec{v}$ (and hence curl $\vec{v}$) is undefined on the z-axis, Stokes' Theorem does not apply on the circle (or on any curve encircling the z-axis).

APPENDIX

Solutions for Section D

1.

$$\int \left(x^2 + 2x + \frac{1}{x}\right) dx = \int x^2\, dx + \int 2x\, dx + \int \frac{1}{x}\, dx$$
$$= \frac{1}{3}x^3 + x^2 + \ln|x| + C,$$

where C is a constant.

2.

$$\int \frac{t+1}{t^2}\, dt = \int \frac{1}{t}\, dt + \int \frac{1}{t^2}\, dt = \ln|t| - \frac{1}{t} + C, \qquad \text{where } C \text{ is a constant.}$$

3.

$$\int \frac{(t+2)^2}{t^3}\, dt = \int \frac{t^2 + 4t + 4}{t^3}\, dt = \int \frac{1}{t}\, dt + \int \frac{4}{t^2}\, dt + \int \frac{4}{t^3}\, dt = \ln|t| - \frac{4}{t} - \frac{2}{t^2} + C,$$

where C is a constant.

4.

$$\int \sin t\, dt = -\cos t + C, \quad \text{where } C \text{ is a constant.}$$

5. Let $2t = w$, then $2dt = dw$, so $dt = \frac{1}{2}dw$, so

$$\int \cos 2t\, dt = \int \frac{1}{2}\cos w\, dw = \frac{1}{2}\sin w + C = \frac{1}{2}\sin 2t + C,$$

where C is a constant.

6.

$$\int \frac{x}{x^2 + 1}\, dx = \int \frac{\frac{1}{2}}{w}\, dw \qquad (x^2 + 1 = w, 2x\,dx = dw, x\,dx = \frac{1}{2}dw)$$
$$= \frac{1}{2} \int \frac{1}{w}\, dw = \frac{1}{2}\ln|w| + C$$
$$= \frac{1}{2}\ln|x^2 + 1| + C,$$

where C is a constant.

7. Let $\cos\theta = w$, then $-\sin\theta\,d\theta = dw$, so

$$\int \tan\theta\, d\theta = \int \frac{\sin\theta}{\cos\theta}\, d\theta = \int \frac{-1}{w}\, dw$$
$$= -\ln|w| + C = -\ln|\cos\theta| + C,$$

where C is a constant.

8. Let $5z = w$, then $5dz = dw$, which means $dz = \frac{1}{5}dw$, so

$$\int e^{5z}\, dz = \int e^{w} \cdot \frac{1}{5}dw = \frac{1}{5}\int e^{w}\, dw = \frac{1}{5}e^{w} + C = \frac{1}{5}e^{5z} + C,$$

where C is a constant.

9. Let $t^2 + 1 = w$, then $2t\,dt = dw$, $t\,dt = \frac{1}{2}dw$, so

$$\int t e^{t^2+1}\, dt = \int e^{w} \cdot \frac{1}{2}\, dw = \frac{1}{2}\int e^{w}\, dw = \frac{1}{2}e^{w} + C = \frac{1}{2}e^{t^2+1} + C,$$

where C is a constant.

10. Since

$$\int \frac{du}{u^2 + a^2} = \frac{1}{a}\tan^{-1}\frac{u}{a},$$

let $u = z, a = 1$, then

$$\int \frac{dz}{1 + z^2} = \frac{1}{1}\tan^{-1}\frac{z}{1} + C = \tan^{-1}z + C, \qquad \text{where } C \text{ is a constant.}$$

11.

$$\begin{aligned}
\int \frac{dz}{1 + 4z^2} &= \frac{1}{4}\int \frac{dz}{\frac{1}{4} + z^2} \qquad \left(u = z, a = \frac{1}{2}\right) \\
&= \frac{1}{4} \cdot \frac{1}{\frac{1}{2}}\tan^{-1}\frac{z}{\frac{1}{2}} + C \\
&= \frac{1}{2}\tan^{-1}2z + C,
\end{aligned}$$

where C is a constant.

12. Let $\sin\theta = w$, then $\cos\theta\, d\theta = dw$, so

$$\int \sin^2\theta \cos\theta\, d\theta = \int w^2\, dw = \frac{1}{3}w^3 + C = \frac{1}{3}\sin^3\theta + C,$$

where C is a constant.

13. Let $\cos 5\theta = w$, then $-5\sin 5\theta\, d\theta = dw$, $\sin 5\theta\, d\theta = -\frac{1}{5}dw$. So

$$\begin{aligned}
\int \sin 5\theta \cos^3 5\theta\, d\theta &= \int w^3 \cdot \left(-\frac{1}{5}\right)dw = -\frac{1}{5}\int w^3\, dw = -\frac{1}{20}w^4 + C \\
&= -\frac{1}{20}\cos^4 5\theta + C,
\end{aligned}$$

where C is a constant.

14.

$$\begin{aligned}
\int \sin^3 z \cos^3 z\, dz &= \int \sin z(1 - \cos^2 z)\cos^3 z\, dz \\
&= \int \sin z \cos^3 z\, dz - \int \sin z \cos^5 z\, dz
\end{aligned}$$

$$= \int w^3 (-dw) - \int w^5 (-dw) \quad \text{(let } \cos z = w, \text{ so } - \sin z \, dz = dw)$$

$$= - \int w^3 \, dw + \int w^5 \, dw$$

$$= -\frac{1}{4} w^4 + \frac{1}{6} w^6 + C$$

$$= -\frac{1}{4} \cos^4 z + \frac{1}{6} \cos^6 z + C,$$

where C is a constant.

15. Let $\ln x = w$, then $\frac{1}{x} \, dx = dw$, so

$$\int \frac{(\ln x)^2}{x} \, dx = \int w^2 \, dw = \frac{1}{3} w^3 + C = \frac{1}{3} (\ln x)^3 + C, \text{ where } C \text{ is a constant.}$$

16. Let $\sin \theta = w$, then $\cos \theta \, d\theta = dw$, so

$$\int \cos \theta \sqrt{1 + \sin \theta} \, d\theta = \int \sqrt{1 + w} \, dw$$

$$= \frac{(1 + w)^{\frac{3}{2}}}{\frac{3}{2}} + C = \frac{2}{3} (1 + \sin \theta)^{3/2} + C,$$

where C is a constant.

17.

$$\int x e^x \, dx = x e^x - \int e^x \, dx \qquad \text{(let } x = u, e^x = v', e^x = v)$$

$$= x e^x - e^x + C,$$

where C is a constant.

18.

$$\int t^3 e^t \, dt = t^3 e^t - \int 3 t^2 e^t \, dt \qquad \text{(let } t^3 = u, e^t = v', 3t^2 = u', e^t = v)$$

$$= t^3 e^t - 3 \int t^2 e^t \, dt \qquad \text{(let } t^2 = u, e^t = v')$$

$$= t^3 e^t - 3(t^2 e^t - \int 2 t e^t \, dt)$$

$$= t^3 e^t - 3 t^2 e^t + 6 \int t e^t \, dt \qquad \text{(let } t = u, e^t = v')$$

$$= t^3 e^t - 3 t^2 e^t + 6(t e^t - \int e^t \, dt)$$

$$= t^3 e^t - 3 t^2 e^t + 6 t e^t - 6 e^t + C,$$

where C is a constant.

19.

$$\int x \ln x \, dx = \frac{1}{2} x^2 \ln x - \int \frac{1}{x} \frac{x^2}{2} \, dx \qquad (\ln x = u, x = v', \frac{1}{x} = u', \frac{1}{2} x^2 = v)$$

546

$$= \frac{1}{2}x^2 \cdot \ln x - \frac{1}{2}\int x\,dx$$

$$= \frac{1}{2}x^2 \ln x - \frac{1}{4}x^2 + C,$$

where C is a constant.

20.

$$\int \frac{1}{\cos^2 \theta}\,d\theta = \int \sec^2 \theta\,d\theta = \tan \theta + C, \qquad \text{where } C \text{ is a constant.}$$

21. Let $x^2 = w$, then $2x\,dx = dw$, $x = 1 \Rightarrow w = 1$, $x = 3 \Rightarrow w = 9$. Thus,

$$\int_1^3 x(x^2+1)^{70}\,dx = \int_1^9 (w+1)^{70}\frac{1}{2}\,dw$$

$$= \frac{1}{2}\cdot\frac{1}{71}(w+1)^{71}\Big|_1^9$$

$$= \frac{1}{142}(10^{71} - 2^{71}).$$

22.

$$\int_0^1 \frac{dx}{x^2+1} = \tan^{-1} x\Big|_0^1 = \tan^{-1} 1 - \tan^{-1} 0 = \frac{\pi}{4} - 0 = \frac{\pi}{4}.$$

23.

$$\int_0^{10} ze^{-z}\,dz = [-ze^{-z}]\Big|_0^{10} - \int_0^{10} -e^{-z}\,dz \qquad (\text{let } z = u, e^{-z} = v', -e^{-z} = v)$$

$$= -10e^{-10} - [e^{-z}]\Big|_0^{10}$$

$$= -10e^{-10} - e^{-10} + 1$$

$$= -11e^{-10} + 1.$$

24. Let $\sin\theta = w$, $\cos\theta\,d\theta = dw$. Then $\theta = -\frac{\pi}{3} \Rightarrow w = -\frac{\sqrt{3}}{2}$, $\theta = \frac{\pi}{4} \Rightarrow w = \frac{\sqrt{2}}{2}$. So

$$\int_{-\pi/3}^{\pi/4} \sin^3\theta\cos\theta\,d\theta = \int_{-\sqrt{3}/2}^{\sqrt{2}/2} w^3\,dw = \frac{1}{4}w^4\Big|_{-\sqrt{3}/2}^{\sqrt{2}/2} = \frac{1}{4}\left[\left(\frac{\sqrt{2}}{2}\right)^4 - \left(\frac{-\sqrt{3}}{2}\right)^4\right]$$

$$= \frac{1}{4}(\frac{4}{16} - \frac{9}{16}) = \frac{1}{4}(-\frac{5}{16}) = -\frac{5}{64}.$$

25. Let $\sqrt{x} = w$, $\frac{1}{2}x^{-\frac{1}{2}}dx = dw$, $\frac{dx}{\sqrt{x}} = 2\,dw$. Then $x = 1 \Rightarrow w = 1$, $x = 4 \Rightarrow w = 2$. So

$$\int_1^4 \frac{e^{\sqrt{x}}}{\sqrt{x}}\,dx = \int_1^2 e^w \cdot 2\,dw = 2e^w\Big|_1^2 = 2(e^2 - e) \approx 9.34.$$

26. Let $y'(t) = \frac{dy}{dt}$. Then y is the antiderivative of y' such that $y(0) = 0$. We know that

$$y(x) = \int_0^x y'(t)\,dt.$$

Thus, $y(x)$ is the area under the graph of $\frac{dy}{dt}$ from $t = 0$ to $t = x$ (note: we interpret "area" to be negative if a region lies below the t-axis). We therefore know that $y(t_1) = 2$, $y(t_3) = 2 - 2 = 0$, and $y(t_5) = 2$.

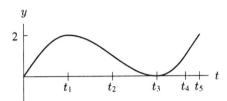

The function y' is positive on the intervals $(0, t_1)$ and (t_3, ∞), so y is increasing on those intervals. y' is negative on the interval (t_1, t_3), so y is decreasing on that interval. y' is increasing on the interval (t_2, t_4), so y is concave up on that interval; y' is decreasing on $(0, t_2)$, so y is concave down there. t_2, the point where the concavity changes, is an inflection point. Finally, since y' is constant on the interval (t_4, ∞), y's graph is linear with positive slope on this interval. $y(t_1) = 2$ is a local maximum, and $y(t_3) = 0$ is a local minimum.

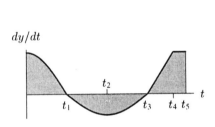

Figure D.1

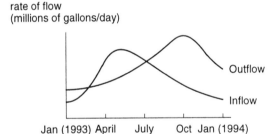

Figure D.2

27. (a) Suppose $Q(t)$ is the amount of water in the reservoir at time t. Then

$$Q'(t) = \begin{array}{c} \text{Rate at which water} \\ \text{in reservoir is increasing} \end{array} = \begin{array}{c} \text{inflow} \\ \text{rate} \end{array} - \begin{array}{c} \text{outflow} \\ \text{rate} \end{array}$$

Thus the amount of water in the reservoir is increasing when the inflow curve is above the outflow, and decreasing when it is below. This means that $Q(t)$ is a maximum where the curves cross in July 1993 (as shown in Figure D.3), and $Q(t)$ is decreasing fastest when the outflow is farthest above the inflow curve, which occurs about October 1993 (see Figure D.3).

To estimate values of $Q(t)$, we use the Fundamental Theorem which says that the change in the total quantity of water in the reservoir is given by

$$Q(t) - Q(\text{Jan'93}) = \int_{\text{Jan93}}^t (\text{inflow rate} - \text{outflow rate})\,dt$$

or

$$Q(t) = Q(\text{Jan'93}) + \int_{\text{Jan93}}^{t} (\text{inflow rate} - \text{outflow rate})dt$$

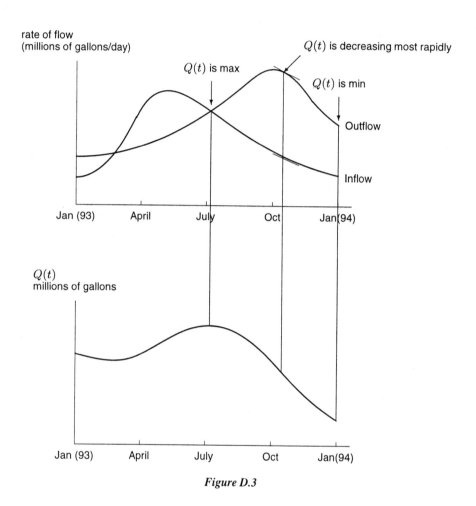

Figure D.3

(b) See Figure D.3. Maximum in July 1993. Minimum in Jan 1994.
(c) See Figure D.3. Increasing fastest in May 1993. Decreasing fastest in Oct 1993.
(d) In order for the water to be the same as Jan'93 the total amount of water which has flowed into the reservoir must be 0, so

$$\int_{\text{Jan93}}^{\text{July94}} (\text{inflow} - \text{outflow})dt = -A_1 + A_2 - A_3 + A_4 = 0$$

giving $A_1 + A_3 = A_2 + A_4$

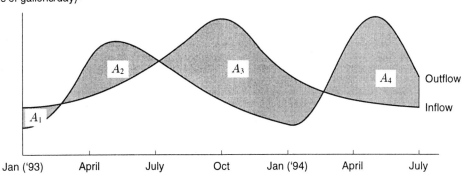

Figure D.4

28. (a) Quantity used = $\int_0^5 f(t)\,dt$.

    (b) Using a left sum, our approximation is

    $$32e^{0.05(0)} + 32e^{0.05(1)} + 32e^{0.05(2)} + 32e^{0.05(3)} + 32e^{0.05(4)} = 177.27.$$

    Since f is an increasing function, this represents an underestimate.

    (c) Each term is a lower estimate of one year's consumption of oil.

29. (a) Suppose we choose an x, $0 \le x \le 2$. If Δx is a small fraction of a meter, then the density of the rod is approximately $\rho(x)$ anywhere from x to $x + \Delta x$ meters from the left end of the rod.

    The mass of the rod from x to $x + \Delta x$ meters is therefore approximately $\rho(x)\Delta x$.

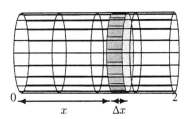

    (b) The definite integral is

    $$M = \int_0^2 \rho(x)\,dx = \int_0^2 (2 + 6x)\,dx = (2x + 3x^2)\Big|_0^2 = 16 \text{ grams.}$$

30. (a)

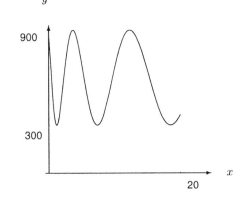

(b) Suppose we choose an x, $0 \leq x \leq 20$. We approximate the density of the number of the cars between x and $x + \Delta x$ miles as $p(x)$ cars per mile. Therefore the number of cars between x and $x + \Delta x$ is approximately $p(x)\Delta x$.

If we slice the 20 mile strip into N slices, we get that the total number of cars is $C \approx \sum_{i=1}^{N} p(x_i)\Delta x = \sum_{i=1}^{N} \left[600 + 300\sin(4\sqrt{x_i + 0.15})\right] \Delta x$, where $\Delta x = 20/N$. (This is a right-hand approximation; the corresponding left-hand approximation is $\sum_{i=0}^{N-1} p(x_i)\Delta x$.)

(c) As $N \to \infty$, the Riemann sum above approaches the integral

$$C = \int_0^{20} (600 + 300\sin 4\sqrt{x + 0.15})\, dx.$$

If we approximate the integral using one of our approximation methods (like Simpson's rule) we find $C \approx 11513$.

We can also find the integral exactly as follows:

$$C = \int_0^{20} (600 + 300\sin 4\sqrt{x + 0.15})\, dx$$

$$= \int_0^{20} 600\, dx + \int_0^{20} 300\sin 4\sqrt{x + 0.15}\, dx$$

$$= 12000 + 300 \int_0^{20} \sin 4\sqrt{x + 0.15}\, dx.$$

Let $w = \sqrt{x + 0.15}$, so $x = w^2 - 0.15$ and $dx = 2w\, dw$. Then

$$\int_{x=0}^{x=20} \sin 4\sqrt{x + 0.15}\, dx = 2 \int_{w=\sqrt{0.15}}^{w=\sqrt{20.15}} w \sin 4w\, dw,$$

using integral table formula 15

$$= 2 \left[-\frac{1}{4}w\cos 4w + \frac{1}{16}\sin 4w \right] \Bigg|_{\sqrt{0.15}}^{\sqrt{20.15}}$$

$$\approx -1.624$$

Using this, we have $C \approx 12000 + 300(-1.624) \approx 11513$, which matches our numerical approximation.

31. (a) We must find where the population density is zero. The density is given by the function

$$10{,}000(3 - r);$$

if $10{,}000(3 - r) = 0$, then we must have $r = 3$. We thus conclude that the radius of Circle City is 3 miles. (Note that for $r > 3$, $10{,}000(3 - r)$ becomes negative, so at that point, our function no longer gives a meaningful representation of population density.)

(b) Let $f(r) = 10{,}000(3 - r)$. The population is approximated by a sum

$$\sum 2\pi r \cdot 10{,}000(3 - r)\Delta r.$$

Since the city radius is 3 miles, r ranges from 0 to 3. Hence as $\Delta r \to 0$, the sum is given by the integral

$$\int_0^3 2\pi r \cdot 10{,}000(3 - r)\, dr.$$

This integral evaluates to $9\pi \cdot 10{,}000 \approx 282{,}743$. So we can say that the population of Circle City is approximately 282,743.

32. (a) Partition $[0, 10{,}000]$ into N subintervals of width Δr. The area in the ith subinterval is $\approx 2\pi r_i \Delta r$. So the total mass in the slick $= M \approx \sum_{i=1}^{N} 2\pi r_i \left(\frac{50}{1+r_i}\right) \Delta r$.

(b) $M = \displaystyle\int_0^{10{,}000} 100\pi \frac{r}{1+r}\,dr$. We may rewrite $\frac{r}{1+r}$ as $\frac{1+r}{1+r} - \frac{1}{1+r} = 1 - \frac{1}{1+r}$, so that

$$M = \int_0^{10{,}000} 100\pi (1 - \frac{1}{1+r})\,dr = 100\pi \left(r - \ln|1+r| \Big|_0^{10{,}000} \right)$$

$$= 100\pi(10{,}000 - \ln(10{,}001)) \approx 3.14 \times 10^6 \text{ kg}.$$

(c) We wish to find an R such that

$$\int_0^R 100\pi \frac{r}{1+r}\,dr = \frac{1}{2} \int_0^{10{,}000} 100\pi \frac{r}{1+r}\,dr \approx 1.57 \times 10^6.$$

So $100\pi(R - \ln|R+1|) \approx 1.57 \times 10^6$; $R - \ln|R+1| \approx 5000$. By trial and error, we find $R \approx 5009$ meters.

33. (a) Partition $0 \le h \le 100$ into N subintervals of width $\Delta h = \dfrac{100}{N}$. The density is taken to be approximately $\rho(h_i)$ on the ith spherical shell, and the volume is approximately the surface area of a sphere of radius $r_e + h_i$ meters times Δh, where $r_e = 6.37 \times 10^6$ meters is the radius of the earth. If the volume of the ith shell is V_i, then $V_i \approx 4\pi(r_e + h_i)^2 \Delta h$, and a left-hand Riemann sum for the total mass is

$$M \approx \sum_{i=0}^{N-1} 4\pi(r_e + h_i)^2 \times 1.28 e^{-0.000124 h_i} \Delta h.$$

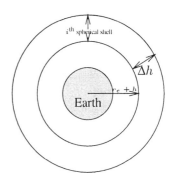

Figure D.5

(b) This Riemann sum becomes the integral

$$M = 4\pi \int_0^{100} (r_e + h)^2 \times 1.28 e^{-0.000124 h}\,dh$$

$$= 4\pi \int_0^{100} (6.37 \times 10^6 + h)^2 \times 1.28 e^{-0.000124 h}\,dh.$$

Evaluating the integral using numerical methods gives $M = 6.48 \times 10^{16}$ kg.

34. We want to take a cross-section of the pipe and cut it up in such a way that the speed of the water is nearly uniform on each slice. We will use thin rings around the pipe's center; if a given ring is narrow enough, all points on it will be roughly equidistant from the center. Since the water speed is a function of the distance from the center, it will be nearly constant on the entire ring.

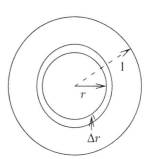

552

Let r be the distance from the center to the inner boundary of the ring, and let Δr be the width of the ring, as in the figure above. By straightening the ring into a thin rectangle, we find that its area is approximately given by the quantity $2\pi r \Delta r$. The speed across any part of the ring is roughly equal to the speed across the inner boundary, $10(1 - r^2)$ inches per second. The flow is defined as the speed times the area; thus on any given ring we have

$$\text{Flow} \approx 10(1 - r^2) \cdot 2\pi r \Delta r.$$

The total flow across the pipe cross-section is approximated by a Riemann sum incorporating all of the rings:

$$\text{Total Flow} \approx 20\pi \sum (1 - r^2) r \Delta r,$$

where r is in between 0 and 1. Letting $\Delta r \to 0$, we obtain the exact solution:

$$\text{Total Flow} = 20\pi \int_0^1 (1 - r^2) r \, dr = 20\pi \left(\frac{r^2}{2} - \frac{r^4}{4} \right) \Big|_0^1 = 5\pi \text{ cubic inches/second.}$$

35. (a) Slice the headlight into N disks of height Δx by cutting perpendicular to the x–axis. The radius of each disk is y; the height is Δx. The volume of each disk is $\pi y^2 \Delta x$. Therefore, the Riemann sum approximating the volume of the headlight is

$$\sum_{i=1}^{N} \pi y_i^2 \Delta x = \sum_{i=1}^{N} \pi \frac{9x_i}{4} \Delta x.$$

(b)

$$\pi \int_0^4 \frac{9x}{4} \, dx = \pi \frac{9}{8} x^2 \Big|_0^4 = 18\pi.$$

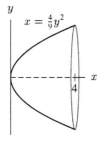

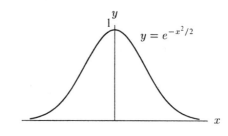

Figure D.6   Figure D.7

36. If $y = e^{-x^2/2}$, then $x = \sqrt{-2 \ln y}$. (Note that since $0 < y \leq 1$, $\ln y \leq 0$.) A typical slice has thickness Δy and radius x.

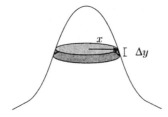

So

$$\text{Volume of slice } = \pi x^2 \, \Delta y = -2\pi \ln y \, \Delta y.$$

Thus,

$$\text{Total volume } = -2\pi \int_0^1 \ln y \, dy.$$

Since $\ln y$ is not defined at $y = 0$, this is an improper integral:

$$\text{Total Volume } = -2\pi \int_0^1 \ln y \, dy = -2\pi \lim_{a \to 0} \int_a^1 \ln y \, dy$$

$$= -2\pi \lim_{a \to 0} (y \ln y - y) \Big|_a^1 = -2\pi \lim_{a \to 0} (-1 - a \ln a + a).$$

By looking at the graph of $x \ln x$ on a calculator, we see that $\lim_{a \to 0} a \ln a = 0$. Thus,

$$\text{Total volume } = -2\pi(-1) = 2\pi.$$

37. We want to approximate $\int_0^{120} A(h) \, dh$, where h is height, and $A(h)$ represents the cross-sectional area of the trunk at height h. Since $A = \pi r^2$ (circular cross-sections), and $c = 2\pi r$, where c is the circumference, we have $A = \pi r^2 = \pi[c/(2\pi)]^2 = c^2/(4\pi)$. We make a table of $A(h)$ based on this:

TABLE D.1

height (feet)	0	20	40	60	80	100	120
Area (square feet)	53.79	38.52	28.73	15.60	2.865	0.716	0.080

We now form left & right sums using the chart:

$$\text{LEFT}(6) = 53.79 \cdot 20 + 38.52 \cdot 20 + 28.73 \cdot 20 + 15.60 \cdot 20 + 2.865 \cdot 20 + 0.716 \cdot 20$$
$$= 2804.42.$$
$$\text{RIGHT}(6) = 38.52 \cdot 20 + 28.73 \cdot 20 + 15.60 \cdot 20 + 2.865 \cdot 20 + 0.716 \cdot 20 + 0.080 \cdot 20$$
$$= 1730.22$$

So

$$\text{TRAP}(6) = \frac{\text{RIGHT}(6) + \text{LEFT}(6)}{2} = \frac{2804.42 + 1730.22}{2} = 2267.32 \text{ cubic feet.}$$

554

38. Although we could work this problem by using the formula for the volume of a right pyramid with a square base, we'll find the volume of the dump by using slices instead. The slices will be squares, and we'll start slicing at the base of the pyramid. The side of a square slice at the base is 100 yards; for every yard above the base, the side of the square slice decreases by 1 yard. Therefore, the side of a slice y yards above the base is $(100 - y)$, and the volume of the slice is $(100 - y)^2 \Delta y$.

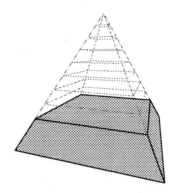

Figure D.8: Garbage Dump

Thus, the volume of the dump is

$$V = \int_0^{20} (100 - y)^2 \, dy$$

$$= \int_0^{20} (100^2 - 200y + y^2) \, dy$$

$$= \left(10{,}000y - 100y^2 + \frac{y^3}{3} \right) \Big|_0^{20}$$

$$\approx 162{,}666.67 \text{ cubic yards.}$$

If 65 cubic yards arrive at the dump every day, then 365(65) cubic yards arrive each year. This means it will take approximately

$$\frac{162{,}667}{365(65)} \approx 6.87 \text{ years}$$

for the dump to fill up.

Solutions for Section F

1.

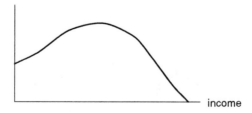

% of population
per dollar of income

income

Figure F.9: Density function

2.

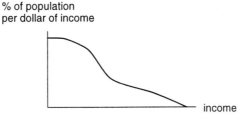

% of population
per dollar of income

income

Figure F.10: Density function

3.

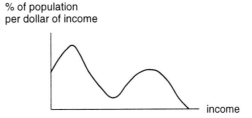

% of population
per dollar of income

income

Figure F.11: Density function

4. No. Though the density function has its maximum value at 50, this does not mean that a large fraction of the population receives scores near 50. The value $p(50)$ can not be interpreted as a probability. Probability corresponds to *area* under the graph of a density function. Most of the area in this case is in the broad hump covering the range $0 \leq x \leq 40$, very little in the peak around $x = 50$. Most people score in the range $0 \leq x \leq 40$.

5. (a) Most of the earth's surface is below sea level. Much of the earth's surface is either around 3 miles below sea level or exactly at sea level. It appears that essentially all of the surface is between 4 miles below sea level and 2 miles above sea level. Very little of the surface is around 1 mile below sea level.

  (b) The fraction below sea level corresponds to the area under the curve from -4 to 0 divided by the total area under the curve. This appears to be about $\frac{3}{4}$.

6. (a)

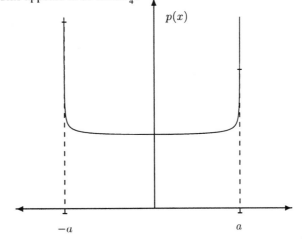

  (b) The graphs should look similar.

(c) We expect $\int_{-a}^{a} \dfrac{dx}{\pi\sqrt{a^2-x^2}} = 1$, since

$$f(x) = \begin{cases} \frac{1}{\pi\sqrt{a^2-x^2}} & -a < x < a; \\ 0 & |x| \geq a, \end{cases}$$

and thus $\int_{-a}^{a} \dfrac{dx}{\pi\sqrt{a^2-x^2}} = 1$ by the definition of a probability density function.
Indeed,

$$\int_{-a}^{a} \frac{dx}{\pi\sqrt{a^2-x^2}} = \frac{1}{\pi} \arcsin \frac{x}{a} \Big|_{-a}^{a}$$
$$= \frac{1}{\pi}(\arcsin 1 - \arcsin(-1))$$
$$= 1.$$

(d) It does make sense, physically speaking. The fact that $f(x) \to \infty$ as $x \to a$ does not mean that the ball spends an infinite amount of time at a, but just that the ratio of the time spent near $-a$ and a to the time spent elsewhere goes to ∞. This makes sense—if we watch a pendulum, we note that more time is spent near the ends of its path (where its velocity is small) than in the middle of the path (where its velocity is largest).

7. (a) Since $\mu = 100$ and $\sigma = 15$:

$$p(x) = \frac{1}{15\sqrt{2\pi}} e^{-\frac{1}{2}\left(\frac{x-100}{15}\right)^2}$$

(b) The fraction of the population with IQ scores between 115 and 120 is (integrating numerically)

$$\int_{115}^{120} p(x)\,dx = \int_{115}^{120} \frac{1}{15\sqrt{2\pi}} e^{-\frac{(x-100)^2}{450}}\,dx$$
$$= \frac{1}{15\sqrt{2\pi}} \int_{115}^{120} e^{-\frac{(x-100)^2}{450}}\,dx$$
$$\approx 0.067 = 6.7\% \text{ of the population.}$$

8. We try to find $\dfrac{1}{\sqrt{2\pi}} \displaystyle\int_{-\infty}^{\infty} e^{-\frac{(x-15)^2}{2}}\,dx$.

Since this has no elementary antiderivative, we must do it numerically. Note that this is a normal distribution with mean 15 and standard deviation 1. Almost all of the area under the graph will lie within 5 standard deviations of the mean, so we can try to find $\dfrac{1}{\sqrt{2\pi}} \displaystyle\int_{10}^{20} e^{-\frac{(x-15)^2}{2}}\,dx$.

Using Simpson's rule with 100 intervals, we get that the above integral is approximately 0.9999994267, which is indeed very close to 1. We can get even closer to 1 by choosing different limits of integration.

9. (a) i.   ii.

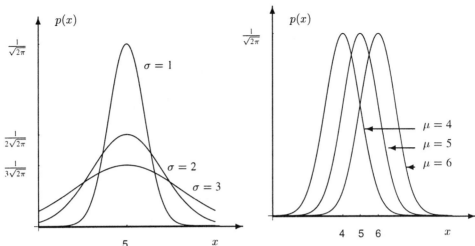

(b) Recall that the mean is the "balancing point." In other words, if the area under the curve was made of cardboard, we'd expect it to balance at the mean. All of the graphs are symmetric across the line $x = \mu$, so μ is the "balancing point" and hence the mean.

As the graphs also show, increasing σ flattens out the graph, in effect lessening the concentration of the data near the mean. Thus, the smaller the σ value, the more data is clustered around the mean.

10. (a) We want to find a such that $\int_0^\infty p(v)\, dv = \lim_{r \to \infty} a \int_0^r v^2 e^{-mv^2/2kT}\, dv = 1$. Therefore,

$$\frac{1}{a} = \lim_{r \to \infty} \int_0^r v^2 e^{-mv^2/2kT}\, dv. \tag{0.1}$$

To evaluate the integral, use integration by parts with the substitutions $u = v$ and $w' = v e^{-mv^2/2kT}$:

$$\int_0^r \underbrace{v}_{u}\, \underbrace{v e^{-mv^2/2kT}}_{w'}\, dv = \underbrace{v}_{u}\, \underbrace{\frac{e^{-mv^2/2kT}}{-m/kT}}_{w} \Big|_0^r - \int_0^r \underbrace{1}_{u'}\, \underbrace{\frac{e^{-mv^2/2kT}}{-m/kT}}_{w}\, dv$$

$$= -\frac{kTr}{m} e^{-mr^2/2kT} + \frac{kT}{m} \int_0^r e^{-mv^2/2kT}\, dv.$$

From the normal distribution we know that $\int_0^\infty \frac{1}{\sqrt{2\pi}} e^{-x^2/2}\, dx = \frac{1}{2}$, so

$$\int_0^\infty e^{-x^2/2}\, dx = \frac{\sqrt{2\pi}}{2}.$$

Therefore in the above integral, make the substitution $x = \sqrt{\frac{m}{kT}}\, v$, so that $dx = \sqrt{\frac{m}{kT}}\, dv$, or $dv = \sqrt{\frac{kT}{m}}\, dx$. Then

$$\frac{kT}{m} \int_0^r e^{-mv^2/2kT}\, dv = \left(\frac{kT}{m}\right)^{3/2} \int_0^{\sqrt{\frac{m}{kT}}\, r} e^{-x^2/2}\, dx.$$

Substituting this into Equation 0.1 we get

$$\frac{1}{a} = \lim_{r \to \infty} \left(-\frac{kTr}{m} e^{-mr^2/2kT} + \left(\frac{kT}{m}\right)^{3/2} \int_0^{\sqrt{\frac{m}{kT}}\, r} e^{-x^2/2}\, dx \right) = 0 + \left(\frac{kT}{m}\right)^{3/2} \cdot \frac{\sqrt{2\pi}}{2}.$$

Therefore, $a = \frac{2}{\sqrt{2\pi}}(\frac{m}{kT})^{3/2}$. Substituting the values for k, T, and m gives $a \approx 3.4 \times 10^{-8}$ SI units.

(b) To find the median, we wish to find the speed x such that

$$\int_0^x p(v)\,dv = \int_0^x av^2 e^{-\frac{mv^2}{2kT}}\,dv = \frac{1}{2},$$

where $a = \frac{2}{\sqrt{2\pi}}(\frac{m}{kT})^{3/2}$. Using a calculator, by trial and error we get $x \approx 441$ m/sec.
To find the mean, we find

$$\int_0^\infty vp(v)\,dv = \int_0^\infty av^3 e^{-\frac{mv^2}{2kT}}\,dv.$$

This integral can be done by substitution. Let $u = v^2$, so $du = 2v\,dv$. Then

$$\int_0^\infty av^3 e^{-\frac{mv^2}{2kT}}\,dv = \frac{a}{2}\int_{v=0}^{v=\infty} v^2 e^{-\frac{mv^2}{2kT}} 2v\,dv$$

$$= \frac{a}{2}\int_{u=0}^{u=\infty} ue^{-\frac{mu}{2kT}}\,du$$

$$= \lim_{r\to\infty}\frac{a}{2}\int_0^r ue^{-\frac{mu}{2kT}}\,du.$$

Now, using the integral table, we have

$$\int_0^\infty av^3 e^{-\frac{mv^2}{2kT}}\,dv = \lim_{r\to\infty}\frac{a}{2}\left[-\frac{2kT}{m}ue^{-\frac{mu}{2kT}} - \left(-\frac{2kT}{m}\right)^2 e^{-\frac{mu}{2kT}}\right]\Bigg|_0^r$$

$$= \frac{a}{2}\left(-\frac{2kT}{m}\right)^2$$

$$\approx 457.7 \text{ m/sec.}$$

The maximum for $p(v)$ will be at a point where $p'(v) = 0$.

$$p'(v) = a(2v)e^{-\frac{mv^2}{2kT}} + av^2\left(-\frac{2mv}{2kT}\right)e^{-\frac{mv^2}{2kT}}$$

$$= ae^{-\frac{mv^2}{2kT}}\left(2v - v^3\frac{m}{kT}\right).$$

Thus $p'(v) = 0$ at $v = 0$ and at $v = \sqrt{\frac{2kT}{m}} \approx 405$. It's obvious that $p(0) = 0$, and that $p \to 0$ as $v \to \infty$. So $v = 405$ gives us a maximum: $p(405) \approx 0.002$.

(c) The mean, as we found in part (b), is $\frac{a}{2}\frac{4k^2T^2}{m^2} = \frac{4}{\sqrt{2\pi}}\frac{k^{1/2}T^{1/2}}{m^{1/2}}$. It is clear, then, that as T increases so does the mean. We found in part (b) that $p(v)$ reached its maximum at $v = \sqrt{\frac{2kT}{m}}$. Thus

$$\text{the maximum value of } p(v) = \frac{2}{\sqrt{2\pi}}\left(\frac{m}{kT}\right)^{3/2}\frac{2kT}{m}e^{-1}$$

$$= \frac{4}{e\sqrt{2\pi}}\frac{m^{1/2}}{kT^{1/2}}.$$

Thus as T increases, the maximum value decreases.

Solutions for Section G

1. $(1,0)$
2. $(0,0)$
3. $(-2,0)$
4. $(-1,-1)$
5. $(\frac{5\sqrt{3}}{2}, -\frac{5}{2})$
6. $(0,3)$
7. $(\cos 1, \sin 1)$
8. $r = \sqrt{1^2 + 0^2} = 1, \quad \theta = 0.$
9. $r = \sqrt{0^2 + 2^2} = 2, \quad \theta = \pi/2.$
10. $r = \sqrt{1^2 + 1^2} = \sqrt{2}.$
    $\tan\theta = 1/1 = 1.$ Since the point is in the first quadrant, $\theta = \pi/4.$
11. $r = \sqrt{(-1)^2 + 1^2} = \sqrt{2}.$
    $\tan\theta = (-1)/1 = -1.$ Since the point is in the second quadrant, $\theta = 3\pi/4.$
12. $r = \sqrt{(-3)^2 + (-3)^2} = 4.2.$
    $\tan\theta = (-3/-3) = 1.$ Since the point is in the third quadrant, $\theta = 5\pi/4.$
13. $r = \sqrt{(0.2)^2 + (-0.2)^2} = 0.28.$
    $\tan\theta = 0.2/(-0.2) = -1.$ Since the point is in the fourth quadrant, $\theta = 7\pi/4.$ (Alternatively $\theta = -\pi/4.$)
14. $r = \sqrt{3^2 + 4^2} = 5, \quad \tan\theta = 4/3.$ The point is in the first quadrant, so $\theta = 0.92.$
15. $r = \sqrt{(-3)^2 + 1^2} = 3.16, \quad \tan\theta = 1/(-3).$ Since the point is in the second quadrant $\theta = 2.82.$
16. For each pair of Cartesian coordinates, there is more than one pair of polar coordinates for that point. For example, if $(x, y) = (1, 0)$ then $(r, \theta) = (1, 0), (r, \theta) = (1, 2\pi),$ and $(r, \theta) = (1, 4\pi)$ all represent the same point.

Notes

Notes

Notes

Notes

Notes

Notes

Notes

Notes

Notes

Notes

Notes